# FOUNDATIONS OF BIOMATERIALS ENGINEERING

# FOUNDATIONS OF BIOMATERIALS ENGINEERING

MARIA CRISTINA TANZI

SILVIA FARÈ

GABRIELE CANDIANI

ACADEMIC PRESS

An imprint of Elsevier

Cover image: From *Prana*, 2017, Rabarama

Academic Press is an imprint of Elsevier
125 London Wall, London EC2Y 5AS, United Kingdom
525 B Street, Suite 1650, San Diego, CA 92101, United States
50 Hampshire Street, 5th Floor, Cambridge, MA 02139, United States
The Boulevard, Langford Lane, Kidlington, Oxford OX5 1GB, United Kingdom

**Notices**
Knowledge and best practice in this field are constantly changing. As new research and
experience broaden our understanding, changes in research methods, professional
practices, or medical treatment may become necessary.

Practitioners and researchers must always rely on their own experience and knowledge in
evaluating and using any information, methods, compounds, or experiments described
herein. In using such information or methods they should be mindful of their own safety
and the safety of others, including parties for whom they have a professional responsibility.

To the fullest extent of the law, neither the Publisher nor the authors, contributors, or
editors, assume any liability for any injury and/or damage to persons or property as a
matter of products liability, negligence or otherwise, or from any use or operation of any
methods, products, instructions, or ideas contained in the material herein.

**Library of Congress Cataloging-in-Publication Data**
A catalog record for this book is available from the Library of Congress

**British Library Cataloguing-in-Publication Data**
A catalogue record for this book is available from the British Library

ISBN **978-0-08-101034-1**

For information on all Academic Press publications
visit our website at https://www.elsevier.com/books-and-journals

  Working together
to grow libraries in
developing countries

www.elsevier.com • www.bookaid.org

*Publisher:* Matthew Deans
*Acquisition Editor:* Sabrina Webber
*Editorial Project Manager:* Leticia Lima
*Production Project Manager:* Maria Bernard
*Cover Designer:* Greg Harris

Typeset by SPi Global, India

# Contents

## 8. Advanced Applications

# Preface

The idea of this textbook is derived from an educational book published in Italian and is now rewritten, expanded, and updated. Although there are currently many textbooks on the subject of biomaterials, we believe that this comprehensive but compact introductory book addresses all the significant aspects of biomaterials science in a balanced way for the first time, providing a global vision with an appropriate balance between depth and broadness in a reasonable number of pages. Conceptual background materials and a broad overview of applications were both envisioned as being integral to this book. Key definitions, equations, and other concepts are concisely pointed out along the text, allowing readers to quickly and easily identify the most important information.

*Foundations of Biomaterials Engineering* is meant to serve as an authoritative tool for training and educating Bachelor students in Biomedical Engineering because it provides them with information generally unavailable in other textbooks. It is also well-suited for students from a wide academic spectrum and other backgrounds who are unfamiliar with the biomedical field. In addition, it can be useful to anyone who wishes to acquire not only a basic knowledge of biomaterials but also of the physiological mechanisms of defense and repair, tissue engineering, and as little as needed for the basis of biotechnology.

The book is divided into eight chapters organized into two major sections. The introductory section (the first three chapters) covers engineering materials, their properties, and traditional and innovative processing methods, and is intended for students who do not yet have a basic knowledge of this subject.

The significant and specific topics of this textbook are addressed in the subsequent section (Chapters 4–7), which is dedicated to "Biomaterials and Biocompatibility," and deals with issues related to the use and application of the various classes of materials in the biomedical field, especially those intended for applications within the human body. It also deals with the mechanisms underlying the physiological processes of defense and repair, and finally the phenomenology of the interaction between the biological environment and biomaterials.

The last part of the book (Chapter 8) concerns two booming sectors: tissue engineering and biotechnology. The chapter introduces the principles and essential technologies for tissue engineering, paying particular attention to scaffolds, their requirements, and methods of fabrication. The last part of the chapter presents the application fields and purposes of current biotechnology, describing the structure and function of nucleic acids, and presenting an outline of current techniques and applications of genetic engineering and gene therapy.

# Acknowledgments

While writing the chapters of this book, each of us three Authors fully expressed his own personal scientific vision and beliefs, and we all are therefore fully and concurrently responsible for the contents. More than a decade of experience teaching the specific topics of this book has helped us select the most relevant information for a fundamental and constructive approach in the field of biomaterials.

Nonetheless, we have been influenced by different readings and discussion with other fellows, and we considered how and where their contributions have impacted our writing. Also, several colleagues have selflessly given us a great deal of help with the artwork and figures used in the different chapters. We are thankful to them all. Each of us was supported in the preparation of this book by postdoctoral and doctoral students of our team, and especially by our families. To all of them, our grateful thanks!

# INTRODUCTION TO MATERIALS

# 1

# Organization, Structure, and Properties of Materials

A philosophical definition of material can be *a substance of which everything is composed or made.* A more scientific definition can be as follows: a material can be defined as *an aggregate of atoms or molecules capable of responding with an appropriate response to a chemical, physical, and mechanical stimulus to allow being used to obtain objects, components, and structures.*

Material properties depend on their microstructure, that is, related to composition and atomic or molecular organization, as well as to chemical and physical treatments to which the material undergoes during its processing (Fig. 1.1). Therefore it is necessary to understand, study, and get knowledge on what and how material is made up; how can it be used, as it can be modified and made better to get more powerful materials; and how new materials can be obtained. For that reason, materials science can be defined as the discipline that studies the relationship between material structure and properties. Furthermore, material technology is the science that studies possible applications starting from material properties.

## 1.1 THE MAIN CLASSES OF MATERIALS

Different classes of materials can be identified based on their chemical structure:

– metals;
– ceramics and glasses;
– synthetic and natural polymers.

A fourth class, called composite materials, is the combination of two or more materials belonging to the three main classes (i.e., metal, ceramics, and polymers).

To choose and use materials consciously, it is crucial to understand that there is a strict bond between the properties and the structure of the material.

### 1.1.1 Structure and Organization of Solids

#### 1.1.1.1 Solid State and Chemical Bonds

Solid state represents something with the adequate characteristics to better fit the previous definition of a material. In fact, materials are

3

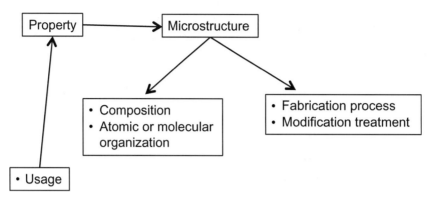

**FIG. 1.1** Scheme of the main relationship among material structure, properties, and processing.

mainly used at a solid state for structural applications, for example, when an adequate response is needed as a reaction to a chemical, physical, or mechanical stimulus.

At the solid state, materials can be classified on the basis of their structure and type of chemical bonds among the atoms. Chemical bonds play an important role in determining chemical, physical, and mechanical properties of a material; hence, it is important to understand the main differences among the type of bonds.

In a material, atoms form bonds with other atoms to reach the energy condition (or configuration) of maximum stability. The electrons of the outer energetic level, named valence electrons, are responsible for the formation of bonds among atoms of a material. The configuration with eight electrons in the outer energetic level is one of maximum chemical stability and is related to noble gases. In all other cases, when a number of electrons lower than eight is present on the outer level, atoms forms bonds with other atoms, so as to reach a more stable configuration. In particular, this is possible by the formation of:

– a covalent bond, which is a bond formed by the sharing of one or more electrons by two atoms;
– an ionic bond, which is an electrostatic bond between two ions formed through the transfer of one or more electrons;

– a metallic bond, which is a bond between atoms in a metallic element, formed by the valence electrons moving freely through the metal lattice.

Among different molecules, other types of bonds can be formed, typically weak electrostatic bonds, such as dipole/dipole, hydrogen bonds, and Van der Waals forces.

## COVALENT BOND

In the covalent bond, atoms of the involved element are able to share electrons (one or more valence electrons) of their outer shell with other atoms to reach a more stable configuration. In fact, this bond is formed when an element has a nearly full outer shell and needs only one more atom to acquire a full outer shell; it then shares their outer electrons with another atom, so that both of them become full and stable (Fig. 1.2). In particular, if the atoms shared one electron, a simple covalent bond is formed (Fig. 1.2A and B); when two electrons are shared, the bond is a double covalent one (Fig. 1.2C); and if three electrons are shared, a triple covalent bond is formed (Fig. 1.2D). In addition, covalent bonds can be formed among different atoms (e.g., C—H, C—O, N—H). In this case, because of the weak difference in electronegativity of the atoms involved in the bond, the shared electrons

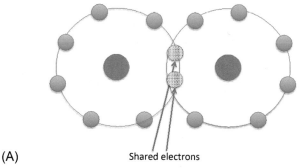

(A)

Shared electrons

**FIG. 1.2** Covalent bonds: (A) and (B) the outer shell of the two atoms shares one electron with the other atom, so that both the atoms become stable (simple covalent bond); (C) two electrons are shared (double covalent bond); (D) three electrons are shared (triple covalent bond).

(B)   $:\!\ddot{F}\!\cdot\; +\; :\!\ddot{F}\!\cdot\; \longrightarrow\; :\!\ddot{F}\!:\!\ddot{F}\!:$     F — F

(C)   $:\!\ddot{O}\!\cdot\; +\; :\!\ddot{O}\!\cdot\; \longrightarrow\; :\!\ddot{O}\!::\!\ddot{O}\!:$     O = O

(D)   $:\!\dot{N}\!\cdot\; +\; :\!\dot{N}\!\cdot\; \longrightarrow\; :\!N\!::\!N\!:$     N ≡ N

are displaced toward the more electronegative atom, forming a dipole.

An atom can also form more simple covalent bonds at the same time; in fact, in the case of carbon that has four valence electrons, it can form up to four covalent bonds to reach a more stable configuration. That is the case of the polymers (see Sections 1.2 and 1.6) that are mainly composed by atoms bonded together in long chains by covalent bonds with lateral bonds with atoms of H, N, O (Fig. 1.3).

Covalent bonds are directional, and, as the electrons are held in place, the materials formed with this bond are generally poor conductors of electricity and heat. The bond a very strong, and typical covalent bond strength (e.g., C—C) is about 350 kJ/mol.

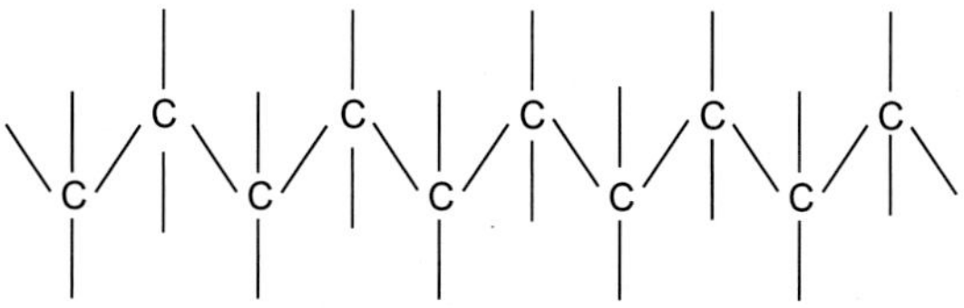

**IONIC BOND**

Ionic bonds (Fig. 1.4) are formed when one atom donates one or more electrons to form a cation, and another atom accepts the electrons to form an anion. In fact, ionic bond is formed between atoms with a high difference in electronegativity values; one of the atoms has, in the outer energetic level, a few electrons (e.g., one or two electrons) and the other atom lacks of the same number of valence electrons to reach eight (i.e., a more stable configuration). In this case, one or more valence electrons are transferred from one atom to another one to regain the most stable configuration (i.e., eight electrons). The atoms that yield electrons become positively charged ions (i.e., cations), and the ones that receive them become negatively charged ions (i.e., anions). The two ions attract

**FIG. 1.3** Covalent bonds: long linear chain of carbon atoms linked together with simple covalent bonds.

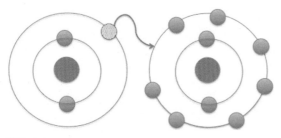

**FIG. 1.4**  Ionic bonds: in this example, the atom on the left has only one electron on its outer shell, and the atom on the right is short one electron. The transfer of one electron from the left atom to the right one gives stability to both atoms, forming a strong ionic bond.

each other. Hence, cations and anions are bonded together via strong electrostatic attraction, forming the ionic bond. In general, this bond is nondirectional and has equal strength in all directions. Bonding energy is generally high, ranging between 600 and 1500 kJ/mol; for example, ionic bond strength for NaCl (i.e., $Na^+Cl^-$) is about 770 kJ/mol. In addition, the electrons are closely held in place and no charge transfer is possible, making ionic materials poor heat and electricity conductors.

## METALLIC BOND

Metals atoms are good donors of electrons, and metallic bonds are characterized by cores formed by packed positive ions surrounded by valence electrons that can form an electron cloud able to float thorough the material (Fig. 1.5). The

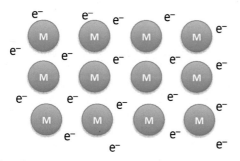

**FIG. 1.5**  Metallic bonds: positive ions (M) are surrounded by an electron cloud.

cores are positively charged, and the electron cloud works as an adhesive for them. The metallic bond is nondirectional, with an energy that can vary depending on the metallic element; in particular, the strength of this bond can be expressed by heat of sublimation, for example, at 25°C, aluminum will have a sublimation heat of 325 kJ/mol and titanium 475 kJ/mol. The loose electron cloud allows good charge transfer, making metals good conductors of electricity and heat.

## SECONDARY BONDS

In addition to strong bonds, weak bonds can be present in the chemical structure of a material; in particular, they can be found as intermolecular or intramolecular bonds. All types of weak interactions are effective only over a short range and require close contact between the reacting groups. Therefore these bonds are based on the attraction between atomic or molecular dipoles, resulting in electrostatic attraction between adjacent atoms or molecules. As weak bonds, the energy involved is lower than that related to strong bonds. In particular, the energy released in the formation of noncovalent bonds is only 1–5 kcal/mol, much less than the bonding energies of single covalent bonds. Because the average kinetic energy of molecules at 25°C is about 0.6 kcal/mol, many molecules will have enough energy to break noncovalent bonds. Secondary bonds do not involve the exchange or sharing of electrons, are less directional, and the strength of the bond is <10% of the one related to covalent bond. However, these bonds are very important as they can strongly influence the properties of materials, in particular polymers. Among weak bonds, four are the main types involved in material and biological systems: hydrogen bonds, hydrophobic interactions, ionic bonds, and Van der Waals interactions.

*Hydrogen bonds* are a type of attractive (dipole-dipole) interaction. Normally, a hydrogen atom (H) forms a covalent bond with only one other atom. However, a H covalently bonded to a donor atom (D) may form an additional weak association, the hydrogen bond, with an acceptor atom (A) (Fig. 1.6).

For a hydrogen bond to form, D must be electronegative, so that the covalent D—H bond is polar. A also must be electronegative, and its outer shell must have at least one nonbonding pair of electrons that attracts the $\delta^+$ charge of the H. As an example, in water molecules, hydrogen bonds form between neighboring water molecules when the hydrogen of one atom comes between the oxygen atoms of its own molecule and that of its neighbor. This happens because the hydrogen atom is attracted to both its own oxygen and other oxygen atoms that come close enough. Oxygen, an electronegative element, attracts electrons better than the hydrogen nucleus with its single positive charge. So neighbor oxygen molecules are capable of attracting hydrogen atoms from other molecules, forming the basis of hydrogen bond formation. Therefore oxygen can distort the binding electron cloud from the hydrogen leaving it with fewer electrons and more plus-charged protons ($\delta^+$). This positive charge will, in turn, interact with the electronegative oxygen ($\delta^-$, Fig. 1.7). Typical hydrogen bond strengths (e.g., O—H $\cdots$ H) are about 20 kJ/mol.

*Hydrophobic interactions* cause nonpolar molecules to adhere to one another. Simply put, *like dissolves like*. Polar molecules dissolve in polar solvents such as water, whereas nonpolar molecules dissolve in nonpolar solvents (e.g., hexane). Nonpolar molecules do not contain ions, possess a dipole moment, or become hydrated. The force that causes hydrophobic molecules

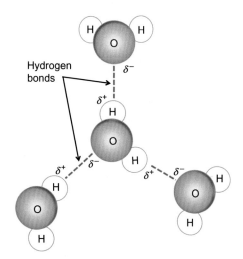

FIG. 1.7 Hydrogen bonds between two molecules of water.

or nonpolar portions of molecules to aggregate together rather than dissolve in water is called a hydrophobic bond. This is not a separate bonding force, nonetheless, it is the result of the energy required to insert a nonpolar molecule into water, because a nonpolar molecule cannot form hydrogen bonds with water molecules, so it distorts the usual water structure, forcing the water into a rigid cage of hydrogen-bonded molecules around it. This situation is energetically unfavorable because it decreases the entropy of the population of water molecules.

*Salt bridges* are present in some compounds (e.g., proteins) when the bonded atoms are so different in electronegativity that the bonding electrons are never shared and the bonds cannot be considered covalent; these electrons are always found around the more electronegative atom. The salt bridge most often arises from an anionic carboxylate (R-COO$^-$) and the cationic ammonium (R-NH$_3^+$) (Fig. 1.8). A salt bridge is generally considered to exist when the centers of charge are 4 Å or less apart.

*Van der Waals forces* are the weak forces that contribute to intermolecular bonding between molecules. These nonspecific interactions result

Hydrogen bond

$$D^{\delta-}-H^{\delta+} + :A^{\delta-} \leftrightarrow D^{\delta-}-H^{\delta+} \cdots :A^{\delta-}$$

FIG. 1.6 Hydrogen bond (highlighted in *blue*): the H in one molecule containing a donor atom (D) is attracted to a pair of electrons in the outer shell of an acceptor atom (A) in an adjacent molecule.

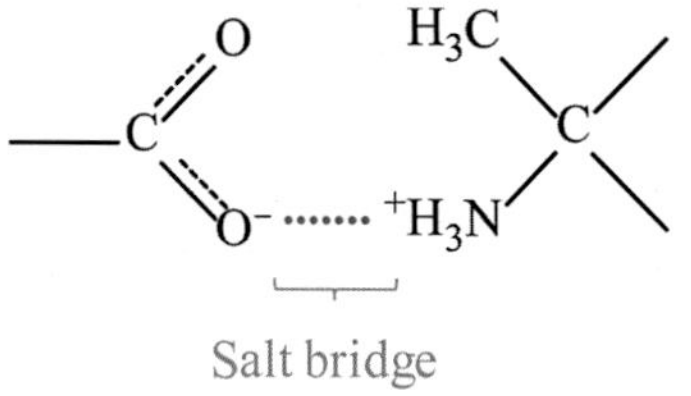

Salt bridge

**FIG. 1.8**   Salt bridge (highlighted in *blue*) between a -COO$^-$ and a $-NH_3^+$ group.

from the momentary random fluctuations in the distribution of the electrons of any atom, which give rise to a transient unequal distribution of electrons, that is, a transient electric dipole. If two noncovalently bonded atoms are close enough together, the transient dipole in one atom will perturb the electron cloud of the other. This perturbation generates a transient dipole in the second atom, and the two dipoles will attract each other weakly. Similarly, a polar covalent bond in one molecule will attract an oppositely oriented dipole in another. Van der Waals interactions, involving either transient induced or permanent electric dipoles, occur in all types of molecules, both polar and nonpolar. Besides, Van der Waals interactions are weaker than the hydrogen bonds. A typical Van der Waals interactive force (e.g., $CH_4 \cdots CH_4$) is about 9 kJ/mol, and it is established only at very close distance (e.g., 2.5–4 Å).

### 1.1.1.2 Solid State and Structural Forms

Atoms can be arranged in defined ratios with covalent bonds to form molecules, or they can combine with metallic, ionic, and/or covalent bonds forming cohesive assemblies of atoms, as in the case of metals and the ceramics. Thus materials can be composed of atoms or molecules. Typical properties of materials, for example, flexibility, elasticity, and hardness, are associated with the way those atoms or molecules are organized in materials. In particular, materials can assume different structural forms, depending on the arrangement of atoms or

molecules in the 3D space. The manner in which atoms or molecules are arranged in 3D space depends on the configuration that allows a more stable condition, minimizing the interatomic energy. For this reason, materials can have different configurations, depending on the degree of order of the chemical elements that compose them:

- crystalline materials: an ordinate and repetitive disposition of the atoms in the 3D space, characterized by a long-range order[1];
- amorphous materials: a disordered spatial disposition of the atoms, possessing a short-range order[2];
- amorphous-crystalline materials: both ordinate and disordered dispositions of the atoms are present in the material; this structure is mainly characteristic of polymeric materials.

Such structures may have defects that are very important in determining the final characteristics of the material.

### CRYSTALLINE MATERIALS

Metals and most of the ceramics are arranged in 3D space in a very regular feature, with atoms[3] that are equidistant one each other (Fig. 1.9A). This disposition can be extended in 3D space assuming different ways in which atoms can be arranged on different planes, with different formats (Fig. 1.9B–E).

---

[1] Long-range order: when atoms or ions are arranged in a manner that is repeated in 3D space, they form a solid that has long-range order, and it is referred to as a crystalline material.

[2] Short-range order: the order exists only in the immediate neighborhood of an atom or a molecule, and atoms are not arranged in a long-range, periodic, and repeatable manner.

[3] We can call them atoms, even if they are ions as previously discussed (see Section Metallic Bonds).

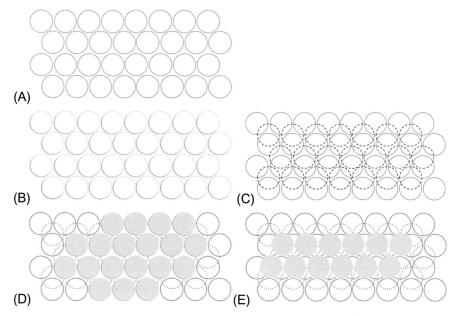

**FIG. 1.9** Crystalline features: (A) A 2D disposition of the atoms: atoms are arranged in a very regular sequence. A second plane is placed on the first one (B) occupying the identic position of the first layer or (C) displacing from the first layer. A third plane can be placed over the previous ones; (D) atoms are located as in the first plane or (E) displaced from the first and the second planes.

A solid in which the atoms are regularly arranged in 3D space is defined as a *crystalline material*. As there are different ways in which the planes can be arranged, different crystal structures can be defined; each of them is characterized by a degree of regularity. Hence, each atom's feature can be identified by a different unit motif, called a *unit cell*, regularly repeated in the 3D environment forming a space lattice (Fig. 1.10). Based on geometrical consideration, there are 14 identified ways in which atoms can

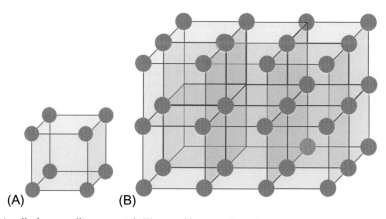

**FIG. 1.10** (A) Unit cell of a crystalline material; (B) crystal lattice, where the unit cell is regularly repeated in the 3D space.

be arranged in 3D, which are called Bravais lattices (see Section 1.3).

In general, crystalline materials are composed of atoms linked together by strong bonds (i.e., metallic, covalent, and/or ionic bonds); hence, these materials are expected to have a high mechanical strength.

## AMORPHOUS MATERIALS

Amorphous materials are characterized by a short-range order, and atoms are bonded in disordered, random spatial positions, because of factors that do not allow the formation of a regular arrangement (Fig. 1.11A). Most polymers and inorganic glasses can have an amorphous structure. In particular in polymers, the secondary bonds among the macromolecules do not allow for the formation of tightly packed chain configurations.

Some polymers (e.g., polyethylene) can have macromolecules efficiently packed in some regions, producing a higher degree of long-range order (i.e., crystalline region), and in other regions macromolecules have a short-range order (i.e., amorphous region), as reported in Fig. 1.11B. For this reason, these polymers are called semicrystalline materials (see Section 1.2).

Inorganic glasses based on silica ($SiO_2$) are an example of ceramic amorphous materials, in which the fundamental unit is a tetrahedron $SiO_4^-$. This structure can be regularly organized in the 3D space to form a long-range order (i.e., a crystalline structure). In fact, when the glass is in the viscous liquid state, crystallization occurs slowly. If the cooling rate increases, the formation of the crystalline structure is not allowed, and tetrahedrons are organized in a short-range order.

In amorphous materials, atoms or molecules are bonded together by weak bonds, determining lower mechanical properties compared to the ones related to crystalline materials.

### 1.1.1.3 Structure of the Different Classes of Materials

Different chemical bonds are involved in the arrangement of the atoms or molecules, so that polymers, metals, and ceramics represent the three main classes of materials used in various fields and applications. Some examples of the three classes of materials are reported in Fig. 1.12.

Polymers are organic materials, composed by long macromolecular chains formed by carbon atoms and other elements (e.g., nitrogen, oxygen, hydrogen). Atoms along the macromolecular chains are bonded by covalent bonds, and secondary bonds are present among the macromolecules; they may assume amorphous, semicrystalline or crystalline structure.

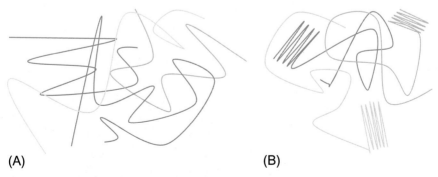

(A)                                                    (B)

FIG. 1.11   (A) Amorphous structure in a polymer: macromolecules are in a random short-range order; (B) semicrystalline material: the material is composed of tight and regular disposition of the macromolecules and random disposition of other macromolecules.

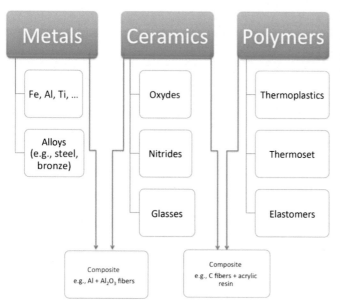

**FIG. 1.12** Three classes of materials: polymers, metals, and ceramics. Composites are produced by combining them together; two possible combinations are reported as representative examples of composites.

Metals are inorganic materials, formed by metallic bonds between metallic elements or metallic and nonmetallic elements (i.e., metal *alloys*); they have, mainly, a crystalline structure. They can be divided between ferrous and non-ferrous metals or alloys (Fig. 1.13).

Ceramics are inorganic materials, composed by metallic (e.g., magnesium, aluminum, iron) and nonmetallic elements (e.g., oxygen) linked together with ionic and/or covalent bonds, commonly in a crystalline structure. Ceramics can have a crystalline structure or an amorphous structure (e.g., inorganic glasses).

Composite materials represent a transversal class of materials (Fig. 1.12); composites are formed by a combination of two or more macro- or microconstituents that are different in shape and chemical composition, and are insoluble one in the other. Polymers, metals, or ceramics can be combined together to obtain a synergic effect of their properties.

## 1.2 POLYMERIC MATERIALS

### 1.2.1 Structure

Polymeric materials (or plastic materials) are organic materials composed of long molecular chains (polymers) formed by many repeat units (monomeric units), chained together by covalent bonds.

Monomeric units derive from simple molecules, called monomers (from Greek: *mono*—one and *meros*—part), which have specific reactive functions capable of reacting repetitively and cumulatively with other monomers to form long polymer chains (*poly*—many—*mers*).

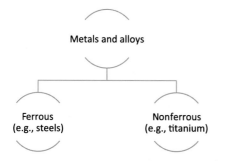

**FIG. 1.13** Classification of metals in ferrous alloys and nonferrous metals and alloys.

Each molecule of a polymer can consist of hundreds, thousands, or even millions of repeat units. Small chains of up to some tens of repeat units are called oligomers (from Greek *oligos*, meaning few), although there is no strict rule about the number of repeat units needed for the transition from oligomers to polymers. The structural units terminating the ends of a macromolecule are known as *end groups*, whereas group of atoms attached to a backbone chain of a macromolecule are called *side groups* or *pendant groups*.

The atoms that constitute the macromolecules (or polymers) are mainly carbon atoms linked to other elements, typically including hydrogen, oxygen, and nitrogen.

The bonds that form the "backbone" of the macromolecule are strong covalent bonds (*primary bonds*, see Section 1.1), whereas intermolecular electrostatic forces of the Van der Waals type and dipole/dipole (*secondary bonds*, see Section 1.1) link together adjacent polymer chains or different segments of the same chain. Ionic bonds may also occur. Valence electrons are predominantly constrained in the formation of simple, directional covalent bonds.

Covalent bonds determine the mechanical, thermal, and chemical properties of a polymeric material. Secondary bonds, instead, regulate the physical characteristics of the material, such as solubility, melting, diffusion, and flow properties (all properties that involve the breaking and forming of these bonds and the relative movements of the macromolecules).

The number of units in the polymeric chains plays a significant role in determining the properties. As the number of units in the chain increases, the product changes from a gas to a liquid and then to a brittle or waxy solid. As the number increases even more, the polymeric chains become long enough to start entangling with each other, leading to the properties more commonly associated with polymers. A typical example is that of polyethylene shown in Fig. 1.30

Depending on how the monomer molecules concatenate, macromolecules may take various *configurations*, with particular reference to *linear, branched,* or *cross-linked (networked)* ones, as shown in Fig. 1.14A. Branched polymers arise due to side reactions in the polymerization process and consist of branches attached to the main

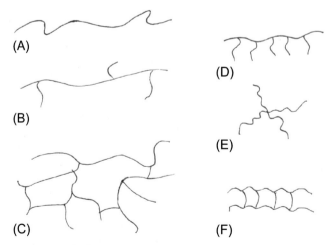

**FIG. 1.14** Schematic representation of polymeric structures (A) linear, (B) branched, (C) cross-linked, (D) comb polymer, (E) star polymer, and (F) ladder polymer. In the case of homopolymers, the monomeric unit is of one type, whereas in copolymers, there are two or more types of monomeric units.

backbone of the macromolecule.[4] When the branches connect with adjacent chains during or after the polymerization process, the product is a networked polymer.

The same basic type of polymer may often exist in different forms depending on the conditions during polymerization.

Other less frequent polymeric structures include *star*, *comb*, and *ladder* polymers. *Star* polymers are structures where multiple polymeric chains originate from one central point (see Fig. 1.14E); in *comb* polymers, multiple side chains are attached to the same side of a linear backbone; and *ladder* polymers are particular structures where two or more independent strands are interconnected in regular distances (similar to a ladder) (see Fig. 1.14F)

Polymers can be classified in different ways, based on the molecular structure (as shown in Fig. 1.14) or on the number and arrangement of repeat units (homopolymers, copolymers, random, block, alternating, and graft, as shown in Fig. 1.15). Other classifications include the thermal behavior (see later in the chapter) or the chemical structure of main bonds in the backbone.

If only one type of monomer (A) is used to form the polymer, the product is called a *homopolymer*; if two types of monomers (A and B) are used in the polymerization reaction, the product is known as a *copolymer*.

There are several possible combinations of copolymers; some of them are shown in Fig. 1.15. If the monomers alternate, then the product is known as an *alternating* copolymer. If they are joined in a random manner, then the result is a *random* copolymer. If they form sequences, then the product is a *block* (*segmented*) copolymer. In the latter case, the placement of the segments in the backbone of the block copolymer can be alternating, random, or branched. The properties of the final product can vary significantly depending

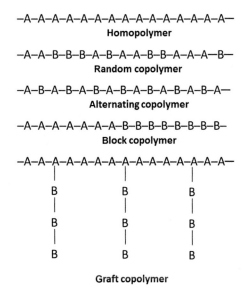

**FIG. 1.15** Considering monomers A and B, their polymerization can form homopolymers (only one monomer is polymerized) or copolymers (when both monomers participate in the polymerization giving different arrangements: alternating, random, segmented, grafted, etc.).

on the relative placement of the monomers or blocks and on the starting ratio of the two monomers. The molecular structure of the chains can also significantly influence the properties.

From the point of view of their thermal behavior, plastic materials can be divided into two major categories: *thermoplastics* and *thermosets*.

Thermoplastic polymers are those retaining permanent plasticity properties. Usually, they are linear or branched in structure and can be melted or dissolved. This makes them easy to fabricate into their final form. Because the molecules do not form covalent bonds with adjacent chains, the chains can flow on each other, and the polymer can behave like a viscous fluid upon heating. These polymers can be repeatedly softened or melted using heat and molded into new shapes.

Thermoset polymers (networked or crosslinked) are those that possess plasticity properties only up to a certain stage in their process of production, after which their physical appearance takes on definitive characteristics, in

---

[4] Branches differ from side groups in their size, which is much higher than a simple group of atoms.

general as a result of heat; they can no longer be remodeled by the effect of heat or pressure.

Thermoplastic materials, in which the macromolecules are linear or branched, can be defined as *multimolecular* materials, whereas thermosetting materials, in which transversal bonds interconnect the macromolecules, can be defined as *unimolecular*.

Cross-linked structures may arise when multifunctional monomers are used in the polymerization reaction. Cross-linking may also take place after polymerization in an existing polymer when subjected to a high-energy source, such as electron beam radiation or gamma rays. In this case, enough energy may be provided to break some bonds within the polymers, and broken bonds may then react with adjacent chains to form a network. As the degree of cross-linking increases, the polymer chains lose their ability to slide on each other, and the polymer structure becomes more rigid and dimensionally stable. Such cross-linked polymers are difficult to melt or dissolve, hence making them difficult to fabricate into products. Usually thermosets are formed into shape by temporarily disabling the cross-linking, then activating it via heat or other means (such as light or chemicals) after they have been put into their final shape.

## 1.2.2 Polymerization Degree and Molecular Weight

In multimolecular polymeric materials, the *degree of polymerization* $(X)$ of a macromolecule can be defined as the total number of concatenated monomeric units, whether in the main chain or in any branching. The synthetic process of polymeric materials always involves the creation of macromolecules having a variable number of chained monomers, for both kinetic and thermodynamic reasons. Thus except in special cases, in a synthetic polymeric material, the degree of polymerization of individual polymer chains is not the same.

It is therefore appropriate to define an average degree of polymerization:

$$\overline{X}$$

which is given by the average value of the degree of polymerization of the individual macromolecules.

Generally, when the average degree of polymerization is low, and therefore the length of the single macromolecules in the polymer is not high, the substance is liquid, waxy, or even solid, but still not very stress-resistant. With the increase of the length of the macromolecules, the properties change gradually, in particular greatly enhancing the mechanical strength and toughness of the material. On the other hand, the processability (machinability) of the material becomes more difficult, so that in the production of some plastic materials, it is necessary to operate with an average degree of polymerization that guarantees a reasonable compromise between tenacity and workability.

To each macromolecule, depending on the atoms that globally compose it, it can be assigned a specific value of molecular weight. In a plurimolecular polymeric material, as the thermoplastics are, for what was previously mentioned, macromolecules of different molecular weight generally coexist. As shown in Fig. 1.16, the typical distribution of the

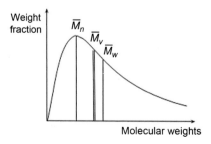

**FIG. 1.16** Typical distribution of molecular weights in a polymer. $\overline{M}_n$ = number average molecular weight, $\overline{M}_v$ = viscosity average molecular weight, $\overline{M}_w$ = weight average molecular weight.

molecular weights of the various macromolecules that form a polymeric material is assimilable to a bell shape whose width is related to the dispersion of molecular weights: the narrower the bell, the more homogeneous the length of the macromolecular chains.

It can therefore be defined as an *average molecular weight*.

There are several ways to express this value. The most interesting values from a practical point of view are the *number average molecular weight* and the *weight average molecular weight*, usually indicated respectively as $\overline{M}_n$ and $\overline{M}_w$.

$\overline{M}_n$ expresses the mean value according to the number, whereas $\overline{M}_w$ expresses the mean value according to the mass of the present macromolecules, that is:

If the number and weight average molecular weights coincided, all present macromolecules would have the same molecular weight; in this case, the polymeric material would be *monodispersed*, and the distribution curve would become a line parallel to the axis of the ordinates. Normally, the ratio of the weight average molecular weight to the number average weight is used as the index of *molecular weight dispersion*. This index, designated as PDI (*polydispersity index*, or *d*, or *p*), is therefore expressed as:

$$\mathrm{PDI} = \frac{\overline{M}_w}{\overline{M}_n} \, (\geq 1)$$

The more this ratio differs from 1, the more the polymer is polydispersed and the distribution

| Number Average Molecular Weight | Weight Average Molecular Weight |
|---|---|
| $M_n = \dfrac{\sum_{i=1}^{\infty} W_i}{\sum_{i=1}^{\infty} N_i} = \dfrac{\sum_{i=1}^{\infty} M_i N_i}{\sum_{i=1}^{\infty} N_i}$ | $M_w = \dfrac{\sum_{i=1}^{\infty} M_i W_i}{\sum_{i=1}^{\infty} M_i N_i} = \dfrac{\sum_{i=1}^{\infty} M_i^2 N_i}{\sum_{i=1}^{\infty} M_i N_i}$ |

Where:

$M_i$ = molecular weight of chains in fraction $i$
$N_i$ = number of chains in fraction $i$
$W_i$ = weight of chains in fraction $i = M_i N_i$

The weight fraction ($W$) of one type of molecule is the weight of that type of molecule divided by the total weight of the sample, as expressed in mathematical form: $M_i N_i / \Sigma M_i N_i$.

The number average molecular weight is strongly influenced by the presence of short chains, whereas the weight average molecular weight is slightly affected by the presence of these small molecules and mainly influenced by the presence of longer chain (high mass) macromolecules.[5] For these reasons, $M_w$ is always higher than $M_n$, as shown in Fig. 1.16.

[5] Note: a bigger molecule contains more of the total mass of the polymer sample than the smaller molecules do.

curve of molecular weights is widened. In commercial polymers, this index is typically between 1 and 3 but can reach up to 10, depending on the synthetic route of the polymeric material.

### 1.2.2.1 *Calculating Average Molecular Weights*

Usually the information about the molecular weight distribution in a polymer is obtained from Size Exclusion Chromatography and may look like the following example (applied to a very small sample and adapted from http://pslc.ws/macrog/average.htm).

| Number of Molecules ($N_i$) | Mass of Each Molecule ($M_i$) | Total Mass of Each Type of Molecule ($M_iN_i$) | Weight Fraction $W_i$ ($M_iN_i/\Sigma M_iN_i$) | $W_iM_i$ |
|---|---|---|---|---|
| 1 | 800,000 | 800,000 | 0.016 | 12,800 |
| 3 | 750,000 | 2,250,000 | 0.045 | 33,750 |
| 5 | 700,000 | 3,500,000 | 0.070 | 49,000 |
| 8 | 650,000 | 5,200,000 | 0.104 | 67,600 |
| 10 | 600,000 | 6,000,000 | 0.120 | 72,000 |
| 13 | 550,000 | 7,150,000 | 0.143 | 78,650 |
| 20 | 500,000 | 10,000,000 | 0.200 | 100,000 |
| 13 | 450,000 | 5,850,000 | 0.117 | 52,650 |
| 10 | 400,000 | 4,000,000 | 0.080 | 32,000 |
| 8 | 350,000 | 2,800,000 | 0.056 | 19,600 |
| 5 | 300,000 | 1,500,000 | 0.030 | 9000 |
| 3 | 250,000 | 750,000 | 0.015 | 3750 |
| 1 | 200,000 | 200,000 | 0.004 | 800 |
| $\Sigma N_i = 100$ | | $\Sigma N_iM_i = 50,000,000$ | | |

For the considered sample: $\overline{M}_n = \Sigma M_iN_i/\Sigma N_i = 50,000,000/100 = 500,000$; $\overline{M}_w = W_iM_i = M_iN_i/\Sigma M_iN_i^2 = 531,600$; and PDI $= 531,600/500,000 = 1.063$.

Another way to calculate the molecular weight comes from the observation that the molecular weight determines the *viscosity* of a dilute solution of a polymer. The relationship between viscosity and molecular weight can be described by the Mark-Houwink-Sakurada equation:

$$[\eta] = KM_v^\alpha$$

where:

$[\eta]$ = viscosity at infinite dilution,
$K$ and $\alpha$ = Mark-Houwink constants (available from published tables),
$M_v$ = viscosity average molecular weight.

The value of $M_v$ is usually positioned between $M_n$ and $M_w$ as shown in Fig. 1.16.

It should be noted that, in the case of networked polymers such as thermosets, we cannot speak either of average degree of polymerization or of weight or number average molecular weights, as the polymer chains are linked together by bridge bonds, so that the whole structure is insoluble and comparable to a single macromolecule with an infinite molecular weight.

### 1.2.3 Production of Polymers

Polymeric materials can be obtained by *chemical isolation*, sometimes followed by *chemical modification*, of natural polymeric substances, or with an entirely synthetic process. Because the first synthetic products obtained by condensation of organic substances were similar to natural resins, the term **resin** was also commonly used as a synonym for plastic material.

### 1.2.3.1 Chemical Isolation

By chemical isolation, we denote obtaining macromolecular materials through appropriate treatment, often followed by chemical modification, of natural substances (see Section 1.6) with a high molecular weight. Among these natural resins, which generally are products of plant origin, are those obtained directly or after elimination of the volatile part by incision of the trunk of various plants, such as copal and rosin (e.g., natural rubber obtained in the form of latex from *Hevea Brasiliensis* or from other exotic plants). Many natural substances have the disadvantage of being formed by macromolecules that are difficult to process, which means they are difficult to transform. A well-known example is *cellulose* that, due to the high number of hydrogen bonds existing between the macromolecular chains, has poor solubility and starts to decompose before reaching the physical state in which it can be processed. For chemical modification reactions, *cellulose nitrate* and *cellulose acetate* can be obtained, as examples, from cellulose. Such reactions that modify only certain groups in the molecule leave its chemical skeleton intact, which leads to products whose properties differ completely from those of the initial compound. In the case of cellulose derivatives, compounds with increased solubility properties are obtained, and therefore more easily processable.

### 1.2.3.2 Synthesis of Polymers (Polymerization)

Synthetic polymers are obtained from small molecules (or monomers) through a reaction called *polymerization*. During the polymerization reaction, many monomer molecules are chained together repetitively and cumulatively, according to different chemical and various kinetic mechanisms, to produce long polymer chains, or macromolecules. To participate in the polymerization reaction, the monomer molecule must possess adequate reactivity characteristics,

that is, it must contain appropriate chemical functions in its structure.

These chemical functions can be a double bond (C=C) or chemical groups able to react to each other (e.g., -OH and -COOH; -NH$_2$ and -COOH; -NCO [isocyanate] and -OH).

Under the term *polymerization reaction* are found all the types of reactions that give rise to the formation of macromolecules. There are two main types of synthetic polymerization methods: *addition* and *condensation*.

During *addition polymerization*, monomers sequentially attach to the growing end of a polymer chain. In this type of polymerization, the atoms of the monomer are directly added to the chain, and consequently the monomer and the repeat unit have the same number of atoms.

Conversely, during *condensation polymerization*, the elimination of some atoms as a by-product takes place during the reaction between the monomer and the growing chain. As such, condensation polymerization results in fewer atoms in the reacted repeat unit than in the monomer.

Another way to classify polymerization is *chain-reaction*, or *chain-growth*, and *step-reaction* or *step-growth*.

### CHAIN-GROWTH POLYMERIZATION

With chain polymerization, what are termed vinyl polymers are obtained, because the monomers from which it starts contain the vinyl group: (CH$_2$=CH—), in which the C=C double bond is present. The polymerization of vinyl monomers occurs by rupture of the double bond and creation of a simple covalent bond with the nearby monomer. Schematically:

$$2\,C=C \rightarrow -C-C-C-C-$$

Usually an *initiator* compound reacts with the monomer to start the reaction, and the mechanism of chain polymerization consists of three phases, called *initiation*, *propagation*, and *termination*.

## INITIATION

The initiator decomposes as a result of light, heat, or a chemical reaction, and generates a reactive species:

$$I \rightarrow I^*$$

This reactive species can be a free radical ($I^·$), a cation ($I^+$), an anion ($I^-$), or a coordination complex (in which metal atoms such as titanium and aluminium form metalorganic complexes with catalytic active centers) that starts the polymerization, either by a radical or ionic mechanism, by adding to the C=C double bond in the monomer (M) and giving rise to a new radical (cation, anion, or complex). The activated monomer (M*) thus obtained becomes the first of the chained (or repeating) units in the polymer chain to be formed:

$$I^* + M \rightarrow IM^*$$

## PROPAGATION

During the propagation step, the activated monomer is added to another molecule of monomer, and with another still, and so on, with the same mechanism seen in the initiation step:

$$IM^* + M \rightarrow IM_2^*$$
$$IM_2^* + M \rightarrow IM_3^*$$
$$\cdots\cdots \rightarrow \cdots\cdots$$
$$IM_{(n-1)}^* + M \rightarrow IM_n^*$$

and the process is repeated until the termination step.

## TERMINATION

The termination step takes place when the growth of the chain is exhausted by reaction with another growing chain:

$$M_m^* + M_n^* \rightarrow M_{m+n}$$

or by reaction with other species present in the system, or by spontaneous decomposition of the active site.

General characteristics of chain polymerization are:

1. Once the initiation step, which is the kinetically slower one, has occurred, the polymer chains are formed very quickly, in times of seconds or fractions of a second. Therefore the molecular weight of the polymer increases quickly, whereas the consumption of monomers occurs at a slow rate.
2. The concentration of active species is very low. For example, in the radical polymerization, the concentration of free radicals is $\sim 10^{-8}$ M, so the polymerization mixture consists mainly of a newly formed polymer and unreacted monomer.
3. Because the double bonds C=C in the monomer are converted into simple bonds, energy is released in the reaction of polymer formation, and the polymerization is therefore *exothermic*.
4. Chain polymerization normally gives rise to polymers with high molecular weight ($10^4 - 10^7$ da).
5. With chain polymerization, it is possible to obtain polymers containing secondary chains (ramifications attached to the main, or primary, chain).
6. With chain polymerization, cross-linked systems can be obtained when secondary chains connect all the primary chains to each other.

An example of chain-growth polymerization: In the reaction shown here, an initiator forms a free radical with an unpaired electron, which then attacks the double bond of a vinyl monomer (i.e., ethylene). The monomer is added, regenerating a free radical that reacts again and continues to add molecules until all the monomers are consumed, or there is a termination step that extinguishes the free radical.

*Initiation*: $R^· + CH_2 = CH_2 \rightarrow RCH_2CH_2^·$
*Propagation*:
$RCH_2CH_2^· + CH_2 = CH_2 \rightarrow RCH_2CH_2CH_2CH_2^·$

*Termination*:

$$RCH_2CH_2CH_2CH_2^\bullet + R^\bullet \rightarrow RCH_2CH_2CH_2CH_2R$$

Other polymers obtained by chain polymerization, with rupture of double covalent bonds and formation of new simple covalent bonds, are the *acrylic polymers*, showing the general formula:

$$-[CH_2 - CR-]_n-$$
$$|$$
$$COOR'$$

Among these is polymethyl methacrylate (PMMA) in which R and R′ are methyl groups ($CH_3$).

## STEP-GROWTH POLYMERIZATION

During the *step-growth* process, reactions can take place between monomers, dimers, trimers, or oligomers because the reactions take place in a multitude of sites (i.e., any two reactive molecules with the correct orientation and energy can react). Therefore reactions take place throughout a multitude of sites (see Fig. 1.17). For that, the increase of molecular weight occurs slowly, although the monomer is consumed rapidly. Increasing viscosity with reaction time prevents the mobility of molecules and reduces the rate of the reaction.

General characteristics of the stage polymerization are:

1. Polymer chains form slowly; sometimes the required time can be many hours or even many days.
2. All monomers are rapidly converted into oligomers, so the concentration of growing chains is high.
3. Because the chemical reactions involved have relatively high activation energies, the polymerization mixture is normally heated because the reaction is triggered.
4. Step polymerization generally results in polymers with moderately high molecular weights (<100,000).

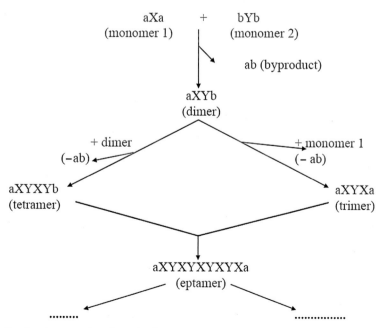

FIG. 1.17  Schematic of a step-growth polymerization.

**5.** No ramifications or cross-links occur unless a monomer with more than two functional groups is used.

Step-growth polymerization generally applies to difunctional monomers (i.e., containing two chemical functions or *functional groups*) able to react reciprocally according to the typical mechanisms of organic chemistry.

For example, a polyester can be obtained by reaction between a monomer containing two hydroxyl groups (i.e., a dialcol) and one containing two carboxylic groups (i.e., a diacid), similarly to what happens in the "traditional" organic chemistry, wherein an alcohol reacts with a carboxylic acid to give an ester.[6]

The reaction of obtaining a polyester can be globally indicated as follows:

[6] The formation of an ester takes place as follows: R-OH + R'COOH = R-O-CO-R' + $H_2O$.

$$\left( n\,HO-R-OH + n\,HOOC-R'-COOH \atop \text{(dialcol)} \qquad\qquad \text{(diacid)} \right)$$
$$\rightarrow (-O-R-O-CO-R'-CO-)_n + \underset{\text{(by product)}}{n\,H_2O}$$
$$\underset{\text{(polyester)}}{}$$

Other polymers obtained by step polymerization are *polyamides, polyurethanes, silicones,* and many others.

Table 1.1 presents the main differences between the two types of polymerization, and Table 1.2 lists the main polymers obtainable with them.

### 1.2.4 Copolymerization

Copolymers can be obtained either by chain polymerization or by step polymerization; in both cases, these processes are called *copolymerization*. When two or more different monomers are used in a chain copolymerization, a copolymer is obtained that contains the corresponding repeating units.

**TABLE 1.1**    Main Differences Between Chain- and Step-Growth Polymerization

| Parameters | Step-Grothi | Chain-Growth |
|---|---|---|
| Reactions | Only one type of reaction is responsible for polymer formation | Initiation, propagation, termination with different rates and mechanisms |
| Polymer growth | Any molecular species present can react; chain growth occurs slowly and randomly | The growth reaction occurs by addition of one unit at a time to the active end of the polymer chain |
| Polymer molecular weight | The molecular weight increases with the progress of the reaction; a high degree of conversion is required to achieve high molecular weights | High molecular weights are formed immediately |
| Monomer concentration during polymerization | The monomer is consumed in the first steps of reaction; at an average degree of polymerization = 10, <1% by weight of monomer is still present | The concentration of monomer decreases progressively throughout the course of the reaction |
| Composition of the reaction product | A relatively broad distribution of molecular species is obtained throughout the course of the reaction | Combinations of polymer chains with high molecular weight, monomer, and only a small amount of growing chains; this occurs both immediately after initiation and at the end of polymerization, since the 100% conversion is never achieved |

TABLE 1.2   Main Types of Polymers Obtained by Chain-Growth and Step-Growth Polymerization

| Chain-Growth | Step-Growth |
|---|---|
| Polyethylene (PE) | Polyesters (e.g., polyethylene terephthalate, PET) |
| Polypropylene (PP) | Polyamides (e.g., Nylon, Kevlar)[a] |
| Polystyrene (PS, PST) | Polycarbonate (PC) |
| Polyvinylchloride (PVC) | Epoxy resins |
| Polytetrafluoroethylene (PTFE) | Phenolic resins |
| Polyisoprene (also: natural rubber) | Acetal resins |
| Polyvinilacetate (PVA) | Polyurethanes[b] |
| Polymethyl acrylate (PMA) | Silicone resins |
| Polymethylmethacrylate (PMMA) | Polysulfones |
| Polyacrylonitrile (PAN) | Polyimides |

[a] *Amide bonds (HN—CO—) are formed from the reaction of amino groups (NH$_2$) with acid groups (-COOH).*
[b] *Urethane bonds (HN—CO—O) are formed from isocyanate (-NCO) and hydroxyl (-OH) groups.*

By varying the copolymerization technique and the quantity of each monomer, even starting from only two monomers, copolymers can be prepared with different properties. The amount of different materials that can be obtained increases enormously with the number of monomers employed, and their properties can be consequently varied; for this reason, most of the current synthetic polymers are represented by copolymers.

It should be underlined that in the case of step polymerization, unless starting from a single monomer containing the two reactive functions in the same molecule,[7] the synthesis reaction actually gives rise to alternating copolymers, as shown in Fig. 1.17. However, because the reciprocal reaction of the starting monomer's reactive functions give rise to new chemical functions, these products are commonly referred to by the name of the new chemical function. For example, if the new bond formed is of the ester type (-COOR-: e.g., -COOCH$_2$), the product is a *polyester*; if it is of the amide type (-CONHR-, e.g., -CONHCH$_2$-), the product is a *polyamide*.

Step copolymerization is currently widely used to obtain block copolymers starting from oligomers provided with the appropriate reactive and separately synthesized functions, instead of monomers:

$n$ aXXXXXXXa + $n$ bYb → a[(XXXXXXXY)$n$]b + $n$ ab or also

$n$ aXXXXa + $n$ bYYYb → a[(XXXXYYY)$n$]b + $n$ ab and so on.

The chain copolymerization can also be used for a cross-linking reaction, that is, the formation of bridging bonds more or less long. It is therefore possible to prepare thermoset products by exploiting the addition mechanism between oligomeric chains previously synthesized (e.g., with the step-growth mechanism) and molecules provided with reactive arms (in this case, double bonds C=C). These products can be referred to as copolymers, even if the most used term is that of resins or foams, besides the appropriate chemical name. A schematic example of this reaction is shown in Fig. 1.18.

[7] When starting from aXb, the homopolymer a[XXX … XX]b is in fact obtained.

**FIG. 1.18** Schematic example of a cross-linking reaction between bifunctional monomers and polyfunctional oligomers.

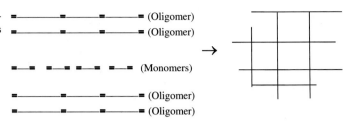

### 1.2.5 Hydrogels

A particular type of polymeric materials is represented by *hydrogels*.

Hydrogels can be defined as: "*Three-dimensional networks of polymer chains that swell, but don't dissolve in water*" (Kopecek, 2002); therefore hydrogels are water-swollen polymer networks.

The *gel* state defines solid, jelly-like materials, which exhibit no flow when in a steady state; in general, hydrogels are structures in which hydrophilic, water-insoluble, polymeric chains are dispersed in water and maintain their shape due to the presence of cross-linking and strong water retention. The cross-linking in hydrogels can be physical (chain entanglements) or chemical (Van der Waals, covalent, ionic, or hydrogen bonds).

Due to their physicochemical properties and high water content that can reach more than 99.9% by weight, hydrogels have found several applications in the pharmaceutical and biomedical fields.

The water-retaining capacity of the hydrogel is an intrinsic property of its structure and depends on the chemical nature of the polymer backbone and, above all, on the chemistry of its functional groups. Although the shape and strength of the hydrogel depend on the type and degree of cross-linking.

Fig. 1.19 exemplifies what happens when a network of polymeric chains changes from dry to hydrated forming a 3D hydrogel structure.

#### 1.2.5.1 Classification of Hydrogels

Hydrogels can be classified into two groups based on their *natural* or *synthetic* origins. Hydrogel-forming natural polymers include proteins such as collagen and gelatine, and polysaccharides such as chitosan, pectin, alginate, and agarose (see Section 1.6). Synthetic

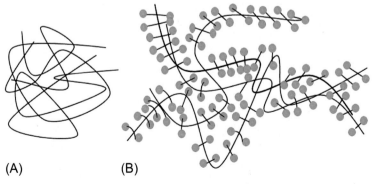

(A)                    (B)

**FIG. 1.19** Schematic representation of (A) a network of polymeric chains in collapsed dry form and (B) swollen polymer chains with water molecules adsorbed to produce a 3D hydrogel structure. *Modified from Agrawal, C.M., et al., 2013. Introduction to Biomaterials: Basic Theory with Engineering Applications. Cambridge Texts in Biomedical Engineering, first ed. Cambridge, Cambridge University Press.*

polymers such as acrylate-based polymers that form hydrogels are traditionally prepared using chemical polymerization methods.

Hydrogels based on natural polymers may vary in their composition (due to their natural origin) and consequently in their properties. Synthetic polymers, on the other hand, can be produced with high fidelity in their molecular weight and composition; therefore their physicochemical properties are more consistent and uniform than natural polymer-based hydrogels.

According to the method of preparation that determines their composition (Ahmed, 2015):

(a) *Homopolymer hydrogels*: Polymer networks derived from a single type of hydrophilic monomeric unit that is cross-linked. Their cross-linked skeletal structure depends on the nature of the monomer and polymerization technique.

(b) *Copolymer hydrogels*: Comprised of two or more different monomer species with at least one hydrophilic component, arranged in a random, block, or alternating configuration along the chain of the polymer network.

(c) *Interpenetrating network hydrogels* (IPN): Made of two independent cross-linked synthetic and/or natural polymer components, contained in a network form. In semi-IPN hydrogel, one component presents a cross-linked structure, and other component is a noncross-linked polymer.

Another possibility of classification is related to the type of cross-linking: chemical or physical. *Chemically cross-linked* networks have permanent junctions, whereas *physical networks* have transient junctions that arise from either polymer chain entanglements or physical interactions such as ionic interactions, hydrogen bonds, or hydrophobic interactions.

According to the electrical charge of the network, hydrogels may be categorized as:

(I) nonionic (neutral)
(II) ionic (including anionic or cationic)

(III) amphoteric electrolyte (ampholytic) containing both acidic and basic groups
(IV) zwitterionic, containing both anionic and cationic groups in each structural repeating unit

In addition, hydrogels' appearance (e.g., bulk, film, membrane, or microsphere) depends on the technique involved in the preparation process.

### 1.2.5.2 Synthesis of Hydrogels

The fabrication of polymers with high water absorption capacity is usually the first step. For this, polymers with desirable functional groups can be synthesized or obtained by modification of an existing polymer. The synthesis of application-specific copolymers or block copolymers is also common.

The next step is the fabrication of cross-linked networks that can be generated using a variety of methods. As an example, short di- or multifunctional linkers can be used to react with long polymer chains, or the cross-linking can occur during the polymerization process if multifunctional monomers are involved. Some of these reactions can be driven by energy provided by UV light as in photopolymerization, electron beams, and radiation such as X-rays or gamma rays to produce free radicals that react to create a cross-linked structure.

Examples of synthetic methods are schematically shown in Figs. 1.20–1.23. All of these take place in an aqueous environment.

## 1.2.6 Physical States of Polymers

### 1.2.6.1 Intermolecular Bonding Forces

Polymers are known to exhibit a wide range of properties. Some are tough and withstand high deformations without breaking, others are rigid and very resistant, others are soft and flexible, and still others considerably resist to impact. All these properties are peculiar to the

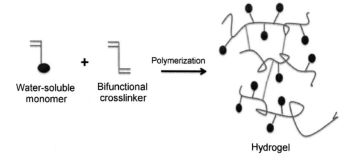

**FIG. 1.20** Synthesis of a hydrophilic polymer network by copolymerization of a water-soluble monomer and a bifunctional cross-linker.

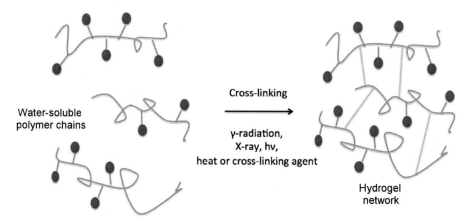

**FIG. 1.21** Synthesis of a hydrogel by cross-linking preobtained water-soluble polymer chains.

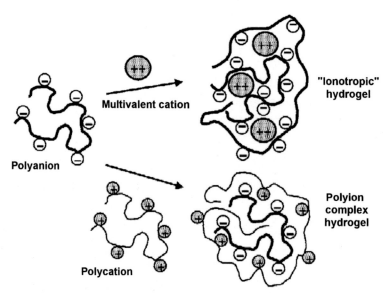

**FIG. 1.22** Schematic of methods for formation of two types of ionic hydrogels. An example of an "ionotropic" hydrogel is calcium alginate, and an example of a polyionic hydrogel is a complex of alginic acid and polylysine.

A. INTRODUCTION TO MATERIALS

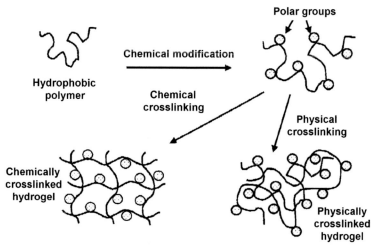

**FIG. 1.23** Synthesis of hydrogels by chemical modification of hydrophobic polymers.

polymer and not characteristics of the starting monomers.

The unusual behavior of polymers is due to the huge amount of interactions among macromolecular chains consisting of various types of intermolecular bonds and physical interconnections. The magnitude of these interactions depends on the nature of the intermolecular bonding forces, on the molecular weight, on the way the chains are reciprocally arranged, and on the flexibility of the polymer chain. For these reasons, the amount of interactions is different in different polymers and very often different even in different samples of the same polymer.

The secondary bonding forces present in the polymers, that is, the electrostatic forces of Van der Waals and dipole-dipole type, are the same as those found in small molecules. In polymers, however, all types of electrostatic forces can be present and act between different portions of the same molecule. Although the individual energies are low, between 0.5 and 10 kcal/mol, the cumulative effect of thousands of these bonds along the polymeric chain results in a large electrostatic attraction field.

In nonpolar linear polymers, such as polyethylene, there are only Van der Waals-type attractions between the chains, and therefore these polymers must have high molecular weights and a very close reciprocal arrangement of the chains to present useful mechanical properties.

Many commercial polymers contain polar functions (e.g., ester: -COOR; nitro-: -NO$_2$; cyano-: -CN groups; halogens: Cl, F, Br, etc.), which allow the formation of strong dipole-dipole interactions. These functions can be either incorporated or lateral (pendent) to the chain. Dipole-dipole interactions obviously are determined by the spatial arrangement and can be increased by suitably orienting the polymer chains.

Polymers exhibiting dipole-dipole interaction, such as polyamides including nylon, and polyurethanes show better mechanical properties.

Another type of polymers, the *ionomers*, have pendent carboxylate groups (COO$^-$) associated with metal cations and display strong ionic interactions between the chains, with consequent increase in the mechanical strength of the material.

Some typical intermolecular bonds in the polymers are shown in Fig. 1.24.

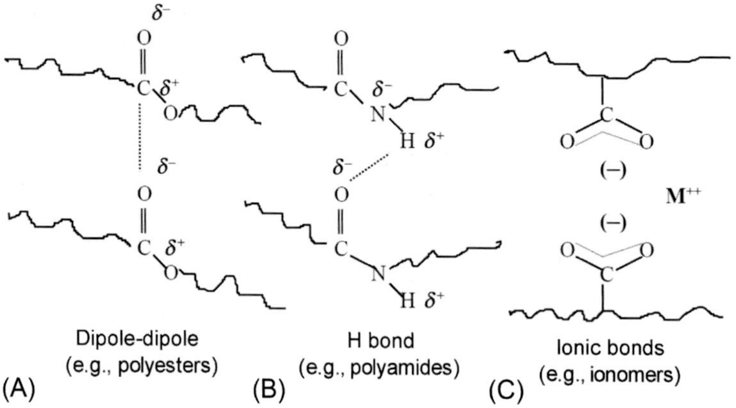

FIG. 1.24   Intermolecular bonds in polymers of the type: (A) dipole-dipole, (B) hydrogen, (C) ionic.

The intensity of the electrostatic forces raises with increasing molecular weight (length) of the chains, because the number of possible interactions increases as well. This also involves changes in the physical state of the polymer (as seen for polyethylene in Fig. 1.30).

It is evident that each type of polymer, based on its chemical structure, can have different types of intermolecular bonding forces. Therefore every type of polymeric material will have a *critical molecular weight*, below which the intensity of the intermolecular forces is not sufficient to impart mechanical properties to the material.

This critical molecular weight, if expressed as the average degree of polymerization ($\overline{X}$), will be, for example, about 40 for polyamides and above 100 for polyhydrocarbons (or polyolefins, such as polyethylene). Above this value, the mechanical properties increase significantly until a second critical value of $\overline{X}$ is reached, above which the mechanical properties no longer vary sensitively. This second value is about 200 for polyamides, whereas it is more than 500 for polyhydrocarbons.

Most of the other polymers require intermediate values between those previously mentioned. Once the value of the second critical point of $\overline{X}$ is reached, the properties become characteristics of that polymer. We also talk about a *molecular weight threshold* to indicate the minimum molecular weight that a polymer must reach to demonstrate the properties required by a particular application. On the other hand, there is also an upper limit of molecular weight above which the processability (workability) of the material becomes practically impossible due to flow resistance phenomena.

It should also be noted that not all the physical properties of a polymer depend on the molecular weight or on the intensity of molecular interactions. For example, the index of refraction of a polymer, the color, and the electrical properties are properties that do not depend on the molecular weight.

### 1.2.6.2 Configuration and Conformation in Polymers

*Configuration* describes the spatial position of atoms within the polymer, namely the geometrical arrangement arising from the order of atoms determined by chemical bonds. The configuration of a polymer cannot be altered unless chemical bonds are broken and reformed.

The regularity and symmetry of the side groups can strongly affect the properties of polymers.

The polymer is *symmetric* when the carbon atoms in the backbone have the same substituents or side groups (Fig. 1.25A). If the

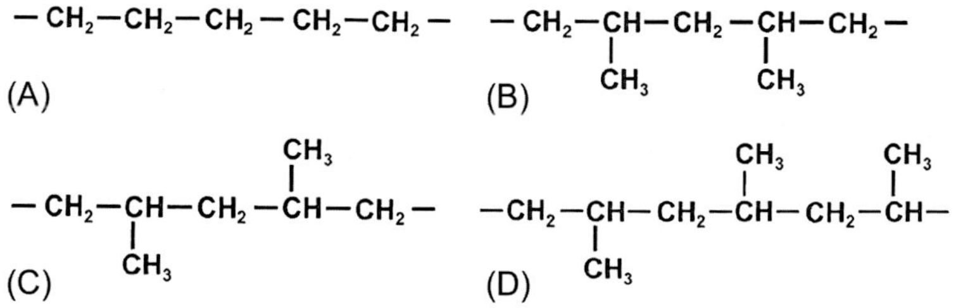

$$— CH_2—CH_2—CH_2—CH_2—CH_2 —$$

(A)

$$— CH_2—CH—CH_2—CH—CH_2—$$
with CH_3, CH_3

(B)

$$— CH_2—CH—CH_2—CH—CH_2 —$$
with CH_3, CH_3

(C)

$$—CH_2—CH—CH_2—CH—CH_2—CH—$$
with CH_3, CH_3, CH_3

(D)

**FIG. 1.25** Simplified different forms for a polymeric molecular chain. (A) no asymmetry, (B) *isotactic*—each pendant group is on the same side of the chain, (C) *syndiotactic*—the pendant groups on alternate sides of the chain in a regular fashion, and (D) *atactic*—the pendant groups on apposite sides of the chain in a random fashion. Example (A) relates to linear polyethylene; examples (B–D) apply to polypropylene.

carbon atoms in the polymer backbone have different side groups, they are called asymmetric; because the carbon atoms are tetrahedral, the asymmetric atoms can exist in different spatial arrangements, and the transition from one to another configuration cannot occur without breaking bonds.

*Tacticity* refers to the order of the succession of configurationally repeating units in the main chain of a polymer molecule. If the side groups are linked in the same order, being all on the same side of the chain, the configuration is called *isotatic* (Fig. 1.25B). Instead, a *syndiotactic* configuration exists when the side groups are at alternative sides of the chain (Fig. 1.25C). In the *atactic* configuration, the side groups are randomly positioned along the polymer backbone (Fig. 1.25D)

*Conformation* refers to order that arises from the rotation of polymer chains around the single covalent bonds. The structure can be changed via simple rotation about a bond.

The different macroscopic properties of polymers correspond to the conformational characteristics: the high viscosity and the rheological characteristics of polymers melted or in solution, the orientability of the macromolecules under mechanical stress or speed gradients, or the possibility of crystallization under stretching, elasticity, and viscoelasticity.

As we have seen, the relative succession of the various atoms constituting the macromolecules, bound to each other by covalent bonds, determines the configuration of the macromolecules. This configuration cannot be changed without changing the distribution of intramolecular bonds.

The real shape that the macromolecule takes in space is, however, influenced by the possibility of rotation around the simple C—C bonds, which assume an angle of 109 degrees in space, as illustrated in Fig. 1.26. This aspect concerns the conformation of the macromolecule.

(A)                              (B)

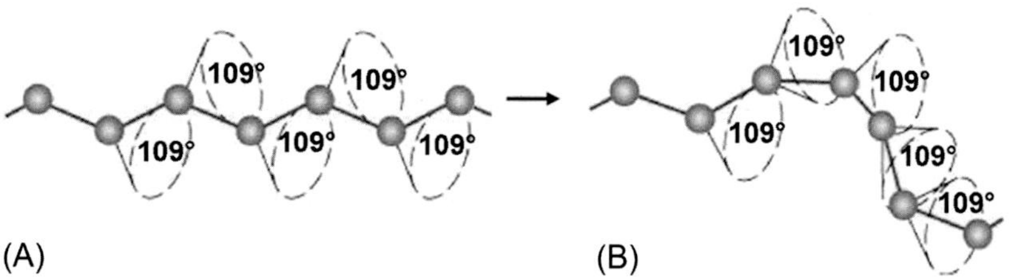

**FIG. 1.26** The possibility of chain bending (A) and twisting (B) of a polymer chain segment is due to the rotation of C atoms around their chain bonds, typically of 109° (tetrahedral sp$^3$ carbon). Because there are hundreds of C—C bonds in a macromolecule, the chains can bend, coil, kink, intertwine, and entangle.

### 1.2.6.3 *Amorphous and Crystalline State in Polymers*

The spaces left free are occupied by other similar molecules, so as to form a tangle that can be imagined as a ball made of several threads. This is true for polymers in a disordered (*amorphous*) state in which there are no constraints other than steric or energetic ones.

However, the macromolecules can also assume regular conformations, as it happens in an ordered state such as the *crystalline* one. In this case, the possible conformations assume an ordered appearance in space. The conformation of the macromolecules of a polymer can therefore be both amorphous and crystalline.

Whereas small (nonpolymeric) molecules are usually either completely amorphous or completely crystalline, polymers can exist in different states depending on the phase, configuration, and arrangement of their molecular chains. It is common for polymers to exist in a *semicrystalline* form wherein both amorphous and crystalline phases coexist.

The conformation of the macromolecules, in the whole, has great influence in determining the characteristics of the specific polymeric material.

**EXAMPLE**

In linear polyethylene macromolecules, the possibility of rotation of the C—C bonds is in principle similar to that of the chain of carbon atoms in Fig. 1.26. A linear chain also has a high probability of being arranged in an ordered manner, for example, if it is oriented in a principal direction by mechanical action (under drawing), so it can assume a crystalline conformation.

However, if the polyethylene molecules are partially branched, as in Fig. 1.27, the freedom of rotation of the individual C—C bonds is restricted by the steric hindrance of the ramifications, and the number of possible disordered or ordered conformations is highly reduced. In particular, there will be less chance of assuming the crystalline conformation.

Many polymers, especially those with a more regular molecular structure and under certain conditions, tend to assume a crystalline structure. This structure is characterized, as in the compounds of low molecular weight, by a

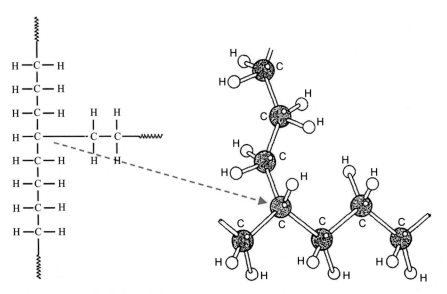

**FIG. 1.27**   Ramification in a polyethylene chain.

geometrically determined arrangement within the *elementary cell* of the constituent atoms, and the elementary cell is repeated with its characteristics of geometrical symmetry in the three dimensions of the space until forming the crystalline lattice that defines the whole crystal. However, in the low molecular weight compounds, the elementary cell generally comprises more molecules, more ions, or more atoms (depending on whether they are molecular, ionic, or metallic solids); in the polymers, the same molecule comprises more elementary cells and sometimes even more crystals.

In any case, the crystallinity of polymers is never complete and the crystalline regions alternate with amorphous zones, and generally the same molecule includes both crystalline parts and amorphous parts, in successive sections (Fig. 1.28). This leads to the presence of marked irregularity of the crystals, which are generally referred to as *crystallites*.

The crystallinity of a polymer is defined as the percentage ratio (C%) between the weight of the substance having a crystalline structure ($w_c$) and the total weight of the material ($w_{tot}$). In polymers with a strongly asymmetric monomer unit, with irregular or strongly branched organization or with frequent transversal bonds, the crystallinity is null or in any case very low, whereas in nonbranched polymers, it can approach 100% but never reaches this value.

$$C\% = \frac{w_c}{w_{tot}} \times 100$$

Crystallinity also depends, for the same type of macromolecular configuration, on the thermal and mechanical history of the sample. Above all, in the case of long linear chains, mechanical processes such as lamination, stretching, and extrusion (especially if performed at temperatures where the mobility of the macromolecules is high) tend to orient the chains parallel to each other, thus favoring the formation of crystallinity.

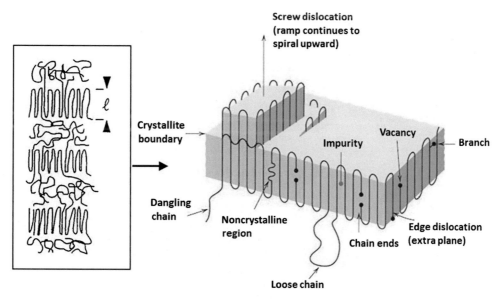

**FIG. 1.28** Crystalline and amorphous regions with schematic representation of imperfection points in the crystalline lattice of a polymer.

## CRYSTALLIZATION PROCESS OF POLYMERS

The process of crystallization of a polymer may be directly followed by an optical microscope; in fact, oppositely to what usually happens for low molecular weight substances, time required for transformation from liquid to solid for polymers is generally quite long, allowing easy observation of the event.

Observation under an optical microscope shows that, starting from a crystallization nucleus, spherical bodies (*spherulites*) are formed, which progressively increase their diameter over time to occupy the whole polymeric mass. Spherulites are made of *fibrils* arranged radially, and fibrils, in turn, are formed by aggregates of lamellar crystallites with a thickness of about 100 Å. Inside the fibrils, *lamellar crystallites* are arranged in such a way that the chain axes are perpendicular to the direction of radial extension of the fibrils, as shown in Fig. 1.29. Among the crystallites, there are amorphous zones constituted by chain fractions with irregular conformations that enter and exit from different crystallites, tying them to each other, as also shown in Fig. 1.28.

The microstructure of semicrystalline polymers can be therefore represented at three levels. In polarized light, they are usually made up of densely packed spherulites, whose diameter is between tens and hundreds of micrometers. In turn, each spherulite is a radial set of narrow crystalline lamellae oriented according to different planes. Within each lamella, the densely packed polymer chains bend back and forth between two limits. The spaces between the lamellae are occupied by amorphous regions, in which the molecules appear in a tangle without order.

Because the crystallites represent more tightly packed chains, crystalline or semicrystalline polymers have a higher density and are usually stiffer, tougher, more resistant to solvents, and opaque. The higher stiffness and lower solubility of the crystalline phase compared to the amorphous one are caused by strong intermolecular forces, which result from the close packing. The higher toughness is a reflection of the intermixing of the amorphous and crystalline regions whereas the opacity is a result of light scattering by the crystallites.

In conclusion, the mechanical properties of a semicrystalline polymer increase by increasing not only the molecular weight, but also crystallinity. As an example, Fig. 1.30 shows the trend of the properties of polyethylene as a function of crystallinity and of molecular weight.

FIG. 1.29 Schematic representation of a spherulite. The magnified area represents the internal structure of a fibril that extends radially in the R direction R. *From Fig. 6.13 in Chapter 6, Characterization of Polymer Biomaterials, Elsevier, 2017.*

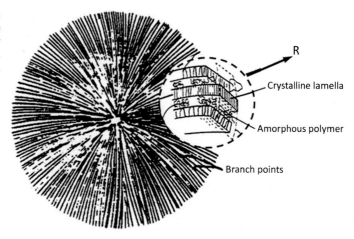

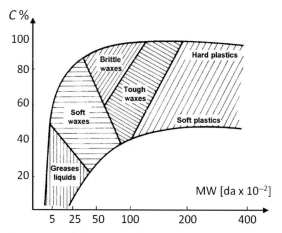

**FIG. 1.30** Trend of polyethylene properties as a function of crystallinity (C) and molecular weight (MW).

## 1.2.7 Thermal Transitions in Polymers: $T_m$ and $T_g$

All polymeric materials undergo thermal transitions as the temperature varies. Depending on the type of polymeric material, there is a temperature in which the amorphous part can pass from a "frozen" state (*glassy state*), where it is rigid and often even fragile, to a plastic-rubbery state, where it acquires flexibility and deformability.

The temperature at which the transition from the glassy state to the rubbery state, solidly indicated with $T_g$, is called the *glass transition temperature*. The $T_g$ is fairly characteristic data of the different polymers, practically independent of the molecular weight above a certain value of the latter, and very important from the practical point of view.

Crystalline polymers have a melting temperature $T_m$. However, usually the melting takes place over a range of temperatures instead of one specific temperature because each specimen contains polymeric chains of various molecular weights and crystallites of many sizes.

Fully amorphous polymers do not have a $T_m$ as they do not contain crystallites, however at temperatures higher than $T_g$, a second transition is observed, when the transition from solid to liquid

(from rubber to viscous liquid) occurs. The temperature range at which this phase transition occurs for amorphous polymeric materials is called **softening** temperature. Anyway, most nonamorphous polymers are semicrystalline, and therefore they exhibit both a $T_g$ and a $T_m$. As a general rule, the ratio of $T_m$ to $T_g$ lies between 1.4 and 2.0 ($T_g$ is lower than $T_m$).

Crystallization in a polymer does not take place until the temperature is lowered below $T_m$, and no effective crystallization occurs below $T_g$. Thus the maximum rate of crystallization takes place between $T_g$ and $T_m$.

On the other hand, an amorphous phase can be achieved by the rapid cooling of a polymer after melting. This fast cooling does not give the chains enough time to organize and orient themselves to form crystallites.

Depending on the structure of the amorphous-crystalline material, the melting temperature may or may not be preponderant on the softening temperature of the rubbery part.

In Table 1.3, the values of $T_g$ and $T_m$ for the most significant polymers are reported.

## 1.2.8 Other Properties of Polymeric Materials

Due to their peculiar properties, in general, polymers display thermal and electrical insulation properties and have low specific weight ($0.9$–$2\,\mathrm{g/cm^3}$), as they consist mainly of elements of low atomic number (carbon, hydrogen, oxygen).

It is also worth noting that polymeric materials soften and decompose at relatively low temperatures (generally <250–300°C), greatly limiting their use at high temperatures.

## 1.3 METALLIC MATERIALS

Metals and metal alloys were not discovered in the last few centuries, as the bronze period fell in 3000 BC, and the iron period was in 1200 BC.

**TABLE 1.3**  Glass Transition ($T_g$) and Melting Temperature ($T_m$) of Some Polymers

| Material | $T_g$ (°C) | $T_m$ (°C) |
|---|---|---|
| Polyethylene (PE) | −30 ÷ −100 | +137−141 |
| Polytetrafluoroethylene (PTFE, Teflon®) | −122, −73 | +335 |
| Polydimethylsiloxane (PDMS, a silicon rubber) | −120 | − 40 |
| Natural rubber (*caoutchouc*) | −73 ÷ −70 | — |
| Polypropylene (PP) | −10 | +174 |
| Polyamide 6,6 (Nylon 6,6) | $\cong 50$ | +255 |
| Polyethylene terephthalate (PET) | $\cong 70$ | +265 |
| Polyvinylchloride (PVC) | $\cong 80$ | ($\leq$ 11% cryst.) |
| Polystyrene (PS, PST) | $\cong 100$ | — |
| Polymethylmethacrylate (PMMA) | $\cong 120$ | — |
| Polyacrylonitrile (PAN) | +87 | +319 |
| Polycarbonate (PC) | +150 | — |

Metals are some of the most widely used materials on the planet, as well as being one of the most extracted. Each extracted metal has different properties important to be investigated, so a specific type of metal can be used in the most appropriate application. Most of the metals are solids in nature except for mercury (Hg), which shows liquid-like motion.

## 1.3.1 Structure

Metal atoms are bonded together by metal bonds, forming regular crystalline structures in 3D space. Atoms are arranged in a network called space lattice that regularly occupies the 3D space. In each space lattice, a unit cell can be identified, where atoms are arranged in a unique motif. A.J. Bravais identified 14 standard unit cells (Table 1.4) that describe the possible lattice networks.

Most of the metals (about 90%) crystallize into three densely packed crystal structures

(Table 1.4), reaching a lower and more stable energy state:

– body-centered cubic (BCC),
– face-centered cubic (FCC),
– hexagonal close-packed (HCP).

### 1.3.1.1 Body-Centered Cubic (BCC) Unit Cell

In a body-centered cubic (BCC) unit cell, the central atom is surrounded by eight atoms (Fig. 1.31A). If a single unit cell is isolated (i.e., highlighting the portion of each atom belongs only to one single cell), the number of atoms per unit cell is two atoms. In fact, one atom is located in the center of the cubic cell and eight of each atom is located at the cube corners:

$$1\,(\text{atom at the center}) + 8$$
$$\times \frac{1}{8}(\text{atoms at the corners})$$
$$= \textbf{2 atoms per BCC unit cell}$$

**TABLE 1.4**  Classification of Bravais Space Lattices

| Crystal System | Space Lattice (Unit Cell) | |
|---|---|---|
| Cubic | Simple cubic | |
| | Body-centered cubic | |
| | Face-centered cubic | |
| Tetragonal | Simple tetragonal | |
| | Body-centered tetragonal | |

*Continued*

**TABLE 1.4**  Classification of Bravais Space Lattices—cont'd

| Crystal System | Space Lattice (Unit Cell) | |
|---|---|---|
| Orthorhombic | Simple orthorhombic | 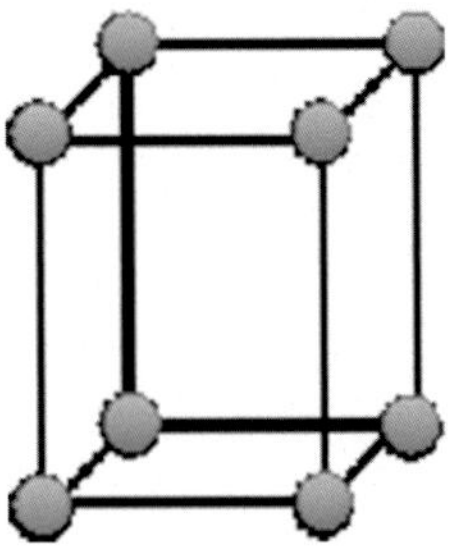 |
| | Body-centered orthorhombic | 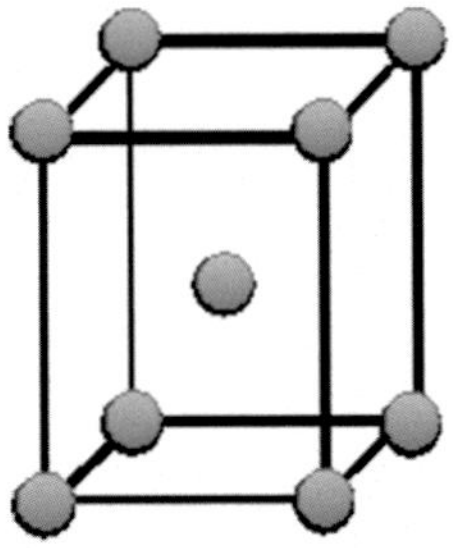 |
| | Base-centered orthorhombic | 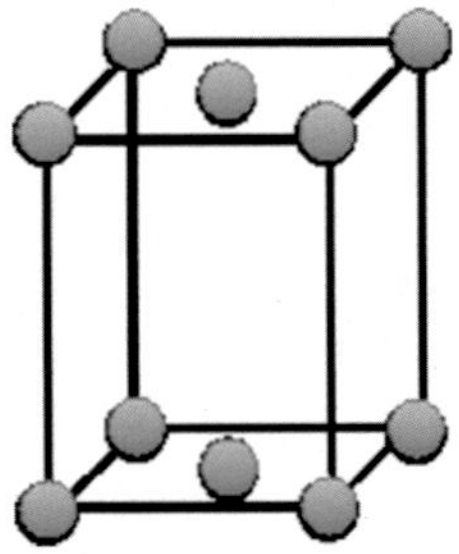 |
| | Face-centered orthorhombic | 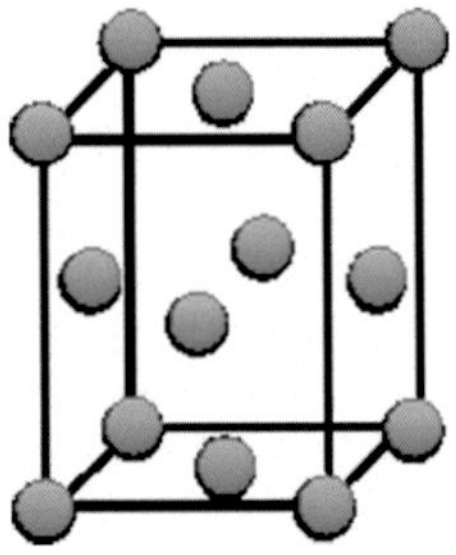 |

**TABLE 1.4**  Classification of Bravais Space Lattices—cont'd

| Crystal System | Space Lattice (Unit Cell) | |
|---|---|---|
| Rhombohedral | Simple rhombohedral | |
| Hexagonal | Simple hexagonal | |
| Monoclinic | Simple monoclinic | |
| | Base-centered monoclinic | |
| Triclinic | Simple triclinic | |

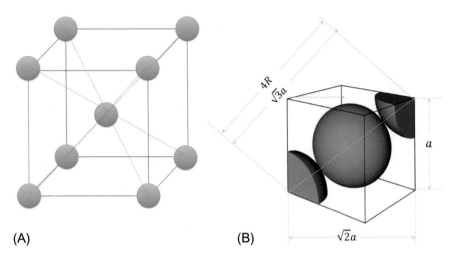

(A)                                              (B)

**FIG. 1.31**  Body-centered cubic (BCC) unit cell: (A) atomic site unit cell; (B) relationship between the lattice constant $a$ and the atomic radius $R$.

To calculate the lattice constant, the atoms are nearer across the cube diagonal (Fig. 1.31B), so that a relation between the length of the cube side $a$ and the atom radius $R$ is as follows:

$$\sqrt{3}a = 4R$$

$$a = \frac{4R}{\sqrt{3}}$$

The atomic packaging factor (APF) can be found for BCC unit cell, based on the definition of APF:

$$APF = \frac{\text{volume of atoms in unit cell}}{\text{volume of unit cell}}$$

Hence, in the case of BCC unit cell, APF is:

$$\text{Volume of atoms in unit cell} = 2 \times \left(\frac{4}{3}\pi R^3\right)$$

$$\text{Volume of unit cell} = a^3$$

Taking into account the definition of APF, it is possible to calculate the APF for the BCC unit cell:

$$APF = \frac{2 \times \frac{4}{3}\pi R^3}{\left(\frac{4}{\sqrt{3}}R\right)^3} = \frac{8.373\,R^3}{12.32\,R^3} = 0.68$$

As the APF is 68%, the volume of the BCC unit cell occupied by atoms is 68% of the entire volume of the cell, and the remaining 32% is empty space. The BCC unit cell is characteristic of metals that possess high hardness and average ductility; examples of these metals are iron (Fe), tungsten (W), chromium (Cr), and tantalum (Ta).

### 1.3.1.2 Face-Centered Cubic (FCC) Unit Cell

In a face-centered cubic (FCC) unit cell, eight atoms are at each corner of the cubic cell, and one atom is at the center of each cube face (Fig. 1.32A). If a single FCC unit cell is isolated, the number of atoms per unit cell is four atoms. In fact, one atom is located in the center of each cubic cell face and eight of each atom is located at the cube corners:

$$6 \times \frac{1}{2}\,(\text{atom at the center of each cube face}) + 8$$

$$\times \frac{1}{8}(\text{atoms at the corners})$$

$$= 4\,\textbf{atoms per FCC unit cell}$$

To calculate the lattice constant, the atoms are closer across the face diagonal (Fig. 1.32B), so that a relation between the length of the cube side $a$ and the atom radius $R$ is:

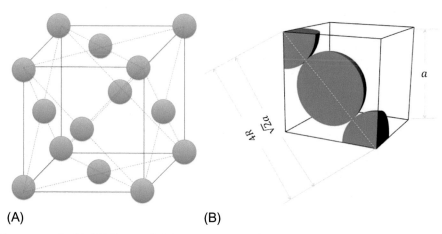

**(A)**                                     **(B)**

**FIG. 1.32** Face-centered cubic (FCC) unit cell: (A) atomic site unit cell; (B) relationship between the lattice constant $a$ and the atomic radius $R$.

$$\sqrt{2}a = 4R$$

$$a = \frac{4R}{\sqrt{2}}$$

The APF for FCC unit cell is:

$$\text{Volume of atoms in unit cell} = 4 \times \left(\frac{4}{3}\pi R^3\right)$$

$$\text{Volume of unit cell} = a^3$$

Taking into account the APF definition, it is possible to calculate the APF for the BCC unit cell:

$$\mathbf{APF} = \frac{4 \times \frac{4}{3}\pi R^3}{\left(\frac{4}{\sqrt{2}}R\right)^3} = 0.74$$

As the APF is 74%, the volume of the FCC unit cell occupied by atoms is 74% of the entire volume of the cell, indicating a higher close-packed structure compared to BCC. The FCC unit cell is characteristic of metals that have high ductility and are good electrical and heat conductors; examples of these metals are copper (Cu), aluminum (Al), lead (Pb), and silver (Ag).

### 1.3.1.3 Hexagonal Close-Packed (HCP) Unit Cell

Hexagonal close-packed unit cell (Fig. 1.33) is one of the most common metallic structures, where atoms are closer together, forming a stable and low-energy condition. The APF of this unit cell is the same as the FCC, that is, 74%, demonstrating the high close-packed condition.

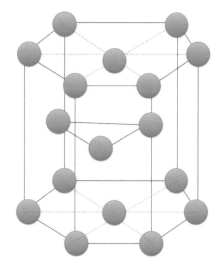

**FIG. 1.33** Hexagonal close-packed (HCP) unit cell: schematic of the crystal structure.

If a single HCP unit cell is isolated, the number of atoms per unit cell is six atoms. In fact, three atoms are located in the middle of the structure, forming a triangle; six atoms are located in the corners of the two hexagonal faces (i.e., bottom and top face); and one atom is at the center of each of the two hexagonal faces:

$$3 \left(\text{atoms in the middle layer}\right) + 2 \times 6$$
$$\times \frac{1}{6} \left(\text{atom at the corners of each hexagonal face}\right)$$
$$+ 2 \times \frac{1}{2} \left(\text{atoms at the center of each}\right.$$

hexagonal face) = **6 atoms** per **HCP** unit cell

The HCP unit cell is characteristic of brittle metals; examples of these metals are titanium (Ti), cobalt (Co), zirconium (Zr), and magnesium (Mg).

### 1.3.1.4 Polymorphism

Each metal assumes a unit cell that is the one in which the metal is at lower energy and stable conditions, but they can reach these conditions assuming different unit cells in different temperature ranges. The phenomenon by which metals and alloys can exist in more than one crystalline unit cell is called *polymorphism*, or *allotropy*.

For example, iron can exist in BCC and FCC in different temperature ranges (Fig. 1.34):

- at lower temperatures (i.e., from $-273°C$ to $912°C$) iron $\alpha$ has a BCC unit cell
- in the temperature range $912–1394°C$, iron $\gamma$ has a FCC unit cell

- over $1394°C$ up to $1539°C$ (i.e., melting temperature for iron), iron $\delta$ has a BCC unit cell.

### 1.3.2 Defects of the Crystalline Structure

Metal crystalline structure can contain different defects that affect the material properties, thus influencing engineering properties. The defects in crystal lattice allow the metal or the alloy to have determinate properties. As an example, defects can influence mechanical, physical, chemical, or electrical properties.

Crystal defects can be classified as follows:

- point defects,
- line defects (i.e., dislocations),
- planar defects (e.g., grain boundaries, precipitates).

In addition to those defects, macroscopic and bulk defects can be included; examples of these defects are pores, cracks, and inclusions.

### 1.3.2.1 Point Defects

Point defects can be *vacancies*, *interstitial atoms*, or *substitutional atoms*. The simplest defect is vacancy, that is, an atom site where the atom is missing (Fig. 1.35). Vacancies can move by exchanging position with their neighbors. They are important in the migration and diffusion of atoms in the solid-state process, in particular at high temperatures, where the mobility of the atoms is greater than at lower temperatures.

When an atom of the same metallic element or another element occupies an interstitial site

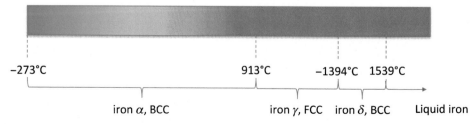

FIG. 1.34    Allotropic forms of iron: iron $\alpha$, iron $\gamma$, and iron $\delta$.

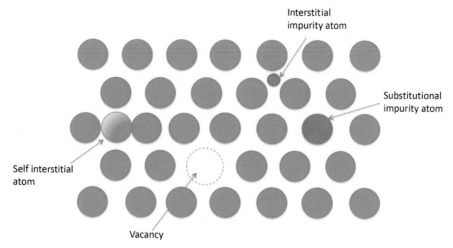

**FIG. 1.35** Point defects: vacancy, self and impurity interstitial atom, substitutional impurity atom.

among surrounding atoms in a normal atom position (Fig. 1.35), the defect is named *self-interstitial atoms*, or *interstitial impurity* atoms. An interstitial atom can determine a structure distortion. In particular, self-interstitial atoms occur only in low concentrations in metals because they distort and highly stress the tightly packed lattice structure. Interstitial impurity atoms are much smaller than the atoms in the bulk matrix. Interstitial impurity atoms fit into the open space between the bulk atoms of the lattice structure.

If an atom of a different element occupies a normal atom site of the crystal cell, the defect is called *substitutional* atom (Fig. 1.35). Hence, a substitutional impurity atom is an atom of a different type than the bulk atoms, which has replaced one of the bulk atoms in the lattice. Substitutional impurity atoms are usually close in size to the bulk atom.

### 1.3.2.2 Line Defects (Dislocations)

Line defects or dislocations are line imperfections, where the atoms are out of position in the crystal structure. Dislocations are generated and move when a stress is applied. The motion of dislocations allows slip—plastic deformation

to occur. There are two basic types of dislocations: edge dislocation and screw dislocation. Actually, edge and screw dislocations are just extreme forms of the possible dislocation structures that can occur. Most dislocations are probably a hybrid of the edge and screw forms.

Understanding the movement of a dislocation is a key issue to understand why dislocations allow deformation to occur at much lower stress than in a perfect crystal.

The dislocations move along the densest planes of atoms in a material, because the stress needed to move the dislocation increases with the spacing between the planes. FCC and BCC metals have many dense planes, so dislocations move relatively easily and these materials have high ductility. Moreover, when a material plastically deforms, more dislocations are produced and they will get into each other's directions and impede further movement, which increases the mechanical strength of the material.

Plastic deformation involves the breaking of a limited number of atomic bonds by the movement of dislocations. Recall that the force needed to break the bonds of all the atoms in a crystal plane all at once is very great, because of the strong nature of the metallic bonds.

However, the movement of dislocations allows atoms in crystal planes to slip past one another at a much lower stress levels. Because the energy required for moving is lowest along the densest planes of atoms, dislocations have a preferred direction of travelling within a grain of the material; this results in slip[8] that occurs along parallel planes within the grain. These parallel slip planes group together to form slip bands, which can be seen with an optical microscope. A slip band appears as a single line under the microscope, but it is in fact made up of closely spaced parallel slip planes.

Hence, dislocations are very important for the deformability of the metal or metal alloy, referring to plastic deformation. When a stress or a strain is applied to the metallic material at low temperature, dislocations move, and their density can significantly increase with cold deformation. In addition, new dislocations are formed by the cold deformation and interact with those already present. As a consequence of that, density of dislocation increases; it becomes more and more difficult for the dislocations to move through them, and the metal strain hardens with increased deformation. Strain hardening is one of the most important methods used for strengthening metal and metal alloys.

If the metal is plastically deformed at high temperature, recrystallization occurs. Recrystallization is a process accomplished by heating, whereby deformed grains are replaced by a new set of grains that nucleate and grow until the original grains have been entirely consumed.

### 1.3.2.3 Planar Defects

Planar defects include external surfaces, grain boundaries, twins, low-angle boundaries, high-angle boundaries, twists, and stacking faults.

Among them, grain boundaries are surface defects in polycrystalline materials that separate grains (i.e., crystals) of different orientation, formed during the solidification of the metal or metal alloy. In particular, grain boundary is a region separating two crystals (grains) of the same phase.[9] These two grains differ in mutual orientations, and the grain boundary thus represents a transition region, where the atoms are shifted from their regular positions as compared to the crystal interior. Shape and dimension of the grain and grain boundaries depend on the restrictions imposed by the growth of neighboring grains. In addition, the atomic packing in grain boundaries is lower than within the grains because of the atomic mismatch. Grain boundaries also have atoms in strained positions that cause an increase in the energy of the grain-boundary region.

The solidification of metal or metal alloy is composed of different steps (Fig. 1.36):

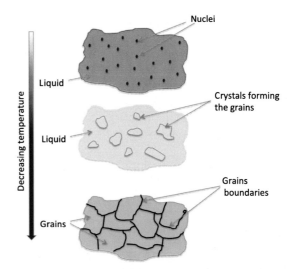

FIG. 1.36   Scheme of metal or metal alloy solidification: when the temperature decreases from melting temperature, nuclei are formed. Decreasing the temperature, nuclei grow into the crystal. At the end of crystal growth, grains and grain boundaries are formed.

---

[8] Slip: the process of atoms moving over each other during the permanent plastic deformation.

[9] Phase: a homogenous portion of a system that has uniform physical and chemical characteristics.

1. formation of stable nuclei in the melting metal/alloy (i.e., nucleation),
2. growth of nuclei into crystals,
3. formation of grain boundaries.

Shape and dimensions of grains are very important because they can influence the mechanism of fracture within the metallic material, depending on the working temperature, as well. In fact, at lower temperature, grain boundaries have more strength than the grains, that is, grains are weaker, and the fracture occurs through the grains (i.e., transgranular fracture; Fig. 1.37A). Indeed, at high temperature, grain boundaries are weaker regions, and the fracture occurs through them (i.e., intergranular fracture; Fig. 1.37B). There is a particular temperature at which transgranular fracture changes to intergranular fracture, and this temperature is called the *equicohesive* temperature.

If a metal component has to work at low temperature, it should be better by reducing the average grain size, to improve its mechanical properties. On the contrary, if the component works at high temperature, the average grain size should be as wide as possible.

## 1.3.3 Typical Properties of Metallic Materials

Metals are inorganic materials with atoms held together by metallic bonds. The cloud of free electrons makes metals good conductors for both heat and electricity. For the same reason, metals are luminescent. They have a high specific weight, for the elements of which they are composed, that have a high atomic number, and for the crystalline compact structure.

The strong strength of the bonds and the crystalline structure of metals make them strong, with high elastic modulus, mechanical strength, and melting point. In addition, they have good ductility and toughness. However, metals can undergo corrosion, especially localized corrosion (see Chapter 5, Section 5.2).

Due to the presence of the dislocations in the crystalline structure, metallic materials can be processed by cold or high temperature deformation.

Due to the presence of point defects, atoms in the crystal lattice can be substituted by different elements, and the metal can form alloys.

## 1.3.4 Metallic Alloys

Although few metals are used in the pure (e.g., copper) or nearly pure form (e.g., titanium), most engineering metals are combined with other metals or nonmetal elements to provide increased strength, higher corrosion resistance, or other specific properties.

Hence, we can define a metal alloy, or an alloy, as a mixture of two or more metals or

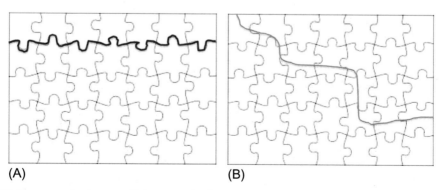

(A)  (B)

**FIG. 1.37**  (A) Transgranular fracture; (B) intergranular fracture.

nonmetal elements; in fact, alloys can have structures that are relatively simple, in which the alloy is composed of two metals (i.e., binary alloys), or can be complex, with about 10 elements in its nominal composition.

### 1.3.4.1 Formation of Metal Alloys

A simple type of alloy is a solid solution, possible for the presence of point defects in the crystal lattice of the metal. In particular, a solid solution is a solid in which two or more elements (metals or nonmetal elements) are dispersed in a single-phase structure (i.e., a single unit cell). In general, two types of solid solutions can be highlighted: substitutional or interstitial solid solutions. In a solid solution, the term *solvent* refers to the more abundant atomic form, and the term *solute* to the less abundant one.

#### SUBSTITUTIONAL SOLID SOLUTION

In the case of a substitutional solid solution, a direct substitution of one type of atom for another occurs so that solute atoms enter the crystal to take positions normally occupied by solvent atoms (Fig. 1.38A).

The fraction of atoms of one element that can dissolve in the solvent can vary, and some conditions can be favorable for higher solubility of solute in the solvent:

- The diameters of the elements must differ by <15%; if the atomic diameters differ, there will be a distortion of the crystal lattice, which can be limited in contraction or expansion.
- The crystal structure of the two elements must be the same.
- There should be no significant difference in the electronegativity of the two elements forming the solid solution; otherwise, the higher electropositive element will lose electrons, the higher electronegative element will acquire electrons, and compound formation will occur.
- The two elements should have the same valence, so that the solid solubility is favored; on the contrary, the binding energy between the atoms of the two elements will be upset, resulting in conditions unfavorable for solid solubility.

#### INTERSTITIAL SOLID SOLUTION

The other type of solid solution is the interstitial one (Fig. 1.38B). Here, the solute atom does not displace a solvent atom but rather enters one of the holes, or interstices, between the solvent atoms. In addition, solute atoms in interstitial alloys must be small in size. Extensive interstitial solid solutions occur only if the solute atom has an apparent diameter smaller than 0.59 of the solvent. The four most important interstitial atoms are carbon (C), nitrogen (N), oxygen (O), boron (B), and hydrogen (H), all of which are small in size.

Small interstitial solute atoms dissolve much more readily in transition metals than in other metals. The ability of transition elements to dissolve interstitial atoms is believed to be due to their unusual electronic structure. All transition elements possess an incomplete electronic shell

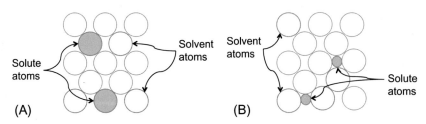

**FIG. 1.38** Mechanisms for solid solution alloying: (A) substitutional solid solutions; (B) interstitial solid solutions.

inside of the outer, or valence, electron shell. The nontransition metals, on the other hand, have filled shells below the valence shell. The extent to which interstitial atoms can dissolve in the transition metals depends on the metal, but it is usually small. Interstitial atoms can diffuse easily through the lattice of the solvent. Diffusion occurs not by a vacancy mechanism but by the solute atoms jumping from one interstitial position to another.

**INTERMEDIATE PHASES**

In many alloy systems, crystal structures or phases are different from those of the components (i.e., pure metals). These homogeneous crystal structures or phases are called *intermediate* phases (Fig. 1.39).

When the new crystal structures occur with simple whole-number fixed ratios of the component atoms, they are intermetallic compounds with stoichiometric compositions. As an example, iron carbide, $Fe_3C$, has an orthorhombic unit cell, where there is a carbon atom for every three iron atoms (Fig. 1.39, Example 2).

If there is no stoichiometric ratio between the elements in the alloy, when the solute is added to the solvent, the two elements can form a solid solution up to the maximum solubility value of solute in the solvent. Above this solubility limit, the solute in excess precipitates, forming a second crystal lattice different from the first one; this second phase can be formed by a substitutional or interstitial solid solution, or it can be an intermediate phase.

### 1.3.4.2 Phase Diagrams

Phase diagrams are important to represent what phases are present in a material system at various temperature, pressures, and compositions. In particular, phase diagrams can be used to study the evolution of an alloy from high to low temperature under slow cooling equilibrium conditions, evidencing different information:

- what phases are present at different compositions and temperatures,
- the equilibrium solid solubility of one element in another,

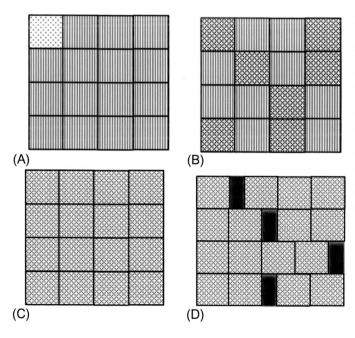

(A)　　　(B)

(C)　　　(D)

FIG. 1.39 Examples of intermediate phases. Example 1: Brass (Cu-Zn): (A) the amount of solute Zn is30% and a single phase $\alpha$ is formed, that is, substitutional solid solution with a CFC unit cell *(green)*; (B) the amount of solute Zn is 40% (i.e., higher than the solubility limit value) and two phases are present, that is, phase $\alpha$ *(green)* and the intermediate phase (CuZn, *blue)* that has a CCC unit cell. Example 2: Steel (Fe-C): (C) the amount of the solute C is 0.01% and a single-phase $\alpha$ is formed, that is, interstitial solid solution (Fe = solvent; C = solute), CCC unit cell *(grey)*; (D) the amount of the solute C is 0.1% and two phases are formed, phase $\alpha$ (CCC unit cell, *grey)* and intermediate phase ($Fe_3C$, *black)*.

- the temperature at which an alloy cooled under equilibrium conditions starts to solidify and the temperature range over which solidification occurs,
- the temperature at which different phases start to melt.

## GIBB'S PHASE RULE

From thermodynamic considerations, the Gibb's phase rule allows us to determine the number of phases that can coexist in equilibrium:

$$P + F = C + 2$$

where

$P$ = number of phases that can coexist in a chosen system,
$F$ = degree of freedom, that is, the number of variables (pressure, temperature, composition) that can be changed independently without changing the phases in equilibrium in the chosen system,
$C$ = number of components in the system.

Considering binary phase diagrams (i.e., two elements in the alloy), pressure is constantly maintained, varying only in composition and temperature, so that the Gibb's phase rule is given by:

$$P + F = C + 1$$

## LEVER RULE

The weight percentages of the two phases in the binary phase diagram can be calculated using the lever rule. For example, for the A–B phase diagram and a considered alloy composition, the weight percent of liquid and solid can be obtained. We can consider the composition $x$, that in Fig. 1.40 is identified by the weight fraction of B in A, $w_0$. In correspondence of temperature $T$, the tie line can be drawn, crossing the *liquidus* and *solidus* line. The alloy $x$ at temperature $T$ is a mix of both phases ($L + S$), and it is possible to identify that the weight fraction of B in A in the liquidus phase is $w_l$ and in the solid phase is $w_s$.

The lever rule can be derived by using weight balances. In fact, the sum of the weight fractions of liquid ($X_l$) and solid state ($X_s$) must be equal to 1:

$$X_l + X_s = 1$$

FIG. 1.40  Binary phase diagram of two metals A and B completely miscible in each other. At temperature $T$, the composition of the alloy is $w_0$, and the composition of liquid and solid phase is, respectively, $w_l$ and $w_s$.

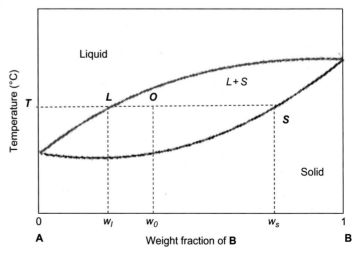

It is possible to write this relationship as:

$$X_l = 1 - X_s$$

or

$$X_s = 1 - X_l$$

Considering $w_0$, $w_l$, and $w_s$ as the weight fraction of B, it is possible to obtain the weight balance of B in the alloy:

$$\left(\frac{\%w_0}{100}\right) = X_s\left(\frac{\%w_s}{100}\right) + X_l\left(\frac{\%w_l}{100}\right)$$

$$w_0 = X_s w_s + X_l w_l$$

considering that

$$X_s = 1 - X_l$$

$$w_0 = (1 - X_l)w_s + X_l w_l$$

$$w_0 = w_s - X_l w_s + X_l w_l$$

Hence:

$$w_0 - w_s = X_l w_s + X_l w_l$$

**weight fraction** of **liquid phase** $= X_l = \dfrac{w_0 - w_s}{w_s - w_l}$

Similarly, it is possible to obtain the weight fraction of the solid phase:

**weight fraction** of **solid phase** $= X_s = \dfrac{w_s - w_0}{w_s - w_l}$

## BINARY ALLOY SYSTEMS: COMPLETE MISCIBILITY

The simplest condition in a binary system is the case in which two components (i.e., each element in the alloy is considered as a separate component) are mutually miscible in any amounts in both solid and liquid phases. In this case, a single crystal structure exists for all compositions, and these systems are called isomorphous systems. An example of an isomorphous system can be represented by a Cu-Ni system; in fact, in that case, the complete solubility occurs because both Cu and Ni have the same crystal structure, FCC, similar radii, electronegativity, and valence.

The phase diagram (Fig. 1.41A) has been obtained for slow cooling equilibrium conditions at atmospheric pressure (i.e., $P =$ constant). The area above the upper line, called the liquidus line, corresponds to the region of stability of the liquid phase; the area below the lower line, called the solidus line, represents the region of stability of the solid phase. The region between the liquidus and solidus line represents the region where both the liquid and solid phase are present. In this region, the amount of each phase depends on the temperature and composition.

To better understand how binary phase diagram works, we can consider an alloy at 1300°C with the composition ($w_0$) of 53 wt% Ni and 47 wt% Cu (Fig. 1.41B). As the two phases are present, neither of them will have the composition 53 wt% Ni and 47 wt% Cu. To determine the composition of each phase, it is necessary to draw a horizontal tie line at 1300°C from the liquidus to the solidus line and draw the vertical line to the composition axis (Fig. 1.41B). Hence, the composition of the liquid phase (named $w_l$) at 1300°C is 45 wt% Ni and 55 wt% Cu, and the one of the solid phase ($w_s$) is 58 wt% Ni and 42 wt% Cu.

Applying the lever rule to the considered alloy composition in the Cu-Ni phase diagram, the weight percent liquid and solid can be obtained. Considering nickel, at 1300°C, the weight percent in the liquid and solid phases are:

$$w_l = 45\% \text{Ni and } w_s = 58\% \text{Ni}$$

$$w_0 = 53\% \text{Ni}$$

Weight fraction of liquid phase $= X_l = \dfrac{w_s - w_0}{w_s - w_l}$

$$= \frac{58 - 53}{58 - 45} = \frac{5}{13} = 0.38$$

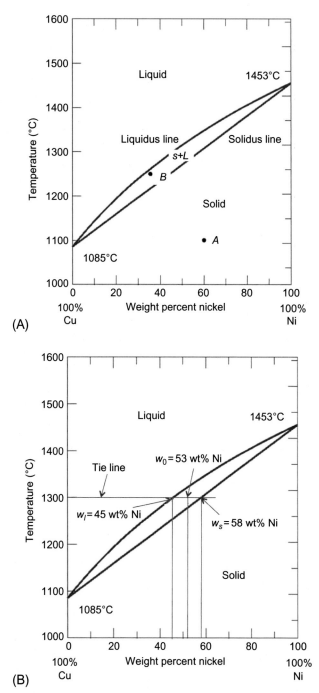

**FIG. 1.41** Copper-nickel phase diagram. (A) Cu and Ni have complete liquid and solid solubility. Point *A* evidences the complete solid solubility of the two elements for a specific composition; point *B* evidences the alloy at a specific composition, in the liquid/solid region. (B) Tie line and the composition of the liquid and solid phase for the alloy 53 wt% Ni and 47 wt% Cu are represented.

Weight% of liquid phase $= X_l = (0.38)(100\%)$
$$= 38\%$$

Weight fraction of solid phase $= X_s = \dfrac{w_0 - w_l}{w_s - w_l}$

$$= \dfrac{53 - 45}{58 - 45} = \dfrac{8}{13} = 0.62$$

Weight% of solid phase $= X_s = (0.62)(100\%)$
$$= 62\%$$

## EUTECTIC ALLOY SYSTEMS: PARTIAL SOLID MISCIBILITY

Many binary alloy systems have components that have limited solid solubility in each other; the lead-tin (Pb-Sn) phase diagram (Fig. 1.42) represents an example of an eutectic alloy system. The regions of solid solubility are named alpha and beta phases, and are called terminal solid solutions. In particular, the alpha phase is a lead-rich solid solution and can dissolve a maximum of 19.2 wt% Sn at 183°C; the beta phase is tin-rich and can dissolve a maximum of 2.5 wt% Sn at 183°C. When temperature decreases below 183°C, the maximum solubility of the two elements decreases, according to the solvus lines in the diagram (Fig. 1.42).

A particular composition can be highlighted in this phase diagram called *eutectic* composition. The eutectic temperature at which there is the eutectic composition is the lowest temperature where liquid phase can exist when cooled slowly. When liquid at the eutectic composition is slowly cooled to the eutectic temperature, the single liquid phase transforms simultaneously into two solid forms (i.e., solid form alpha and beta). This transformation is named an eutectic reaction, and it is an invariant reaction because it occurs under equilibrium conditions at a specific temperature and alloy composition that cannot be varied (i.e., $F = 0$). In fact, applying the Gibb's rule, it is demonstrated that the number of degrees of freedom is zero:

$$F = C + 1 - P$$

$P = 3$, that is, liquid, solid solution alpha, solid solution beta

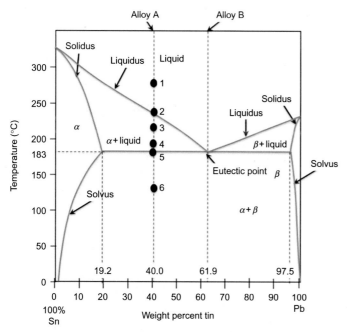

**FIG. 1.42** Lead-tin phase diagram. The diagram is characterized by a partial solid solubility of each terminal phase (named alpha and beta). The eutectic invariant reaction at 61.9 wt% Sn and 183°C is evidenced; at the eutectic point, three phases coexist: alpha (19.2% Sn), beta (97.5% Sn), and liquid (61.9% Sn).

$C = 2$, that is, it is a binary alloy; hence, two elements compose the system.

Considering alloy A that has the eutectic composition, nothing can be varied. During the temperature decrease, down to 183°C the alloy remains liquid, but at 183°C (i.e., eutectic temperature), the alloy solidifies by the eutectic reaction, forming a mixture of solid solution $\alpha$ and solid solution $\beta$. After completion of the eutectic reaction, decreasing the temperature down to room lower temperature, the two solid solutions are still distinguished in the microstructure of the alloy. The compositions to the right of the eutectic point are named *hypoeutectic*, and the ones on the left are called *hypereutectic*.

Considering alloy B, from the liquid state (point 1), where the alloy is in the liquid phase, at point 2, the liquidus line is reached. After that, solid solution alpha begins to precipitate; this first formed solid is called *proeutectic* alpha. Cooling down the alloy from the liquidus line to slightly above eutectic temperature (i.e., 183°C), the alloy is in the alpha + liquid region. At the eutectic temperature, all the liquid solidifies by the eutectic reaction. When the eutectic reaction is completed, the alloy consists of proeutectic alpha and the eutectic mixture formed by alpha and beta solid solution.

It is noted that it is possible to apply the lever rule either in the regions alpha + liquid and beta + liquid, and in the regions alpha + beta, alpha, and beta. For example, we can make phase analysis at the eutectic point, and at points 3, 4, and 5.

At eutectic composition (i.e., 61.9% Sn), just below 183°C:

| Phases Present | Alpha | Beta |
|---|---|---|
| Phase composition | 19.2% Sn | 97.5% Sn |
| Phase amount | wt% alpha phase $=\frac{97.5-61.9}{97.5-19.2}(100\%)$ $=45.5\%$ | wt% beta phase $=\frac{61.9-19.2}{97.5-19.2}(100\%)$ $=54.5\%$ |

At point 3 at 40.0% Sn:

| Phases Present | Liquid | Alpha |
|---|---|---|
| Phase composition | 48.0% Sn | 15.0% Sn |
| Phase amount | wt% liquid phase $=\frac{40-15}{48-15}(100\%)$ $=76.0\%$ | wt% alpha phase $=\frac{48-40}{48-15}(100\%)$ $=24.0\%$ |

At point 4 at 40.0% Sn:

| Phases Present | Liquid | Alpha |
|---|---|---|
| Phase composition | 61.9% Sn | 19.2% Sn |
| Phase amount | wt% liquid phase $=\frac{40-19.2}{61.9-19.2}(100\%)$ $=49.0\%$ | wt% alpha phase $=\frac{61.9-40}{61.9-19.2}(100\%)$ $=51.0\%$ |

At point 5 at 40.0% Sn:

| Phases Present | Alpha | Beta |
|---|---|---|
| Phase composition | 19.2% Sn | 97.5% Sn |
| Phase amount | wt% alpha phase $=\frac{97.5-40}{97.5-19.2}(100\%)$ $=73.0\%$ | wt% beta phase $=\frac{40-19.2}{97.5-19.2}(100\%)$ $=27.0\%$ |

## EXAMPLE: FE-C PHASE DIAGRAM

In their simplest form (i.e., plain-carbon steels), steels are alloys of iron and carbon. The Fe-C phase diagram is a fairly complex one; only compositions up to 6.67% carbon are shown in Fig. 1.43. In particular, Fe-C alloys can be of two types:

1. Steels: alloys of iron and carbon containing up to 2.06% C. Other alloying elements may also be present in steels.
2. Cast irons: alloys of iron and carbon containing more than 2.06% C, up to 6.67%. Other alloying elements may also be present in cast irons.

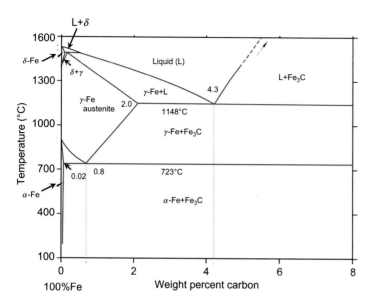

**FIG. 1.43** Iron-carbon phase diagram. The main structures are represented: liquid phase, $\delta$ ferrite (1393–1536°C), max solubility C = 0.1%, austenite, $\gamma$ phase (723–1493°C), max solubility C = 2.06%, $\alpha$ ferrite (up to 723°C), max solubility C = 0.02%, Fe$_3$C (iron carbide), intermetallic compound, C = 6.67%.

In the Fe-Fe$_3$C phase diagram, different solid phases in the steel region are present:

– $\alpha$ ferrite: it is an interstitial solid solution of C in BCC iron crystal lattice. It has very limited solubility for carbon (maximum 0.022% at 723°C, and 0.005% at room temperature). Alpha ferrite is soft and ductile.
– austenite ($\gamma$ phase): it is an interstitial solid solution of C in gamma iron, with a FCC iron crystal lattice. In FCC, C has a higher solubility than in BCC (i.e., $\alpha$ ferrite). Austenite can have maximum 2.08% at 1148°C, and decreases to 0.8% at 723°C. It is normally not stable at room temperature. Austenite is nonmagnetic and soft.
– cementite or iron carbide (Fe$_3$C): it is an intermetallic compound of iron and carbon, that is very hard and brittle. This intermetallic compound is a metastable phase, and it remains as a compound indefinitely at room temperature.
– $\delta$ ferrite: it is an interstitial solid solution of C in delta iron, with a BCC structure like a ferrite, but it has a greater lattice constant. The maximum solid solubility of C in $\delta$ ferrite is

0.09% at 1465°C. It is stable at high temperatures.

**TERNARY PHASE DIAGRAMS**

Phase diagrams help us interpret the structures of metal and alloys, as most commercial alloys have more than two components. Ternary diagram is used to represent the stability of different phases if the alloy consists of three components. Usually ternary phase diagrams are represented by an equilateral triangle. Compositions of ternary systems are represented on this base, with the pure content of each element at each edge of the triangle (Fig. 1.44A). Normally, they are constructed at a constant pressure ($P = 1$ atm). Temperature is uniform in the whole diagram, because this phase diagram is a isothermal section. Hence, to show a range of temperatures, a figure with temperature on a vertical axis with a triangular base should be constructed.

The three sides of the triangle represent the compositions of the three binary alloys. At every point on the line, BC (the side opposite the vertex A) has 0% A. Each side of the triangle can be

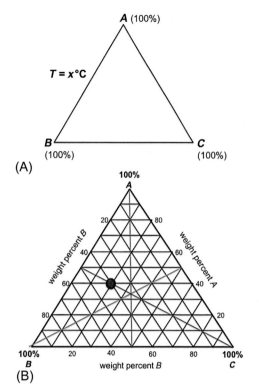

**FIG. 1.44** Ternary phase diagram. (A) Composition base for a system with three components (A, B, and C); (B) determination of the composition for the alloy evidenced by the red point.

period of time, and cooling at some specified rate. The aim is to obtain a desired microstructure to achieve certain predetermined properties (physical, mechanical, magnetic, or electrical).

Heat treatment is any one of a number of controlled heating and cooling operations used to bring about a desired change in the physical properties of a metal. Its purpose is to improve the structural and physical properties for some particular use or for future work of the metal. The major objectives of metal and alloys heat treatments can be defined as follows:

– to increase strength, harness, and wear resistance,
– to increase ductility and softness,
– to increase toughness,
– to obtain fine grain size,
– to remove internal stresses induced by differential deformation by cold working, nonuniform cooling from high temperature during casting and welding,
– to improve machinability,
– to improve cutting properties of tool steels,
– to improve surface properties,
– to improve electrical properties,
– to improve magnetic properties.

There are five basic heat-treating processes: hardening, tempering, annealing, normalizing, and case hardening. Although each of these processes brings about different results in metal, all of them involve three basic steps: heating, soaking, and cooling (Fig. 1.45).

*Heating* is the first step in a heat-treating process. Many alloys change structure when they are heated to specific temperatures.

Once a metal part has been heated to the temperature at which desired changes in its structure will take place, it must remain at that

subdivided into 10 parts by a set of points, so that the entire space is now divided into a set of small equilateral triangles. Let us consider an alloy represented by the point X in Fig. 1.44B, we have to obtain the composition of this alloy. For example, %A is read by measuring the distance from the side (BC) opposite vertex A: it is equal to 30%. The %B is read from the side AC, and it is equal to 30%. Finally, %C is read from the side AB, and it is equal to 40%.

## 1.3.5 Thermal Treatments

Heat treatment is an operation or combination of operations involving heating at a specific rate, soaking at a specified temperature for a

**FIG. 1.45** Steps in heat treatments: heating, soaking, and cooling.

temperature until the entire part has been evenly heated throughout. This step is called *soaking*.

The third step is *cooling* the metal component. The structure may change from one chemical composition to another, it may stay the same, or it may revert to its original form. Many metals can be made to conform to specific structures to increase their hardness, toughness, ductility, and tensile strength.

For ferrous materials, mainly for steel, different heat treatments are performed to change the crystalline structure, hence the properties, of the materials. Among the possible heat treatments, the principal ones are briefly described here.

### 1.3.5.1 Hardening

A ferrous metal is normally hardened by heating the metal to the required temperature and then cooling it rapidly by plunging the hot metal into a quenching medium (e.g., oil, water, or brine). Most steels must be cooled rapidly to harden them. The hardening process increases the hardness and strength of metal, but also increases its brittleness.

### 1.3.5.2 Tempering

Steel is usually harder than necessary and too brittle for practical use after being hardened. Severe internal stresses are set up during the rapid cooling of the metal. Steel is tempered after being hardened to relieve the internal stresses and reduce its brittleness. Tempering consists of heating the metal to a specified temperature and then permitting the metal to cool. The rate of cooling usually has no effect on the metal structure during tempering. Therefore the metal is usually permitted to cool in still air. Temperatures used for tempering are normally much lower than the hardening temperatures. The higher the tempering temperature used, the softer the metal becomes.

### 1.3.5.3 Annealing

Metals are annealed to relieve internal stresses, soften them, make them more ductile, and refine their grain structures. Metal is annealed by heating it to a prescribed temperature, holding it at that temperature for the required time, and then cooling it back to room temperature. The rate at which metal is cooled from the annealing temperature varies greatly. Steel must be cooled very slowly by burying the hot part in sand, ashes, or some other substance that does not conduct heat readily (packing), or by shutting off the furnace and allowing the furnace and part to cool together (furnace cooling).

### 1.3.5.4 Normalizing

Ferrous metals are normalized to relieve the internal stresses produced by machining, forging, or welding. Normalized steels are harder and stronger than annealed steels. Steel is much tougher in the normalized condition than in any other condition. Parts that will be subjected to impact and parts that require maximum toughness and resistance to external stresses are usually normalized. Normalizing prior to hardening is beneficial in obtaining the desired hardness, provided the hardening operation is performed correctly. Normalizing is achieved by heating the metal to a specified temperature (i.e., higher than either the hardening or annealing temperatures), soaking the metal until it is uniformly heated, and cooling it in still air.

## 1.4  CERAMIC MATERIALS

A ceramic is defined as an inorganic material that consists of metallic and nonmetallic elements. Usually ceramics are nitrides, carbides, and oxides. However, carbons are included in ceramics, as well.

### 1.4.1  Structure

Ceramics are solid materials characterized by ionic bonds or a combination of covalent and ionic bonds. Depending on the atomic feature, ceramics can either have a crystalline or amorphous structure. For example, amorphous ceramics are glass and silicate, and crystalline ceramics are porcelain and alumina.

When bonds in ceramics are a combination of both ionic and covalent bonds, the percentage between ionic and covalent character determines the type of crystalline structure (Table 1.5). Taking into account the difference in electronegativity values of the atoms that compose the ceramic material, it is possible to obtain approximate values of the percentage of ionic and covalent character in the chemical bond. Based on Pauling's equation, the percentage of ionic character can be found applying the following equation:

$$\%\text{ionic character} = \left[1 - e^{(0.25)(X_A - X_B)^2}\right] \times 100$$

where $X_A$ and $X_B$ are the electronegativity values of the elements in the ceramic material.

The percentage of ionic and covalent bonds between the ions in ceramic materials is important because it can influence the crystalline unit cell they can form, hence the properties of the material. In Table 1.5, some examples of ceramics materials are reported, and it is possible to note that the percent of ionic and covalent character varies considerably among them.

In ionic ceramics, crystalline structures are formed by a combination of cations and anions, where the electrical neutrality is needed, balancing electrostatic charges; in addition, it is important for the relative size of the ions to have a stable configuration. The energies of the atoms involved in the ionic bonds are lowered by the formation of the ions and their bonding in an ionic solid. Hence, the ions are packed together as densely as possible to lower the overall energy. An example of ionic ceramic material is NaCl, where a face-centered unit cell is composed of $Cl^-$ anions, and $Na^+$ cations fill into the voids between adjacent anions, maintaining the neutral charge (Fig. 1.46).

## 1.4.2 Typical Properties

Covalent and ionic bonds in ceramics are responsible for their properties, and significant differences in properties can be evidenced among ceramics, depending on the percentages of the two bonds. In general, ceramics have dimensional stability even at high temperatures, due to their relatively high melting temperature. They possess high chemical inertness in hostile environments due to the stability of their strong bonds. For the nature of chemical bonds, ceramics are strong in compression, and weak in tension, with high hardness and elastic modulus; moreover, they are difficult to deform and have low ductility. As a result of their strong bonds, they are brittle, causing them to fracture rapidly and without plastic deformation, with low fracture toughness. In addition, because of

**TABLE 1.5**   Examples of Ceramic Materials With Different Percentages of Ionic and Covalent Bonds

| Material | $X_A$ | $X_B$ | Δ Electronegativity | % Ionic Character | % Covalent Character |
|----------|-------|-------|---------------------|-------------------|----------------------|
| $ZrO_2$ | 1.2 | 3.5 | 2.30 | 73.0 | 27.0 |
| $TiO_2$ | 1.54 | 3.5 | 1.96 | 62.0 | 38.0 |
| MgO | 1.3 | 3.5 | 2.20 | 69.0 | 31.0 |
| $Al_2O_3$ | 1.5 | 3.5 | 2.00 | 63.0 | 37.0 |
| $SiO_2$ | 1.8 | 3.5 | 1.70 | 51.0 | 49.0 |
| $Si_3N_4$ | 1.8 | 3.1 | 1.30 | 34.5 | 65.5 |
| SiC | 1.8 | 2.5 | 0.70 | 11.0 | 89.0 |

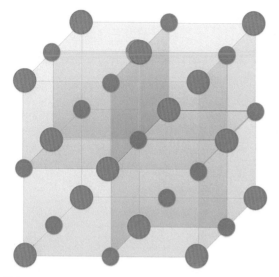

**FIG. 1.46** Structure of NaCl, with $Cl^-$ anions forming a CFC unit cell and $Na^+$ cations occupying the void between them.

the absence of conduction electrons in ionic and/or covalent bonds, they are good heat and electricity insulators (i.e., due to the ionic and covalent bonds), exhibiting electrical resistivity in the range of $10^{14}$ $\Omega \times$ cm compared to copper in the range of $10^{-6}$ $\Omega \times$ cm. Specific properties can be conferred to ceramic materials, for example, ferromagnetism or semiconductivity.

## 1.4.3 Traditional Ceramics and Advanced Ceramics

There are different ways to classify ceramics; in particular, they can be classified based on general form, composition, or engineering applications. These classifications are discussed here.

### 1.4.3.1 Classification Based on Form

Depending on the specific function, ceramics are classified as follows:

– powders;
– coatings;
– bulk shapes.

Powders are dry, solid particles of various size that are not fused together, so that they can flow when shaken or tilted. Coatings, in general, refer to films or deposits on substrates, whereas bulk shapes refer to the densified form of the ceramics.

### 1.4.3.2 Classification Based on Composition

Ceramics can be classified as oxides or non-oxides, and are chemically inert. However, oxide ceramics (e.g., alumina, zirconia, magnesium oxide) are oxidation-resistant, electrically insulating, and have generally low thermal conductivity. In contrast, nonoxide ceramics (e.g., carbides, nitrides, silicates) possess low oxidation resistance, are electrically conducting, and have high thermal conductivity. Among ceramics, carbon can exists in many allotropic forms, such as graphite, noncrystalline carbon, and pyrolitic carbon.

### 1.4.3.3 Classification Based on Applications

Ceramic materials can be divided in two groups (Fig. 1.47):

– traditional ceramics;
– advanced ceramics.

*Traditional* ceramics are made from three basic components: clay, silica, and feldspar; some examples of traditional ceramics are glasses, bricks, tiles, and electrical porcelain. In particular, clay confers ductility and hardness, silica determines high stability at high temperature and an elevated melting point, and feldspar produces the glass phase when the ceramic is cooked.

*Advanced* ceramics generally consist of pure or nearly pure (i.e., >99.9% purity) compounds, such as aluminum oxide ($Al_2O_3$) and silicon carbide (SiC). In this group, materials composed of only carbon are included.

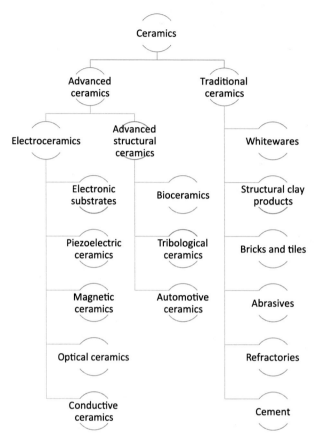

FIG. 1.47 Classification of ceramics based on their applications: traditional and advanced ceramics.

## 1.4.4 Carbon and Its Allotropes

Carbon materials can exist in different forms that have significant different characteristics due to the specific structure that carbon can assume in the different allotropes. In this section, the structure and properties of graphite, turbostratic, and glassy carbon will be briefly discussed, as they can be used for different biomedical applications.

### 1.4.4.1 Graphite

Graphite has a layered structure where the carbon atoms in each layer are bonded in hexagonal arrays with covalent bonds; the layers are bonded to each other by secondary bonds, that is, Van der Waals interactions (Fig. 1.48). The weak bonds among the layers determine weak shear strength, so that they can slide one each other by applying low force. Hence, graphite is anisotropic, as properties depend on the direction of force application. It is a good thermal and electrical conductor along each layer of graphite but not perpendicular to it. The reason for the good electrical conductivity is due to the structure of graphite. In fact, each carbon atom is bonded into its layer with three strong covalent bonds; this leaves each atom with a spare electron, which together form a delocalized sea of electrons loosely bonding the layers together.

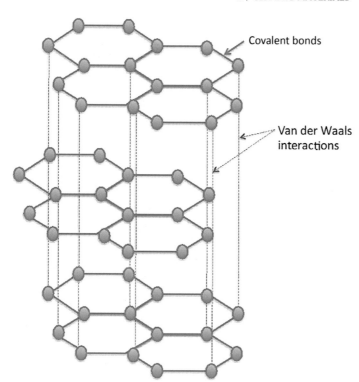

**FIG. 1.48** Structure of crystalline graphite: carbon atoms form hexagonal arrays bonded in layers by strong covalent bonds. Among the layers, there are weak Van der Waals interactions.

Covalent bonds

Van der Waals interactions

These delocalized electrons can all move along together on each layer, making graphite a good electrical conductor. In addition, due to the sliding of the layers, graphite can be used as lubricant.

### 1.4.4.2 Turbostratic Carbon

Turbostratic carbon has a structure in which the layers are disordered, resulting in wrinkles or distortions within layers (Fig. 1.49). Therefore the crystalline structure has a distorted lattice structure with random unassociated carbon atoms unlike graphite. This structure provides it with isotropic properties (i.e., similar in all directions), giving turbostratic carbon improved durability compared to graphite. In addition, it exhibits excellent dimension stability, chemical inertness, mechanical strength, wear resistance, fatigue resistance, and biocompatibility.

### 1.4.4.3 Glassy Carbon

This type of carbon has a glass-like appearance and is referred to as glassy carbon. This carbon structure is visually a highly disordered structure. Glassy carbon does not revert to graphite form at high temperatures; this characteristic makes glassy carbon nongraphitizable, because thermal treatments at high temperature ($T > 2500°C$) do not significantly modify the structural disorder. The most important properties are high temperature resistance, low density, electrical resistance, friction, thermal resistance, high resistance to chemical attack, and impermeability to gases and liquids. It is hard, brittle, and isotropic but with mechanical properties lower than those of turbostratic carbon. Glassy carbon is widely used as an electrode material in electrochemistry, as well as for high temperature crucibles and as a component of some prosthetic devices.

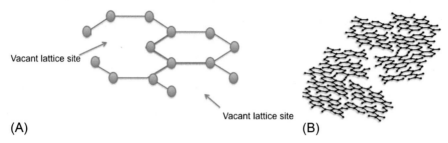

Vacant lattice site

Vacant lattice site

(A)                                                                        (B)

FIG. 1.49   (A) Structure of single layer with defects (vacant lattice site); (B) turbostratic carbon structure, where different orientation of the single layers can be observed, causing isotropic properties.

## 1.5  COMPOSITE MATERIALS

It is very difficult to find a widely accepted definition of what a composite material is. A dictionary defines a composite as something made up of distinct parts or constituents. In engineering design, a composite material usually refers to a material consisting of constituents in the micro- to macroscale range. A composite material can be defined as a material system composed of two or more different micro- or macrophases or constituents with distinct interfaces among them that differ in form and chemical composition, and are insoluble in each other. Typically, the idea is that, by combining two or more distinct materials at micro- up to macroscale, one can engineer a new material with the desired combination of properties (e.g., light, strong, corrosion-resistant), that are superior, or more important, to the properties of individual components.

Usual composites have just two phases:

− matrix (continuous),
− dispersed phase (particulates, fibers).

Examples of composite materials are represented by "old" composites, known for many thousands of years, and "new" composites that are high-tech materials, engineered to specific applications.

One early example is mud bricks. Mud can be dried out into a brick shape to form a building material; it is strong when squashing it (i.e., it has good compressive strength), but it breaks quite easily if it bends (i.e., it has poor tensile strength). Straw seems very strong when stretched, but it is possible to crumple it up easily. By mixing mud and straw together, it is possible to make bricks that are resistant to both squeezing and tearing, making excellent building blocks.

Another ancient composite is concrete. Concrete is a mix of aggregate (i.e., small stones or gravel), cement, and sand. It has good compressive strength (i.e., it resists squashing). In more recent times, it has been found that adding metal rods or wires to the concrete can increase its tensile (i.e., bending) strength. Concrete containing such rods or wires is called reinforced concrete.

Composites are also present in nature, in plants and animals. Wood is a composite made from long longitudinal cellulose fibers held together by a much weaker matrix, composed of lignin, hemicellulose, and other natural components. Bone in the human body is also a natural composite. It is made from a hard but brittle material, that is, hydroxyapatite, and a soft and flexible material, that is, collagen. On its own, collagen would not be useful in the skeleton, but when combined with hydroxyapatite, it gives bone the properties needed to support the body.

## 1.5.1 Properties of Composite Materials

Properties of composites depend on the characteristics of the components (i.e., matrix and reinforcement), dimension of reinforcement, morphology of the entire system, and nature of the interface between the two phases. In addition, properties of the composite can vary, changing material composition and orientation of the reinforcement in the matrix. Summarizing, composite can be designed *ad hoc* for a specific application, taking into account:

– properties of phases,
– geometry of dispersed phase (particle size, distribution, orientation),
– amount of phases.

A characteristic of composite is the balancing between a low density, a favorable weight-to-performance ratio, and high mechanical properties (Fig. 1.50). Among the properties of a composite, the following ones can be highlighted:

– decrease of the coefficient of thermal expansion,
– increase in stiffness,
– high mechanical strength, also fatigue behavior,
– high dimensional stability,
– wear resistance,

**FIG. 1.50** In composite materials, an appropriate balance between mechanical properties and density should be found as represented in the schema.

– long life,
– design flexibility,
– cost reduction.

The main disadvantages of the composites are the difficulty in fabrication and reproducibility, and the reduction of some mechanical properties for the increase of other ones.

## 1.5.2 Classification

Composite materials can be classified depending on the reinforcement used following the classification here, schematically shown in Fig. 1.51:

– particle-reinforced (large-particle and dispersion-strengthened),
– fiber-reinforced (continuous, aligned) and short fibers (aligned or random),
– structural (laminates and sandwich panels).

### 1.5.2.1 Particle-Reinforced Composite

These are the cheapest and most widely used composites. They fall into two categories depending on the size of the particles:

– large-particle composites, which act by restraining the movement of the matrix, if well bonded,
– dispersion-strengthened composites, containing 10–100 nm particles, similar to what was discussed under precipitation hardening. The matrix bears the major portion of the applied load, and the small particles hinder dislocation motion, limiting plastic deformation.

### 1.5.2.2 Fiber-Reinforced Composite

In many applications, such as aircraft parts, there is a need for high strength per unit weight (specific strength). This can be achieved by composites consisting of a low density and soft matrix reinforced with stiff fibers. The strength depends on the fiber length and orientation with respect to the stress direction. The efficiency of

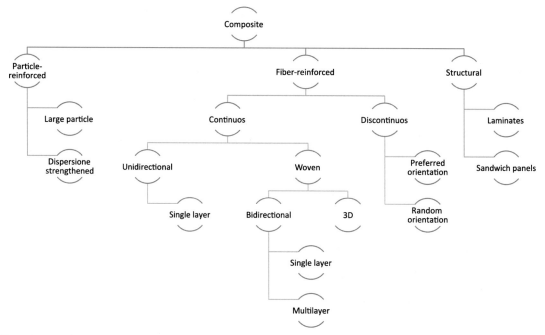

**FIG. 1.51** Classification of composite materials based on the characteristics of the reinforcement.

load transfer between matrix and fiber depends on the interfacial bond.

Particle-reinforced composites (Fig. 1.51) consist of a matrix reinforced with a dispersed phase in the form of particles. The effect of the dispersed particles on the composite properties depends on their dimensions. In fact, very small particles (diameter lower than 0.25 μm) finely distributed in the matrix impede deformation of the material, so that the resulting composite exhibits high mechanical strength. Depending on the material used for the particulate reinforcement, a dispersed phase of this kind of composite can be stable at high temperatures, so the strengthening effect is retained (i.e., using ceramic particles). When large dispersed phase particles are used as reinforcement phase, the composite material gains a low strengthening effect, so that an increase of stiffness and decrease of ductility are characteristics of those materials. Moreover, hard particles dispersed in a softer matrix increase wear and abrasion resistance. Indeed, soft dispersed particles in a harder matrix improve machinability (e.g., lead particles in a steel or copper matrix) and reduce coefficient of friction (e.g., tin in an aluminum matrix or lead in a copper matrix).

When the dispersed phase of these materials consists of two-dimensional flat platelets (flakes) laid parallel to each other, the material exhibits anisotropy. In the case of flakes oriented parallel to a particular plane, the material demonstrates equal properties in all directions parallel to the plane and different properties in the direction normal to the plane.

If the reinforcement is in form of fibers (Fig. 1.51), the composite shows an improvement in strength, stiffness, and fracture toughness, impeding crack growth in the direction normal to the fiber. The effect of the strength increase becomes much more significant when the fibers are arranged in a particular direction (preferred orientation; Fig. 1.51) and a stress is applied along the same direction.

The strengthening effect is higher in long-fiber (continuous-fiber)-reinforced composites than in short-fiber (discontinuous-fiber)-reinforced composites. In fact, short-fiber (length $<100 \times$ diameter)-reinforced composites have a limited ability to share load. On the contrary, load applied to a long-fiber-reinforced composite is carried mostly by the fibers.

Laminate composites (Fig. 1.51) consist of layers with different anisotropic orientations or of a matrix reinforced with a dispersed phase in the form of sheets. When a fiber-reinforced composite consists of several layers with different fiber orientations, it is called multilayer (angle-ply) composite. Laminate composites provide increased mechanical strength in the directions in which the fibers are aligned, but in the direction perpendicular to the preferred orientations of the fibers or sheet, mechanical properties of the material are low.

## 1.5.3 The Role of the Components

As described in the previous sections, composite materials are composed of a matrix and a reinforcement; their main properties and materials will be described in the following sections.

In addition to matrix and reinforcement, the bonding between the two phases plays an important role in bonding them together, ensuring an adequate transfer of load through the composite. In addition, when fibers are used as reinforcement, bonding between the two phases is required so fibers do not separate from the matrix during load application. For these reasons, the interfacial bond should be chemically and physically stable during use. In fact, if the adhesion between matrix and reinforcement is weak, an unexpected mechanical behavior of the composite should be shown, due to a noncorrect load transfer. When the direct bonding between matrix and reinforcement does not guarantee the formation of a good interface, particles or fibers can be coated with special agents

to improve bonding and moisture resistance (i.e., in this case, the coating represents the bonding between the two composite phases).

### 1.5.3.1 Matrix

Matrix in such materials serves only as a binder of the fibers, keeping them in a desired shape and protecting them from mechanical or chemical damages.

The three classes of matrix materials are polymers, metals, and ceramics (Fig. 1.52). The properties of these types differ substantially and have profound effects on the properties of the composites using them.

*Polymeric matrix.* Polymer matrices generally are relatively weak, low-stiffness, viscoelastic materials; in fact, mechanical strength and stiffness come primarily from the reinforcing fibers or particles. There are two major classes of polymers used as matrix materials: thermosets and thermoplastics. At this time, for industrial applications, thermosets are the most widely used matrix resins for structural applications, although thermosets are making steady gains. Thermosets tend to be more resistant to solvents and corrosive environments than thermoplastics. Thermoplastics and elastomers have been deeply investigated for biomedical applications and will be further detailed later (see Chapter 4, Section 4.5).

*Thermosetting resins.* The key types of thermosetting resins used in composites are epoxies, thermosetting polyimides, cyanate esters, thermosetting polyesters, unsaturated polyesters, vinyl esters, silicones, and phenolics; it should be noted that this list is continually expanding. Epoxies are used to produce composites with excellent structural properties, for example, airframe structures and other aerospace applications. Epoxies tend to be rather brittle materials, but toughened formulations with greatly improved impact resistance are available; they cost more than other thermosetting polymeric matrices, but their advantages make them a primary choice for carbon- and aramid-fiber

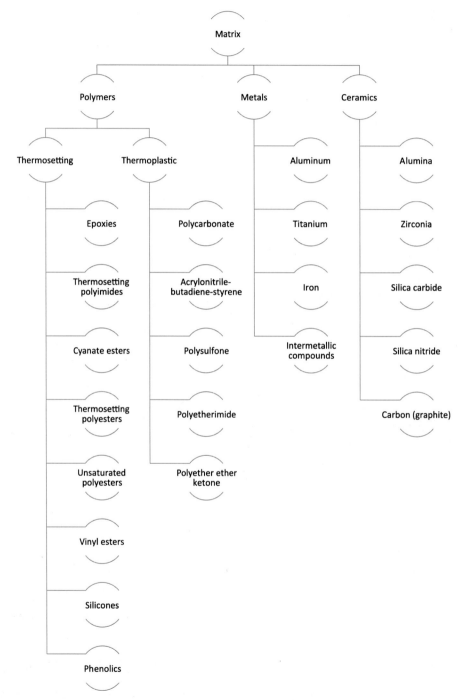

**FIG. 1.52**   Main polymers, metals, and ceramics used as matrices in composite materials.

composites. Polyester resins are the most-used resins in commercial applications, because they are relatively inexpensive, easy to process, and corrosion-resistant. Unsaturated polyesters are widely used as matrices for fiber-reinforced composites, in boat hulls, building panels, and structural panels for automotive and aerospace applications.

*Thermoplastic resins.* Among thermoplastics, both amorphous and semicrystalline polymers are usable as matrices in composites. In particular, amorphous thermoplastics tend to have poor solvent resistance, whereas crystalline materials tend to be better in this respect. Relatively inexpensive thermoplastics like nylon are extensively used with chopped E-glass fiber[10] reinforcements in countless injection-molded parts. There is an increasing number of applications using continuous fiber-reinforced thermoplastics.

*Metallic matrix.* Metals initially used as matrix materials were traditional alloys. Now, innovative metallic materials are tailored for use in composites. Main metallic matrix materials are alloys of aluminum, titanium, and iron; other metals used as matrix include magnesium, cobalt, silver, and superalloys. There was a significant amount of research on composites using intermetallic compound matrix materials, such as titanium aluminides, but these were largely unsuccessful.

*Ceramic matrix.* The key ceramics used as matrices are silicon carbide, alumina, silicon nitride, mullite, and various cements. Ceramics are very flaw-sensitive, resulting in a decrease in strength with increasing material volume. As a result, there is no single value that describes the tensile strength of ceramics. In fact, because of the very brittle nature of ceramics, it is difficult to measure tensile strength, and flexural strength is often reported. For that reason, monolithic ceramics are rarely used in applications where they are subjected to significant tensile stresses.

Carbon is an incredible ceramic material and includes materials ranging from lubricants to diamonds to structural fibers. The forms of carbon matrices resulting from the various carbon-carbon manufacturing processes tend to be rather weak, brittle materials. Thermal conductivities range from very low to high, depending on precursor materials and processes.

### 1.5.3.2 Reinforcement

Reinforcement is added to the matrix mainly to improve mechanical properties, for example, stiffness, mechanical strength, and toughness. The four key types of reinforcements used in composites are continuous fibers, discontinuous fibers, whiskers (elongated single crystals), and particles made of different materials (Fig. 1.53). Most of these materials (i.e., glasses, ceramics, or carbon) are brittle, but particles or fibers are embedded in the matrix, so that stress concentration is reduced. Continuous, aligned fibers are the most efficient reinforcement form and are widely used, especially in high-performance applications. However, for ease of fabrication and to achieve specific properties, such as improved impact resistance, continuous fibers are processed using textile technology.

### 1.5.3.3 Fibers

The great importance of composites, and the revolutionary improvements in properties that they can offer compared to conventional materials, are derived to a great extent from the development of fibers with extraordinary properties. The key fibers for mechanical engineering applications are glasses, carbons, several types of ceramics, and high-modulus organics.

*Glass fibers.* Glass fibers are used primarily to reinforce polymers. The leading types of glass fibers for mechanical engineering applications are E-glass and S-glass[11] fibers. E-glass fibers are the most widely used among all fibrous

---

[10] E-glass: E (electrical) glass; it is a borosilicate glass, and it is very commonly used for fibers in fiberglass-reinforced plastic composites.

[11] S-glass: S-glass is a magnesia-alumina-silicate glass, and it is used for fibers in fiberglass-reinforced plastics composites when extra-high-strength fibers are required.

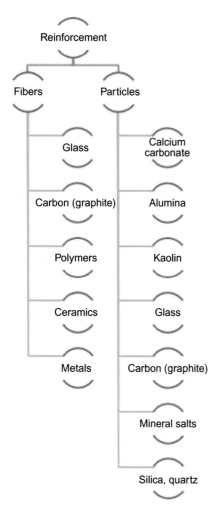

FIG. 1.53 Main reinforcements used in composite materials.

characterized by high stiffness and strength. In addition, carbon fibers have excellent resistance to creep, break stress, fatigue, and corrosive environments. An important characteristic, together with a low density, is a low coefficient of thermal expansion (CTE). In particular, carbon fibers generally have small, negative axial CTE (i.e., they get shorter when heated), and they can be used for the production of carbon-reinforced composites for aerospatial components, in which the temperature difference between the part exposed to the sun and the part in the shade can reach 400–500°C. Most carbon fibers are highly anisotropic. Axial stiffness, tension, compression strength, and thermal conductivity are typically much greater than the corresponding properties in the radial direction. Carbon fibers are made primarily from three key precursor materials: polyacrylonitrile (PAN), petroleum pitch, and coal tar pitch. PAN-based fibers are the most widely used type of carbon fibers. A key advantage of pitch-based fibers is that they can be produced with much higher axial moduli than those made from PAN precursors. However, composites made with pitch-based carbon fibers generally are somewhat weaker in tension and shear, and much weaker in compression than those produced using PAN-based reinforcements. The characteristics of carbon fibers make them adequate as reinforcement of composites with polymeric matrix for aerospace applications.

*Polymer fibers.* Among polymeric fibers, the aramid fibers are the most used; for that reason, in this textbook, only these fibers will be described. *Aramid fibers* are a high modulus organic reinforcement, primarily used to reinforce polymers and cement, and for ballistic protection. Aramids are aromatic polyamide polymers with a very rigid molecular structure so their mechanical properties are very high. There are a number of commercial aramid fibers produced by several manufacturers; Kevlar 29, Kevlar 49, and Twaron are examples. The high properties of Kevlar are caused by the fact that

reinforcements, because of their low cost and relatively low elastic moduli compared to other reinforcements. S-glass fibers are stiffer and stronger than E-glass ones, and have better resistance to fatigue and creep.

*Carbon fibers.* Carbon fibers are used as reinforcements for polymers, metals, ceramics, and carbon. There are different carbon fibers, with a wide range of strength and modulus values. As a class of reinforcements, carbon fibers are

hydrogen bonds maintain macromolecular chains together in the transverse direction (Fig. 1.54). Consequently, these fibers have high strength in the longitudinal direction and weak strength in the transverse direction. In addition, the aromatic ring structure gives high stiffness to the macromolecular chains, so that they have a rod-like structure. These fibers, for their low density, high strength, and modulus, are used as reinforcement in composites with polymeric matrix (e.g., epoxy) for aerospace, marine, and automotive applications.

*Other fibers.* Boron fibers are primarily used to reinforce polymers (e.g., epoxy) and metals. Boron fibers are produced as monofilaments of boron by chemical vapor deposition (CVD) on a tungsten wire or carbon fiber. The properties of boron fibers are influenced by the ratio of overall fiber diameter to that of the tungsten core. Silicon-carbide-based fibers are primarily used to reinforce metals and ceramics. There are different types: monofilament and multifilament silicon-carbide-based fibers. Some of these fibers are far from pure silicon carbide (SiC) and contain varying amounts of silicon, carbon and oxygen, titanium, nitrogen, zirconium, and hydrogen.

#### 1.5.3.4 *Particles*

Particle reinforcement in composites is less effective in strengthening than fiber reinforcement. Particulate-reinforced composites mainly achieve gains in stiffness, but they also can achieve increases in strength and toughness. In any case, the improvements are lower than would be achieved in a fiber-reinforced composite. The main benefit of particle-reinforced composites is their low cost and ease of production and forming compared to fiber-reinforced ones.

In particular, particulate-reinforced composites find applications where high levels of wear-resistance are required such as road surfaces. In fact, gravel added as a reinforcing filler significantly increases the hardness of cement.

### 1.5.4 Design of Composite Materials

Fiber-reinforced composites are composed by continuous (long fibers) or discontinuous (short fibers). When the fibers (long or short) are aligned, they provide maximum strength but only along the direction of alignment (Fig. 1.55A and B). The composite is considerably weaker along other directions and is therefore highly anisotropic. This anisotropy can be

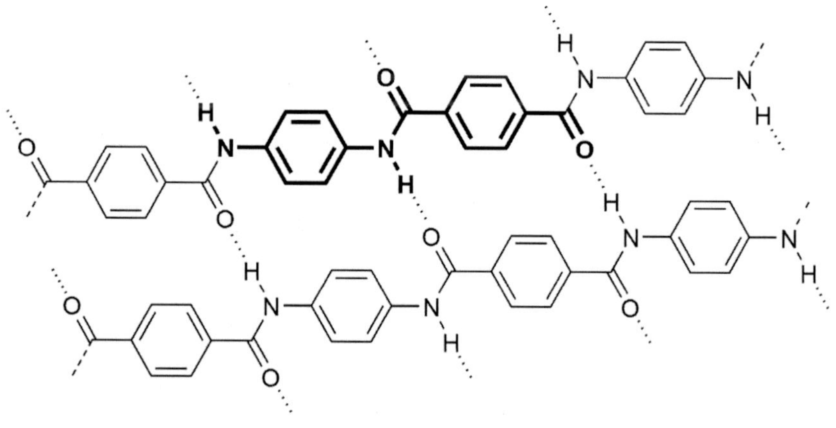

**FIG. 1.54** Structure of aramid, in particular the structure is referred to as Kevlar fiber. The interaction caused by hydrogen bonds between two macromolecular chains is evidenced.

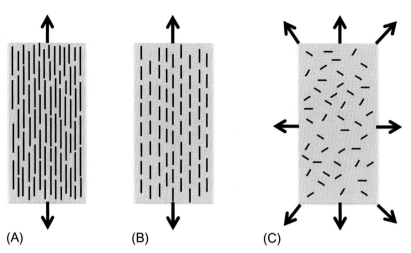

**FIG. 1.55**   Different representations of (A) continuous, long and aligned fibers; (B) discontinuous, short and aligned fibers; and (C) discontinuous, short and randomly oriented fiber-reinforced composites. In (A) and (B), the composite has an anisotropic behavior, whereas in (C), the composite exhibits an isotropic behavior.

overcome by fibers in all directions; in that case, usually short fibers are used (Fig. 1.55C).

As previously discussed, optimum strength and stiffness can be achieved in a composite by aligning the fibers parallel to the direction of loading. However, in this case, the composite can perform very poorly when the load is applied perpendicular to the fibers (Fig. 1.56A). To overcome this problem, some composite structures are designed to obtain a more isotropic composite material, using multiple plies of continuous fibers with the direction of the fibers differing in each ply. In this case, for example, fibers can vary by 0-degree, 90-degree, +45-degree, or −45-degree angles in each different ply, to accommodate for the direction of the loads applied to the laminated structure (Fig. 1.56B).

When a laminate is designed, it is important to answer two different questions: (1) What is the relationship of the composite material properties to the properties of the constituents (Fig. 1.57A)? (2) What is the relationship between the properties of the composite and the properties of the single plies (Fig. 1.57B)?

Micromechanics and macromechanics study the mechanical behavior at the two levels to optimize the design of the composite laminate to obtain the expected behavior. The mathematical treatment related to them is not within the scope of this book, and micromechanics and macromechanics will be briefly exposed in general terms. Micromechanics is the study of composite behavior encompassing the interaction of constituent materials, taking into account the design of single plies of laminate in terms of volumetric composition, geometrical properties, fiber disposition, and properties of reinforcement and matrix material. In addition, the percentages of the constituent materials (i.e., matrix vs reinforcement) can be varied to arrive at the desired composite stiffness and strength.

On the contrary, macromechanics is related to the study of composite material behavior wherein the material is assumed homogeneous, and the effects of the constituent materials are detected only as averaged apparent properties of the composite ply and how arrange different plies together to obtain specific properties of the composite.

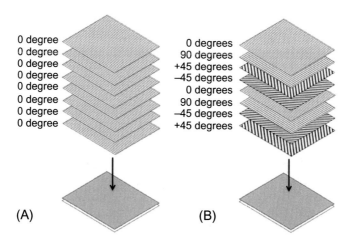

FIG. 1.56  Possible design of laminated composites: (A) unidirectional and (B) multidirectional laminated plies.

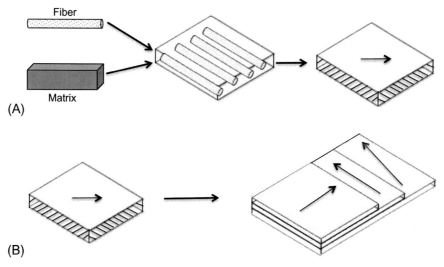

FIG. 1.57  (A) Micromechanics and (B) macromechanics in a laminated composite material.

## 1.6 NATURAL POLYMERS

Future technologies will be inspired by nature (Adikwu, 2012). Nature itself has made good use of structural natural polymers, as in crustacean chitin, and produced strong polymeric fibers, as in spider webs, long before synthetic chemists arrived on the scene to complicate matters (Shogren and Bagley, 1999). Natural polymers by themselves are a class of polymers that refers to polymers derived from natural sources, such as plants, animals, and microorganisms, through extraction from their bulk form in nature (Singh, 2016). Such natural polymers also include polymers produced by biological processes such as bacteria synthesis or fermentation (Olatunji, 2016). Natural polymers possess wide scope in drug, food, and cosmetic industries. Natural polymers are biogenic, and their biological properties, such as cell recognition and interactions, enzymatic degradability, semblance to the extracellular matrix (ECM), and their chemical flexibility, make them materials of choice for multiple applications (Singh, 2016).

Many of the applications of natural polymers lost out in the competitive world to synthetic polymers, in which properties could be tailored to needs (Shogren and Bagley, 1999). Processing of natural polymers has been taking place since the early humans who have long woven and dyed fibers of silk, wool, and carbohydrates from flax and cotton.

In comparison to synthetics, natural polymers remain attractive primarily because (Alam Md et al., 2014; Singh, 2016):

1. They are inexpensive, or at least cheaper, and their production cost is generally lower than synthetic materials. Besides, they are cheaper to utilize and do have low disposal costs.
2. They are readily available through environmentally friendly processing. There are many types of natural polymers obtained from different sources and collected in immensely large quantities due to the simple production processes involved.
3. They are (bio)compatible, typically devoid of side or adverse effects, or at least safer than synthetic polymers on human beings due to their natural origin. Patient tolerance as well as public acceptance are therefore key strengths.
4. They are potentially biodegradable.

Conversely, some major drawbacks about the use of natural polymers include (Singh, 2016):

1. Batch to batch variability. Synthetic manufacturing is a controlled procedure with fixed quantities of ingredients, whereas production of natural polymers is dependent on environmental factors. Due to differences in the collection of natural materials at different times, as well as differences in region, species, and climate conditions, the percentage of chemical constituents present in a given material may sometimes vary greatly.
2. Slow rate of production. As the production rate depends upon the environment and many other factors, it cannot be changed at will.
3. Heavy metal and microbial contamination, as they are naturally sourced materials.

Although these problems should cause natural polymers to be comparatively less attractive at the commercial level, they have been widely used in biomedical research and medicine so far. In the following text, we are going to discuss the three main classes of natural polymers, namely proteins, polysaccharides, and nucleic acids.

## 1.6.1 Proteins

Proteins are biopolymers present in and vital to every living organism that serve an astonishing variety of functions. The human body contains about $10^5$ different proteins, playing important roles in all aspects of cell structure and function. For instance, some proteins contribute to the structural elements of a cell, whereas others are used to bind cells into tissues or catalyze metabolic reactions, DNA replication, and transport molecules from one location to another. Besides, proteins take the form of antibodies to protect from diseases and interferon to fight off microorganisms. Other proteins are silk, hair, and collagen (used in plastic surgery, and a protein in our body used for strengthening and packing). On the other hand, the deadly properties of protein toxins and venoms are less widely appreciated. Botulinum toxin A, from *Clostridium botulinum*, is regarded as the most powerful poison known. Meat, fish, eggs, milk, cheese, peas, beans, nuts, and many other kinds of food contain large amounts of proteins.

From a chemical point of view, proteins are by far the most structurally complex and functionally sophisticated polymers known. Despite the variety of physiological and pathological functions and differences in physical properties, proteins are sufficiently similar in molecular structure to warrant treating them as a single chemical family.

### 1.6.1.1 Building Blocks

Proteins are macromolecules variable in length and structure, consisting of one or more chains of building blocks called amino acids, so named because an amino group (-NH$_2$) is bound to the C atom, next to the carbonyl moiety (-COOH). Otherwise, they are amphoteric (having both acidic and basic properties) (Fig. 1.58). Some common features of such amino acids warrant special mention. Amino acids are crystalline solids that dissolve in water. Besides, every amino acid, that is an $\alpha$-aminocarboxylic acid in chemistry, consists of a central hydrocarbon moiety (i.e., a C$_\alpha$-H), to which are bonded three other chemical moieties. Two of these groups are the same for every amino acid. Indeed, with the exception of proline (Pro — P), they display a -NH$_2$ and a -COOH group (Fig. 1.58). The third moiety, known as side chain or R group, differentiates one amino acid from another. The chemical properties of the side chain thus determine which of four categories the amino acid falls into

(see the classification herein). The simplest $\alpha$-amino acid is the aminoacetic acid, also called glycine (Gly — G). It bears a hydrogen side chain (i.e., R = H) (Table 1.6). With the exception of Gly, the $\alpha$-amino acids are all *chiral* (see Annex 1 — Chirality). The hydrolysis of proteins, that is, the breakdown of proteins through the addition of water, boiling aqueous acid or base, yields an assortment of small building blocks called *amino acids*. The physical and chemical properties of a protein are determined by its constituent amino acids.

The predominant form of the amino acid depends on the pH of the solution they are solubilized in. Amino acids are less acidic than most carboxylic acids and less basic than most amines. Contrary to your expectations, the acidic part of the amino acid molecule is the $-NH_3^+$ group, not the -COOH moiety, and the basic part is the -COO$^-$ group, and not the free -NH$_2$ moiety. Because amino acids contain both acidic and basic groups, they are said to be *amphoteric* because they do display both acidic and basic behaviors. Noteworthy, by varying the pH of the solution, we can control the charge on every amino acid. In an acidic solution, the -COO$^-$ is protonated to a free -COOH group, and the molecule has an overall positive charge. As the pH is raised, the -COOH loses the hydrogen ion (H$^+$) at

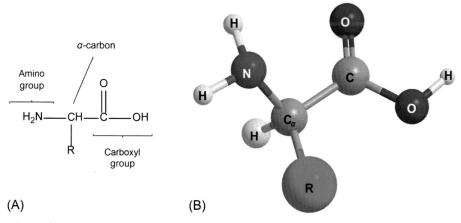

(A)                                                        (B)

**FIG. 1.58** Planar (A) and 3D chemical structure (B) of a general amino acid. Every amino acid contains the same amino (-NH$_2$) and carboxyl (-COOH) groups, but the side chain, or R group, makes them different from each other.

**TABLE 1.6**  Amino Acid Classification Between Nonessential and Essential

**Amino Acid Backbone**

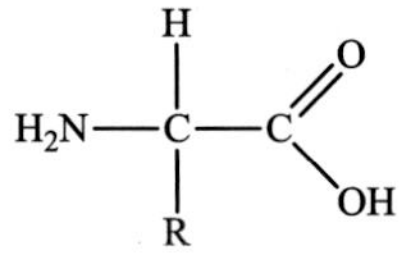

R group of nonessential amino acids

Alanine (Ala—A)

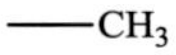

—CH$_3$

Arginine (Arg—R)

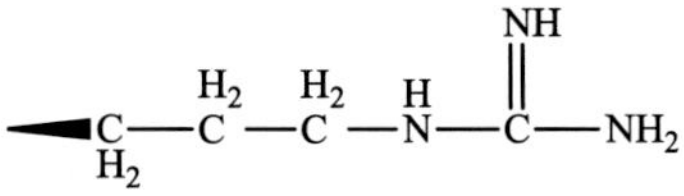

Asparagine (Asn—N)

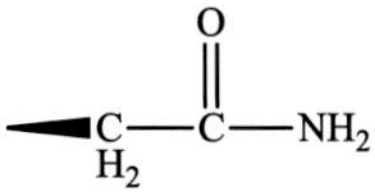

Aspartic acid (Asp—D)

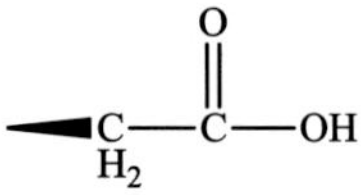

Cysteine (Cys—C)

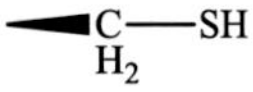

Glutamic acid (Glu—E)

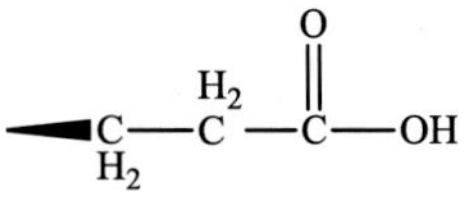

Glutamine (Gln—Q)

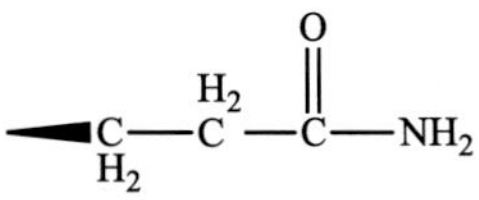

R group of essential amino acids

Isoleucine (Ile—I)

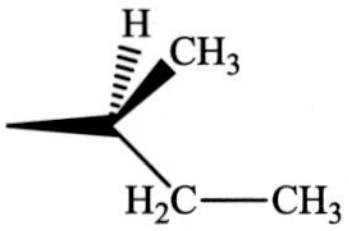

Leucine (Leu—L)

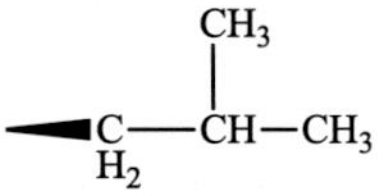

Lysine (Lys—K)

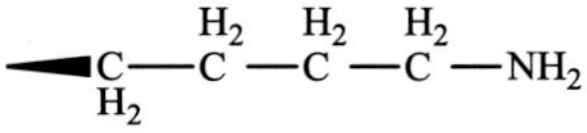

Methionine (Met—M)

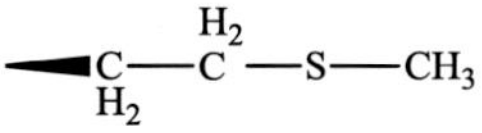

Phenylalanine (Phe—F)

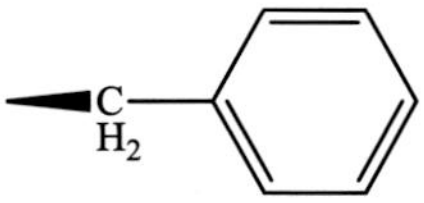

Threonine (Thr—T)

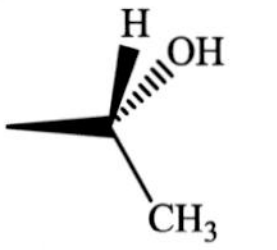

Tryptophan (Trp—W)

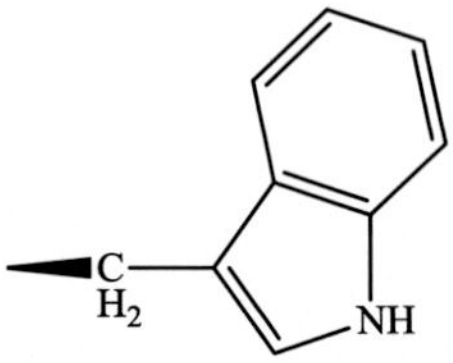

**TABLE 1.6**   Amino Acid Classification Between Nonessential and Essential—cont'd

**Amino Acid Backbone**

Glycine (Gly—G)

—H

Valine (Val—V)

$CH_3$
|
CH—$CH_3$

Histidine (His—H)

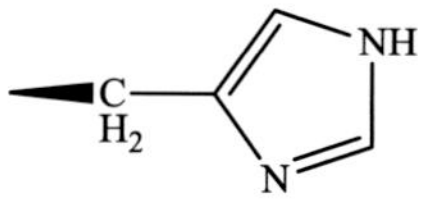

Proline (Pro—P)—Whole amino acid structure

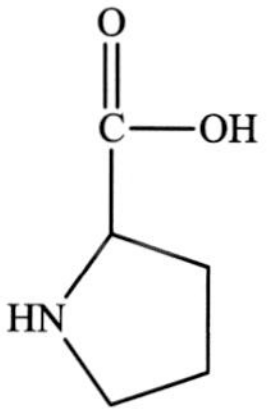

Serine (Ser—S)

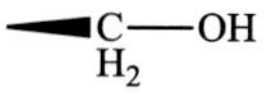

Tyrosine (Tyr—Y)

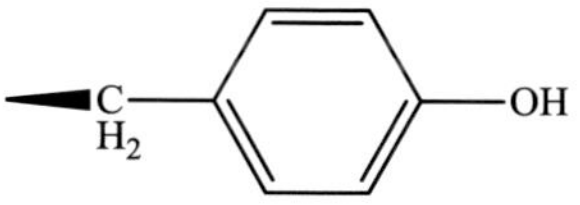

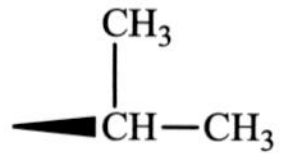

Each amino acid is also assigned a three- or one-letter abbreviation. Sometimes there are difficulties in using the three-letter system, particularly when the amino acid sequence is long. A single-letter code has been designated for convenience, which is helpful in summarizing large amounts of data and for aligning and comparing homologous sequences. Single-letter codes are also used to label residues in 3D pictures of proteins. However, this system is less easily understood than the three-letter abbreviation.

about pH 2 (Fig. 1.59, left). This point is named $pK_{a1}$, the first acid-dissociation constant. As the pH is raised further, the $-NH_3^+$ group loses its $H^+$ at pH $\approx$9–10 (Fig. 1.59, right). This point is called $pK_{a2}$, the second acid-dissociation constant. Above this pH, the amino acid has an overall negative charge. Therefore there must be an intermediate pH where the amino acid is evenly balanced between the two forms, as the dipolar zwitterion displays a net charge of zero. This pH value is called isoelectric pH or isoelectric point (pI) (Fig. 1.59, middle).

Although more than 100 amino acids occur naturally, only 20 are commonly used in protein synthesis that the human body needs to function properly. For such reasons, they are also called standard amino acids. Interestingly, they are the same in every living organism, from protozoa to plants and animals. Scientists classify amino acids based on lots of different features.

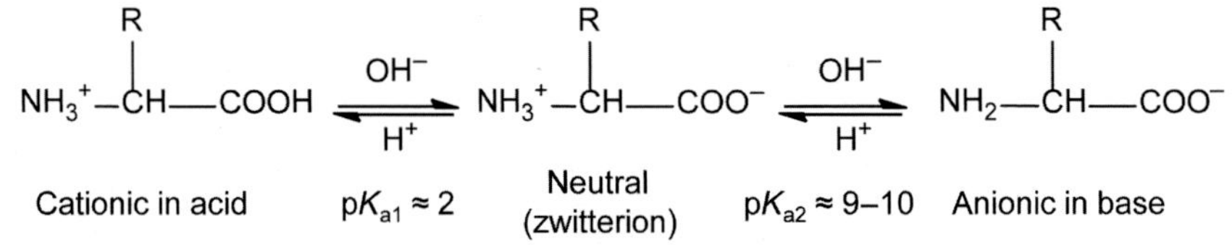

**FIG. 1.59** When an amino acid is dissolved in water, it exists predominantly in the isoelectric form. The isoelectric point (pI) is the pH of an aqueous solution of an amino acid at which the molecules have no net charge because the cationic groups are exactly balanced by their anionic counterparts.

For instance, Table 1.6 displays a widely accepted classification that relies on a dichotomous key to identify essential and nonessential amino acids. However, it is worth noting that the classification as essential or nonessential does not actually reflect their importance, as all 20 of them are necessary for human health. Eleven out of 20 standard amino acids are called nonessential because they are produced naturally in the body, made from other basic chemicals found within it, whereas the other nine are called essential amino acids because they have to be supplied by food. The digestive system breaks down consumed proteins into amino acids that are absorbed by the intestine and enter the bloodstream. Cells next utilize the amino acids provided as building blocks to rebuild various proteins. If an individual does not acquire enough essential amino acids from food consumption to create the necessary proteins, the body is susceptible to many shortcomings such as a weakened immune system, a decrease in hormone production, and the breakdown of muscle tissues.

A protein differs from another primarily in the amino acid sequence, which is dictated by the succession of nucleotides of a nucleic acid sequence called a gene, and which typically results in a protein, such that it folds into a specific 3D structure determining its activity. Notice that the chemical versatility that the 20 standard amino acids provide is vitally important to the function of every protein.

Most of the amino acids found in naturally occurring proteins exhibit the same absolute steric configuration as (L)-glyceraldehyde and thus appear to be of the (L)-type (see Annex 1 — Chirality). Although (D)-amino acids occasionally occur in nature, such as those found in small peptides that are part of the bacteria cell wall, and are often found in polypeptide antibiotics, we usually assume the amino acids under discussion are typically (L)-amino acids. We would like to remind you that the (D) and (L) nomenclature, like the (R) and (S) designation, gives the configuration of the asymmetric carbon atom. It does not imply the sign of the optical rotation or which must be determined experimentally (see Annex 1 — Chirality).

In actuality, all amino acids do display acid-base properties, as they each contain an amino group and a carboxyl moiety. However, from an acid-base viewpoint, these groups neutralize each other, leaving the side chain to determine whether an amino acid is ultimately acidic or basic. It is worthy of note that the collective properties of the amino acid side chains underlie all the diverse and sophisticated functions of proteins. Based on the propensity of the side chain to be in contact with a polar solvent like water, amino acids are classified as:

1. *nonpolar and neutral.* A general chemical, and specifically an amino acid, is nonpolar if it is uncharged, that is, it has a low propensity to be in contact with water. This state occurs either because its atoms do not carry a charge at all or because its positively and negatively charged atoms are present in equal amounts and cancel each other out. Amino acids containing uncharged hydrocarbon groups or benzene rings as side chains are nonpolar.

Neutral side chains exhibit neither acidic nor basic behaviors. Although the bulk of amino acids is neutral, the ones that are both nonpolar and neutral are Ala, Gly, Ile, Leu, Met, Phe, Pro, and Val;

2. *polar and neutral.* Polar amino acids do have side chains with either a net-positive or a net-negative charge, though the degree of polarity varies depending on the specific chemistry of the side chain. For instance, Ser exhibits greater polarity, or it is less nonpolar, than Thr and Tyr, because its side chain is shorter and lacks a ring group. In addition to Ser, Thr, and Tyr, Asn, Cys, Gln, and Trp are polar and chemically neutral at once;

3. *polar and acidic.* Acidic amino acids contain a side chain that functions as an acid and is capable of becoming ionized by donating a $H^+$ to the surrounding environment. The only two acidic amino acids are Asp and Glu. These molecules are polar as well;

4. *polar and basic.* The remaining amino acids, namely Arg, His, and Lys, are polar and basic. Each of their side chains contains an additional amino group beyond what is found in the core structure of the monomer. They therefore function as basic molecules that can accept a $H^+$ from the existing environment. The polarity of these and other charged amino acids makes them more water-soluble than the eight nonpolar amino acids. This chemical feature means that they are typically located on the exterior of a protein in an aqueous solution.

A different classification criterion relies on the side chain structure of each specific amino acid. Scientists recognize five types in this classification:

1. containing sulfur (Cys and Met);
2. neutral (Asn, Ser, Thr, Gln, Pro);
3. acidic (Asp and Glu) and basic (Arg, His, and Lys);
4. aliphatic (Leu, Ile, Gly, Val, and Ala);
5. aromatic (Phe, Trp, and Tyr).

## 1.6.2 Structure and Function of Proteins

Proteins are single, unbranched (i.e., linear) repeats of amino acid monomers. Amino acids are joined covalently by peptide bonds on the ribosome during protein synthesis, or translation, to yield the astonishing variety of proteins. A linear chain of amino acid residues is called a *polypeptide*. A polypeptide is a peptide containing many amino acid residues but usually having a molecular weight of <5,000. Notice that a protein contains at least one long polypeptide. Even though there is no clear distinction of when a polypeptide might become a protein, proteins contain more amino acid units, with molecular weights ranging from ≈5,000 to about 40,000,000. The term *oligopeptide* is occasionally used for peptides containing about 4–10 amino acid residues. In naming peptides, the names of acyl groups ending in "yl" are used. Hence, if Gly and Cys were to condense so that Gly acylates Ala, the dipeptide formed (the product is called dipeptide because it consists of two amino acids) is named glycylcysteine. However, oligopeptides and polypeptides of biological origin often have trivial names because their sequences are usually described more conveniently by symbols rather than by constructing lengthy names. The peptide bond, also called amide bond or linkage in organic chemistry (the same bond is found in synthetic fibers such as Nylon), is formed by a condensation (water-loss) reaction between the carboxyl group of one amino acid and the amino group of the next amino acid occurring in a protein (Fig. 1.60). Thus proteins are formed by the linear arrangement of amino acids in a particular order. The free $-NH_3^+$ is invariably drawn on the left of the chain and is called the *N*-terminal end or the *N*-terminus. On the other hand, the right of the polypeptide chain ends with the free -COO⁻, which is called the *C*-terminal end or the *C*-terminus. This is available for the formation of a peptide bond with another amino acid.

Crucial to the understanding of protein structure is the knowledge of the structure of the

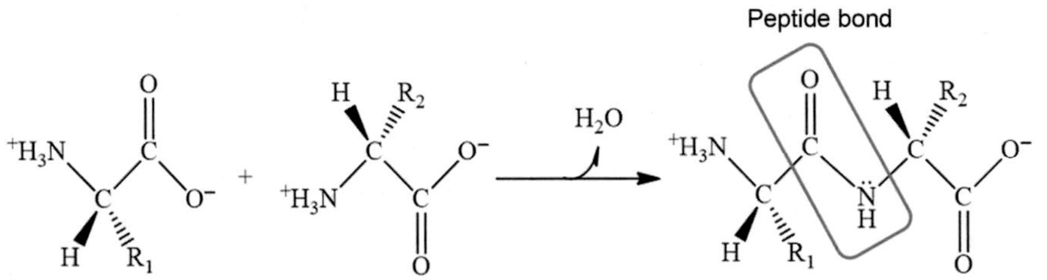

Peptide bond

**FIG. 1.60** Two amino acids form a dipeptide by means of a condensation reaction, giving rise to a peptide bond. *N*-terminus and *C*-terminus are invariably drawn to the left and the right, respectively, of the neosynthesized peptide.

peptide bond. Linus Pauling, in the 1930s, used X-ray diffraction to examine the nature of the peptide bond formed between two amino acids. He reported that the peptide bond has a rigid planar structure because the lone pair of electrons on the N is delocalized into the C=O, thus forming a partial double bond between the N and the C=O. This effect is an example of resonance that can be thought of as a sharing of electrons between bonds. For that very reason, it is estimated that a peptide is described by resonance structure A for 62% and by B for 28%, which does not sum to 100% because there are additional resonance forms not depicted in Fig. 1.61. As can be seen in Fig. 1.61, steric hindrance between the functional groups attached to the $C_\alpha$ atoms will be greater in the *cis* configuration. Therefore, the peptide bond nearly always has the *trans* configuration because it is

more favorable than *cis*, which is sometimes found to occur with Pro residues.

Because the bond between the C-O and the C-N has a partial double bond character, rotation around this bond is restricted. Therefore the peptide unit is a planar, rigid structure and rotation in the peptide backbone is restricted to the bonds involving the $C_\alpha$.

The preferred location of different amino acids in protein molecules can be characterized by calculating the extent by which an amino acid is buried in the structure or exposed to solvent. Although membrane proteins follow a different pattern, most proteins have a hydrophobic core that is not accessible to solvent, which is mostly water, and a polar surface in contact with the environment. Although hydrophobic amino acid residues are mostly buried within the core of the structure, polar and charged amino acids

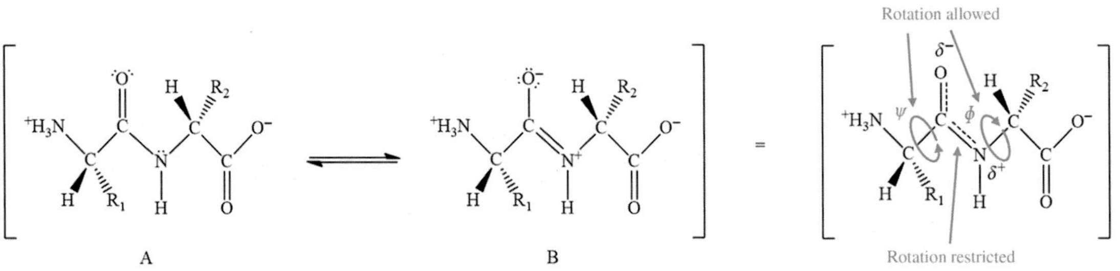

**FIG. 1.61** Resonance interactions between electrons in the C=O bond and the C—N bond of the peptide group mean that there is "sharing" of electrons between these bonds. Note the charges on the N and O atoms. Therefore the O-C-N-H atoms in the peptide bond are usually considered to be coplanar. The rigid, planar nature of the peptide unit has implications for the detailed 3D structure of peptides. Besides, note that though rotation is not permitted about the peptide bonds, rotation around the $C_\alpha$—N and $C_\alpha$—C bonds can occur and is defined by the torsion angles $\phi$ and $\psi$, respectively.

preferentially cover the surface of the molecule and are exposed to and in contact with solvent to a much higher degree due to their ability to form hydrogen bonds. For a hydrogen bond to be formed, two electronegative atoms (for instance, in the case of an α-helix, the amide N and the carbonyl O) have to interact with the same H. In proteins, essentially all groups capable of forming hydrogen bonds (both main chain and side chain, independently of whether the residues are within a secondary structure or some other type of structure) are usually hydrogen-bonded to each other or to water molecules. Due to their electronic structure, water molecules may accept two hydrogen bonds and donate two, thus being simultaneously engaged in a total of four hydrogen bonds. Water molecules may also be involved in the stabilization of protein structures by making hydrogen bonds with the main chain and side chain groups in proteins and even linking different protein groups to each other. In addition, water is often found to be involved in ligand binding to proteins, mediating ligand interactions with polar or charged side chain or main chain atoms. Positively and negatively charged amino acids often form so-called salt bridges. These interactions may be important for the stabilization of the protein 3D structure, for instance, proteins from thermophilic organisms (i.e., organisms that live at high temperatures, such as 80–90°C or higher) often have an extensive network of salt bridges on their surface, which contributes to the thermostability of these proteins, preventing their denaturation at high temperature.

Protein structure is the 3D arrangement of atoms in an amino acid-chain molecule. There are four distinct levels of structure, or hierarchical levels of organization, which determine the final shape of proteins, namely primary, secondary, tertiary, and quaternary structure.

1. The *primary structure* of a protein refers to the ordered sequence of amino acids of which it is composed (Fig. 1.62). Amide linkages (peptide bonds) form the backbone of the amino acid chains. This sequence ultimately determines the shape that the protein adopts (Anfinsen's dogma), according to the spatial limitations on the arrangement of the atoms in the protein, the chemical properties of the component amino acid residues, and the protein environment. Bearing in mind the planar nature of the peptide group, a polypeptide chain can be seen to have a backbone that consists of a series of rigid planar peptide groups linked by the $C_\alpha$ atoms. The angles of rotation, termed *torsion angles*, about these bonds specify the conformation of a polypeptide backbone. The torsion angles about the $C_\alpha$—N and $C_\alpha$—C bonds are referred to as $\phi$ (phi) and $\psi$ (psi) (Fig. 1.61), respectively, and they are defined as 180 degrees when the polypeptide is in the extended planar conformation, that is, in the primary structure.

2. The conformation, or local folding, adopted by the polypeptide backbone of a protein is referred to as *secondary structure*. Although it is true to say that all proteins have a unique 3D structure, or conformation, specified by the nature and the sequence of their amino acids, there are certain structural elements, or types of secondary structure, that are readily recognized in many different proteins. These secondary structural elements include *helices, pleated sheets,* and *turns.* As well as conforming to allowed torsion angles for component residues, secondary structures are stabilized by noncovalent interactions between atoms and groups in the polypeptide. The polypeptide may fold and turn many times, and such interactions are often between residues some distance apart in terms of the primary structure. Roughly half of an average globular protein (see the section *Classification of Proteins*) consists of regular repetitive secondary structures (helices and pleated sheet) whereas the remainder has an irregular so-called coil or loop conformation.

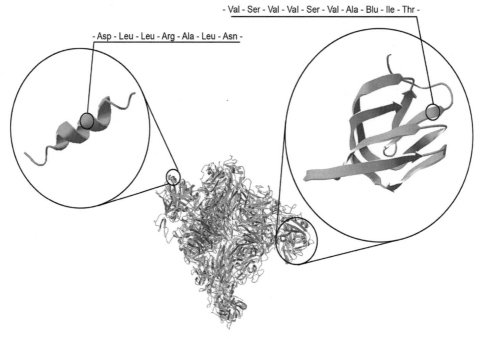

- Val - Ser - Val - Val - Ser - Val - Ala - Blu - Ile - Thr -

- Asp - Leu - Leu - Arg - Ala - Leu - Asn -

**FIG. 1.62**    Picture generated by Swiss-PdbViewer (http://www.expasy.org/spdbv/) of the human cell adhesion protein integrin β-2 (lower central panel) displaying a tertiary structure. Inset pictures depict two different secondary structures. Specifically, an α-helix and some β sheets are shown on the left and the right of the figure, respectively. Coil conformations are in green. Part of the primary structures, which is the amino acid sequence, specific to each secondary structure is displayed as well.

An important point to note is that a helix has a handedness; that is, if viewed along its axis, the chain turns either in a clockwise direction (right-handed helix) or in an anticlockwise direction (left-handed helix). A number of different helical structures have been identified in proteins. The most common is the α-helix (Fig. 1.62). The α-helix structure is stabilized by hydrogen bonds between the peptide CO and the peptide NH groups that are four residues along. In this way, the full hydrogen-bonding capacity of the polypeptide backbone is utilized. Note that the side chains (Rs) all project outward and backward from the helix as it rises; thus steric interference with the backbone or with other side chains is avoided. The helix core is tightly packed and stabilized by Van der Waals interactions. Because the hydrophilic polypeptide backbone is optimally hydrogen-bonded to itself and hidden away at the core of the α-helix, such regions of secondary structure are commonly seen in proteins that traverse the cell membrane, such as transmembrane receptors and transport proteins. In such cases, the side chains, which project into the lipid environment, are typically nonpolar. Another common secondary structure is the β-pleated sheet (Fig. 1.62), which contains extended stretches of polypeptide chain with hydrogen bonds between neighboring strands. In a parallel β-pleated sheet, polypeptide strands run in the same direction (i.e., from N-terminus to C-terminus) whereas in antiparallel β-pleated sheet, neighboring strands extend in opposite directions. Strands are not fully extended but have a zig-zag shape, which gives the sheet formation, in both parallel and antiparallel structures, a pleated appearance when viewed edge-on. The $C_\alpha$ of successive residues are at, alternately, the top and bottom of each pleat, with the side chains pointing away from the sheet. Antiparallel β sheets appear to be more stable than parallel sheets. The relative instability of parallel β sheets may be due to the offset in hydrogen-bonding groups between neighboring strands. This offset causes some distortion, and hence weakening, in the hydrogen bonds compared to those between antiparallel strands. Mixed parallel and antiparallel β sheets also occur. Within the context of the entire peptide chain, regions of β sheet are connected by linking peptides. In addition to the repetitive helical and pleated sheet structures, there are other nonrepetitive, and therefore more varied, elements of secondary structure called coil conformations (Fig. 1.62). However, the lack of regular repetitive order in coil conformations does not mean that these structures are disordered or unstable. Sometimes referred to as *random coil*, such regions may play an important part in the protein function. For instance, random coils

may be involved in the binding of a *ligand*, with consequent changes in the conformation and activity of the protein.

3. *Tertiary structure* is the arrangement of the polypeptide as a whole, that is the spatial relationship between its elements of secondary structure (Fig. 1.62). Though it may not be immediately obvious, proteins do follow certain recognizable folding patterns common to many different proteins. Proteins that are secreted by the cell, or are attached to the extracellular surface of the plasma membrane (the so called *exofacial* proteins), are usually subject to more extreme conditions than those experienced by intracellular proteins. Apart from the peptide bond, a second kind of covalent bond is typically possible between any Cys residues present. The sulfhydryl groups (-SH) of two Cys residues in close proximity in the folded protein can form disulfide bridges (also called disulfide linkages, -S-S-) that can join two chains or link a single chain into a ring (Fig. 1.63). This disulfide-linked dimer of Cys is called *cystine*. Disulfide bonds do not affect the conformation of the protein and are only formed when the folding is complete. Therefore they act to secure the conformation and increase the stability of the protein.

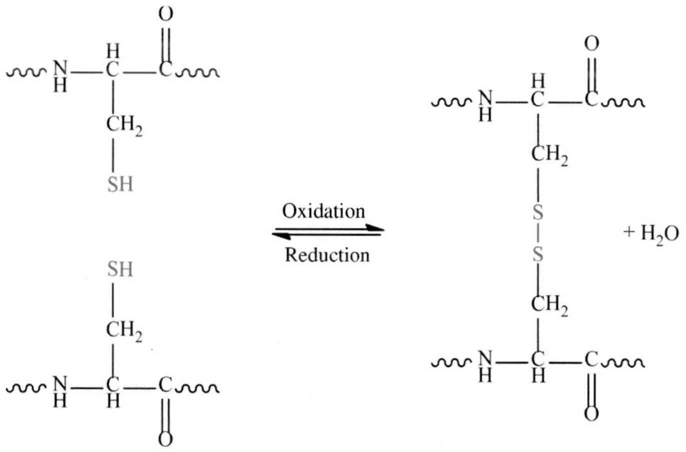

FIG. 1.63 Formation of a cystine disulfide bridge linking two peptide chains. Mild oxidation of two Cys residues joins two molecules of a thiol into a disulfide. A disulfide bridge (highlighted in *blue*) is easily cleaved by reducing them to the thiol (Cys) form. These reduced Cys residues have a tendency to reoxidize and reform disulfide bridges, however.

4. *Quaternary structure* applies only to those proteins that consist of more than one polypeptide chain, termed subunits, and describes their arrangement. In such proteins, sometimes referred to as multisubunit proteins, the same kinds of noncovalent interactions that stabilize the folded polypeptides also specify the assembly of complexes of subunits. Quaternary structure refers to the way in which the subunits of such proteins are assembled in the finished protein. Multisubunit proteins can have a number of identical (*homomeric*) or nonidentical (*heteromeric*) subunits. The simplest multisubunit proteins are homodimers, which are two identical polypeptide chains that are independently folded but held together by noncovalent interactions.

Protein folding is the physical process through which a protein chain acquires its native, characteristic, and functional 3D structure from a random coil in an expeditious and reproducible manner. Conversely, failure to fold into a native structure generally produces inactive proteins, but in some instances, misfolded proteins have modified or toxic functionality. In this regard, many allergies are caused by incorrect folding of some proteins, because the immune system does not produce antibodies for certain protein structures. The energy landscape describes the folding pathways in which the unfolded protein is able to assume its native state. Folding is a spontaneous process that is mainly guided by hydrophobic interactions, formation of intramolecular hydrogen bonds, and Van der Waals forces, and it is opposed by conformational entropy. Nevertheless, protein folding is thermodynamically favorable within a cell for it to be a spontaneous reaction. The process of folding often begins cotranslationally, so that the *N*-terminus of the protein begins to fold while the *C*-terminal portion of the protein is still being synthesized by the ribosome. Denaturation is a process in which proteins lose the quaternary, tertiary, and secondary structure present in their native state, by application of some external stress or compound such as a strong acid or base (e.g., acetic acid, trichloroacetic acid, sulfosalicylic acid, or sodium bicarbonate), a concentrated inorganic salt (e.g., urea 6–8 mol/L, guanidinium chloride 6 mol/L, lithium perchlorate 4.5 mol/L), an organic solvent (e.g., ethanol, chloroform), cross-linking reagents (e.g., formaldehyde, glutaraldehyde), radiation, or heat. If proteins in a living cell are denatured, this results in disruption of cell activity and possibly cell death. Protein denaturation may be also a consequence of cell death. Denatured proteins can exhibit a wide range of characteristics, from conformational changes and loss of solubility to aggregation due to the exposure of hydrophobic groups. Denatured proteins lose their 3D structure and therefore lose their function. For instance, when food is cooked, some of its proteins become denatured because they become thermally unstable. This is why boiled eggs become hard and cooked meat becomes firm. In this case, enzymes lose their activity, because the substrates can no longer bind to the active site, and because amino acid residues involved in stabilizing substrates transition states are no longer positioned to be able to do so. Proteins from *thermophiles* (i.e., microorganisms with optimal growth temperatures above 45°C, generally ranging between 50°C and 110°C) have more salt bridges than proteins from *mesophiles* (i.e., microbes whose optimal development temperatures are moderate, ranging from 15°C to 45°C). These additional salt bridges contribute to stability, resisting denaturation by high temperature.

Denaturation is reversible in many cases, in other words, proteins can regain their native state when the denaturing influence is removed. This process can be called renaturation.

## 1.6.3 Classification of Proteins

Different methods of protein classification have been proposed, but currently none of them is universally valid. For now, we briefly survey their general classifications. Proteins may be classified according to their chemical composition, their shape, or their biological function.

On the basis of their chemical composition, proteins are divided into two classes:

1. *simple proteins* (or unconjugated proteins) are also known as homoproteins because they are made up of only amino acids. Examples are plasma albumin, collagen, and keratin;

2. *conjugated* (or complex) *proteins*. Sometimes also called heteroproteins, they contain in their structure a nonprotein portion, called a *prosthetic group*. Good examples of this are glycoproteins, phosphoproteins, chromoproteins, and lipoproteins. *Glycoproteins* are proteins that covalently bind one or more carbohydrate units to the polypeptide backbone. Typically, the branches consist of not more than 15–20 carbohydrate units, where you can find arabinose, fucose (6-deoxygalactose), galactose, glucose, mannose, and *N*-acetylglucosamine (GlcNAc, or NAG). Examples of glycoproteins are: *fibronectin*, which anchors cells to the ECM through interactions on one side with collagen or other fibrous proteins, whereas on the other side with cell membranes; all blood plasma proteins, except albumin; and immunoglobulins or antibodies. *Phosphoproteins* are proteins that bind phosphoric acid to Ser and Thr residues. Generally speaking, they have a structural function, such as tooth dentin, or reserve function, such as milk caseins (α, β, γ, and δ). Reversible post-translational modifications, especially phosphorylation, are important in regulation and cell signaling. *Chromoproteins* contain at least one colored prosthetic group. Typical examples are *hemoglobin* and *myoglobin* that bind respectively one and four heme groups, *chlorophylls* that bind a porphyrin ring with a magnesium atom at its center, and *rhodopsins* that bind retinal. *Lipoproteins* are complex particles with a central core containing cholesterol esters and triglycerides surrounded by free cholesterol, phospholipids, and apolipoproteins. Because cholesterol and triglycerides are insoluble in water, these lipids must be transported in association with proteins. Lipoproteins play a key role in the absorption and transport of dietary lipids by the small intestine, in the transport of lipids from the liver to peripheral tissues, and in the transport of lipids from peripheral tissues to the liver and intestine (reverse cholesterol transport). A secondary function is to transport toxic foreign hydrophobic and amphipathic compounds, such as bacterial endotoxin, from areas of invasion and infection.

According to their shape, proteins are classified into two groups:

1. **Fibrous** or **structural proteins**. They have primarily mechanical and structural functions, providing support to the cells as well as the whole organism. These proteins are insoluble in water as they contain, both internally and on their surface, many hydrophobic amino acids. The presence on their surface of hydrophobic amino acids facilitates their packaging into very complex supramolecular structures. In this regard, it should be noted that their polypeptide chains form long filaments or sheets, whereas in most cases, only one type of secondary structure, that repeats itself, is found. In vertebrates, these proteins provide external protection, support, and shape; in fact, thanks to their structural properties, they ensure flexibility and/or strength. Typical examples are *fibroin, collagen, elastin,* and *keratin.* Some fibrous proteins, such as α-keratins, are only partially hydrolyzed in the intestine.

- Spiders and insects, such as the silkworm *Bombyx mori*, produce fibroin. *Fibroins* are large complex proteins, the specific different structural details of those making up different types of silk that constitute, with sericin, silk fibers. On the other hand, all fibroins have some common features. The primary structure of all fibroins mainly consists of the recurrent amino acid sequence $(Gly-Ser-Gly-Ala-Gly-Ala)_n$. The high Gly (and, to a lesser extent, Ala) content allows for tight packing of the sheets, which contributes to the rigid structure of silk and tensile strength. Besides, fibroin proteins consist of layers of antiparallel $\beta$ sheets.

- *Collagen* is a family of at least 28 different structurally related proteins found exclusively in animals. Types I to V are the major collagen types in existence, being type I collagen the most abundant collagen of the human body. Collagen types I, II, III, and V are fibrillar, whereas type IV is nonfibrillar and forms the basic two-dimensional (2D) network of all basal laminae, a thin sheet-like network of ECM components that underlies most animal epithelial layers and other organized groups of cells (e.g., muscle), separating them from the connective tissue. Collagen is the extracellular scaffolding of multicellular organisms that can be thought of as the glue that holds cells together in a tissue. As the main component of connective tissue, it is the most abundant protein in mammals. In vertebrates, it makes up about 25%–30% (w/w) of the total protein content. Collagen is found in different tissues and organs, such as tendons and the organic matrix of bone, where they are present in very high percentages, and also in cartilage and in the cornea of the eye. In different tissues, specific types of collagen form different structures, each capable of

satisfying a particular need. Collagens therefore differ in their ability to form fibers and to organize the fibers into networks. For example, type I collagen fibrils are used as the reinforcing rods in construction of bone, and type II is the major collagen in cartilage.

Of note, the different types of collagen have low nutritional value as deficient in several amino acids. In fact, they contain no Try and low amount of the other essential amino acids. The gelatin used in food preparation is a derivative of collagen. Like the other collagens, a single type I collagen molecule consists of a 300 nm-long, 1.5 nm-thick triple helix made of three polypeptide chains or strands wound together. Collagen fibers are rope-like structures consisting of many of these overlapped units assembled into long cross-linked fibers, organized in a hierarchical order from collagen molecules, to microfibrils, fibrils, and collagenous fibers. The COL1A1 gene produces the pro-$\alpha$1 (I) chain. This chain combines with another pro-$\alpha$1(I) chain and also with a pro-$\alpha$2 (I) chain (produced by the COL1A2 gene) to make a molecule of type I procollagen. These triple-stranded, rope-like procollagen molecules must be processed by enzymes outside the cell (Fig. 1.64). Once these molecules are processed, they arrange themselves into long, thin fibrils that cross-link to one another in the spaces around cells. Many three-stranded type I collagen molecules pack together side-by-side, displaced from one another by about one-quarter of their length, forming fibrils with a diameter of 50–200 nm. The unique properties of the fibrous collagen types I, II, III, and V are due to the ability of the rod-like triple helices to form such side-by-side interactions.

The triple-helical structure of collagen arises from an unusual abundance of three

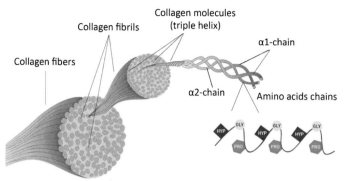

**FIG. 1.64** Collagen structure. Collagen molecules display a triple-stranded helical structure composed of three polypeptide chains (α-chains) and pack together to form long thin fibrils arranged in bundles.

amino acids that constitute the characteristic repeating motifs Gly-Pro-X and Gly-X-hydroxyproline (Hyp), where X is any amino acid other than Gly, Pro, or Hyp. Covalent aldol cross-links, catalyzed by *lysyl oxidase* (LOX) enzyme, form between two Lys or hydroxylysine (Hyl) residues at the *C*-terminus of one collagen molecule with two similar residues at the *N*-terminus of an adjacent molecule (Fig. 1.65). These cross-links stabilize the side-by-side packing of collagen molecules and generate a strong fibril. Type I collagen fibers take a wavy course that probably provides resilience to the fibers themselves, which also serves as a cushion against the direct tension.

- *Elastin* protein is a major *bioelastomer* characterized by the occurrence of the pentapeptide sequence (Val-Pro-Gly-Val-Gly)$_n$ where $n$ is between 10 and 13. Elastin is a highly elastic protein in connective tissue and allows many tissues and organs in the body to resume their shape after stretching or contracting. Elastic ligaments, large arteries, lungs, and skin are made of 70%, 50%, 30%, and 2%–4% elastin, respectively. Elastin protein is made by linking together many small soluble precursor tropoelastin molecules of 50–70 kDa, to make the final massive insoluble, durable complex. To

make mature elastin fibers, the tropoelastin molecules are cross-linked via their Lys residues with desmosine and isodesmosine cross-linking molecules. The enzyme that performs the cross-linking is LOX, as for the generation of collagen fibers (Figs. 1.65 and 1.66).

The unlinked tropoelastin molecules are not normally available in the cell, as they become cross-linked into elastin fibers immediately after their synthesis and during their export into the ECM. Elastin, which is produced as an ECM protein, provides skin and blood vessels with elastic properties because of its random coiled structure. It is encoded by the ELN gene in humans.

- *α-keratins* constitute almost the entire dry weight of nails, claws, beak, hooves, horns, hair, wool, and a large part of the outer layer of the skin. Their different stiffness and flexibility is a consequence of the number of disulfide bonds that contribute, together with other noncovalent forces, to stabilize the protein structure. This is the reason why wool keratins, which have a low number of disulfide bonds, are flexible, soft, and extensible, unlike claw and beak keratins that are rich in disulfide bonds.

2. **Globular** or **functional proteins**. Most of the proteins belong to this class. They have a compact and more or less spherical structure,

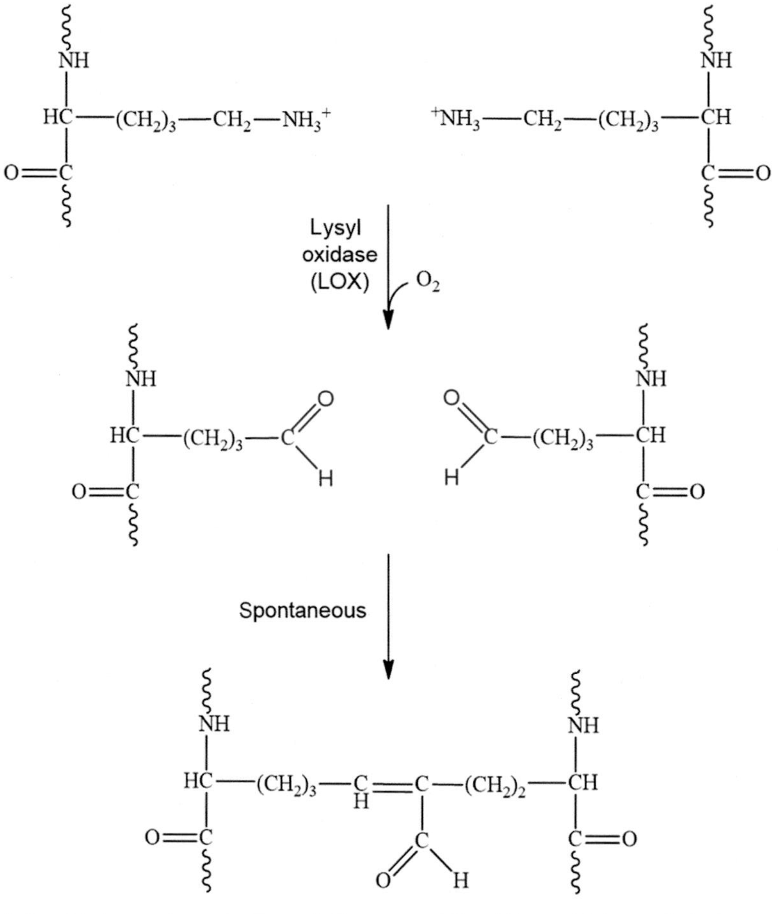

**FIG. 1.65** The side-by-side interactions of collagen helices are stabilized by an aldol cross-link (lower panel) between two Lys (or Hyl) side chains (upper panel). The extracellular copper-dependent enzyme lysyl oxidase (LOX) catalyzes the formation of the aldehyde groups (highlighted in *blue*) from amino moieties in collagen and elastin precursors.

more complex than fibrous proteins. In this regard, tertiary and quaternary structures are found, in addition to the secondary structures. They are generally soluble in water but can also be found inserted into biological membranes (transmembrane proteins), thus in a hydrophobic environment. Unlike fibrous proteins that have structural and mechanical functions, globular proteins act as enzymes, hormones, membrane transporters and receptors, transporters of triglycerides, fatty acids and

oxygen in the blood, immunoglobulins or antibodies, and storage proteins. Examples of globular proteins are myoglobin, hemoglobin, and cytochrome c.

From the functional point of view, proteins are divided into at least seven groups:

1. *Enzymes* (biochemical catalysts). In living organisms, almost all reactions are catalyzed by enzymes because they have a high catalytic power, increasing the rate of the reaction in which they are involved at least by

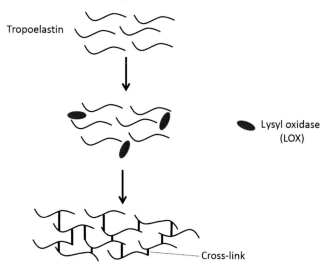

Tropoelastin

Lysyl oxidase
(LOX)

Cross-link

**FIG. 1.66** Schematic of the stages of elastogenesis. Tropoelastin aggregates are oxidized by lysyl oxidase (LOX) leading to cross-linked elastin.

a factor of $10^6$. Therefore life as we know could not exist without their facilitating action. Except some catalytic RNA molecules called *ribozymes* or *ribonucleic acid enzymes*, almost all known enzymes are proteins.

2. *Transport proteins*. Many small organic and inorganic molecules are transported through the bloodstream and extracellular fluids, across the cell membranes, and inside the cells from one compartment to another, by specific proteins. A typical example is given by the *hemoglobin*.

3. *Storage proteins*, which serve as biological reserves of metal ions and amino acids used by the organism. Examples are ferritin, which stores iron intracellularly in a nontoxic form, and milk caseins, which act as a reserve of amino acids for the milk.

4. *Mechanical support proteins*. They have a pivotal role in the stabilization of many structures. Examples are α-keratins, collagen, and elastin. The same cytoskeleton, the scaffolding element of the cell, is made up of proteins that generate movement. They are

responsible, among others, for the contraction of the muscle fibers (of which myosin is the main component), the propulsion of spermatozoa and microorganisms with flagella, and the separation of chromosomes during mitosis.

5. *Communication proteins*. They are regulatory molecules involved in the control of many cellular functions, from metabolism to reproduction, such as hormone proteins. Examples are insulin, glucagon, and thyroid-stimulating hormone (TSH).

6. *Protection proteins* against the action of harmful agents. The *antibodies* or *immunoglobulins* are glycoproteins that recognize antigens expressed on the surface of viruses, bacteria, and other infectious agents. *Interferon, fibrinogen,* and *factors of blood coagulation* are other members of this group.

7. *Energy storage proteins*. It is worthy of note that proteins, and in particular the amino acids that constitute them, act as energy storage, second in size only to the adipose tissue, that in particular conditions, such as prolonged

fasting, may become essential for survival. However, their reduction of more than 30% leads to a decrease of the contraction capacity of respiratory muscle, immune function, and organ function, which are not compatible with life. Therefore proteins are an extremely valuable energy source.

## 1.6.4 Polysaccharides

In scientific literature, the chemical term "carbohydrate" has many synonyms, such as "sugar," "saccharide," "glucide," "hydrate of carbon," or "polyhydroxy compounds with aldehyde or ketone." *Carbohydrates* are the most abundant organic compounds in nature. The term carbohydrate or hydrates of carbon is most common in biochemistry. It arises because most sugars have molecular formulas $C_n(H_2O)_m$, where $n$ could be different from $m$, suggesting that C atoms are combined in some way with water. The saccharides are divided into four chemical groups: *monosaccharides, disaccharides, oligosaccharides,* and *polysaccharides*. Disaccharides are sugars whose molecules contain two monosaccharide residues. Small polysaccharides, containing about 3–10 monosaccharide units, are sometimes called oligosaccharides.

Nearly all plants and animals synthesize and metabolize carbohydrates, using them to store energy and deliver it to their cells. Most living organisms oxidize glucose to carbon dioxide and water to provide the energy needed by their cells. Plants can retrieve the glucose units from *starch* when needed. In effect, starch is a plant's storage unit for solar energy for later use. Almost every aspect of human life involves carbohydrates in one form or another. In fact, animals can store glucose energy by linking many molecules together to form glycogen, another form of starch. Cellulose makes up the cell walls of plants and forms their structural framework. Cellulose is the major component of wood, a strong yet supple material that supports the great weight of the oak, yet allows the willow to bend with the wind.

### 1.6.4.1 Building Blocks

Monosaccharides are classified according to three different features:

1. the placement of their carbonyl group (Fig. 1.67). If the carbonyl group is a ketone (R-CO-R') (usually at C2), the monosaccharide is a *ketose*, whereas if the carbonyl group is an aldehyde (R-CHO) at one end, the monosaccharide is an *aldose*.

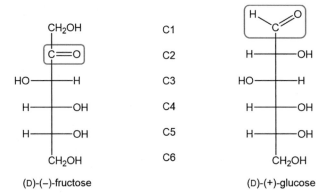

(D)-(−)-fructose          (D)-(+)-glucose

**FIG. 1.67**  Open-chain representations of a ketohexose, that is, a ketone-containing six-carbon monosaccharide, (D)-(−)-fructose, and an aldohexose, that is, an aldehyde-containing six-carbon monosaccharide, (D)-(+)-glucose. The carbonyl group is circled in *blue*. C atom numbering is also displayed in the middle.

2. The number of C atoms they contain, with $3 \leq C \leq 7$ (Figs. 1.68 and 1.69). Monosaccharides with three C atoms are called *trioses*; those with four are called *tetroses*; five are *pentoses*; six are *hexoses*; and seven are *eptoses*.

3. Their chiral handedness (Fig. 1.69). The (D) and (L) designation refer to the asymmetric C atom farthest from the carbonyl group in monosaccharides with more than one chiral center. Notice that the (D) or (L) configurations do not tell you which way a sugar rotates the plane of polarized light. This must be determined experimentally. Some (D) sugars have (+) rotations, and others have (−) rotations (see Annex 1 — Chirality). Notice that most naturally occurring sugars have the (D) configuration, that is, with the hydroxyl group (-OH) group of the bottom asymmetric C on the right in the Fischer projection, and most members of the (D) family of aldoses (up through six C atoms) are found in nature. At the time the (D) and (L) system of relative configurations was introduced, chemists could not determine the absolute configurations of chiral molecules. They

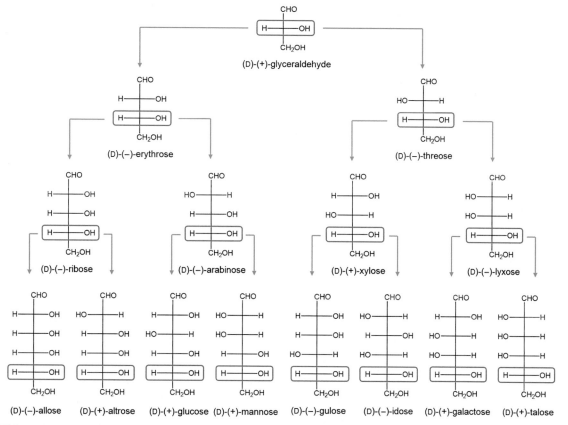

FIG. 1.68 The family tree of (D)-aldoses represented in the open chain form, generated by starting with (D)-(+)-glyceraldehyde and adding other one to three carbons at the top to generate aldotetroses, aldopentoses, and aldohesoses. All these monosaccharide occur naturally except for threose, lyxose, allose, and gulose. Notice that the (D) or (L) configuration does not tell us which way a sugar rotates the plane of polarized light. This must be determined by experimentation. Some (D) sugars have (+) rotations, and others have (−) rotations.

A. INTRODUCTION TO MATERIALS

**FIG. 1.69** Epimers are monosaccharides differing only by the stereochemistry at one single C atom. If the number of the C atom is not specified, it is assumed to be C2. Therefore mannose and glucose are C2 epimers or simply epimers, and glucose and galactose are C4 epimers. The stereochemistry of specific C is highlighted.

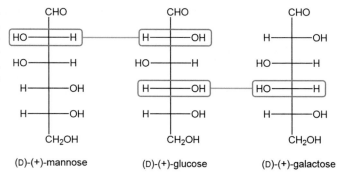

(D)-(+)-mannose          (D)-(+)-glucose          (D)-(+)-galactose

decided to draw the (D) series with the -OH group of glyceraldehyde on the right, and the (L) series with it on the left. This guess later proved to be correct, so it was not necessary to revise all the old structures.

The latter two systems of classification are often used in combination. For example, *glucose* is an aldohexose (a six-carbon aldehyde), *fructose* is a ketohexose (a six-carbon ketone) (Fig. 1.68), and *ribose* is an aldopentose (a five-carbon aldehyde) (Fig. 1.68).

*Erythrose* is the aldotetrose with the -OH groups of its two asymmetric C situated on the same side of the Fischer projection, and *threose* is the diastereomer, or distereoisomer, with the -OH groups on opposite sides of the Fischer projection. These names have evolved into a shorthand way of naming diastereoisomers with two adjacent asymmetric C atoms. A diastereoisomer is called *erythro* if its Fischer projection shows similar groups on the same side of the molecule. It is called *threo* if similar groups are on opposite sides of the Fischer projection.

Many common sugars are closely related, differing only by the stereochemistry at a single C atom. For instance, *glucose* and *mannose* differ only at C2, the first asymmetric C atom. Sugars that differ only by the stereochemistry at a single C are called *epimers*, and the C atom where they differ is generally stated. If the number of a carbon atom is not specified, it is assumed to be C2 (Fig. 1.69).

As an aldehyde (R-CHO) may react with one molecule of alcohol (R-OH) to give a hemiacetal, and with a second R-OH molecule to give an acetal; the hemiacetal is not as stable as the acetal, and most hemiacetals decompose spontaneously into R-CHO and R-OH. However, if the R-CHO and the R-OH are part of the same molecule, a cyclic hemiacetal results. Even though aldose monosaccharides exist as an equilibrium mixture of the cyclic hemiacetal and the open-chain form in solution, for most sugars, the equilibrium favors cyclic hemiacetals. That is because in this form they are often more stable than in their open-chain form. Cyclic monosaccharide structures are named according to the number of atoms constituting their rings. Five- and six-membered cyclic hemiacetal are called *furanose* and *pyranose*, derived from the name of the five- and six-membered cyclic ethers furan and pyran, respectively. For instance, the five-membered ring of fructose is called *fructofuranose* and the six-membered ring of glucose is called *glucopyranose*.

Aldohexoses such as glucose can form cyclic hemiacetals containing either five-membered or six-membered rings. For most common aldohexoses, the equilibrium favors six-membered rings with a hemiacetal linkage between the aldehyde carbon and the C5-OH. For instance, the solid, crystalline form of the best-known aldohexose (D)-(+)-glucose is the six-membered ring (pyranohexose) hemiacetal, well represented by the Haworth projections (Fig. 1.70). The prefixes α and β refer to the configurations of the anomeric C1 in the Haworth projections. The α-anomer is the isomer with the C1-OH group that lies on the opposite side of the plane with respect to the C5-CH$_2$OH substituent, that is, pointing down, whereas the β-anomer has the C1-OH group pointing up. Notice that the prefixes α and β identify the stereochemical configurations at C1, whereas (D) and (L) highlight those at C5. The hemiacetal C atom is called the *anomeric* carbon, easily identified as the only C atom bonded to two oxygens. Because anomers are diastereomers, they generally have different properties. For instance, α-(D)-glucose has a melting point of 146°C and a specific rotation of +112.2°, whereas β-(D)-glucose has a melting point of 150°C and a specific rotation of +18.7°.

The Haworth projections are widely used in biology texts because they are useful for comparing stereochemical features of pyranoses. In fact, the cyclic structure is often drawn initially in the Haworth projection, which depicts the ring as being flat, but, of course, it is not. That is why most chemists prefer to use the more realistic chair conformation (Fig. 1.71).

α and β anomers of monosaccharides slowly interconvert in an aqueous solution (Fig. 1.69). In fact, the concentration of a pure sample of α-(D)-glucose in water slowly decreases at the same rate that β-(D)-glucose appears in the solution. Ultimately, the solution contains an equilibrium mixture of α-(D)-glucose and β-(D)-glucose where the concentration of the two anomers is identical to the initial concentration of α-(D)-glucose. The same equilibrium mixture

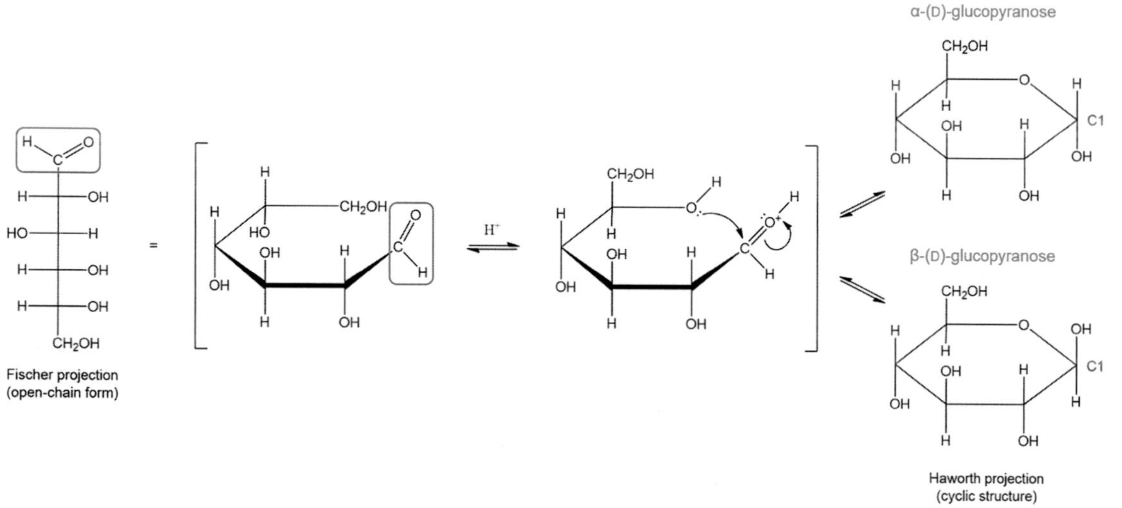

**FIG. 1.70** (D)-glucose exists almost entirely as its cyclic hemiacetal form called glucopyranose. Glucose is an aldohexose, as shown by the Fischer projection that displays the skeleton of the acyclic monosaccharide. This means that the top of the Fischer projection of glucose contains an aldehyde group (-CHO) (circled in *blue*) and that there are six C atoms in the polyhydroxy chain. Because the -OH group on the bottom-most asymmetric C5 is on the right side, the notation is (D). Every -OH substituent on the right side of the vertical line in a Fischer projection is pointing down in the Haworth projection, whereas the C5-CH$_2$OH group points up. The α or β configuration is determined by looking at the anomeric C1: the anomeric carbon is the carbonyl carbon in the acyclic form. In an α-monosaccharide, the C1-OH group is on the right side of the Fischer projection and points down in the Haworth projection. In a β-monosaccharide, the -OH group attached to the anomeric C1 is on the left side of the Fischer projection and points up in the Haworth projection.

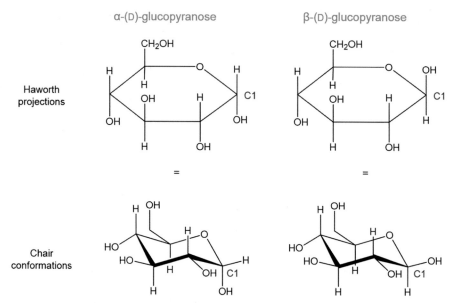

**FIG. 1.71** The difference between the α-(D)-glucose and the β-(D)-glucose. In an α monosaccharide, the -OH group attached to the anomeric C1 points down in the Haworth projection and is axial in the chair conformation. In a β monosaccharide, the C1-OH group points up in the Haworth projection and is equatorial in the chair conformation.

arises when we place a pure sample of β-(D)-glucose in water, and analogous equilibria do exist for α and β anomers of any other monosaccharide in solution. *Anomerization* is the process of conversion of one anomer to the other, and this equilibrium of anomers is called *mutarotation*, as it causes the optical rotation of a water solution of a pure anomer to change. This is because hemiacetals (or hemiketals) undergo a reaction in water to give the carbonyl -CHO (or a ketone) and an alcohol (Fig. 1.72). In the solution, the two anomers are in equilibrium through a small amount of the open-chain forms, and this equilibrium continues to supply more of the anomer crystallizing out of the solution. Such reaction is the reverse of hemiacetal formation, as displayed in Fig. 1.72. At room temperature, the equilibrium mixture of (D)-glucopyranose is approximately 36% α and 64% β, whereas the acyclic (D)-glucose intermediate represents only a trace of the total (D)-glucose. The two anomers differ in concentration because they have different stabilities ($\triangle\triangle G \approx$

3.6 kcal/mol) due to the fact that all substituents on the β-(D)-glucopyranose are equatorial, whereas the C1-OH of α-(D)-glucopyranose is axial when the other groups are equatorial.

Not all sugars exist as six-membered rings in their hemiacetal forms. Many aldopentoses and ketohexoses form five-membered rings (Fig. 1.73). Five-membered rings are not puckered as much as six-membered rings, so they are usually depicted as flat Haworth projections. The five-membered ring is customarily drawn with the ring oxygen in back and the hemiacetal C2 atom (the one bonded to two O) on the right. The C5-CH$_2$OH is in the up position for (D)-series ketohexoses (Fig. 1.73).

Most monosaccharides exist in nature in polysaccharides such as *cellulose, starch, chitin, alginate,* and many others. To provide a basis for understanding structures of large polysaccharides, we have to first learn about structural features of oligosaccharides with two or three monosaccharide units, called *disaccharides* and *trisaccharides,* respectively. The

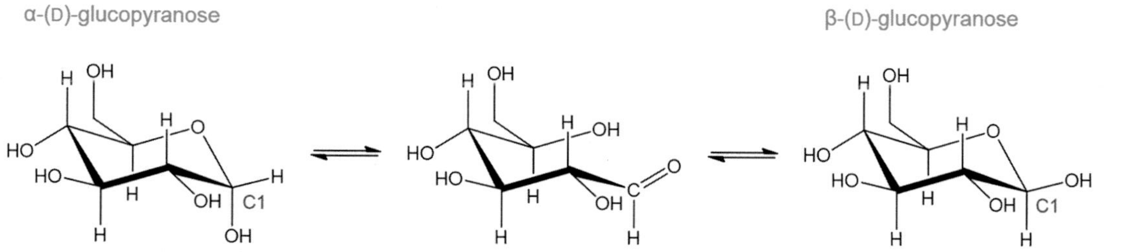

α-(D)-glucopyranose

β-(D)-glucopyranose

**FIG. 1.72** Chair conformations of the two C1 anomers (highlighted in *red*) of the (D)-glucopyranose. An aqueous solution of (D)-glucose contains an equilibrium mixture of α-(D)-glucopyranose, β-(D)-glucopyranose, and the intermediate open-chain form called simply acyclic (D)-glucose. The two anomers interconvert in solution through a process called mutarotation.

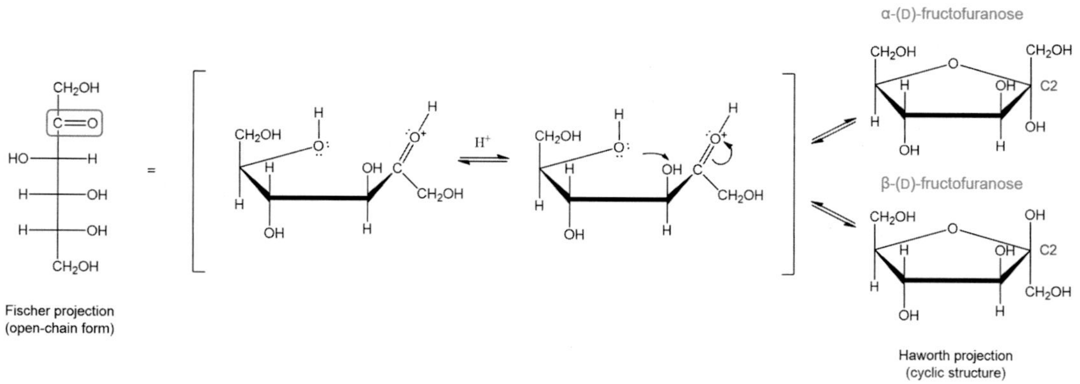

**FIG. 1.73** (D)-fructose exists almost entirely as its cyclic hemiacetal form called fructofuranose. Fructose is a ketohexose, as shown by the Fischer projection that displays the skeleton of the acyclic monosaccharide. This means that the C2 of the Fischer projection of fructose contains a ketone group (*circled in blue*) and that there are six C atoms in the polyhydroxy chain. Because the C5-OH is on the right side, the notation is (D). The configuration of α or β anomers is determined by looking at the anomeric C2: the anomeric carbon is the carbonyl carbon in the acyclic form. In an α-monosaccharide, the C2-OH group is on the right side of the Fischer projection and points down in the Haworth projection, whereas in a β-monosaccharide it is on the left side of the Fischer projection and points up in the Haworth projection.

joining of monosaccharides into a double sugar, or biose, happens by a condensation reaction between the hemiacetal or hemiketal group of a saccharide and the -OH group of another saccharide to give a glycosidic bond. In principle, the anomeric C1 can react with any of the -OH groups of another sugar to form a disaccharide. However, in naturally occurring disaccharides, there are three common glycosidic bonding arrangements:

1. *(1,4′) glycosidic bond.* The anomeric C1 is bonded to the O atom on C4′ of the second sugar. The prime symbol (′) indicates that C4 is on the second sugar. This is by far the most common glycosidic linkage. It is found in cellobiose, maltose, and lactose disaccharides.

2. *(1,6′) glycosidic bond.* The anomeric C1 is bonded to the O atom on C6′ of the second sugar.

3. *(1,1′) glycosidic bond.* The anomeric C1 of the first sugar is bonded through an O atom to the C1′ of the second sugar. It is found in the sucrose disaccharide. Because fructose is a ketose and its anomeric carbon is C2, this is actually a (1,2′) linkage.

Cellobiose, maltose, sucrose, and lactose illustrate the structural diversity of disaccharides.

*Cellobiose*, the disaccharide obtained by partial hydrolysis of cellulose, contains a β(1,4′) glycosidic linkage. In cellobiose, the anomeric C1 of one β-(D)-glucopyranose unit is linked through an equatorial (β) C—O bond to C4′ of another β-(D)-glucopyranose ring (Fig. 1.74A).

*Maltose* is a disaccharide formed when starch is treated with sprouted barley, called malt. This malting process is the first step in brewing beer, converting polysaccharides to disaccharides and monosaccharides that ferment more easily. Maltose contains a α(1,4′) glycosidic linkage between two α-(D)-glucopyranose units (Fig. 1.74B). It means that, although maltose and cellobiose give identical mixture of α-(D)-

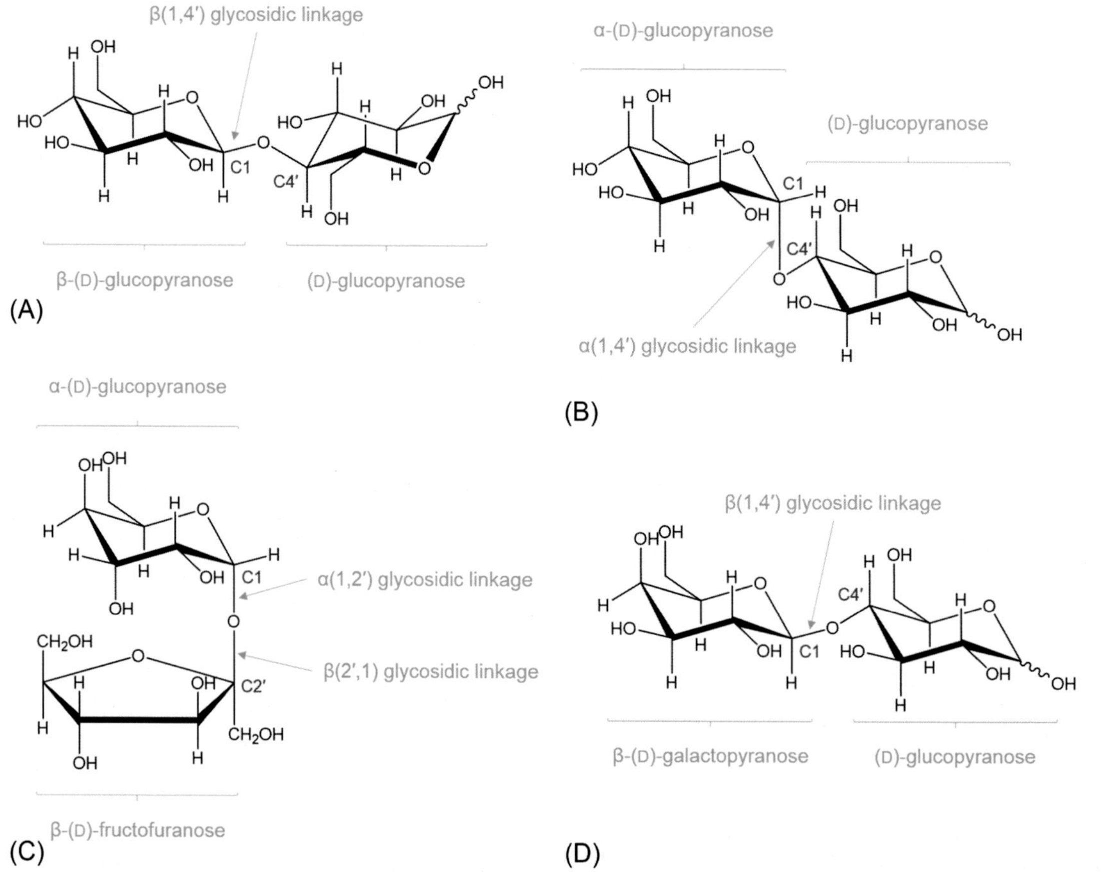

**FIG. 1.74** Most common naturally occurring disaccharides. (A) In cellobiose, the anomeric C1 of one β-(D)-glucopyranose is linked through an equatorial (β) C—O bond to C4′ of another β-(D)-glucopyranose. (B) Maltose contains a α(1,4′) glycosidic linkage between two α-(D)-glucopyranose units. (C) Sucrose is a disaccharide in which the glycosidic linkage is in the α position with respect to the (D)-glucopyranose and in the β position with respect to the (D)-fructofuranose ring. In contrast to the other three disaccharides in the figure, mutarotation does not occur in either anomeric C of sucrose because both anomeric C do have glycosidic bonds. (D) Lactose is composed of one β-(D)-galactopyranose and one (D)-glucose linked through a β(1,4′) glycosidic linkage.

glucopyranose and β-(D)-glucopyranose on hydrolysis, they are structurally different. Indeed, their anomeric C1 have different configurations.

Some monosaccharides are joined by a direct glycosidic linkage between their anomeric C1. An example is given by the *sucrose*, also known as common table sugar. Sucrose is composed of one glucose unit and one fructose unit bonded by an O atom linking their anomeric C1 that are both present as acetals. Notice that the linkage is in the α position with respect to the (D)-glucopyranose ring and in the β position with respect to the (D)-fructofuranose ring (Fig. 1.74C).

*Lactose* is composed of one β-(D)-galactopyranose ring and one (D)-glucose unit linked together through a β(1,4′) glycosidic linkage. The bond is between the acetal C1 of β-(D)-galactopyranose and the C4′ on the (D)-glucose (Fig. 1.74D). Lactose occurs naturally in the milk of mammals, and its hydrolysis requires the *lactase* or β-galactosidase enzyme. Some humans synthesize the lactase, but others do not. This enzyme is present in humans since birth. Unfortunately, once the child stops drinking milk, the production of the enzyme gradually stops, therefore he can no longer digest lactose. Consumption of milk or milk products can cause digestive discomfort in lactose-intolerant people who lack the β-galactosidase enzyme.

Hydrolysis of maltose gives an equilibrium mixture of only (D)-(+)-glucopyranose anomers, so both of its monosaccharide units are (D)-(+)-glucose. Because the same is true of cellobiose, maltose and cellobiose are both classified as *homo-oligosaccharides*. In contrast, sucrose and lactose are called *hetero-oligosaccharides* because each gives a mixture of two different monosaccharides upon hydrolysis.

### 1.6.4.2 Classification of Polysaccharides

Polysaccharides are biopolymers, or naturally occurring polymers, of carbohydrates that consist of more than 2 monosaccharides linked together covalently by glycosidic linkages in a condensation reaction and hydrolyzed to give rise to the constituent monosaccharides or oligosaccharides. Most polysaccharides, being comparatively large macromolecules with hundreds or thousands of simple sugar units linked together into long polymer chains, are most often insoluble or slightly soluble in water. Polysaccharides are generally white and amorphous products (glassy) not sweet in taste that form colloidal solutions or suspensions in water. As reported herein, at least five major structural variables that are not mutually exclusive allow classifying polysaccharides.

1. The presence of the same or different specific monosaccharide units (Fig. 1.75). When they are formed by the same kind of monosaccharides, they are called *homopolysaccharides* or *homoglycans*, like *starch*, *glycogen*, and *cellulose*, formed each of them by hundreds of molecules of (D)-glucose linked by glycosidic bonds. If the polysaccharides molecules are formed by different kinds of monosaccharides, they are considered heteropolysaccharides or heteroglycans. An example of heteropolysaccharide is represented by *hyaluronic acid*, which is formed by thousands of alternative units of N-acetyl glucosamine (NAG) and glucuronic acid.

2. The different kinds of glycosidic bonds joining monosaccharide building blocks together in linear and branched chains. There are mainly (1,4′) and (1,6′) glycosidic bonds, and rarely (1,3′) linkages found primarily in fungal and yeast β-glucans. (1,4′) bonds may exist as α(1,4′) and β(1,4′) and give rise to linear chains. The former is formed when the -OH on C1 is below the glucopyranose ring (such as the one present in maltose, the disaccharide constituting the polysaccharide amylose, and the straight backbone of amylopectin and glycogen), whereas the

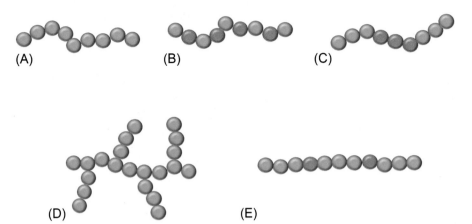

**FIG. 1.75** The structural variables 1 and 2 reported herein may be combined to define specific polysaccharide classes. (A) Linear homopolysaccharides with only one kind of monosaccharide unit (e.g., amylose, cellulose, chitin); (B) Linear polysaccharides with an alternation of two different monosaccharides (e.g., hyaluronic acid); (C) Linear block polysaccharides (e.g., alginate); (D) Branched homopolysaccharides (e.g., amylopectin); and (E) Linear block polysaccharides interrupted by single different monosaccharides (e.g., chitosan).

latter is formed when the -OH is above the plane (e.g., it is found in lactose and cellobiose, which is the constituent of the main chain of the polysaccharide *cellulose*). These types of bonds invariably form straight chains. There are also α(1,6′) glycosidic bonds where C1 on one sugar is linked to C6′ on another one; these are branch points in the main chain of polysaccharide molecules, such as those found in *glycogen* and *amylopectin*. Of note, the rate at which these bonds appear may vary. Taken together, the combination of α, β, (1,4′), and (1,6′) glycosidic bonds in a polysaccharide makes it unique and a target of specific enzymes and receptors.

3. Differences in molecular weight, which may vary greatly depending on the source of extraction, from one class of polysaccharides to another and even in a given class, from 0.5 kDa to 10 MDa.

4. Differences in superstructures that look remarkably like the secondary structures found in proteins. For instance, there are close similarities in the secondary structure of chitin and cellulose. The chitin of insect and crustacean cuticle occurs in the form of microfibrils of typically 10–25 nm in diameter and 2–3 m in length. Instead, amylose displays a helical structure.

5. Differences in overall charge at neutral pH allows classifying polysaccharides as **anionic** (e.g., alginates, hyaluronic acid, heparin, and other glycosaminoglycans (GAG)), **neutral** (e.g., cellulose, starch), and **cationic** (e.g., chitosan).

All these structural variables together do deeply affect the extraordinary functional variability of polysaccharides in terms of structural, protective, lubricity, and energetic behaviors, among others. Specifically, polysaccharides serve a number of crucial biological functions in organisms such as cell wall support (structural polysaccharides), food (energy) storage (storage polysaccharides), and as the ECM surrounding connective tissue (mucopolysaccharides). Polysaccharides are also present in many proteins called glycoproteins.

*Cellulose* is a straight chain, neutral homopolysaccharide that contains thousands of β-(D)-glucose monosaccharides chemically bonded by β(1,4′) glycosidic linkages (Fig. 1.76). No

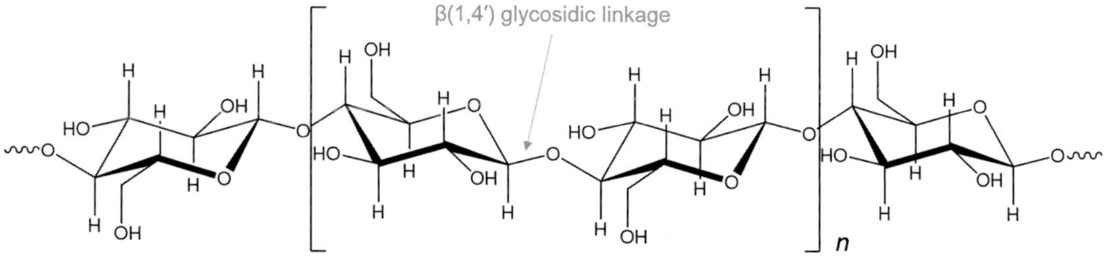

**FIG. 1.76** Partial molecular structure of cellulose. Cellobiose is a disaccharide (in square brackets), product of cellulose breakdown. It consists of two β-(D) glucose molecules linked by a β(1,4') glycosidic bond.

coiling or branching occurs because such bonding arrangement, unlike the α(1,4') bonds of starch, is rather rigid and very stable, and forces cellulose to form long and sturdy straight chains. Besides, compared to starch, cellulose is also much more crystalline. The multiple -OH groups on the (D)-glucose from one chain form hydrogen bonds with O atoms on the same or on a neighbor chain, holding the chains firmly together side-by-side and forming microfibrils with high tensile strength. This confers tensile strength in cell walls, where cellulose microfibrils, which are held in bundles to give fibers, are meshed into a polysaccharide matrix. For this reason, cellulose is insoluble in water. However, individual strands of cellulose are not very hydrophobic as compared to other polysaccharides. Cellulose is synthesized mainly by plants, and it is responsible for a structural role such as withstanding the weight of the plant itself. In nature, cellulose almost never purely occurs. Its main other components are *hemicellulose*, *pectin*, and *lignin*. In fact, dry wood is largely cellulose and lignin, whereas paper and cotton are nearly pure (≈95%) cellulose. Humans and many animals lack the *β-glucosidase* enzyme (or cellulase) to break the β-linkages, so they are not able to digest cellulose. Some ruminants like cows and sheep instead contain certain symbiotic bacteria in the flora of the rumen, and these bacteria produce cellulases that help the microorganism digest cellulose; the breakdown products are then used by the bacteria for proliferation.

*Starch* is a neutral homopolysaccharide produced by most green plants as energy storage. Plants store starch within specialized organelles called amyloplasts. It is the most common carbohydrate in human diets, and it is contained in large amounts in staple foods like potatoes, wheat, corn, and rice. Pure starch is a white, tasteless and odorless powder. About 20% of the mass of starch granules in plants is water-soluble *amylose*, and the remaining 80% is water-insoluble *amylopectin* polysaccharide. Amylose is a much smaller molecule than amylopectin, although there are about 150 times more amylose than amylopectin molecules. Amylose is basically a linear polymer of α-(D)-glucose units linked with α(1,4') glycosidic bonds (Fig 1.77A). Although amylose was thought to be completely unbranched, it is now known that some molecules may contain very few α(1,6') branching points. The "subtle" stereochemical difference between cellulose and amylose results in some striking physical and chemical differences. The linkage in amylose kinks the polymer chain into a helical structure. This bending increases hydrogen bonding with water and lends additional solubility. As a result, amylose is soluble in water yet cellulose is not. Besides, unlike cellulose, amylose is an excellent food source for all animals because the α(1,4') glycosidic linkage is easily hydrolyzed enzymatically. Instead, amylopectin is a highly branched polysaccharide formed of 2000–200,000 α-(D)-glucose units. Its inner chains are formed of 20–24 glucose subunits linked in a

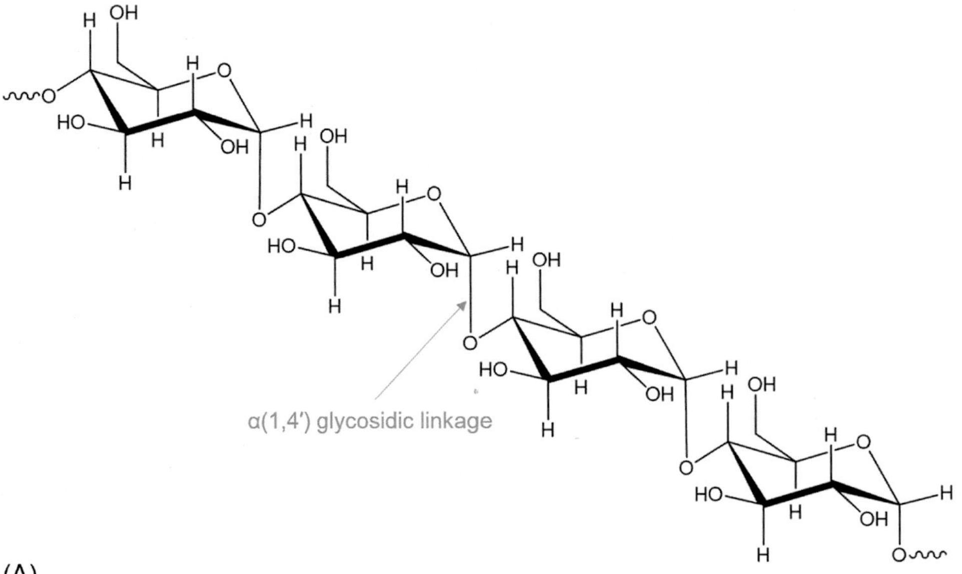

(A)

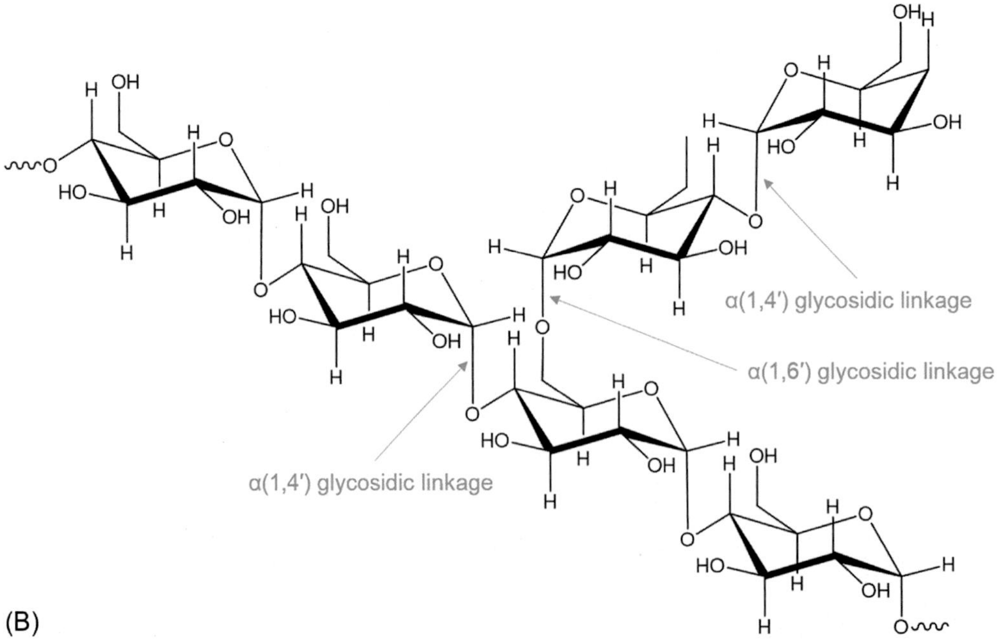

(B)

FIG. 1.77   Partial molecular structure of starch. Maltose is a disaccharide product of starch breakdown. (A) Amylose is basically a linear homopolysaccharide made of α-(D)-glucopyranose units linked through α(1,4') glycosidic bonds. (B) Amylopectin is a highly branched polysaccharide, instead. Branching takes place with α(1,6') bonds on the main chain that is the same as for amylose.

linear way with α(1,4′) glycosidic bonds. Branching takes place with α(1,6′) bonds occurring every 24–30 glucose units along the main chain (Fig 1.77B), resulting in a soluble molecule that can be quickly degraded as it has many end points onto which enzymes can attach.

In digesting such branched polysaccharides, *α-amylase* is the relevant catalyst. α-amylase, however, only digests α(1,4′) glycosidic bonds, leaving disaccharide and polysaccharide fragments known as dextrins, containing α(1,6′) bonds. Its counterpart in animals is *glycogen*, which has the same composition and structure but with more extensive branching that occurs every 8–12 α-(D)-glucose units. A large amount of glycogen is stored in the muscles themselves, ready for immediate hydrolysis and metabolism. Additional glycogen is stored in the liver, where it can be hydrolyzed to glucose for secretion into the bloodstream.

*Chitin* is a neutral homopolysaccharide composed by the repetition of amino-monosaccharide (actually an amide) called β(1,4′)-N-acetyl-β-(D)-glucosamine units (Fig. 1.78). In N-acetyl-glucosamine, the -OH group on C2′ of glucose is replaced by an amino group (forming glucosamine), and that amino group is acetylated (CH$_3$-CO-). Therefore chitin is bonded like cellulose, except using N-acetylglucosamine instead of β-(D)-glucose. It is the main component of the exoskeleton of insects and crustaceans, such as crab and shrimp shells, as well as the cell wall of yeasts and fungi where its relative amounts are in the range of 30%–60% in weight. It possesses an amino group covalently linked to acetyl group as compared to some free amino groups in *chitosan*. Chitosan is produced commercially by deacetylation of chitin (Alam Md et al., 2014). Chitosan is the best kenned natural polysaccharide utilized for its multifarious applications in pharmaceutical industry.

*Alginate*, also called algin or *alginic acid* in some literature, or E400 as given by the International Numbering System (INS) for Food Additives, is an anionic heteropolysaccharide comprising from 30% to 60% of the dry weight of various species of brown seaweeds. Alginate has dietary fiber properties. It is a linear copolymer with homopolymeric blocks of β-(D)-mannuronic acid (M) and α-(L)-gulurononic acid (G) residues, respectively, covalently linked through (1,4′) linked together in different sequences or blocks (Fig. 1.79). The blocks are composed of consecutive G residues (GGGGGG, G-blocks), consecutive M residues (MMMMMM, M-blocks), and alternating M and G residues (GMGMGM, MG-blocks), each of which has different conformational preferences and behavior. For instance, the G-blocks do form stiffer two-fold screw helical chains, and they are the only ones believed to participate in intermolecular cross-linking with

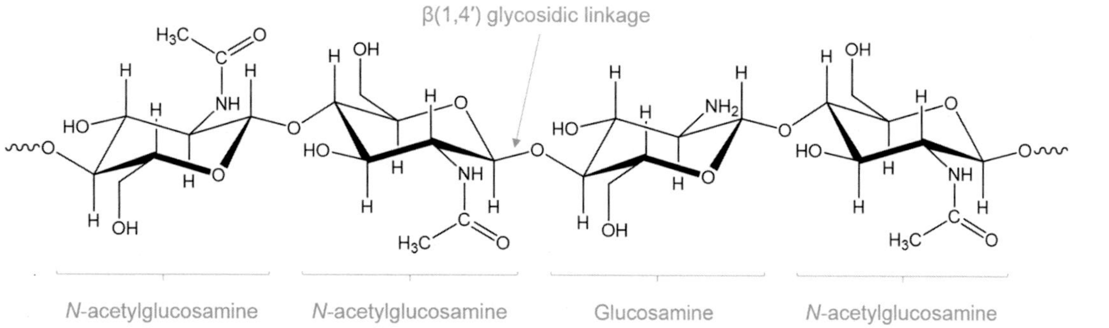

**FIG. 1.78** Partial molecular structure of chitosan. It consists of a random repetition of N-acetylglucosamine and glucosamine molecules linked by a β(1,4′) glycosidic bond.

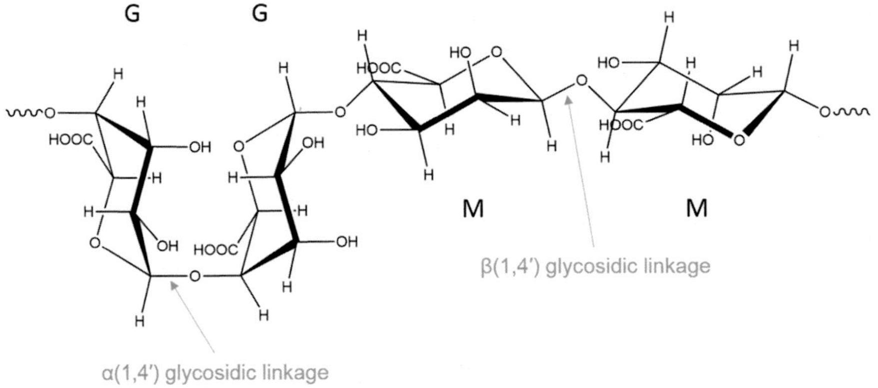

**FIG. 1.79** Alginates are basically linear unbranched polymers comprising α(1,4′)-linked (L)-guluronic acid (G) residues (G-blocks) and β(1,4′)-linked (D)-mannuronic acid residues (M-blocks).

divalent cations (e.g. $Ca^{2+}$), with the exception of $Mg^{2+}$, to form hydrogels. The composition (i.e., M/G ratio), sequence, G-block length, and molecular weight are thus critical factors affecting the physical properties of alginate and its resultant hydrogels. The molecular weight of commercially available sodium alginates range between 32 and 400 kDa.

*Hyaluronic acid*, also called *hyaluronan*, is an anionic glycosaminoglycan (GAG) distributed widely throughout connective, epithelial, and neural tissues, being one of the chief components of the ECM. It is unique among GAGs in that it is nonsulfated. Hyaluronic acid is a linear heteropolysaccharide constituted by the alternation of β-(D)-glucuronic acid residues and N-acetyl-β-(D)-glucosamine units linked together through β(1,4′) and β(1,3′) glycosidic bonds (Fig. 1.80). The molecular weight of hyaluronic acid preparations varies with purification procedures, that is, the extent of degradation as well as the source. In some cancers, hyaluronic acid levels correlate well with malignancy and poor prognosis. It is often used as a tumor marker for prostate and breast cancer, and it may be used to monitor the progression of the disease as well.

*Heparin*, is a linear heteropolysaccharide that consists of a variably sulfated repeating disaccharide units. Although others may be present, the main monosaccharides occurring in heparin and present in decreasing amounts are 2-deoxy-2-sulfamino-α-(D)-glucose-6-sulfate, α-(L)-iduronic acid 2-sulfate, 2-acetamido-2-deoxy-α-(D)-glucose, β-(D)-glucuronic acid, and α-(L)-iduronic acid joined by glycosidic linkages. It is a highly sulfated GAG that displays the highest anionic charge density of any known biological molecule. In Nature, heparin is a polymer of varying chain size and a molecular weight ranging typically from 3 to 40 kDa, although the fractionated version commercially available is about 12–15 kDa. Heparin is usually stored within the secretory granules of *mast cells* and released only into the vasculature at sites of tissue injury. In therapeutic doses, it acts as an anticoagulant, preventing the formation of clots and extension of existing clots within the blood. In fact, heparin binds to the enzyme inhibitor antithrombin III (AT), causing its activation (see Section 6.2.1 ). The activated AT then inactivates thrombin, factor Xa and other proteases. Heparin is used in medical practice to treat and prevent deep vein thrombosis, pulmonary embolism, and arterial thromboembolism. Given by intravenous injection, heparin is also used in the treatment of heart attacks and unstable angina.

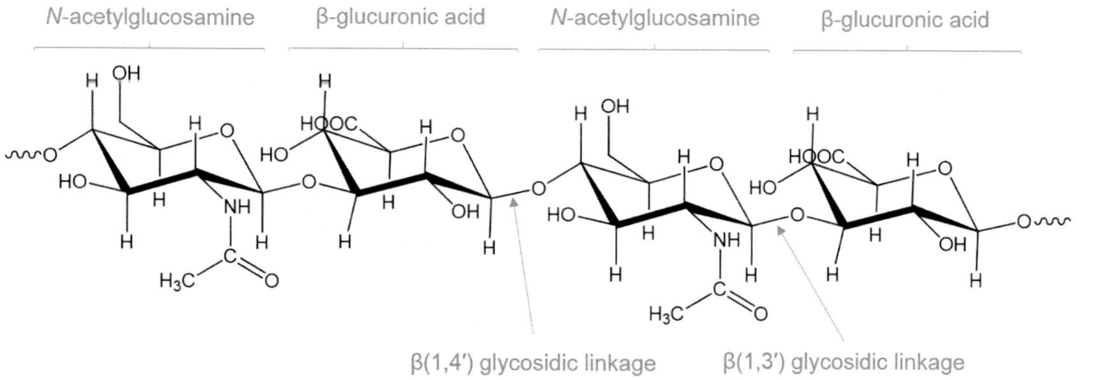

N-acetylglucosamine          β-glucuronic acid          N-acetylglucosamine          β-glucuronic acid

β(1,4′) glycosidic linkage          β(1,3′) glycosidic linkage

**FIG. 1.80**   Hyaluronic acid is a linear heteropolysaccharide characterized by a highly polymerized chain of β-(D)-glucuronic acid residues and N-acetyl-β-(D)-glucosamine units linked through β(1,4′) and β(1,3′) glycosidic bonds.

## 1.6.5 Nucleic Acids

Nucleic acids are the main information-carrying molecules of the cell, and, by directing the process of protein synthesis, they determine the inherited characteristics of every living organism. Nucleic acids are biopolymers that come in two naturally occurring varieties. *Deoxyribonucleic acid* (DNA) and *ribonucleic acid* (RNA). DNA contains the genetic information in the form of genes that dictate the specific sequence of amino acids found in all proteins. RNA is essential in various biological roles in coding, decoding, regulation, and expression of genes. In eukaryotes, such as plants and animals, DNA is as long as several hundred million nucleotides, and it is found in the nucleus, a specialized, membrane-bound vault in the cell, as well as in certain other types of organelles (such as mitochondria and the chloroplasts of plants). In prokaryotes, such as bacteria, the DNA is not enclosed in a membranous envelope, although it is located in a specialized cell region called the nucleoid. Instead, RNAs range in length from <100 to many thousands of nucleotides. Like DNA, RNA is assembled as a chain of nucleotides, but unlike DNA, it is more often found in nature as a single-strand folded onto itself, rather than a paired double-strand.

### 1.6.5.1 Building Blocks

A single nucleic acid strand is therefore a phosphate-pentose polymer (a polyester) with purine and pyrimidine bases as side groups. In their primary structures, both DNA and RNA are linear polymers, which are long chain-like molecules composed of a series of nearly identical building blocks composed of monomers called *nucleotides*. When such monomers combine, the resulting chain is called a *polynucleotide*. All nucleotides have a common structure: a phosphate group linked by a phosphodiester bond to a pentose (a five-carbon atom monosaccharide) that, in turn, is covalently linked to an organic base (Fig. 1.81). A nucleoside is a nitrogenous base and a five-carbon sugar instead. Nucleotides found in polynucleotides are nucleosides that have one phosphate group esterified at the 5′-OH (Fig. 1.81).

The organic base components of nucleic acids are heterocyclic rings, or nitrogenous bases containing nitrogen and carbon. *Adenine* (A) and *guanine* (G) are **purines**, which contain a pair of fused rings; *cytosine*, (C) *thymine* (T), and *uracil* (U) are **pyrimidines**, which contain a single ring. However, T is found only in DNA (Fig. 1.82), whereas U is an exclusive basic ingredient of RNA (Fig. 1.83). DNA nucleotides are therefore composed of a phosphate-deoxyribose sugar

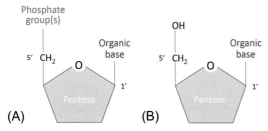

(A)                    (B)

**FIG. 1.81**   (A) Chemical structure of a generic nucleotide and (B) a nucleoside. All nucleotides are composed of a phosphate moiety (in *green*), which may contain up to three phosphate groups, linked to the 5′-OH of a pentose sugar (in *light blue*), whose C1′ is linked to an organic base (in *red*). By convention, the C atoms of pentoses are numbered with primes. In natural nucleotides, the 1′-carbon is joined by a β-linkage to the base, which is in the plane above the furanose ring, as is the phosphate.

backbone and one of the four nitrogenous bases: A, G, C, and T (Fig. 1.82). Deoxy-, a chemical prefix designating a compound containing one less O atom than the reference substance, specifically means that *deoxyribose* found in DNA nucleotides lacks the 2′-OH, which is instead present in the ribose of RNA nucleotides (Figs. 1.82 and 1.83). Nevertheless, for the sake of simplicity and by convention, DNA deoxyribonucleotides are also referred simply to as nucleotides or "nt" in short.

### 1.6.5.2 Structure and Function of Nucleic Acids

DNA is the master blueprint for life and constitutes the genetic material in all free-living organisms and most viruses. In fact, DNA contains the genetic information that dictates the specific sequence of amino acids found in all

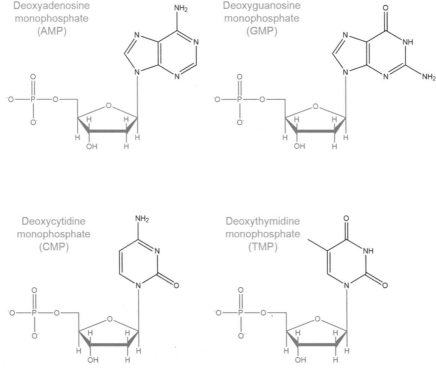

**FIG. 1.82**   The four DNA nucleotides. Every nucleotide invariably consists of a pentose monosaccharide, 2′-deoxyribose or 2′-deoxyribofuranose (in *light blue*), 5′-linked to a phosphate group (in *green*), and 1′-linked to a variable nitrogen-containing aromatic base (in *red*) that are adenine (A), guanine (G), cytosine (C), and thymine (T) for AMP, GMP, CMP, and TMP, respectively. A and G are called purines, and C and T are categorized as pyrimidines.

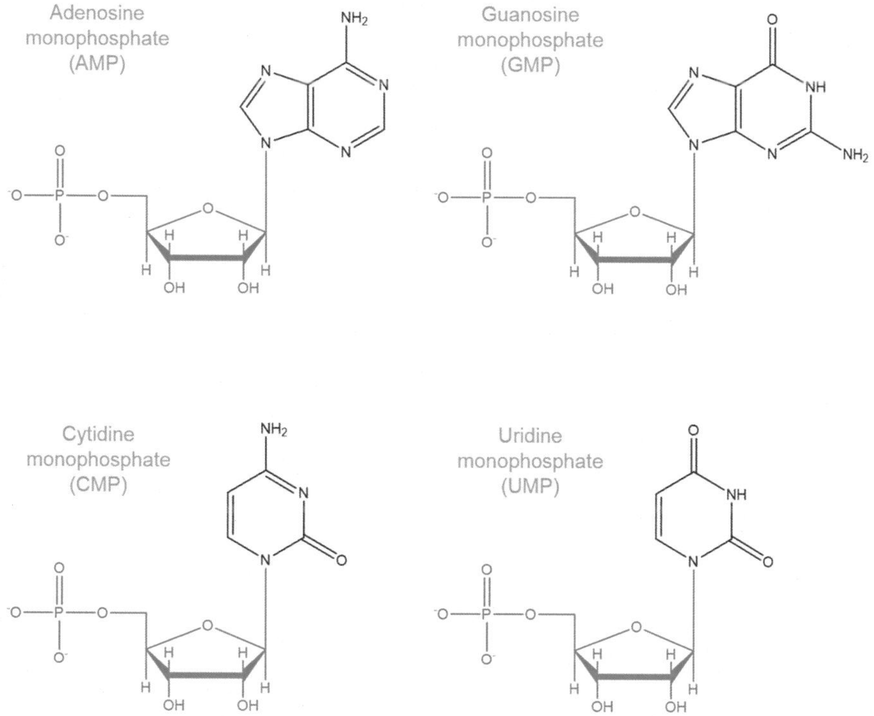

**FIG. 1.83** The four RNA nucleotides. Every nucleotide invariably consists of a pentose monosaccharide, ribose or ribofuranose (in *light blue*), 5'-linked to a phosphate group (in *green*), and 1'-linked to a variable nitrogen-containing aromatic base (in *red*) that are adenine (A), guanine (G), cytosine (C), and uracil (U) for AMP, GMP, CMP, and UMP, respectively. A and G are categorized as purines, and C and U are called pyrimidines.

proteins. RNA is the genetic material of certain viruses (e.g., Ebola, SARS, the common cold, influenza, hepatitis C, West Nile, etc.), but it is also found in all living cells, where it plays an important role in certain processes such as the making of proteins. RNA molecules are typically much smaller than DNA, and they are more easily hydrolyzed and broken down. But nevertheless, in their primary structures, both DNA and RNA are linear polymers that are long chain-like molecules composed of a series of nearly identical nucleotides. Nucleotides are nucleosides that have one, two, or three phosphate groups esterified at the 5'-OH. Nucleoside monophosphates have a single esterified phosphate (Pi), diphosphates contain a pyrophosphate group (*PPi*), and triphosphates have a third phosphate. During DNA replication and RNA synthesis (i.e., in a process called

*transcription*), when generic nucleotides polymerize to give rise to nucleic acids, the 3'-OH group of one nucleotide forms a phosphodiester bond with the phosphate attached to the C5' of another nucleotide. The deoxynucleoside triphosphates (dNTPs) are the precursors used in the synthesis of DNA, according to the reaction reported in Fig. 1.84. Thus a single nucleic acid strand is a phosphate-pentose polymer (a polyester) with purine and pyrimidine bases as side groups. Because the synthesis proceeds from the 5' to the 3'-end, sequences are written by convention from in the 5'→3' direction, therefore from left to the right in the text. DNA polymerases catalyze the nucleophilic attack of the 3'-OH terminus and the elongation of polynucleotide chains through the addition of new nucleotides from incoming dNTPs. The energy required to drive the reaction comes from

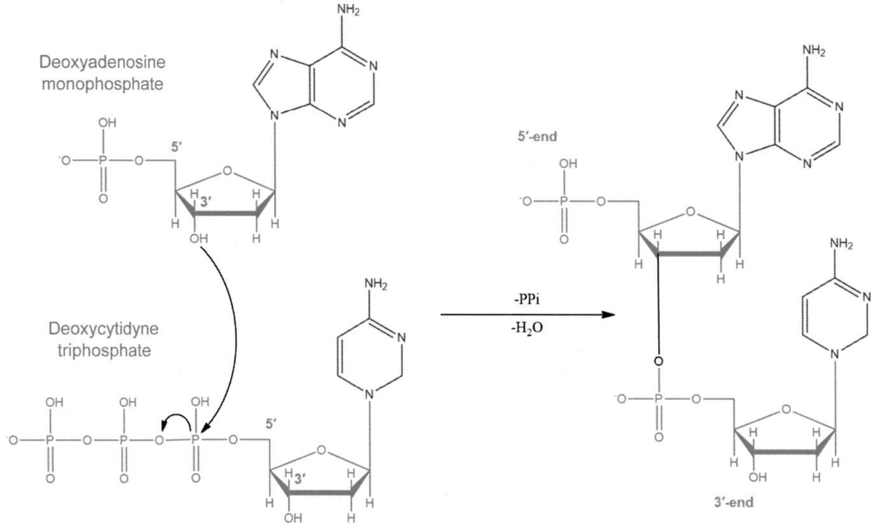

**FIG. 1.84**   Phosphodiester bond formation between A and C deoxynucleotides during DNA polymerization (synthesis). Similarly to other biological molecules, nucleotides are joined together by a condensation reaction in which two nucleotides are joined by a phosphate linkage or phosphodiester bond between the 3′-OH of one nucleotide and the 5′ phosphate of the other. Unlike proteins and carbohydrates, however, the molecule that is released is pyrophosphate (two phosphate groups bound together, PPi). When pyrophosphate is cleaved by the addition of water, a great deal of free energy is released, ensuring that the reverse process (hydrolysis of the phosphodiester bond to give free nucleotides) is very unlikely to occur.

cutting high energy phosphate bonds on the dNTPs used as the source of the nucleotides needed in the reaction.

The same process happens for RNA synthesis, apart that building blocks are nucleoside triphosphates (NTPs) and polymerization is catalyzed by a RNA polymerase.

Unless it is in a ring-like form, every nucleic acid molecule has invariably two ends. One end has a free 5′ phosphate group, that we refer to as 5′-end, and the other end has a free 3′-OH moiety, that is called 3′-end (Figs. 1.84 and 1.85). The linear sequence of nucleotides linked by phosphodiester bonds constitutes the primary structure of nucleic acids. Nevertheless, polynucleotides can twist and fold into 3D conformations stabilized by noncovalent bonds; in this respect, they are similar to polypeptides.

In nature, DNA contains two complementary polynucleotide chains, or regions thereof, held together and winded together through space to form a structure described as DNA double helix or secondary structure. In principle, two polynucleotide strands can form either a right-handed or a left-handed helix. Because the geometry of the sugar-phosphate backbone is more compatible with the former, natural DNA is a right-handed helix. The two strands are antiparallel: one strand runs 5′→3′, whereas the other runs in the opposite direction. Fig. 1.85 shows a stretch of a double strand of DNA (dsDNA), with every base paired with the complementary. Each pyrimidine base belonging to one strand forms a stable hydrogen-bonded pair with only one of the two purine bases. By definition, a base pair (bp) is a unit consisting of two nucleobases bound to each other by hydrogen bonds. To maintain the geometry of the double-helical structure shown in Fig. 1.85, a larger purine must pair with a smaller pyrimidine. Specifically, G forms a base pair, joined by three hydrogen bonds, with C. A forms a base pair with T (or U in RNA), joined by two hydrogen bonds (Fig. 1.85). Therefore G is said to be complementary to C, and A is complementary to T (or U in RNA). In this way, hydrogen bonding contributes to the folded structure of

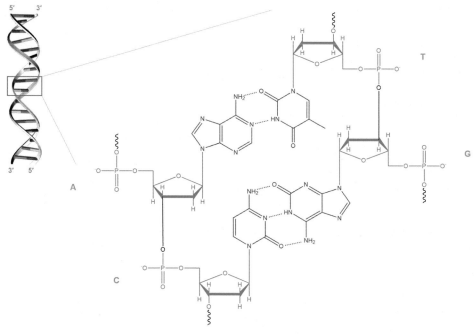

**FIG. 1.85** Stretch of double-stranded DNA (dsDNA) helix typically constituted by two antiparallel nucleic acid strands. The inset displays base pairing between 5′-AC-3′ and 5′-GT-3′. Notice that adenine pairs with thymine (A=T) through two hydrogen bonds *(violet dotted lines)* and guanine pairs with cytosine (G≡C) with three hydrogen bonds *(violet dotted lines)*. Although the linear sequence of nucleotides linked by phosphodiester bonds constitutes the primary structure of nucleic acids, the double helix is the secondary structure.

both DNA and, less frequently, RNA. Although the multitude of hydrogen and hydrophobic bonds between the polynucleotide strands provides stability to DNA, the double helix is somewhat flexible about its long axis. Besides, in DNA replication and in the copying of RNA from DNA, the strands of the helix must separate at least temporarily through a process called denaturation or melting.

The acidic character of nucleic acids is due to the presence of C5′-phosphates, which dissociate at the pH found within cells, freeing H⁺ and leaving phosphates negatively charged (Fig. 1.85). Because these charges attract proteins, most nucleic acids in cells are associated with proteins, such as histones, highly alkaline proteins found in eukaryotic cell nuclei that pack and order the DNA into structural units called *nucleosomes*.

DNA and RNA have minor structural differences that lead to major functional differences.

All living cells use DNA as the primary genetic material passed from one generation to another. In eukaryotes, like animals and plants, nuclear DNA directs the synthesis of messenger RNA (mRNA), which leaves the nucleus to serve as a template for the construction of protein molecules in the ribosomes. After it has served its purpose, the mRNA is enzymatically hydrolyzed to nucleotides, which become available for assembly into new RNA molecules to direct other syntheses. Notice that not all genes encode protein products. For instance, some genes specify *ribosomal RNAs* (rRNAs), which serve as structural components of ribosomes, or *transfer RNAs* (tRNAs), cloverleaf-shaped RNA molecules that bring amino acids to the ribosome for protein synthesis. Still other RNA molecules, such as tiny *microRNAs* (miRNAs), act as regulators of other genes, and new types of nonprotein-coding RNAs are being discovered all the time.

## ANNEX 1. CHIRALITY

An *isomer* is a molecule with the same molecular formula as another molecule but with a different chemical structure; therefore they are structurally different in some way. Remember that isomerism is a property between a pair or more molecules, thus a molecule is an isomer of another molecule. There are two main forms of isomerism: structural isomerism, sometimes referred to as constitutional isomerism, and stereoisomerism (or spatial isomerism). In structural isomers, the atoms and functional groups are joined together in different ways. In stereoisomers, the bond structure is the same, but the geometrical positioning of atoms and functional groups in space differs.

Stereochemistry is the study of the three-dimensional (3D) structure of molecules. The understanding of stereochemistry is fundamental to the comprehension of organic chemistry and biochemistry. What is the difference between your left and right hand? They look similar, yet a left-handed glove does not fit the right hand. The relationship between our two hands is that they are nonsuperimposable (nonidentical) mirror images of each other. No matter how we twist and turn our hands, they never look exactly the same. Chemists use the term chirality (from chiral, the Greek word for "handed") to refer to the property of handedness when it applies to molecules. Differences in spatial orientation might seem unimportant, but stereoisomers often have remarkably different physical and chemical properties. Notice that biological systems are exquisitely selective, and they are often able to discriminate between molecules with subtle stereochemical differences.

A molecule is chiral if it cannot be superimposed on its mirror image, whereas it is called achiral, meaning "not chiral," if it is superimposable on its mirror image. A tetrahedral carbon atom (C) with four different atoms or groups bonded is chiral. A stereocenter, or stereogenic atom, is any C at which the interchange of two moieties gives a stereoisomer. Nonsuperimposable mirror-image molecules are called enantiomers. Notice that a chiral compound always has an enantiomer. To draw the mirror image of a structure, keep up-and-down and front-and-back aspects, as they are in the original structure, but reverse left and right. If two or more of the groups or atoms on a tetrahedral C are identical, the C is achiral, and the molecule are superimposable. Two molecules are said to be superimposable if they can be placed on top of each other and the 3D position of each atom of one molecule coincides with the equivalent atom of the other one. An achiral compound always has a mirror image that is the same as the original molecule, so any molecule that is achiral cannot have an enantiomer. To distinguish two enantiomers, we use additional nomenclature that we describe herein.

Students who take chemistry and biochemistry are exposed to an outmoded, confusing, and often wrong method of specifying configurations at chiral centers such as (D) or (L). The enantiomer that rotates the plane of the polarized light clockwise (+) was arbitrarily labeled (D) whereas the other enantiomer (−) that rotates the plane of the polarized light counterclockwise became (L). The source of the (D) and (L) labels was the Latin words *dexter* (toward the right, or clockwise) and *laevus* (toward the left, or counterclockwise). Glyceraldehyde ($HOCH_2CH(OH)CHO$, Gly) enantiomers were chosen as the standards for defining molecular configurations (Fig. A1). Instead, (R) and (S) come from rectus (right-handed) and sinister (left-handed).

The assignments in modern notation are (R) and (S), respectively, but it does not always work out that (D) = (R) and (L) = (S). In fact, this happened by chance, and there is no general correlation between these two labels. Using International Union of Pure and Applied Chemistry (IUPAC) notation, the direction of

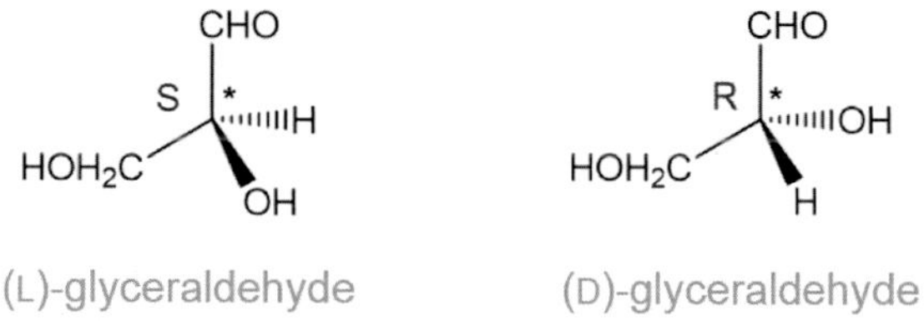

(L)-glyceraldehyde (D)-glyceraldehyde

**FIG. A1** Schematic representation of (L)-glyceraldehyde (Gly) and (D)-glyceraldehyde enantiomers. The anomeric C is highlighted with an asterisk (*).

rotation is specified by the (+) or (−) sign of the rotation. Any other molecule containing a single chiral center was to be assigned as (D) or (L) by imagining a resemblance between the ligands on its chiral center and those in Gly. Differences in enantiomers become apparent in their interactions with other chiral molecules, such as enzymes. Still, we need a simple method to distinguish between enantiomers and measure their purity in the laboratory. Polarimetry is a common method used to distinguish between enantiomers, based on their ability to rotate the plane of polarized light in opposite directions. Most of what we see is unpolarized light, vibrating randomly in all directions. When unpolarized light passes through a polarizing filter, the randomly vibrating light waves are filtered so that most of the light passing through is vibrating in one direction. The direction of vibration is called the axis of the filter. Plane-polarized light is composed of waves that vibrate in only one plane. When polarized light passes through a solution containing a chiral compound, this causes the plane of vibration to rotate. Rotation of the plane of polarized light is called optical activity, and substances that rotate the plane of polarized light are said to be optically active. Before the relationship between chirality and optical activity was known, enantiomers were called optical isomers because they seemed identical except for their opposite optical activity. Two enantiomers have identical physical properties, except for the direction they rotate the plane of polarized light.

Thus enantiomeric compounds rotate the plane of polarized light by exactly the same amount but in opposite directions. Again, do not confuse the process for naming a structure (R) or (S) with the process for measuring an optical rotation. That is to say, both (R) or (S) stereocenters can be dextrorotatory or levorotatory. The specific rotation [α] depends on the rotation observed (α) (expressed in °) at 25°C in a polarimeter divided by the concentration of the sample solution (c) (expressed in g/mL), and the length of the cell (l) (expressed in dm), according to the following equation:

$$[\alpha] = \frac{\alpha}{c \cdot l}$$

A solution of equal amounts of two enantiomers, so that the mixture is optically inactive, is called a racemic mixture. Sometimes a racemic mixture is called a racemate, a pair, or a (d, l) pair. A racemic mixture is symbolized by placing (±), or (d, l), in front of the name of the compound. Often we unfortunately deal with mixtures that are neither optically pure (all one enantiomer) nor racemic (equal amounts of two enantiomers). In these cases, you must specify the optical purity (o.p.) (expressed in %) of the mixture, in accordance to the following equation:

$$\text{o.p.} = \frac{\text{observed rotation}}{\text{rotation of pure enantiomer}} \times 100\%$$

If the direction of rotation of polarized light were the only difference between enantiomers, one might ask whether the difference would be important. Biological systems, such as enzymes, commonly distinguish between enantiomers. In fact, two enantiomers may have completely different biological properties, and just one enantiomer produces the characteristic effect whereas the other either produces no effect or has a different effect.

We have been using dashed lines and wedges to indicate perspective in drawing the stereochemistry of stereocenters. When we draw molecules with several asymmetric C atoms, perspective drawings become time-consuming and cumbersome. In addition, the complicated drawings make it difficult to see the similarities and differences in moieties of stereoisomers. At the turn of the 20th century, Emil Fischer developed a symbolic way of drawing asymmetric C atoms, allowing them to be drawn rapidly. Asymmetric carbons are at the centers of crosses: the vertical lines project away from the viewer, the horizontal lines toward the viewer. The carbon chain is placed along the vertical, with the IUPAC numbering from the top to the bottom. The Fischer projection facilitates comparison of stereoisomers, such as those of glyceraldehyde in Fig. A2, holding them in their most symmetric conformation and emphasizing any differences in stereochemistry.

The (R) or (S) configuration can also be determined directly from the Fischer projection, without having to convert it to a perspective

drawing. As an example, consider the Fischer projection formula of the (R)-glyceraldehyde enantiomer (Figs. A2 and A3). According to the Cahn-Ingold-Prelog (CIP) priority rule, to completely and unequivocally name a stereoisomer of a molecule, after the substituents of a stereocenter have been assigned their priorities, the molecule is oriented in space so that the group with the lowest priority is pointed away from the observer. If the substituents are numbered from 1 (highest priority, that is, the group having the atom of higher atomic number) to 4 (lowest priority), then the sense of rotation of a curve passing through 1, 2, and 3 distinguishes the stereoisomers. A center with a clockwise sense of rotation is (R) and a center with a counterclockwise sense of rotation is (S). First priority goes to the -OH 1, followed by the carbonyl group (-CHO) 2 and the methanol group (-CH$_2$OH) 3. The hydrogen (-H) receives the lowest priority, as usual. The arrow from group 1 to 2 to 3 appears counterclockwise in the Fischer projection. If the molecule is turned over so the -H is in back, the arrow is clockwise, so this is the (R) enantiomer of glyceraldehyde.

To properly draw the mirror image of a molecule drawn in Fischer projection, the rule is to reverse left and right yet keeping the other directions (up and down, front and back) in their same positions. Thus testing for enantiomerism is particularly simple using Fischer projections. If the Fischer projections are drawn, and if the mirror image cannot be made to look the same

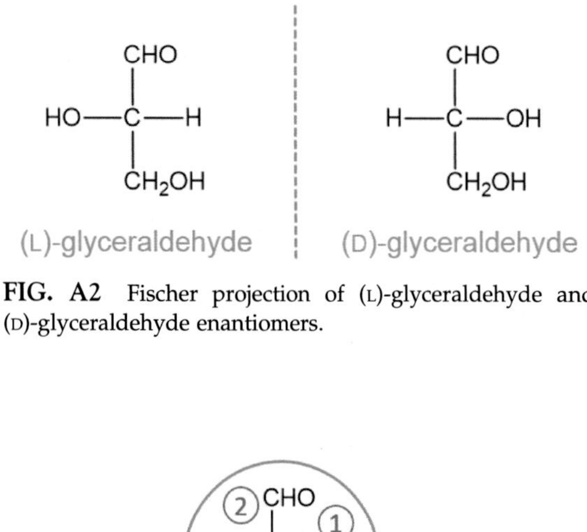

FIG. A2   Fischer projection of (L)-glyceraldehyde and (D)-glyceraldehyde enantiomers.

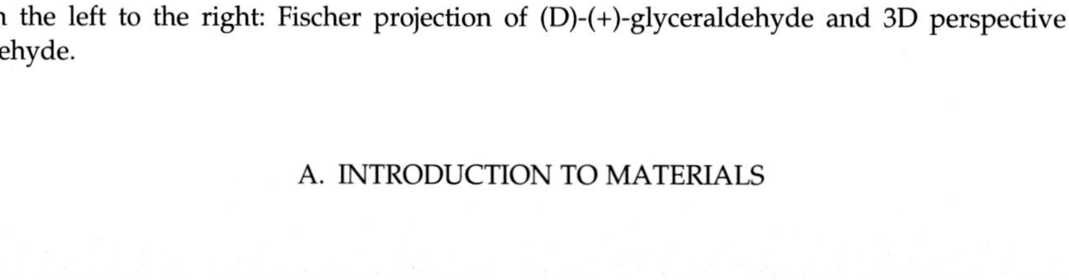

FIG. A3   From the left to the right: Fischer projection of (D)-(+)-glyceraldehyde and 3D perspective drawing of the (R)-(+)-glyceraldehyde.

as the original structure with a 180 degree rotation in the plane of the paper, the two mirror images are enantiomers. All the other stereoisomers are classified as diastereoisomers, which are stereoisomers that are not mirror images.

We now know that the enantiomer of glyceraldehyde has its -OH group on the right in the Fischer projection, as shown in Fig. A2. Therefore sugars of the (D) series have the -OH group of the bottom asymmetric C (C5) on the right in the Fischer projection (see Fig. 1.67). Sugars of the (L) series have the -OH group of the bottom asymmetric C on the left.

## References

Adikwu, M.U., 2012. Introduction. In: Biopolymers in Drug Delivery: Recent Advances and Challenges. Bentham Science Publishers, pp. 1–6.

Ahmed, E.M., 2015. Hydrogel: preparation, characterization and applications. a review. J. Adv. Res. 6, 105–121.

Alam Md, T., Parvez, N., et al., 2014. J. Pharm., 5, 1–6.

Kopecek, J., 2002. Nature 417, 388.

Olatunji, O., 2016. Nat. Polym. 1–17 (Chapter 1).

Shogren, R.L., Bagley, E.B., 1999. Biopolymers, 2–11 (Chapter 1).

Singh, J., 2016. World J. Pharm. Pharm. Sci. 5 (4), 805–816.

## Further Reading

Agrawal, C.M., 2013. Introduction to Biomaterials: Basic Theory With Engineering Applications. Cambridge Texts in Biomedical Engineering, Cambridge University Press, Cambridge.

Alberts, B., Johnson, A., Lewis, J., Raff, M., Roberts, K., Walter, P., 2002. Molecular Biology of the Cell, fourth ed. Garland Science Inc., New York.

Angeletti, R.H., 1998. Proteins: Analysis and Design. Academic Press, Cambridge, MA.

Barton A, States of Matter, State of Mind., 1997. Bristol, Institute of Physics Publishing.

Callister, W.D., 2007. Materials Science and Engineering: An Introduction. John Wiley & Sons Inc., New York City, NY

Lodish, H., Berk, A., Zipursky, S.L., Matsudaira, P., Baltimore, D., Darnell, J., 2000. Molecular Cell Biology, fourth ed. WH Freeman, New York.

Ratner, B.D., Hoffman, A.S., Schoen, F.J., Lemons, J.E., 2012. Biomaterials Science: An Introduction to Materials in Medicine, third ed. Academic Press, Cambridge, MA.

Smith, W., Hashemi, J., 2009. Foundations of Materials Science and Engineering, fifth ed. McGraw-Hill Education, Europe.

Tanzi, M.C., Farè, S., 2017. Characterization of Polymeric Biomaterials, first ed. Woodhead Pub./Elsevier Ltd, Duxford, UK.

Wade Jr., L.G., 2013. Organic Chemistry, eighth ed. Pearson Education Inc., London.

Williams, D., 2014. Essential Biomaterials Science. Cambridge Texts in Biomedical Engineering, Cambridge University Press, Cambridge.

# 2

# Mechanical Properties of Materials

## 2.1 INTRODUCTION

Mechanical properties are important for every object we have in everyday life. We can define an object as soft or hard, stiff or flexible, but from an engineering point of view, it is important to quantify these terms. For every material, properties have to be studied quantitatively following standardized methods that define test conditions and outputs. A variety of properties need to be investigated, depending on the external stimulus applied to the material:

- mechanical properties: they quantify the response of a material under different force systems;
- chemical properties: they characterize the material behavior in a particularly aggressive environment (e.g., human body);
- physical properties: they are related to the material behavior under particular conditions (e.g., temperature, electrical and/magnetic field, UV light).

It is important to investigate all these properties, as the material behavior can change based on independent variables (e.g., temperature, rate of application of force). This is particularly important when a biomedical application has

to be considered. In fact, the in vivo environment is very complex, and the detailed selection of appropriate biomaterials and materials properties is essential.

In this chapter, the mechanical properties and behavior of metallic, polymeric, and ceramic materials will be described, as well as the failure mechanisms under ductile or brittle conditions and the time-dependent mechanical behavior of materials, especially polymeric ones.

## 2.2 THE MECHANICAL BEHAVIOR OF MATERIALS

Mechanical properties determine the behavior of the materials when forces or loads are applied. The response of the material to the applied force depends on type of bonds, structural arrangement of atoms or molecules, and type and number of defects, which are always present in solids except rare situations (i.e., perfect monocrystalline structures). For this reason, the mechanical properties are influenced by the processing and transformation procedures that can affect the characteristics of materials having the same chemical composition. Furthermore, the type of stress and the way in which it is

*Foundations of Biomaterials Engineering*
https://doi.org/10.1016/B978-0-08-101034-1.00002-5

applied can influence the material's behavior even more than chemical composition, heat treatments, or temperature.

The materials can exhibit an *elastic* behavior (i.e., typical of brittle materials, such as ceramics), an *elastoplastic* behavior (i.e., typical of most metallic materials), or a *viscoelastic* behavior (i.e., typical of polymers). This classification is related to the different deformation type in response to the applied force: elastic, plastic, and viscous deformation.

In addition, *isotropic* materials exhibit uniform properties in all directions; on the contrary, in *anisotropic* materials, properties vary in different directions. For example, amorphous materials, such as glass and many polymers, are isotropic, both at the microscopic and macroscopic scale. Materials such as wood, laminated fibrous composites, and biological tissues are highly anisotropic.

Mechanical properties of the materials are measured by standardized mechanical tests, using specimens subjected to well-defined load conditions. To ensure a better comparison among mechanical tests carried out in different laboratories, specimens of standard dimensions are obtained from the material under evaluation, taking into account International Standards (e.g., ISO, CEN, ASTM, DIN). In particular cases, some specific working conditions can be simulated (e.g., accelerated aging, wear, fatigue) to better study the real behavior of the material. When the result of laboratory tests has to be moved to the behavior in situ (i.e., working application), many warnings are required, as the actual conditions are generally more complex than those simulated during the normalized tests performed in lab.

Later in this chapter, the different behaviors will be described.

## 2.2.1 Stress and Strain

Each force or load applied to a material produces a stress and a deformation in the material. The *stress* is the intensity of the reaction force in each point of the sample subjected to loads in service, under different processing and manufacturing conditions and after thermal variations. The stress, $\sigma$, is measured as the force acting per unit area, and it is expressed as the ratio between the applied force and the area of the material cross-section on which the force acts (Eq. 2.1):

$$\sigma = \frac{F}{A} \qquad (2.1)$$

The forces or loads acting on the material can be static or dynamic, depending on the mode of application. Static force remains essentially constant or slowly changes; dynamic force can be classified as impact force, alternating force, or inverse force. Impact force is produced when the kinetic energy that hits the material specimen is absorbed by deflection in the material. Alternating force oscillates between two limits, typically in a sinusoidal manner such as during a vibration state. The inverse force is a particular alternating force when the two limits are the same in magnitude but opposite in sign.

The static stress can be a compressive or tensile stress (Fig. 2.1A, B, D), or a shear stress (Fig. 2.1C and E). In particular, stress is called compressive when it tends to keep the material compact, and tensile when it tends to separate the material.

The change in shape or dimensions caused by a stress on a material is called *deformation* (or *strain*, $\varepsilon$). The deformation is expressed in a dimensional units such as mm/mm or as a percentage. Because there are three main types of stress (i.e., tensile, compressive, and shear), strain can be expressed in different ways.

The tensile strain $\varepsilon_t$ is expressed as an elongation per length unit (Fig. 2.2A), as described in Eq. (2.2):

$$\varepsilon_t = \frac{L - L_0}{L_0} = \frac{\Delta L}{L_0} \qquad (2.2)$$

where $L_0$ is the original length.

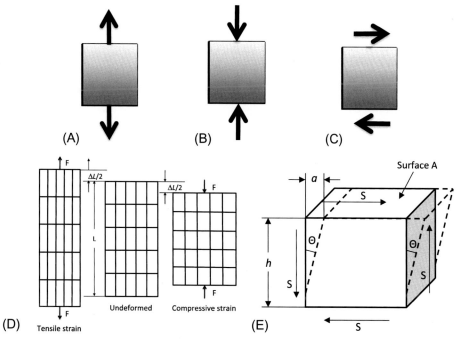

**FIG. 2.1** (A) Tensile stress; (B) compressive stress; (C) shear stress; (D) tensile or compressive stress and connected deformation; (E) shear stress (S=shear force).

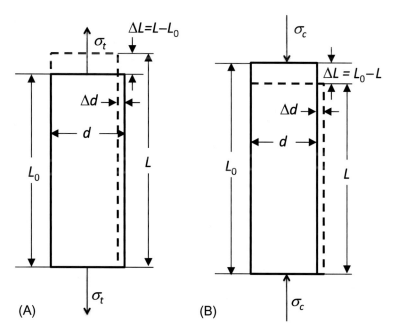

**FIG. 2.2** (A): Tensile or longitudinal deformation $\Delta L/L_0$ corresponding to lateral deformation $\Delta d/d$. (B): compressive deformation $\Delta L/L_0$ and corresponding lateral elongation $\Delta d/d$.

A. INTRODUCTION TO MATERIALS

Compressive deformation $\varepsilon_c$ is measured by the ratio of contraction (Fig. 2.2B) and initial length (Eq. 2.3):

$$\varepsilon_c = \frac{L_0 - L}{L_0} = \frac{\Delta L}{L_0} \qquad (2.3)$$

It can also be noted that the tensile stress causes a contraction perpendicular to its main direction, whereas the compressive stress produces an elongation perpendicular to the main direction.

The shear deformation is expressed by the amplitude of the angle $\theta$ resulting from the variation in the inclination of a particular plane subjected to the shear stress, $\tau$ (Fig. 2.1B). The deformation is measured by the tangent of the angle caused by the slippage of the two planes and corresponds to the angle itself for small elastic deformations.

## 2.2.2 Elasticity

A material exhibits an elastic behavior when the deformation produced under an applied force is completely recovered after its removal. The relationship between stress and the corresponding strain in the elastic region of a material is governed by the Hooke's law (Eq. 2.4a), which states that the stress is proportional to the strain, independently of time:

$$\sigma = \text{constant} \times \varepsilon \qquad (2.4a)$$

This law can be generally applied to all materials for very small deformations, as described in the next section.

### 2.2.2.1 Elasticity Modulus

From Hooke's law, the relationship between stress and strain is a constant characteristic of a material; it is called *elastic modulus* (*E*), and its measure unit is usually GPa. The modulus of elasticity is a measure of the ability of a material to undergo deformation under an applied stress, that is, its rigidity or stiffness (susceptibility to break without appreciable deformation) or

its flexibility (ability to deform before breaking). The higher the elastic modulus, the more stiff the material is.

Because there are two main types of stress in uniaxial condition (i.e., tensile and shear stress), there will be two corresponding modulus of elasticity, as described here.

The modulus of elasticity in tensile condition, or Young's modulus, indicated with *E*, is expressed as in Eq. 2.4b:

$$E = \frac{\sigma_t}{\varepsilon_t} \qquad (2.4b)$$

where $\sigma_t$ is the tensile stress and $\varepsilon_t$ the tensile strain, or longitudinal deformation (expressed as $\Delta L/L_0$). Table 2.1 reports the elastic modulus values of some materials.

The shear modulus, indicated with *G*, is expressed by Eq. 2.4c:

$$G = \frac{\tau}{\gamma} \qquad (2.4c)$$

where $\tau$ is the shear stress and $\gamma$ is the shear strain.

### 2.2.2.2 Poisson's Ratio

The dimensionless relationship between lateral deformation and axial deformation is another constant, named *Poisson's ratio*, and describes the elastic deformations that a material can undergo in the three spatial directions.

If $\sigma$ is the stress needed for the deformation $\varepsilon$ along an axis, the same stress simultaneously produces two minor and opposite strains along the two normal axes. If the applied load is in tensile mode, the $\varepsilon_z$ is an axial deformation and there are two lateral contractions on the orthogonal planes (i.e., $-\varepsilon_x$ and $-\varepsilon_y$). If the applied load is in compression mode, there will be traction strain in the other two normal directions. The Poisson's ratio is expressed by Eq. 2.5:

$$\nu = \frac{\varepsilon \text{ (lateral)}}{\varepsilon \text{ (longitudinal)}} = -\frac{\varepsilon_x}{\varepsilon_z} = -\frac{\varepsilon_y}{\varepsilon_z} \qquad (2.5)$$

**TABLE 2.1** Values of Elastic Modulus, E (evaluated in tensile test) for Different Materials at Room Temperature

| Material | E (GPa) | Material | E (GPa) |
|---|---|---|---|
| Diamond | 1000 | Wood | 6–15 |
| Silicon carbide, SiC | 450 | Polyvinyl chloride, PVC | 3.5 |
| Alumina, $Al_2O_3$ | 400 | Polystyrene, PS | 3 |
| Carbon fibers | 300 | Polyamide, Nylon 6,6 | 2 |
| Steel | 210 | Polypropylene, PP | 1.5 |
| Titanium | 110 | Low density polyethylene, LDPE | 0.3 |
| Inorganic glass | 70 | Rubber | 0.005÷3 |
| Aluminum | 70 | | |

For isotropic behavior, $-\varepsilon_x$ and $-\varepsilon_y$ are equal, because the isotropic material exhibits properties with the same values when measured along axes in all directions.

If the lateral deformations do not involve volume variations (i.e., ideal materials), the Poisson's ratio is equal to 0.5, as happens with liquids. When $\nu$ is lower than 0.5, the material undergoes a volume increase (i.e., expansion) when subjected to a tensile force, and undergoes a volume decrease (i.e., shrinking) if subjected to a compression load. Some representative values of the Poisson's ratio are reported in Fig. 2.3.

The natural rubber has a Poisson's ratio $\nu = 0.5$, caused by an elastic deformation without volume variations. The Poisson's ratio for metals is close to 0.35; for ceramics, it ranges from 0.17 to 0.27. Moreover, the stronger the interatomic bond is (e.g., diamond), the higher the volume increase during tensile deformation is. For most structural materials, Poisson's ratio $\nu$ is in the range 0.2–0.3; for cork it is almost null. Many

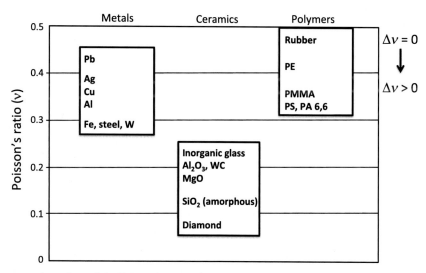

**FIG. 2.3** Representative values of the Poisson's ratio $\nu$ for some materials (values at room temperature).

biological materials have values higher than 0.5, because of the anisotropy characteristic of natural tissues.

Recently, negative Poisson's ratio values have been measured, ranging between $-0.3$ and $-0.1$, in microporous solids and composites (e.g., metal and polymeric porous materials). These materials expand laterally when subjected to a tensile stress and contract laterally under compression load.

### 2.2.2.3 *Enthalpic and Entropic Elasticity*

All materials exhibit an elastic behavior, even if partially, during their deformation. Sometimes, the elastic behavior is macroscopically observable (e.g., natural rubber in a rubber band) or not visible at all, for example, in ceramic materials (e.g., porcelain). The difference among the elastic behavior of ceramic materials, metals, and polymers is mainly due to the different chemical bonds involved in the structure of the material and to atomic arrangement.

When a solid is exposed to an uniaxial stress, it undergoes a deformation in the direction of the applied stress, and therefore an elongation is detected. This elongation causes the appearance of a retraction force, in the opposite direction to the applied force. In general, the elastic retraction force is caused by the displacement of the atoms from their equilibrium position.

In metallic and ceramic materials (metals, crystalline ceramics, mineral or organic glass), thermoset polymers, and polymers in glassy state, the elastic deformation determines a minimum displacement of the atoms of the solid from their equilibrium position (Fig. 2.4A). In this case, internal energy (i.e., enthalpy) considerably increases, yet the order of the system (i.e., entropy) does not significantly change, because atoms move very little from their equilibrium position (Fig. 2.4B); therefore metals and ceramics are called materials exhibiting *enthalpic elasticity*. This behavior is related to the high energy of the bonds in the structure, that is, metallic bonds in metals and ionic/covalent

bonds in ceramic materials. For that reason, the elastic retraction force is due to a minimum displacement of the atoms of the solid from their equilibrium position; the cohesion energy is very high and the elastic retraction forces are very intense. Hence, due to the high energy of these solids, high forces are needed to determine small deformations. For that reason, the elastic modulus values are high (70–2000 GPa) and the maximum elastic strain (elastic region of the stress/strain curve) is limited (about 1%–0.5%).

On the other hand, elastomers are made up of long macromolecular chains bound together by a reduced number of covalent bridge bonds (at about one bridge unit for every 100 structural units), and the cohesion among the chains is very low, because it depends on intermolecular bonds with weak forces. An applied stress causes a deformation that brings to a strong modification in the order of the system and a small variation of the internal energy (Fig. 2.4C); therefore small forces are needed for large deformations. These polymers are solids with *entropic elasticity*. In elastomers, the retraction forces are very low, explaining the reason for the low elastic modulus ($E \cong 1$–$10$ MPa) and large elongation, for example, up to 1000% of elastomers (Fig. 2.4D).

Most of the materials are solid with enthalpic elasticity, and the elastic modulus can vary on three orders of magnitude, from diamond ($E = 10^3$ GPa) to organic glass ($E \cong 2$ GPa). This variation depends on the structure; amorphous materials generally have an elastic modulus lower than that of crystalline solids and mineral glass, exhibiting a much higher modulus.

The amorphous/crystalline polymers have mechanical properties intermediate between those of organic glass and elastomers. In fact, if the amorphous phase is glassy (i.e., $T < T_g$), their elastic modulus is of the same order of magnitude of the organic glass ($E \cong 2$–$3$ GPa); on the contrary, if it is in the rubber state (i.e., $T > T_g$), the value of the elastic modulus varies between 0.2 and 1.5 GPa depending on the

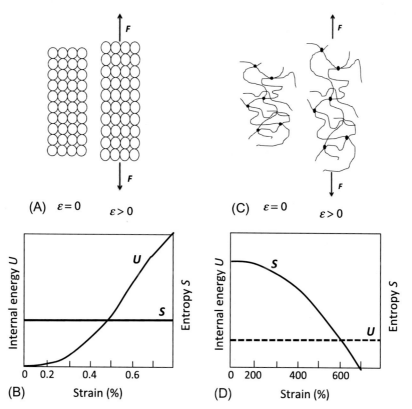

**FIG. 2.4** Mechanism of the elasticity in solids versus the variation of internal energy, $U$, and entropy, $S$. (A) Deformation mechanism and (B) internal energy variation in solids with enthalpic elasticity (e.g., metals and ceramics); (C) deformation mechanism and (D) entropic variations in the entropic elastic solids (i.e., elastomers); note the difference in the strain scale between (B) and (D).

percentage of crystallinity. Partially crystalline polymers are, therefore, a class of materials in which the elastic retraction force can simultaneously have both an enthalpic component and an entropic one. In particular, at temperatures close to the glass transition and to the melting temperature, the mechanical behavior is frequently influenced by viscoelastic effects (see the following section).

Non-elastomeric polymeric materials have an intermediate behavior between the two cases, with partially enthalpic and partially entropic elasticity.

### 2.2.3 Viscoelasticity

The viscoelastic materials exhibit mechanical properties intermediate between those of viscous liquid and those of elastic solid. Polymeric materials, and in particular the thermoplastic ones, are viscoelastic materials.

When a viscoelastic material is subjected to a stress, the response is composed by an elastic deformation (which stores energy) and a viscous flow (which dissipates energy).

For an elastic material, the stress is directly proportional to the strain according to Hooke's law (Eq. 2.1).

In a viscous fluid (i.e., Newtonian fluid), the shear stress, $\tau$, is directly proportional to the strain rate $(d\varepsilon/dt)$ and the relationship is described in Eq. 2.6:

$$\tau = \text{constant} \times \frac{d\varepsilon}{dt} \qquad (2.6)$$

where *constant* is represented by the viscosity $\eta$.

In a viscoelastic material, stress is a function of deformation and time, so it can be described by Eq. 2.7:

$$\sigma = f(\varepsilon, t) \qquad (2.7)$$

Eq. 2.7 can be reduced to a simpler form, described in Eq. 2.8, as follows:

$$\sigma = \varepsilon \times f(t) \qquad (2.8)$$

Eq. 2.8 is the basis of linear viscoelasticity and indicates that, for fixed time values, the stress is directly proportional to the strain. In general, the polymeric materials, and in particular the thermoplastic materials, are viscoelastic, that is, their mechanical properties simultaneously show the characteristics of viscous liquids and elastic solids. In general, the properties of viscoelastic materials depend on time, temperature, and strain rate.

Briefly, an elastic deformation can be described by the spring model (i.e., the elongation is null when the load is removed), whereas a viscous deformation is described by the shock absorber model (i.e., all deformation occurring by viscous sliding of the materials structure, and it is not spontaneously reversible). Different models can be used to describe the viscoelastic behavior (Fig. 2.5); among them, Maxwell model (i.e., spring and shock absorber in series) and Voigt-Kelvin model (spring and shock absorber in parallel) describe the simpler ones.

In the series model (i.e., Maxwell model), adaptable to different thermoplastic polymers, the elastic deformation is immediate and the viscous sliding takes time; after load removal, the elastic deformation is immediately reversible, whereas the viscous sliding is not reversible.

In the parallel model (i.e., Voigt-Kelvin model), adaptable to the behavior of rubber tires, the deformation of the system requires both an elastic deformation and a viscous sliding; after load removal, the elastic component eliminates the deformation of the system causing the inversion of the sliding.

The behavior of the different types of polymeric materials can be more complex than the one described by the previous models, and in most cases an appropriate combination of the various models will be closer to behavior of each polymer depending on its structure. For example, highly cross-linked thermosetting materials will tend to follow Hooke's law, vice-versa thermoplastic materials will tend to follow, at a temperature higher than the melting one, Newton's law (i.e., viscous fluid state), and at low temperature, the Maxwell model (Fig. 2.5). In addition, materials with a typically rubber behavior will be essentially described by the Voigt-Kelvin model (Fig. 2.5). However, both the Maxwell model and the Voigt-Kelvin model are very simplified approximations. The viscoelastic behavior of the thermoplastic materials is better represented, for example, by a combination of the Maxwell and Voigt-Kelvin models (Fig. 2.5).

## 2.2.4 Materials Behaviour in Tensile Test

The most performed test to evaluate the mechanical behavior of a material is the tensile test; in this test, a tensile force is applied to a specimen of standardized dimensions until its break, following a loading process at a constant strain rate. In a standard apparatus for tensile tests, both ends of the material specimen are fixed between clamps; the lower clamp is held fixed in place, and the upper one is attached to the crosshead via a load cell, which measures the applied load (Fig. 2.6A). The material specimens can be bar-shaped with straight sides and uniform rectangular or round cross-section. On the contrary, test specimens can have varying cross-sections and be shaped into dog-bone or

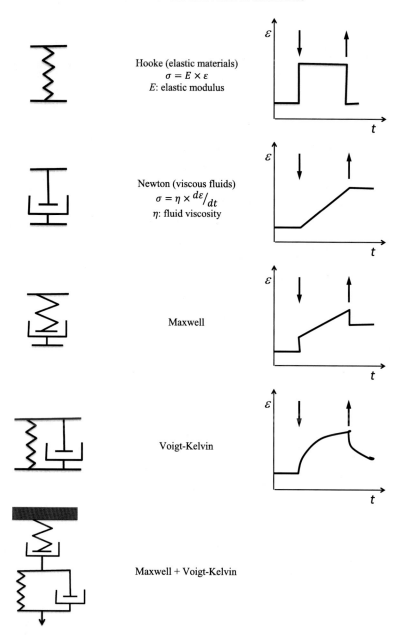

**FIG. 2.5** Main models for the viscoelastic behavior of polymeric materials. The load is applied in ↓ and is removed in ↑.

dumbbell shapes with a rectangular or circular cross-section (Fig. 2.6B).

By recording the force applied to the specimen by the tensile apparatus and the progressive elongation of the specimen, important mechanical characteristics are determined. The force $F$ and the elongation $\Delta L$ are compared to the initial dimensions of the specimen.

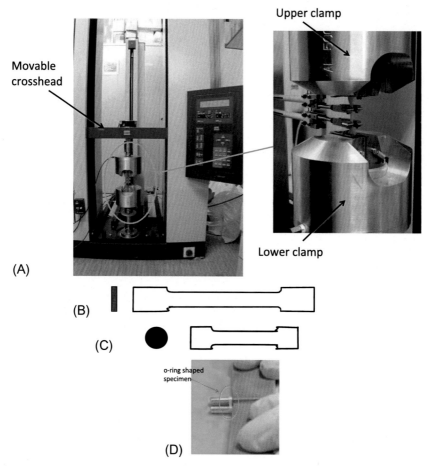

**FIG. 2.6**   (A) Tensile testing instrument; (B) dog-bone-shaped specimen with a rectangular cross-section; (C) dumbbell-shaped specimen with a circular cross-section; (D) o-ring-shaped specimen.

The *nominal stress*, $\sigma$ (measured in MPa), is thus obtained (Eq. 2.9):

$$\sigma = \frac{F}{A_0} \qquad (2.9)$$

where $A_0$ is the area of the initial cross-section.

The nominal deformation $\varepsilon$ is also defined (Eq. 2.2):

$$\varepsilon = \frac{\Delta L}{L_0}$$

where $L_0$ corresponds to the initial length of the specimen. The value of $\varepsilon$ is usually expressed as a percentage.

The application of a force first causes an *elastic deformation*; for many materials, such as metals and some polymers, reversible elastic deformation is followed by an irreversible (permanent) deformation called *plastic deformation*. On the contrary, ceramics exhibit only an elastic deformation.

In general, in the tensile curve obtained by performing a uniaxial test, two types of stress can be detected:

– *yield stress* (or yield strength, yield point), $\sigma_y$: it indicates the stress reached by the material at the end of the elastic deformation region;

– *stress at break* (or ultimate tensile strength), $\sigma_b$: it indicates the stress supported by the material before break.

For stiff materials such as ceramics and many organic polymers in a glassy state (i.e., $T < T_g$), which break without plastic deformation, the yield stress is not detectable. Hence, in fact, the distinction between the two stress values is useless; in fact, yield stress corresponds to the stress at break.

Moreover, yield stress determines the stress that must not be exceeded if permanent deformations have to be avoided. A large gap between the $\sigma_y$ and $\sigma_b$ values, together with a high value of $\varepsilon_r$, should guarantee safety for the use of the material when a localized overcoming of the yield stress occurs in the stressed part of the material.

However, even when the applied load is lower than the yield stress, prolonged periodic deformation can cause the material to break due to a phenomenon called *fatigue* (see Section 2.2.11).

Even a surface degradation of the materials can often be overlapped to the effect of a static or cyclic load, decreasing the real mechanical resistance of the material.

Other important parameters generally determined from the tensile stress/strain curve are the following ones:

– *ductility*: characterized by the level of permanent deformation reached at break; it is mainly measured in terms of elongation or necking of the cross-section area;
– *toughness*: it represents the amount of energy absorbed by a material until it breaks (see Section 2.2.6).

Table 2.2 reports the main mechanical parameters that can be derived from a tensile test and their measure units.

Organic polymers (at $T > T_g$), metals and ceramics (at elevated temperatures) can have viscoelastic behavior. In this case, the instantaneous elastic response is followed by a viscous deformation, which causes an important

**TABLE 2.2**  Mechanical Parameters Derivable From a Tensile Test Performed on a Material

| Parameter | | International Units |
|---|---|---|
| Elastic modulus ($E$) | $\frac{F}{A}$ | N/m$^2$ (Pa) generally expressed in GPa |
| Yield strength ($\sigma_y$) | $\frac{F}{A}$ | N/m$^2$ (Pa) generally expressed in MPa |
| Stress at break ($\sigma_b$) | $\frac{F}{A}$ | N/m$^2$ (Pa) generally expressed in MPa |
| Ductility | % | % |
| Toughness | $F \times l / V$ | J/m$^3$ |

Note: $F$ = force; $A$ = cross-section area; $l$ = length; $V$ = volume.

variation in the mechanical properties as a function of the load application rate.

Relating to the thermodynamic study of the uniaxial deformation, as previously discussed (see Section 2.2.2.3), it is possible to distinguish two large classes of materials: those with enthalpic elasticity (e.g., metals, organic and mineral glasses, semicrystalline polymers) and the elastomers, which have an entropic elasticity. Hence, different tensile behavior can follow the same classification. For that reason, in the next sections, the different general behavior of metals, polymers, and ceramics will be detailed, highlighting the mechanical parameters and characteristics for each class of materials.

### 2.2.4.1 Metals

The tensile stress/strain curve of a ductile metal is reported in Fig. 2.7; from the analysis of this curve, different mechanical parameters can be detected, as described here.

As observable in Fig. 2.7, the stress/strain curve can be divided into two different regions. In the first part of the curve, that is, elastic region, there is proportionality between stress and deformation. In the second part of the curve, that is, plastic region, the stress/strain curve deviates from a linear behavior; in fact, when the specimen is unloaded, a residual plastic

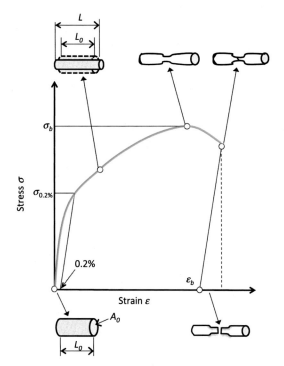

**FIG. 2.7**   Stress/strain curve of a cylindrical specimen of a ductile metal subjected to tensile test.

deformation remains on the specimen. At the microstructural level in metallic materials, there is an irreversible flow among the atoms of crystalline lattice as a result of the displacement of dislocations; thus a permanent residual deformation occurs in the sample.

Analyzing the curve reported in Fig. 2.7, the following characteristics can be highlighted:

- The initial part of the metal deformation is elastic: the specimen recovers its dimensions when the load is removed; there is a total and instant reversibility of the deformation. In metals, elastic deformation generally has a linear behavior and, in the case of materials with a high modulus of elasticity, it usually does not exceed a $\varepsilon$ value equal to 0.1%. In the elastic region, the contraction of the cross-section is very small and can be neglected in the evaluation of the stress.

- In the elastic region, the *elastic modulus*, $E$ (or Young's modulus) can be detected. $E$ is given by the slope of the elastic part of the $\sigma/\varepsilon$ curve and is calculated as reported in Eq. 2.4b; it is expressed in GPa. As already explained, the elastic modulus is a function of the energy of the bonds between the atoms in the structure of the material.
- The *yield stress*, $\sigma_{0.2\%}$ or $R_y$, gives the value of the nominal stress from which the material begins to plastically deform. Because plastic deformation often occurs increasingly, it is difficult to accurately determine the yield stress. For metal materials, a conventional yield strength $\sigma_{0.2\%}$ is generally used; it represents the nominal stress corresponding to a permanent deformation of 0.2%. A detailed description of the method to get information on the yield point is reported later in this section.
- When the yield stress is exceeded, the length of the specimen is permanently increased when the load is removed; in fact, there is an irreversible deformation (i.e., plastic deformation). The plastic deformation mechanisms in metals and metal alloys cause movement of dislocations, increasing the mechanical resistance of the metal/metal alloy.
- During the initial elastic deformation, and for small plastic deformations, the elongation of the specimen is accompanied by a homogeneous contraction along the whole length. Starting from a certain percentage of critical deformation, the contraction of the cross-section stops to be homogeneous and becomes more and more accentuated in a single random point (i.e., theoretically in the middle of the specimen). In this case, we talk about the phenomenon of *necking* (Fig. 2.8): from the beginning of this phenomenon, the local cross-section decreases rapidly and the mechanical resistance of the specimen, proportional to the decreasing in cross-section, decreases as

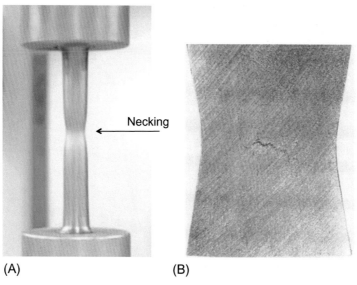

(A)                              (B)

**FIG. 2.8** Necking in a ductile metal specimen. (A) Specimen under tensile test; (B) magnification of the necking zone. The localized decrease in the cross-section can be observed.

well, as observable in the last part of the stress/strain curve after the maximum point (Fig. 2.7). *Percentage necking Z%* can be defined and characterizes the relative decrease of the cross-section measured after the break with the following relation (Eq. 2.10):

$$Z\% = \frac{A_0 - A_b}{A_0} \times 100 \qquad (2.10)$$

where $A_0$ is the area of the initial cross-section and $A_b$ is the cross-section area at break. The percentage necking $Z$ represents a measure related to the ductility of the material. In fact, $Z$ parameter can range from zero (brittle materials) up to 100 (very ductile materials).

- In the final part of the stress/strain curve, the applied stress apparently decreases, that is, the material apparently opposes a lower resistance to the deformation. This is due to the phenomenon of the necking, as previously discussed.
- The break stress, $\sigma_b$ or $R_m$, is defined as the maximum nominal stress value (Fig. 2.7).

- The deformation at break, $\varepsilon_b$, corresponds to the nominal plastic deformation at which the break occurs, due to tensile force applied to the specimen. The value of the deformation at break represents a characteristic related to ductility.

Determination of yield stress is strongly related to the shape of the stress/strain curve. When the curve shows a discontinuous trend, as reported in Fig. 2.9A, the yield stress value is assumed as the stress corresponding to the zone of discontinuity of the curve. A maximum $\sigma_{up}$ value (*upper yield point*) is detectable at the maximum stress value, and a minimum $\sigma_{low}$ value (*lower yield point*) corresponds to the minimum value assumed by the nominal stress in the oscillating portion of the curve. The first corresponds to the stress necessary to start movement of dislocations; the second one is related to the stress necessary to allow them moving within the crystalline lattice.

In the case reported in Fig. 2.9B, in which the stress/strain curve shows a continuous trend,

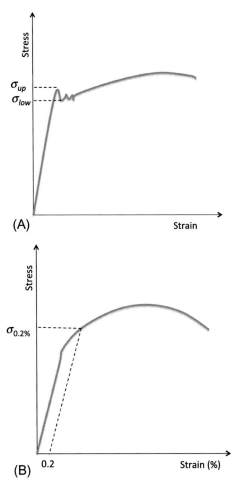

(A)

(B)

FIG. 2.9  Yield strength detection in the case of stress/strain curve: (A) with a discontinuous trend and (B) with a continuous trend.

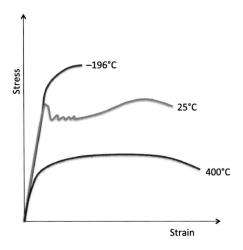

FIG. 2.10  Stress/strain curves of a low carbon steel at different test temperature.

### 2.2.4.2 Polymers

In the case of polymers, we observe a great variety of behaviors depending on the molecular structure, temperature, and deformation rate. A typical tensile stress-strain curve of a thermoplastic polymeric material is shown in Fig. 2.11.

Different information can be detected from this stress/strain curves, as reported here:

- the elastic modulus $E$: it will be described in further detail in the following section;
- the *yield stress*, $\sigma_y$: in general, it corresponds, in the stress-strain curve, to the peak following the first pseudolinear curve of the curve;
- after yield stress, necking occurs, and a phase of drawing is exhibited by the polymer; in that step, macromolecular chains are aligned in the direction of the applied load, with a homogenous decrease in the cross-section area;
- when macromolecules reach the maximum elongation, an increase in the applied load to get a higher deformation of the specimen is required, due to the fact that it is necessary, in this last part of the curve, playing on the strong bonds on the macromolecular chains (i.e., covalent bonds);

the nominal stress corresponding to a predetermined permanent deformation, generally equal to 0.2% strain, is taken as the conventional yield strength point.

For metals, stress/strain behavior is not significantly influenced by the temperature (Fig. 2.10); in fact, a different mechanical behavior can be detected only for high variation of temperature. In particular, when the test temperature increases, resistance to loads decreases and ductility increases.

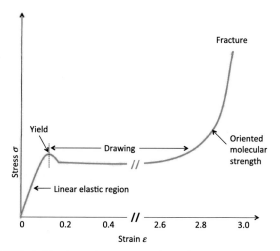

**FIG. 2.11** Stress/strain curve for a thermoplastic polymeric material.

- the *break stress*, $\sigma_b$ and the corresponding strain at break, $\varepsilon_b$, are detected as last point of the stress/strain curve.

The tensile response of polymeric materials when a uniaxial force is applied can be correlated to the different types of material; in a test performed at the same temperature and stress rate different behavior can be detected (Fig. 2.12).

### ELASTIC MODULUS

The design of an object or a component based on the elastic modulus obtained from a tensile test does not predict long-term behavior, because plastics are viscoelastic materials, that is, modulus, mechanical strength, ductility, and friction coefficient are affected by the deformation rate, history and type stress, temperature, time, etc.

Most of the elasticity modulus values reported in the technical literature of plastics represent the slope of the tangent at the origin of the $\sigma/\varepsilon$ curve. However, because this value cannot always be determined with precision and does not have great significance for materials that bear high deformations before breaking, the *secant modulus* can be considered,

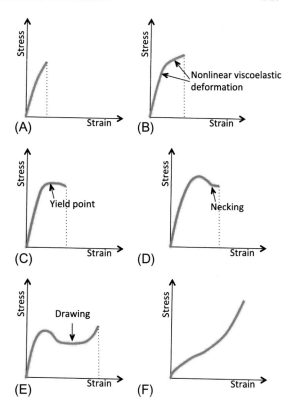

**FIG. 2.12** Mechanical behavior of polymeric materials under tensile test. (A) brittle behavior; (B) quite brittle behavior; (C) quite tough behavior; (D) tough behavior; (E) high tough behavior; and (F) rubber or elastomeric behavior.

choosing a certain deformation (e.g., C' in Fig. 2.13), identifying the corresponding point on the curve $\sigma/\varepsilon$ (e.g., C in Fig. 2.13) and tracing the line that connects this point to the origin. The slope of this line represents the secant modulus at the chosen deformation value and does not take into account the true trend of the $\sigma/\varepsilon$ curve.

### TENSILE BEHAVIOR AS A FUNCTION OF TEMPERATURE

The value of the modulus varies with the elongation rate and with the test temperature, because of the presence of a viscous component in addition to the elastic one. Only for lower temperature (e.g., $T < T_g$) for amorphous

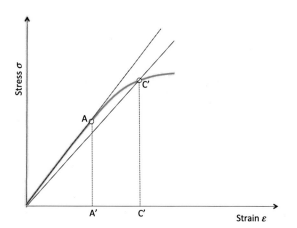

**FIG. 2.13** Elastic modulus and secant modulus for a polymeric material.

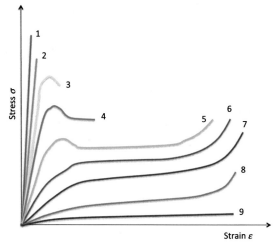

**FIG. 2.14** Influence of temperature in tensile tests for polymeric materials.

polymeric materials, the effect of the deformation rate can be neglected and there is an elastic linear behavior (i.e., brittle failure) until break of the polymeric material (curves 1 and 2, Fig. 2.14).

For higher temperatures, but still lower than $T_g$, curves characterized by the presence of a maximum corresponding to the yield stress $\sigma_y$ can be observed for amorphous polymers (curves 3 and 4, Fig. 2.14). The maximum evidenced in the curves corresponds to a ductile behavior of the polymer, which at a certain point is subjected to a necking of the cross-section, with an increase in the stress referred to the surface unit, following by a decrease of the stress due to the viscous deformation that occurs. After necking, the curve (curve 4, Fig. 2.14) exhibits a section almost parallel to the strain axis and more or less prolonged until it breaks (i.e., ductile failure).

Increasing the temperature at $T > T_g$, there are curves (curves 5, 6, and 7, Fig. 2.14) in which the yield point disappears (mostly in curves 6 and 7, Fig. 2.14) and very high elongations occur, in large part recovered at the end of the test (i.e., rubbery state). In addition, the high elongation is due to the increasing flexibility of the macromolecules at relatively high temperatures, after which there is a final ascending section of the

curve that occurs when the maximum extensibility of the macromolecules has been reached, and therefore the stress necessary to produce a further elongation of the specimen increases more quickly until break.

Curves 8 and 9 (Fig. 2.14) are observable at high temperature, reaching the maximum elongation, typical of the rubbery state.

For crystalline polymeric materials (i.e., high percentage of crystallinity) at temperatures lower than the $T_g$ of the amorphous component, the stress/strain curves are similar to those of amorphous polymers under the same conditions (curve 1, Fig. 2.15). At higher temperatures, great elongations, practically irreversible, due to the orientation of the macromolecular chains and the crystallites can be observed (curve 2, Fig. 2.15).

### TENSILE BEHAVIOR AS A FUNCTION OF DIRECTION AND RATE OF DEFORMATION

The tensile behavior of polymeric materials considerably changes depending on the direction of the applied load (i.e., parallel or perpendicular) with respect to the orientation direction of the polymeric chains. As an example, in

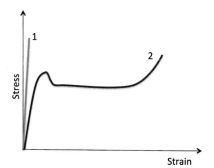

**FIG. 2.15** Stress/strain behavior of crystalline polymeric materials: (1) brittle behavior; (2) ductile behavior.

Fig. 2.16A, the behavior of brittle polymers is reported, and Fig. 2.16B shows the behavior of ductile polymers when the tensile load is applied in different directions with respect to the orientation of the chains.

Moreover, by varying the stress rate, the response of the material changes (Fig. 2.17). In general, at low elongation rate, the polymeric material has a more ductile behavior because the macromolecular chains have time to align under the effect of the applied stress. Hence, the material is able to flow at the same rate as the one it is stressed. On the contrary, when the elongation rate is high, the polymeric chains are not able to align in the direction of the applied load, and the break of the covalent bonds in the macromolecular chains occurs.

### TENSILE BEHAVIOR OF ELASTOMERS

Elastomers are a particular class of polymeric materials characterized by an elastic retraction force of almost exclusively entropic origin (i.e., materials with entropic elastic behavior). They have a great ability for elastic deformation, recovering the majority of the deformation once the force is removed.

Elastomers are made up of three-dimensional networks in which the macromolecular chains are bonded together by short and not very numerous bridge bonds. They are used at a temperature higher than their $T_g$ (e.g., $-70°C$), and

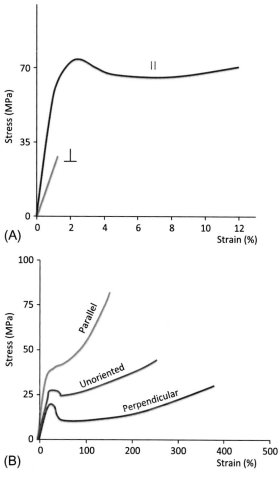

**FIG. 2.16** (A) Behavior of brittle polymers. ($\parallel$): tensile load parallel to the orientation direction of the polymeric chains, and ($\perp$): tensile load perpendicular to the direction of the polymeric chains. (B) Tensile behavior of ductile polymers: *unoriented*, with load applied parallel to the direction of the chains orientation (*parallel*) and perpendicular to the direction of the chains orientation (*perpendicular*).

the cohesion forces between the segments are remarkably low. Hence, the chain segments can move one on each other without producing a noticeable change in the internal energy, breaking at a very high deformation value ($\varepsilon_b$ at about 700%–1000%) without any plastic deformation (Fig. 2.18).

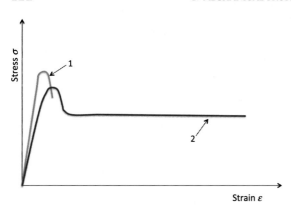

FIG. 2.17    Tensile behavior of polymer tested at different deformation rate: (1) high deformation rate; (2) low deformation rate.

FIG. 2.19    Tensile stress/strain curve for a typical ceramic material.

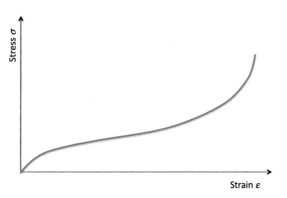

FIG. 2.18    Tensile stress/strain curve for an elastomeric material.

The stress/strain curve for an elastomer (i.e., tensile test) is reported in Fig. 2.18; the slope of the curve at the origin makes it possible to calculate the elastic modulus $E$, which is very low. In addition, $E$ is proportional to the temperature at which load is applied as well as to the number of chain segments per unit volume of the elastomer.

### 2.2.4.3 CERAMICS    In the case of ceramics, no plastic deformation can occur, because the movement of the crystalline planes is strongly inhibited, at room temperature, due to the high

energy and stiffness of the ionic/covalent bonds, or by the presence of ions of opposite signs that prevent their displacement. For that reason, the ceramics are essentially brittle materials that break without appreciable plastic deformation (Fig. 2.19); in fact, for these materials, the yield stress and the stress at break coincide.

### 2.2.4.4 COMPARISON BETWEEN THE TENSILE BEHAVIOR OF THE DIFFERENT CLASSES OF MATERIALS    To compare the tensile behavior of the different types of materials, values of $E$ and $\sigma_b$ are reported in Table 2.3. The plastic deformation mechanisms of metals and metal alloys cause movement of dislocations. In the case of ceramics, this movement is strongly inhibited, at room temperature, due to the great energy and rigidity of the bonds, or by the presence of ions of opposite sign that prevent their displacement. The ceramics are essentially fragile materials that break without appreciable plastic deformation: in practice, the yield stress and the breaking stress coincide.

In the case of polymers, we observe a great variety of behaviors depending on the molecular structure and temperature.

**TABLE 2.3** Tensile Mechanical Properties of the Main Classes of Materials

| | Elastic Modulus, $E$ (GPa) | Stress at Break, $\sigma_b$ (MPa) |
|---|---|---|
| Rubber | 0.005–3 | 10–30 |
| Plastic materials | 1–8 | 10–120 |
| Reinforced plastic materials | 45–200 | 800–1500 |
| Wood, timber | 6–15 | 30–100 |
| Metals | 70–250 | 170–1000 |
| Traditional ceramics | 10–100 | 1–30 |
| Advanced ceramics | 200–400 | 150–210 |
| Metal fibers | 200–350 | 1000–4000 |
| Synthetic polymeric fibers | 10–500 | 300–3000 |
| Natural polymeric fibers | 2–20 | 150–1000 |
| Whiskers | 400–1000 | 4000–40,000 |
| Diamond | ~1000 | ~2000 |

## 2.2.5 True Stress and Strain Versus Engineering (Nominal) Stress and Strain

The stress/strain curve reported in Fig. 2.7 for a ductile metal indicates that the stress at break is lower than the maximum stress. This apparent anomaly is due to the fact that the stress is calculated with respect to the area of the starting cross-section of the specimen (i.e., nominal stress) and does not take into account the variations in cross-section area occurring during the mechanical test. The same observation can be made for the deformation, where the variation in length does not take into account the plastic deformation. True stress and strain values for the area and length of the specimen at each moment of the tensile test will be very different

from the nominal ones, especially near the breaking point.

True deformation ($\varepsilon_T$) can be calculated by integrating the increments of deformations from the first point of the curve. If a differential deformation element is indicated with $d\varepsilon$ (Eq. 2.11):

$$d\varepsilon = \frac{dl}{l} \qquad (2.11)$$

where $l$ is the instantaneous length of the test specimen.

By integrating this equation between the starting, original length $l_0$ and the instantaneous length $l$, the total deformation during the test will be given by Eq. 2.12:

$$\varepsilon_T = \int_{l_0}^{l} \frac{dl}{l} = \ln\frac{l}{l_0} \qquad (2.12)$$

Assuming that there are no variations in volume (i.e., Poisson's ratio $\nu = 0.5$) as in Eqs. (2.13a), (2.13b):

$$A_0 l_0 = Al \qquad (2.13a)$$

and

$$\frac{l}{l_0} = \frac{A_0}{A} \qquad (2.13b)$$

the true stress can be calculated with Eq. (2.14):

$$\sigma_T = \ln\frac{A_0}{A} \qquad (2.14)$$

where $A_0$ is the initial area and $A$ is the instantaneous area of the specimen cross-section.

Fig. 2.20 shows the tensile curve of a ductile metal expressed in nominal and true values of stresses and strains. It can be observed that the difference between the true and the nominal deformation is small for a deformation lower than 10%, and it can be detected that the maximum nominal stress does not correspond to the maximum in the true values of the intrinsic resistance of the material.

Indeed, the true tensile strength of the material increases until the specimen breaks. The maximum of the tensile curve under nominal stress is due to the phenomenon of necking, which

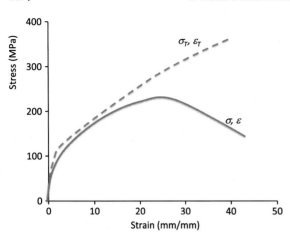

**FIG. 2.20** True and nominal stress/strain curves for a ductile metal.

involves a considerable reduction of the cross-section of the sample. The tensile force necessary for sample deformation, after necking, is lower compared to the increase in mechanical strength of the material. Hence, the nominal stress for a metal is a measure of the resistance of the specimen and not of the material.

Comparing the nominal stress/strain curve with the true curve for an elastomer, no significant differences can be detected up to 500% strain (Fig. 2.21). When strain is higher than 500%, the difference between the two curves is considerable, and it is determined by the fact that when the macromolecular chains are extended and oriented, the tensile force is progressively applied to the covalent bonds present in the polymeric chains, considerably increasing the retraction force.

## 2.2.6 Hysteresis

For a perfectly elastic material the tensile curve is reversible, and the energy absorbed during the load step is entirely recovered when force is removed (i.e., behavior similar to that of a spring, Fig. 2.22A).

Most materials do not have an ideal elastic behavior, and part of the energy absorbed in

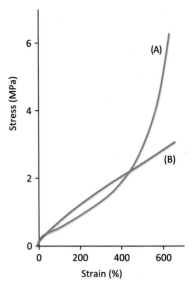

**FIG. 2.21** Stress/strain curve for natural rubber: (a) experimental curve (i.e., nominal stress and strain); (b) true curve.

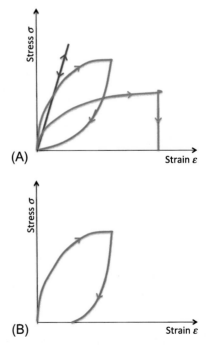

**FIG. 2.22** (A) Hysteresis curves for the different types of behavior: elastic, viscoelastic, and plastic; (B) hysteresis cycle with residual deformation.

the load phase is dissipated in the specimen bulk by internal friction mechanisms (e.g., movement of dislocations in metals or macromolecular chains in polymers). In this case, there is a difference between the energy provided to the material and the recovered energy; hence, the unloading curve is not equivalent to the load curve, even if the sample recovers its initial length (Fig. 2.22A). Elastomers can form hysteresis ring and give rise to internal heating phenomena due to the friction of the elastomer chains on each other (e.g., heating of the tires).

During a deformation cycle, a hysteresis ring is formed due to an inelastic effect. The hysteresis (and consequently the energy dissipation) is greater in polymeric materials (with viscoelastic behavior) than in ceramic or metal materials, which are closer to an ideally elastic or elastoplastic behavior, respectively. Fig. 2.22A shows the various types of behavior under load and unloading conditions of the applied force.

In the case of the viscoelastic behavior, as shown in Fig. 2.22B, it can be verified that, under unloading conditions, the hysteresis ring has a residual deformation (point A in Fig. 2.22B). This deformation can be recovered over time, or it can be permanent. In particular, increasing time, a viscoelastic material with residual deformation can recover the initial conditions or permanently maintain a deformation state.

## 2.2.7 Toughness and Resilience

Among the quantities examined so far, some of them, such as $\sigma_y$ or $\sigma_b$, give information on the ability of the material to resist to applied loads; others, such as $\varepsilon\%$ and $Z\%$, provide the material's capacity to deform (i.e., ductility). To have a more complete parameter that simultaneously takes into account the mechanical resistance and the ductility of the material, we must consider parameters such as the energy absorbed during the mechanical test.

*Toughness* is the ability of a material to absorb energy before break. In a static tensile test, this

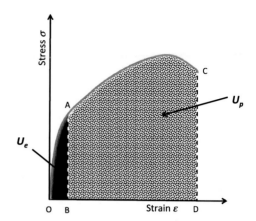

**FIG. 2.23** Schematic representation of toughness and resilience. The OACD area measures the toughness, and the OAB area measures resilience.

energy is measured from the area under the stress/strain curve, and represents the work required to break the specimen under test (Fig. 2.23). The energy of volumetric deformation $U$ is given by the Eq. 2.15:

$$U = \int_0^\varepsilon \sigma \, d\varepsilon \qquad (2.15)$$

As reported in Fig. 2.23, an energy related to the elastic volumetric deformation, $U_e$ (area subtended between points O, A and B) and an energy related to the plastic volumetric deformation, $U_p$ (area subtended between B, A, C, and D) can be detected.

The ability of a material to absorb energy in the elastic region is called *resilience*.

Materials with high toughness must have high yield strength and high ductility. Brittle materials have low toughness, because they exhibit only small plastic deformations before break.

Moreover, toughness is required in objects or components subjected to dynamic loads (e.g., shock and impact load).

However, toughness is not usually measured starting from the tensile stress/strain curve, but with dynamic tests, such as impact resistance test, or by using the fracture mechanics theory, that investigates the mode of propagation of cracks in a material. The latter will not be discussed in the present chapter because it is outside the scope of this book.

## 2.2.8 Brittle Fracture and Ductile Fracture

In most materials, there are cracks (internal or superficial cracks) or other defects that constitute fracture beginnings by concentrating the loads (which locally reach a value higher than the applied load).

The study of the propagation conditions of a crack by action of a stress is the subject of fracture mechanics that will not be discussed in this book. Here, the mechanisms of fracture in different materials will be briefly discussed.

### 2.2.8.1 Metals and Ceramics

In the case of a material with a brittle behavior, such as ceramics, the failure of the material is of brittle type and occurs by breaking the primary bonds without significant deformations on the specimen (Fig. 2.24).

In the case of metallic materials with a ductile behavior (Fig. 2.24), at a microstructural level an irreversible flow occurs between the atoms of the crystalline lattice through the displacement of dislocations, resulting in a permanent residual deformation. An important characteristic of the ductile fracture compared to the brittle one is the large amount of energy necessary to produce it.

### 2.2.8.2 Polymers

The fracture of the polymeric materials can be brittle, ductile, or intermediary between them,

FIG. 2.24 Schematic representation of fracture mechanisms at the tip of a crack. (A) and (B) Macroscopic appearance, in the case of a brittle (ceramic) and ductile fracture material with plastic deformation zones (metal), respectively. (C) and (D) Microscopic appearance, schematic representation of the shape of the plastic deformation zone absent in the brittle material (C) and due to the movement of the dislocations in the metals (D). (E) and (F) Macroscopic images of a specimen that underwent a brittle and ductile breaking, respectively.

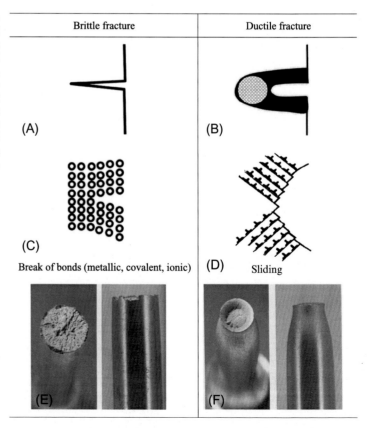

as already discussed for metals. In general, thermosetting plastics undergo brittle failure, whereas thermoplastics may undergo fracture in a brittle or ductile manner.

If the fracture of a thermoplastic occurs below its glass transition temperature, it will fracture in a brittle manner. When the fracture occurs above the $T_g$ of the polymeric material, a ductile failure can be detected. Hence the temperature can greatly influence the way the thermoplastics fail.

Cross-linked plastics (i.e., thermoset), if heated just below their degradation temperature, become weaker and fail at lower stresses, but the fracture is always mainly brittle because the network of covalent bonds remains intact until the decomposition temperature.

Deformation rate is also an important factor in thermoplastic fracture behavior: low deformation rate promotes ductile fracture, because the macromolecular chains are able to draw and align.

The elastomeric materials are deformed essentially in the same way, but there is a much higher elongation of the polymeric chains in the elastic region; however, when the stress applied to the material becomes too high and the elongation of the macromolecular chains is maximum, the covalent bonds of the macromolecular chains break and cause the ductile fracture of the material.

The surface energy required for the fracture of a glassy and brittle amorphous polymeric material, such as polystyrene or polymethyl methacrylate (PMMA), is much higher than that required if the fracture involved only the break of C—C bonds on a fracture plane. Hence, glassy polymeric materials such as PMMA are tougher than inorganic glass. The energy required to cause fracture of glassy thermoplastics is much higher because localized distorted regions, called *crazes,* are formed before failure (i.e., cracking) occurs. A craze in a glassy thermoplastic is formed in a highly stressed region of the material and exists in the alignment of the molecular chains combined with a high density of voids. Fig. 2.25A schematically shows the variation in the molecular structure in presence of the craze in a glassy thermoplastic, such as PMMA. If the applied stress is high enough, a crack is formed through the craze, as shown in Fig. 2.25B.

As the crack propagates along the craze, the stress concentration at the top of the crack spreads throughout the craze. The work necessary for aligning the polymer macromolecules in the craze is the cause of the relatively high amount of work required for the fracture of glassy polymeric materials.

Thermoplastics above glass transition temperature may have plastic deformation before the fracture. During plastic deformation, the macromolecular chains stretch out, slide on each other, and gradually align compacting themselves in the direction of the applied load (Fig. 2.26). When the stress applied to the macromolecular chains becomes too high, the covalent bonds of the macromolecular chains break and the material fracture occurs.

## 2.2.9 Deformation Mechanisms of Ceramic Materials

The lack of plasticity in crystalline ceramic materials is due to their ionic and covalent chemical bonds. In ceramic materials mainly covalently linked, the bond between atoms is specific and directional. As a result, when covalent crystals are stressed to a sufficient level, they show a brittle fracture caused by the separation of the bonds between the electron pairs without their subsequent reformation. In addition, covalently bonded ceramics are brittle both in the monocrystalline state (i.e., assuming they can exist in practice) and in the polycrystalline state.

The deformation of ceramics mainly composed by ionic bonds is different. Let us try to briefly examine some conditions in which an ionic crystal can be deformed, as reported in Fig. 2.27. The sliding of one ion plane over

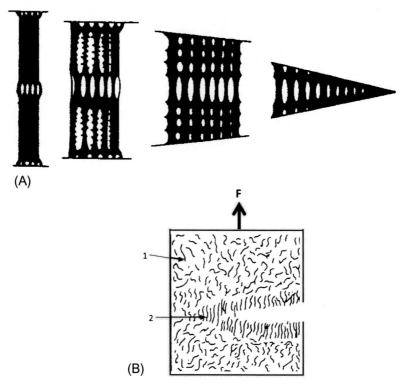

(A)

(B)

**FIG. 2.25**   (A) Scheme of the variation of the microstructure of a craze in a glassy thermoplastic under stress; (B) Scheme of the structure of a craze near the end of a crack in a glassy thermoplastic: F=direction of the applied stress; 1=randomly folded macromolecules; 2=macromolecules oriented in the craze region.

another involves different charged ions coming into contact, and therefore attraction and repulsion forces can be introduced. The sliding that occurs in most of the ion-bonded crystals having a NaCl-type structure occurs along the plane family {110} as it involves only different charged ions, and therefore the sliding planes remain linked to one another by secondary forces during the sliding process. On the other hand, sliding on the plane family {100} is rarely observed because the ions of the same charge coming into contact tend to separate the planes of the ions flowing one on top of the other. In polycrystalline ceramic materials, adjacent grains must vary their shape during deformation. Because there are few sliding systems in the solids with ionic bonds, cracks are formed at the grain edges and consequently the brittle fracture occurs.

## 2.2.10 Impact Test

Toughness, that is the amount of energy absorbed before break, can be determined by means of the impact test under dynamic loads, in which a certain load is applied instantaneously to a specimen.

*Impact stress* is related to the load force that acts at high speed on the product or on the specimen, in contrast to the static load.

The impact test is aimed at evaluating the amount of energy absorbed by a standardized specimen when hit by a hammer with a known value of kinetic energy. The absorbed energy is used as a parameter for the susceptibility to ductile or brittle fracture. The impact resistance of a material is very sensitive to the notch, which can cause a state of triaxial tension near the bottom of the notch.

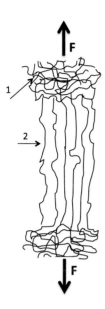

**FIG. 2.26** Plastic failure of a thermoplastic polymeric material under stress. Macromolecular chains draw and flow on each other to align themselves in the direction of applied load. When the stress is too high, the macromolecular chains are broken causing the material fracture. F = direction of applied stress; 1 = randomly oriented macromolecular chains; 2 = oriented macromolecular chains.

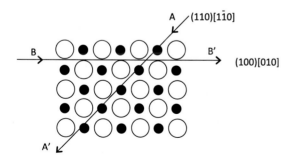

**FIG. 2.27** Conditions in which a ionic crystal (e.g., NaCl) can be deformed: sliding on the plane (110), in the direction AA′ is possible because it involves only ions with different charge, whereas the sliding on the plane (100), in the direction BB′ is rarely observed because ions with the same charge that tend to repel come in contact.

For the impact tests, the Charpy pendulum, shown in Fig. 2.28A, is used. The specimens are fixed at either end (Charpy method, Fig. 2.28B) or at one end (Izod method, Fig. 2.28C).

The resilience test also allows to investigate the modifications of the material behavior as a function of the temperature, or to determine the representative curve of the ductile-brittle transition of a material. In fact, by repeating the tests at different temperatures, a critical temperature can be detected so that below that value the material exhibits a ductile behavior (i.e., high absorbed energy) and a brittle one above that value (i.e., low absorbed energy).

By repeating the impact tests at different temperatures, it is possible to identify a critical temperature above which the material is able to absorb a high quantity of energy (i.e., ductile behavior) and below that a low quantity of energy is absorbed (i.e., brittle behavior). The ductile-brittle transition temperature obtained by a resilience test must be taken into account as an indicative value for the selection of a materials and as comparison value among different materials; in addition, it is an indication of a temperature below which it is certainly dangerous to use the material (i.e., brittle behavior). On the contrary, it is theoretically incorrect to consider the resilience value obtained at room temperature as a consistent criterion for selecting a material for applications at low temperatures. This is clearly shown in Fig. 2.29 for two different steels: steel A has a higher toughness than steel B at room temperature, but if the working temperature of the structure or component decreases, the behavior is opposite.

## 2.2.11 Hardness

Hardness is a measure of the material resistance to a permanent localized plastic deformation. This property is most relevant at the surface of the materials.

Hardness is often measured using an indentation technique. It is in general calculated by

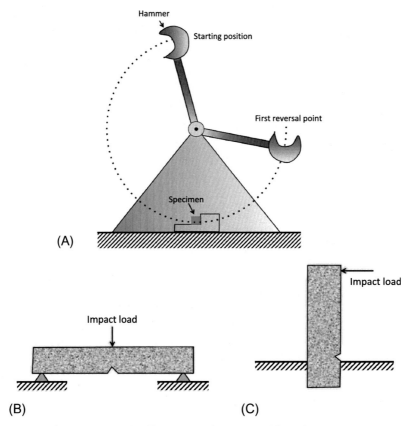

**FIG. 2.28**   (A) Apparatus for impact tests; (B) Charpy v-notch specimen; (C) Izod specimen.

measuring the surface morphology or the depth of the impression obtained by compressing a penetrator on the surface of the material to be investigated. The penetrator (e.g., sphere, pyramid, or cone) is made of a material much harder than the one under evaluation (e.g., hardened steel, tungsten carbide, diamond). There are various scales for indentation hardness, depending on the used method (e.g., Brinell, Vickers, Rockwell, Shore). However, these different scales use variations of the same measuring protocol, wherein a standard material of a known geometry is used to indent the surface of the material under a known standard load (Fig. 2.30).

The Rockwell scales (i.e., $R_A$, $R_C$) use a variety of indenters (spheres or cones, Fig. 2.30) of known

dimensions, and the hardness value is related to the inverse of the depth of the indentation. Thus the more shallow the indentation, the higher the hardness. Other scales include the Vickers test that uses a square pyramidal indenter (Fig. 2.30B), and the Brinell hardness test that uses sphere for indentation (Fig. 2.30A). Unless there are conversion tables, it is not possible to switch automatically from one scale to another.

The hardness test, also for the simplicity and speed of its execution, represents a useful and precious method for a rough measurement of the mechanical resistance of a material.

Variations in hardness also allow detecting changes in the structure of a material due, for example, to a different production batch,

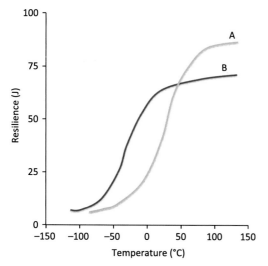

**FIG. 2.29** Scheme of the ductile-brittle transition of two steels: steel A is more ductile than B at room temperature, but not at a lower temperature rather than at higher temperature.

thermal history, etc. This allows control of the production (e.g., quality and reproducibility of the product) or to verify the success of particular processes, such as welding.

In the case of polymers, that is, viscoelastic materials, the hardness measurements should take place according to preestablished times, to allow viscoelastic recovery and therefore measure only the permanent plastic deformation.

## 2.2.12 Fatigue

A material subjected to a time-varying stress in a cyclic manner can be broken at a stress level much lower than that required to have a break in case of constant stress over time. This kind of break is called *fatigue*.

Fatigue failures are particularly insidious, because they arise after a certain operating time

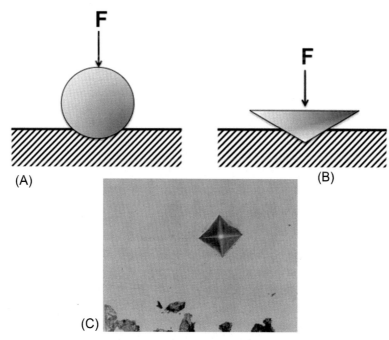

**FIG. 2.30** Hardness test. (A) Brinell hadrness test; (B) Vickers hardness test; (A) and (B) Rockwell hardness test; (C) example of hardness test performed with Vickers penetrator (pyramidal).

without any premonitory sign, such as, for example, macroscopic plastic deformation.

In a fatigue failure, a number of factors are involved, such as:

- factors related to the type and mode of application of the stress (e.g., maximum stress level, load oscillation law, total number and frequency of oscillations, action of any overloads);
- factors related to the material (e.g., presence of notches, surface finish, material structure);
- environmental factors, such as temperature and aggressive actions.

It must be stressed that fatigue failure can occur not only with the simplest types of alternating stress (i.e., tensile, compression, torsion, and flexion) but also with more complex types of applied stress, with an irregular pattern, such as the one reported in Fig. 2.31.

When a fatigue test is performed in laboratory, the specimen is subjected to the simplest and time-varying types of stress in a regular manner; examples of applied stress are represented by the symmetrical alternating, wavy pulsated stress, and pulsatory (or oscillating) stress. More rarely, the real state of stress to which the component is subject under working condition is replicated.

The tests are carried out by stressing a certain number of specimens at different stress levels and measuring the number of cycles after which they break. The maximum stress is then reported in a semilogarithmic diagram versus the number of cycles at break, obtaining curves as the ones shown in Fig. 2.32. Such curves are often referred to as the $\sigma$-N curves, or *Wohler curves*.

Analyzing the Wohler curves reported in Fig. 2.32, some consideration could be highlighted:

- as the maximum applied fatigue stress ($\sigma_{max}$) decreases, the number of cycles necessary for specimen failure increases;
- for many materials, including ferrous ones, there is a threshold stress below which, although the number of cycles is increased, the specimen does not break;
- for other materials, such as aluminum alloys, the threshold stress does not exist, and the material breaks, however, after a sufficiently high number of cycles.

The threshold stress below that there is no break, is called *fatigue limit*, or endurance limit (Fig. 2.32A). The presence of the fatigue limit is important because it allows stating that, if a component made of steel is stressed under the fatigue limit, it does not break whatever the operating time of the component.

FIG. 2.31   Evolution over time of an irregular stress acting on a machine element.

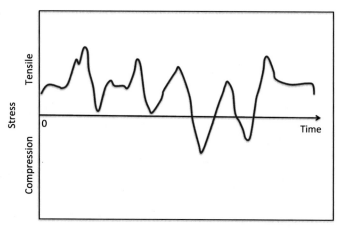

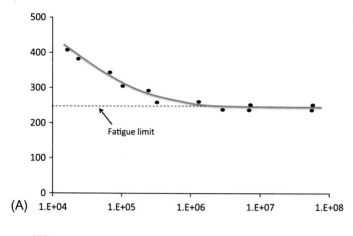

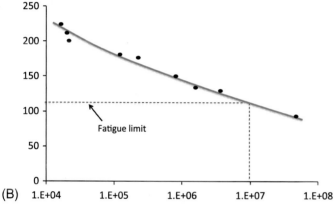

**FIG. 2.32** Representative $\sigma_{max}$—number of cycles curves in the case of: (A) steel, (B) aluminum alloy.

In case there is not a fatigue limit (Fig. 2.32B), a conventional fatigue limit can be fixed, defined as the stress that determines the fatigue failure in correspondence with a prefixed number of cycles (to be indicated), for example, $10^7$ or $10^8$ cycles.

### 2.2.12.1 Fatigue in Polymeric Materials

Plastic materials undergo fatigue failure when subjected to cyclic stresses, more or less in the same way as other materials. In addition, plastics are subject to thermal softening phenomena if the cyclic stress and its rate are high.

The fatigue failure of ceramic and metal materials is always brittle, and there is no visible evidence of the moment in which it will occur. On the contrary, the molecular structure of polymeric materials makes impossible the same type of fracture initiation detectable for metallic or ceramic materials, although it is expected that, once initiated, the fracture propagates in a similar way.

The viscoelastic behavior of polymers indicates that the deformation rate (or the frequency) plays an important role in fatigue characteristics. The increase in temperature due to the low thermal conductivity and the damping capacity of the plastic materials can cause deterioration of the mechanical properties, or softening so that the material is no longer able to bear loads. Before the fatigue failure, in these cases, the "thermal" yielding occurs.

**FIG. 2.33** Typical fatigue behavior for an acetal resin at a frequency of 5 Hz.

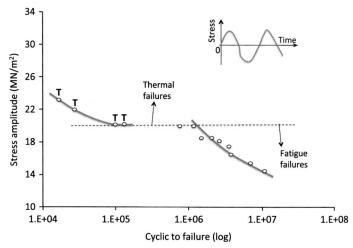

Typical fatigue behavior is shown in Fig. 2.33 (e.g., acetal resin).

In crystalline polymers, the surroundings of spherulitic areas (i.e., crystalline domains) are considered weakness areas that may undergo cracks during deformation, so that they can act as sites for fracture propagation.

In amorphous polymers, it is possible that the cracks arise in the voids formed during the viscous flow.

In each sample of polymeric material, there is a random distribution of microcracks, internal flows, and localized residual stresses. These defects result from structural defects (e.g., variations in molecular weight) or from the used processing method. The breaking process begins at these defects; therefore the fracture propagation depends on a series of random events.

Printed plastic materials may have defects (e.g., junction lines or presence of additives particles, such as pigments and stabilizers) acting as crack initiation sites. A concentration of stresses due to acute geometric discontinuity will be the main source of fatigue failure beginning.

As for metals, once the stress type and frequency (or rate) have been defined, a typical curve of fatigue behavior for plastic materials is made by placing the maximum stress applied during each cycle in the $y$ axis, and the corresponding number of cycles sustained by the material before breaking on the $x$ axis, as reported in Fig. 2.34. The fatigue strength is defined by the number of cycles the material supports before breaking versus the maximum load applied during each cycle. It is possible to recognize the presence of a fatigue limit as a threshold stress under which the break does not occur.

### 2.2.12.2 Factors Influencing Fatigue Resistance

The fatigue behavior of machine components and devices generally depends on the intrinsic properties of the materials' resistance, that is, the fatigue limit, and above all on a correct design of the object. In particular, among the main factors that can influence the fatigue behavior of a material, the following ones are the most important:

– *geometric shape* controls the average nominal level of the stress, as well as the intensification of the stress in presence of notches;
– *processing* involves the presence or the absence of residual internal stresses that can be added to the external forces already present on the component;

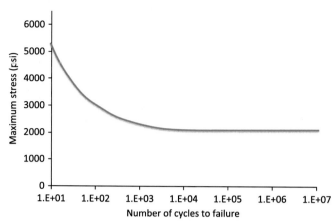

FIG. 2.34 Typical fatigue behavior curve of polymeric material.

— *surface roughness* affects the probability of crack development.

Therefore a fatigue problem can be solved in some cases by controlling shape, processing and finishing of the object, or the aggressiveness of the environment in which it will work, rather than changing the material.

## 2.2.13 Time-Depending Properties

The main characteristic of viscoelastic materials is that of presenting time-dependent deformations under constant stress (i.e., *creep*) and time-dependent stresses under constant deformation (i.e., *relaxation*). In addition, when the applied load is removed, the material has the ability to recover (i.e., *recovery*) slowly toward the initial dimensions for a definite period of time.

### 2.2.13.1 Creep and Strain Recovery

Plastics exhibit time-dependent deformations in response to an applied constant stress. This behavior is called *creep*. Similarly, if the load is removed, the material exhibits a time-dependent recovery, toward the original dimensions (Fig. 2.35).

*Creep break*: when a polymeric material is subjected to a constant tensile stress, its deformation increases

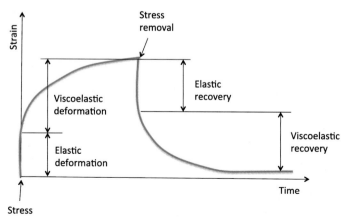

FIG. 2.35 Typical creep *(blue curve)* and recovery behavior *(red curve)* of plastic materials.

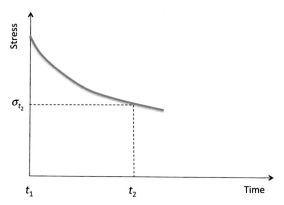

FIG. 2.36　Typical stress relaxation behavior of a plastic material (strain = constant).

untilitreachesthebreakpoint(i.e.,breakbycreepor, occasionally, static fatigue). It is important that the designer is aware of this, to avoid of assuming that a material capable of withstanding a static short-term load, will continue to behave in the same way for indefinite times.

### 2.2.13.2 Creep in Metallic Materials

Metallic materials subjected to constant stress over time, at a temperature higher than about a half of their melting temperature $(T \geq \approx \frac{1}{2} T_m)$, undergo to progressive plastic deformation phenomena over time, which can cause them to break. Hence, also in metallic material, creep can occur.

It should be noted that on metallic materials, the temperature at which the viscous sliding phenomenon begins is in any case relatively high (for example, about 450°C for steels), unlike polymeric materials for which the viscous sliding phenomena also occur at room temperature.

### 2.2.13.3 Stress Relaxation

Another important consequence of the viscoelastic nature of polymeric materials is that, if subjected to a particular level of deformation and the deformation is kept constant for a long time, over time the stress necessary to maintain that deformation decreases. This phenomenon is called *stress relaxation* and is very important in the design, for example, of gaskets, springs, and sealed closures. Fig. 2.36 shows a typical trend of the stress as a function of time under constant deformation.

## Further Reading

Agrawal, C.M., Ong, J.L., Appleford, M.R., Mani, G., 2013. Biomaterials: Basic Theory With Engineering Applications. Cambridge University Press, Cambridge, UK.

Callister, W.D., Rethwisch, D.G., 2007. Materials Science and Engineering: An Introduction. John Wiley & Sons, Berlin, Germany.

Draghi, L., 2017. Static and uniaxial characterization of polymer biomaterials. In: Tanzi, M.C., Farè, S. (Eds.), Characterization of Polymeric Biomaterials. Woodhead Publishing, New York City, NY, USA.

Meyers, M.A., Chawla, K.K., 2009. Mechanical Behavior of Materials. Cambridge University Press, Cambridge, UK.

Smith, W.F., Hashemi, J., 2006. Foundations of Materials Science and Engineering. McGraw-Hill—Series in Materials Science, McGraw-Hill, New York City, NY, USA.

Ward, I.M., Sweeney, J., 2004. An Introduction to the Mechanical Properties of Solid Polymers. Wiley, Berlin, Germany.

Williams, D., 2014. Essential Biomaterials Science. Cambridge University Press, Cambridge, UK.

# 3

# Manufacturing Technologies

## 3.1 PRODUCTION AND PROCESSING OF MATERIALS

The sequence of technological processes (i.e., sequence of operations) used to create an object follows a path Raw Materials → Finished Product that depends on the intrinsic characteristics of the material used (i.e., polymeric, metallic, or ceramic material), and the shape and dimensions of the object itself.

### 3.1.1 Involved Sectors

The raw materials are handled by the primary processing industry (e.g., casting processing), which makes semifinished products available (i.e., "raw" materials or products, which generally must undergo further processing before their final use).

The manufacturing industry is involved in transforming semifinished products into objects or components of preestablished shape and size. The processing, after eventual and opportune assembly and finishing steps, leads to obtaining finished products (Fig. 3.1).

Both the primary processing industry and the manufacturing industry make use of the energy industry involved in transforming energy into work (e.g., electrical or mechanical type) necessary for the operation of production plants.

### 3.1.2 Classification of Material Processing

It can be described, with a diagram similar to the one shown in Fig. 3.2, how the material or materials of origin (without shape or random shape, raw or semifinished) are transformed by suitable procedures, treatments, and processing into a finished object (product or component).

This scheme is more suitable for the processing of metallic materials but can be generalized to polymeric materials as well, considering, for instance, thermoplastics or thermosetting polymers, and processing techniques typical for polymers.

This classification refers to the concept of material cohesion, meaning:

- an assembly of submicroscopic particles (e.g., atoms or molecules);
- an assembly of microscopic particles (e.g., grains in a metallic or ceramic material);
- an assembly of macroscopic components joined or connected to each other (e.g., matrix-reinforcement of a composite material).

**137**

**FIG. 3.1** Path for obtaining finished products starting from raw materials.

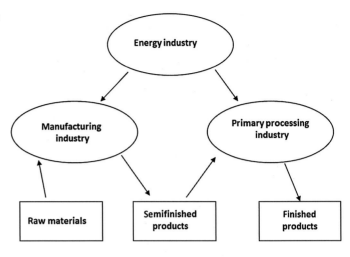

## 3.2 POLYMERIC MATERIALS (PLASTICS)

### 3.2.1 Traditional Technologies

The processes used to transform granules, pellets, and tablets (preshaped forms) into finished products such as sheets, bars, extruded objects, tubes, or molded parts are many and different. The process used depends to a certain extent on whether it is a thermoplastic or a thermosetting material. In general, *thermoplastics* are heated until softened and then molded before cooling. Instead, for *thermosetting* materials, which are not completely polymerized before being molded into the final shape, a process is adopted by which a chemical reaction takes place so that the polymer chains are bonded together to form a three-dimensional network (i.e., cross-linked material). Cross-linking can be achieved by heat

**FIG. 3.2** Classification of material processing.

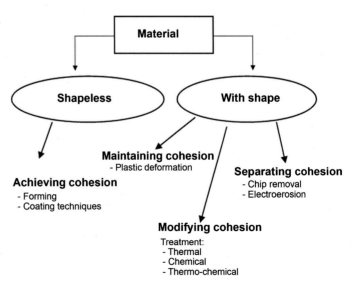

and pressure, or by action of a catalyst at room temperature or at higher temperatures.

This section will describe some of the most important processes used for thermoplastic and thermosetting materials.

### 3.2.1.1 Compression Molding and Transfer Molding

One of the most widely used methods for molding plastics is *compression molding*, through which the thermoplastic polymeric material is introduced in the form of powder, flakes, granules, or pellets, into the cavity of an appropriate mold and mechanically compressed by the relative counter-mold (Fig. 3.3). The mold, generally made of stainless steel, is heated by means of gas, steam, or by electrical resistors, sometimes even with high frequency or infrared systems. In the mold, the polymeric material softens and, under the action of pressure, completely fills the cavity, assuming the desired shape. Afterward, the mold is cooled and opened, and the object is extracted.

Many thermosetting resins are processed in solid forms by compression molding. With this method, the uncross-linked components of the resin (starting compounds such as oligomers,

monomers, prepolymers, additives, and various catalysts) are introduced into the heated mold, containing one or more cavities, depending on the number of objects to be obtained at the same time. The applied pressure and heat cause the mixture to liquefy, pushing it into the cavity. Continuing the heating process (usually 1 or 2 min), the complete cross-linking of the thermosetting resin is obtained. The piece is then ejected from the mold. The flashes are removed afterward from the piece.

A variant to the compression molding process is that of *transfer molding*, in which the mold is closed in advance, the material is fluidized into a separate cavity, then introduced into the mold by the action of a piston. In this way, precisions higher than with the simple heating and compression molding are achieved in a single phase.

All operations in modern equipment occur automatically, one after the other, according to continuous cycles. The detachment and expulsion of the objects from the mold is generally managed by the introduction of special bars, and it is often facilitated by introducing, in the molding step, particular additives known as *detaching agents*.

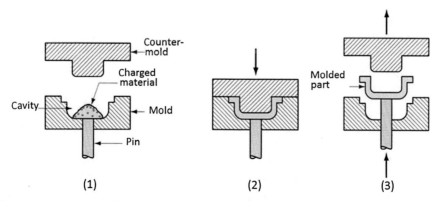

**FIG. 3.3** Schematic representation of compression molding: (1) mold and counter-mold opened, the material is introduced into the cavity; (2) the mold is closed, heat and pressure are applied, and the material melts and fills the cavity; (3) the mold is cooled and opened, and the molded part is ejected.

A. INTRODUCTION TO MATERIALS

### 3.2.1.2 Casting

In some cases, molding is carried out without the use of pressure, that is, by simply pouring the molten material into a mold (Fig. 3.4). In the case of a thermosetting resin, once the mold is completely filled, the same mold is transferred to an oven where cross-linking will take place.

Advantages of casting over different processes such as injection molding include: (a) the mold is simpler and less costly, (b) the cast piece is comparatively free of residual stresses and viscoelastic memory, and (c) the process is suitable for low production quantities.

### 3.2.1.3 Extrusion

Extrusion is one of the essential shaping processes for thermoplastic polymers and elastomers, as well as for metals and ceramics.

**FIG. 3.4** Example of processing by casting: a machined pattern is reproduced with silicone. *Courtesy of Prof. Michal Zalewski.*

Extrusion is a process in which the material is forced to flow through a die orifice to provide long continuous products with cross-sectional shape determined by the shape of the die. As a shaping process, it is extensively used to produce tubing, pipes, sheet and film, continuous filaments, and all kind of shapes. For these types of products, extrusion is carried out as a continuous process; the extrudate (i.e., the extruded product) is subsequently cut into desired lengths.

In the extrusion process, the thermoplastic resin is introduced through a hopper into a heated cylinder (i.e., the barrel with the screw), then the molten plastic material is pushed by a rotating screw through one or more openings in a die of precise shape. A schematic representaion of an extruder is provided in Fig. 3.5.

When exiting the die, the extruded piece has to be cooled below its glass transition temperature to ensure dimensional stability. Cooling is usually achieved with an air-blowing or water-cooling system.

The extrusion machine is also used for the production of raw plastic forms, such as pellets and granules, and for the recovery of waste thermoplastic materials.

Extrusion lends itself to the covering of wires, metal sheets, sheets of paper, and other materials with synthetic resins for *coextrusion*, by extruding together both the support and the covering material.

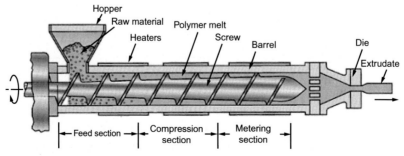

**FIG. 3.5** Schematic representation of an extruder and its various functional parts.

### 3.2.1.4 Injection Molding

Injection molding is one of the most used processing methods for forming thermoplastic materials. Modern injection molding machines use a reciprocating screw mechanism that allows the plastic to be melted and injected into a mold (Fig. 3.6). The rotation of the screw forces the molten polymer and allows the homogeneous fusion of the substance to be injected. The mold may contain more than one cavity (i.e., depending on the number of replicates to be obtained), so that several objects are produced each cycle. Complex and sophisticated shapes are possible with injection molding: the challenge in these cases is to fabricate a mold whose cavity is the same geometry as the part and that also allows for part removal.

### 3.2.1.5 Blow Molding

Blow molding is used for the production of bottles or other empty objects in one piece with thermoplastic materials. This process consists in the hot expansion of a hollow portion of thermoplastic material (e.g., a tube) by means of air pressure, until it adheres to the walls of a mold, from which, after cooling, it is extracted by opening the mold itself (Fig. 3.7).

The semicast hollow plastic material can be introduced into the mold by means of an injector (injection blow molding) or by extrusion (extrusion blow molding).

### 3.2.1.6 Thermoforming

In thermoforming (Fig. 3.8), a sheet of heated plastic is forced to press against the walls of a mold. Mechanical pressure, vacuum, or compressed air can be used to drag the heated plastic sheet into an open mold.

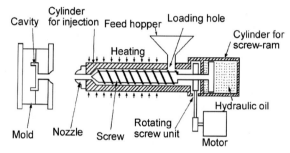

**FIG. 3.6**  Diagram of an injection molding machine, reciprocating screw type.

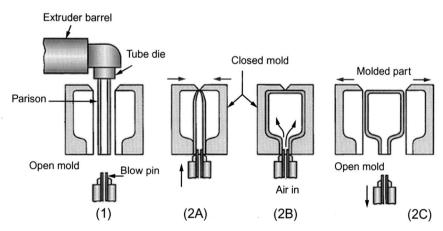

**FIG. 3.7**  Blow molding is done in two steps: (1) fabrication of a starting tube of molten plastic, called a parison (same as in glass-blowing); and (2A–C) inflation of the tube to the desired final shape. Forming the parison is done by either extrusion or injection molding.

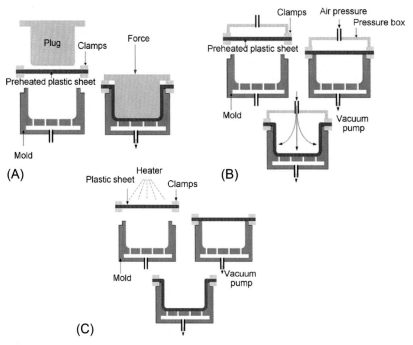

**FIG. 3.8** Schematic of thermoforming process via (A) mechanical pressure, (B) air pressure, and (C) vacuum (http://www.substech.com/dokuwiki/doku.php?id=thermoforming).

### 3.2.1.7 Calendering or Rolling

Calendering or rolling is the process of smoothing and compressing a thermoplastic material during production by passing a single continuous sheet through a number of pairs of heated rolls to obtain the desired thickness. The rolls in combination are called calenders. Calender rolls are usually constructed of steel with a hardened surface.

Calendering frequently produces thin and continuous sheets. In this type of processing, the material, in the form of a hot plastic mass, is passed through a series of internally heated pairs of calender rolls, with parallel axes rotating in the opposite direction, increasingly closer together. In this way, the advancement of the "sheet" of plastic material is caused, which is gradually made homogeneous and thinned to the desired thickness. Eventually, this sheet is made to adhere to a support made of paper or fabric, thus creating a *composite calendering*. Cut and preheated sheets can then be press-molded or decompressed, making a vacuum in the molds.

Calendering is also widely used in the manufacture of textile fabrics, coated fabrics (Fig. 3.9), and plastic sheeting to provide the desired surface finish and texture.

### 3.2.1.8 Rotational Molding

Rotational molding (rotomolding) uses gravity surrounded by a rotating mold to accomplish a hollow form. It represents an alternative to blow molding for making very large, hollow shapes. It is used mainly for thermoplastic polymers, but applications for thermosets and

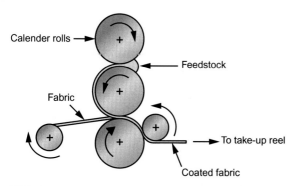

**FIG. 3.9** Coating of a fabric with plastic (rubber) using a calendering process (https://me-mechanicalengineering.com/shaping-processes/).

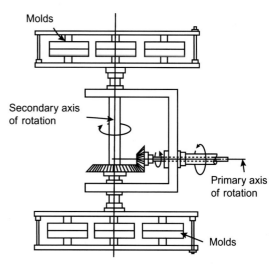

**FIG. 3.10** Mechanism of biaxial rotation in rotational molding.

elastomers are becoming common. Rotomolding allows more difficult external geometries, larger parts, and low production quantities than blow molding. In this process, a fixed amount of polymer powder or pellets is loaded into the cavity of a split mold, then the mold is heated and concurrently rotated on two perpendicular axes, so that the powder impinges on all internal surfaces of the mold, slowly forming a fused layer of uniform thickness. While still rotating, the mold is cooled so that the plastic skin solidifies. Then the mold is opened, and the part is unloaded.

Rotational speeds used in the process are moderately slow. The uniform coating of the mold surfaces is caused by gravity, not by centrifugal force. Molds in rotational molding are easy to use and cheap compared to injection molding or blow molding, but the production cycle is greatly longer. Therefore rotational molding is often performed on a multicavity machine, such as the three-station machine shown in Fig. 3.10. The products also do not require finishing and are of higher precision than other molding techniques.

### 3.2.1.9 Machining With Machine Tools

By use of the processing technologies previously described, it is possible to obtain semifinished or finished products having a defined geometry.

If it is not possible to directly obtain the object of desired shape and size, a semifinished product must be machined with machine tools (for example, by turning, milling, or drilling; see Section 3.3.4). In general, plastics can be easily processed with these technologies, but the tool geometry (for example, a drill tip with a more or less sharp cutting edge) and the processing speed (high or low) more suitable for the specific polymeric material have to be selected. When drilling a hole on a plastic material such as PMMA, running at high speed implies the risk that the material around the hole gets too hot, softening, and leading to a hole having neither the geometry nor the desired size. Therefore it will then be necessary to choose a lower speed that allows the hole to be made without material degradation.

## 3.2.2 Foams, Fibers, Filament Winding

### 3.2.2.1 Foams and Foamed Plastics

Another particular field of considerable importance in polymer processing is the manufacture of expanded products or foams that have achieved great importance in domestic applications, in thermal and acoustic insulation, and in packaging. These products consist of spongy materials, obtained by incorporating into the starting material, before solidification, bubbles of volatile gases or nonsolvent liquids or chemical products capable of developing them. Foamed plastics present very low apparent densities (up to $10 \, kg/m^3$) and very low thermal and electrical conductivity. Depending on the type of polymer, flexible or rigid products can be obtained. When the cavities (pores) are in communication with one another (i.e., "interconnected"), they can also be used as filter plates or porous baffles, but generally, the cavities (pores) are separated from one another (i.e., "closed porosity"). The dimensions of the pores present in the solidified mass depend on the methods and conditions of preparation.

Using different technologies, the expansion process can be carried out in such a way as to form blocks of material (of very large dimensions), which can be subsequently reduced and shaped, or small-shaped articles. In many cases, the process can be carried out on site.

The polymers used in the production of foams are primarily polystyrene and polyurethanes.

### EXPANDED POLYSTYRENE

Polystyrene for the expansion process is produced in beads of appropriate size, depending on the final density of the article to be obtained.

The starting polymer (as beads), containing about 6%–8% of a volatile blowing agent (e.g., pentane), is introduced from below into a preexpansion chamber, in which the beads are heated to atmospheric pressure (e.g., injecting steam and air) and then under vacuum. This is how the material is expanded: the beads become progressively lighter, rise into the chamber and, once they have reached the upper end, after encountering a sieve, they are conveyed by an air flow from the expansion chamber to bags or barrels where they are left to condition.

The molding operation is carried out by loading the pre-expanded beads into the mold, blowing in steam, allowing the beads to expand into the free spaces, and causing welding between them. The pressure inside the mold, due to steam, air, and blowing agent, is discharged, for example, by applying vacuum. The basic expanded polystyrene (EPS) product is white, although it can be colored otherwise.

Fig. 3.11 illustrates the manufacturing process for expanded polystyrene objects.

### POLYURETHANE FOAMS

Polyurethane foams (PUF) are produced by casting, injection, or extrusion, bringing into contact the various components (i.e., polyols, isocyanates, and catalysts) at the time of use by means of suitable mixing devices. The expansion occurs normally by the action of carbon dioxide, which develops in the reaction between the isocyanates and OH groups, or by the effect of added blowing agents (e.g., methylene chloride).

In Fig. 3.12, an example of PU foam production using water as an expanding reagent, in lab scale, is provided. This type of PU foam was designed with the aim to be used as scaffold for supporting cell growth in tissue engineering, and therefore to present, besides adequate mechanical and biocompatibility properties, an open (interconnected) pore structure (Bertoldi et al., 2010; Tanzi et al., 2003).

### 3.2.2.2 *Technology of Fibers*

To be suitable for the formation of fibers, a polymer must possess a set of properties that allow its use in the foreseeable sectors of application. The two major sectors of fibrous material use are textile and industrial.[1]

---

[1] A *fiber* can be defined as a long, thin strand of material whose length is at least 100 times its cross-sectional dimension. A *filament* is a fiber of continuous length. Fibers can be made of natural or synthetic polymers. Synthetic fibers constitute about 75% of the total fiber market today, polyester being the most important, followed by nylon, acrylics, and rayon. Natural fibers are about 25% of the total produced, with cotton by far the most important staple (wool production is significantly less than cotton).

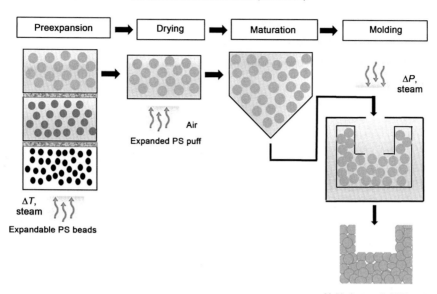

**FIG. 3.11** Process for producing EPS (Expanded PolyStyrene) articles: PS beads are fed to a vertical tank containing a stirrer and a controlled steam input; the expanded beads are then stored in the open air for few hours as a drying stage; during storage they are allowed to reach the ambient temperature; this step is called "maturation," or "stabilization" process; finally, expanded beads are poured in a mold, steam and pressure are applied and beads fusion takes place.

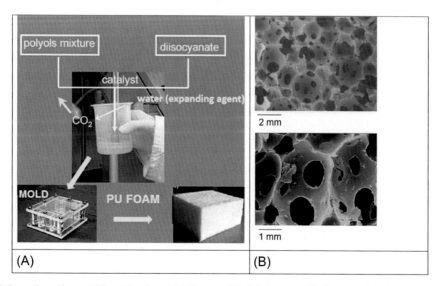

**FIG. 3.12** Polyurethane foam (PU) production at the Biomaterials Laboratory of Politecnico di Milano (Italy) using water as expanding agent. (A) Water is added to the appropriate reagents (polyols, diisocyanate and catalysts) under mixing, and $CO_2$ is produced from the reaction of water with isocyanate groups; (B) SEM images of the obtained PU foam, showing a high fraction of pores with highly open interconnection (BioMatLab, Politecnico di Milano, Italy).

From an application point of view, the fibrous materials are used in the form of continuous filaments, for which high tenacity, modest or low elongation, and high modulus are usually requested (e.g., silk), or in the form of a bow, with a longer or shorter cut, for which higher elongation values are required (e.g., wool- and cotton-type flakes).

## SPINNING TECHNOLOGIES

The process of forming a synthetic fiber basically consists of the extrusion of a fluid thread through the orifice of a spinneret (actual spinning operation) in the solidification of this thread and in the subsequent plastic elongation of the consolidated thread along its longitudinal axis, a certain number of times its initial length (*drawing* process).

Depending on the type of polymer and the required material, three spinning processes can be used: spinning in the molten state (*melt spinning*), spinning of a solution with evaporation of the solvent (*dry spinning*), and spinning of a solution for coagulation in the bath (*wet spinning*), as schematically shown in Fig. 3.13.

*Melt spinning* is used when the starting polymer can best be processed by heating to the molten state and pumping through the spinneret, much in the manner of conventional extrusion. The process is applicable to polymers, which can be maintained above the melting temperature for the generally short time necessary to perform the extrusion. After suitably drying the polymer in the form of granules (chips), a multizone extruder is fed under a nitrogen stream, where the material is melted, transported, and pushed into a spinning head. At the exit (i.e., die), the fluid threads descend vertically and begin to cool. The wire is then collected and wound onto cylindrical reels, then subjected to drawing, by which an orientation of the macromolecules along the fiber axis is achieved, with an increase in crystallinity and an improvement of the mechanical characteristics of the thread itself.

The melt process is used, for instance, for spinning nylon 6, nylon 6-6, polyethylene terephthalate, polyethylene, and polypropylene (http://www.tikp.co.uk/knowledge/ technology/fibre-and-filament-production/ melt-spinning/).

In *dry spinning*, the starting polymer is in solution, and the solvent can be separated by evaporation. The extruded part is pulled through a heated chamber that removes the solvent; besides this, the sequence is similar to melt spinning. By this process, fibers of cellulose acetate and acrylic polymers are produced.

In *wet spinning*, the polymer is also in solution, however, the solvent is nonvolatile. To separate the polymer, the extrudate must be passed through a liquid chemical that coagulates or precipitates the polymer into coherent strands that are then collected onto bobbins.

A technique similar to wet spinning is that leading to the production of *microporous tubes* of variable thickness, studied in the past as small-caliber vascular prostheses. The polymer solution is extruded in a tubular form and coagulated in a coagulation bath (represented by a nonsolvent).

### 3.2.2.3 Filament Winding

This process involves producing tubular forms by winding fibers on a mandrel.

The equipment consists of a rotating spindle regulated by a motor system and a fiber supply unit (Fig. 3.14). The compaction of the wall thickness is guaranteed by the degree of tension applied to the fibers and the orientation of the fibers that can be varied by adjusting the deposition angle. Multilayered shapes can also be obtained, depending on the thickness and mechanical properties to be obtained.

This method of processing is also used to produce tubular forms in composite material (see Section 3.6.1.3). In this case, the fibers are incorporated into a polymeric matrix that increases their cohesion.

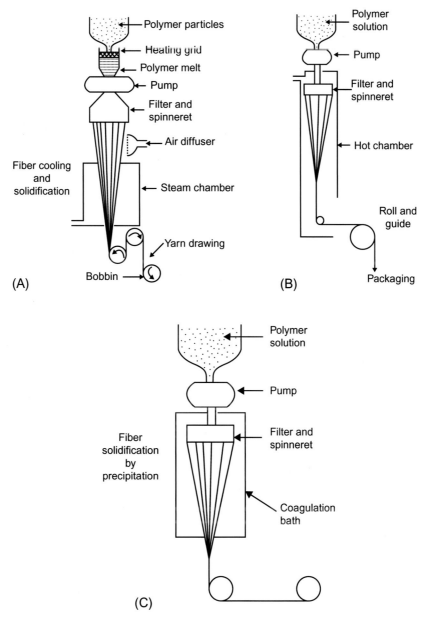

**FIG. 3.13** Schematic of the three main types of fiber spinning: (A) melt spinning, (B) dry spinning, (C) wet spinning. *From Carraher, C.E. Jr., 2002. Polymer News, 27 (3), 91, cited by Chanda, M., Roy, S.K., 2009. In: Plastics Fabrication and Recycling. CRC Press.*

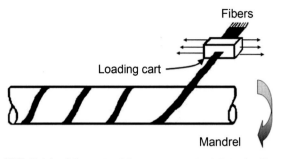

**FIG. 3.14** Schematic of the operating principle in the filament winding process.

## 3.2.3 Forming From Solution

This type of processing is widely used for the construction of devices or part of them for biomedical applications, where, for example, thin thicknesses or particular geometries are required.

The most used techniques involve preparing a more or less concentrated solution of the polymeric material in a suitable solvent and obtaining the object of appropriate geometric shape by subsequent evaporation of the solvent.

Forming from solution includes *solvent casting*, *dip casting*, *solution coating*, and *spinning* of microporous tubular structures.

### 3.2.3.1 Solvent Casting and Solution Coating

These methods are used, for example, to obtain films and membranes of variable thickness (from μm to mm) for deposition or casting of the solution on a smooth or shaped support or mold, and subsequent evaporation of the solvent. Very homogeneous surfaces are obtained, but the surface structure or topography on the support side can be influenced by the support itself (e.g., surface composition and morphology).

For subsequent soakings (*dipping*) in the solution of the polymeric material of a shaped object, it is possible to obtain tubes, bags, and various objects. The concentration of polymer solution, hence its viscosity, may vary according to the choice of polymer and stage of fabrication.

With the technique of casting in a mold, flaps (*leaflets*) of artificial biomimetic heart valves or artificial ventricles (*ventricular chambers*) can be produced.

Polymeric prosthetic heart valve leaflets can be fabricated by use of a specifically designed mandrel, which undergoes repeated cycles of dipping in a polymer solution and subsequent curing in air or in a dry air oven until the desired thickness is attained (Mackay et al., 1996). The usual process of dip coating typically involves multiple dips into a low viscosity polymer solution. The major disadvantage of this method is that it is difficult to accurately control the leaflet thickness distribution. In a proposed technique, the leaflets could be made by just a single dip into a concentrated polymer solution. Accordingly, this would allow more accurate reproducibility and would minimize operator dependency (Mackay et al., 1996).

Recent advancements led to the development of a robotic droplet deposition technique of a polyurethane solution to fabricate the leaflets of aortic and mitral valves (ADIAM life sciences, Fig. 3.15) (Ghanbari et al., 2009).

## 3.2.4 Advanced Technologies

### 3.2.4.1 Electrospinning

Among several techniques developed to obtain nanostructured fibrous structures, *electrospinning* seems to be one of the most promising. The interest in this technique for biomedical purposes was limited to a few research groups until 1995, when the potential of electrospinning in the fabrication of nanofibers and nanofibrous structures was demonstrated using different polymers.

Electrospinning is a spinning process based on electrostatic interactions, which produces long solid nanofibers with a uniform diameter, based on the stretching of polymeric materials from solution or melt. This is made possible by the application of an external electric field that, due to the electrostatic repulsions between the

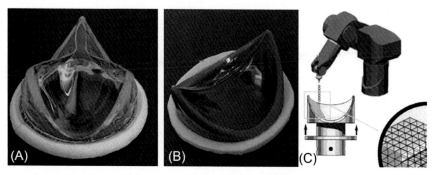

**FIG. 3.15**   (A) Aortic trileaflet and (B) mitral bileaflet polycarbonate urethane valve developed by ADIAM life sciences, Erkelenz, Germany, produced by a robotic droplet deposition technique, schematically shown in (C). *From Bezuidenhout, D., Williams, D.F., Zilla, P., 2015. Biomaterials 36, 6–25.*

charged surfaces (i.e., collector and spinneret) and the evaporation of the solvent, allows a continuous spinning of the polymer (Fig. 3.16).

Different types of polymeric materials, both synthetic and natural, are processed to obtain fibers with diameters ranging from a dozen nm up to a few μm. Thanks to the simplicity and versatility of the technique and the possibility of making changes according to different needs, electrospinning is used in the biomedical field for the production of nanostructured networks, gauzes, membranes, and scaffolds that mimic the morphology of the extracellular matrix of human tissues as much as possible.

**ELECTROSPINNING SET-UP**

A typical electrospinning set-up is composed of three basic elements: a high voltage DC (or AC) power supply, a capillary (including a solution container and a spinneret), and a grounded metal collector (usually aluminum foil). One electrode of the high voltage power supply is connected with the spinneret containing the spinning solution or melt and the other attached to the collector, which is usually grounded, as shown in Fig. 3.16.

How does the electrospinning process transform the fluid solution through a millimetric capillary tube into solid micro/nanofibers, which are four to five orders of magnitude smaller in diameter? Theoretical and experimental studies have demonstrated that the electrospinning process generally consists of three stages:

**(1)** jet initiation and elongation of the charged jet along a straight line;
**(2)** growth of electrical bending instability (also known as *whipping instability*) and further elongation of the jet, which may or may not be accompanied with the jet branching and/ or splitting;
**(3)** solidification of the jet into micro/nanofibers and deposition on collector.

In particular, the polymeric solution is extruded from the spinneret to form a small droplet in the presence of an electric field, and then the charged solution jets are extruded from the cone formed at the tip of the spinneret. Generally, the fluid extension occurs first in uniform straight filament (step 1), and then the straight flow lines undergo vigorous whipping and/or splitting motion due to fluid and electrically driven bending instabilities (step 2). Finally, the continuous as-spun fibers are deposited, commonly as a 2D nonwoven random web on the collector (step 3, Fig. 3.17B). Besides fiber mesh without orientation, as shown in Fig. 3.17C–I, other fiber morphologies and structures (e.g., parallel and crossed fiber arrays,

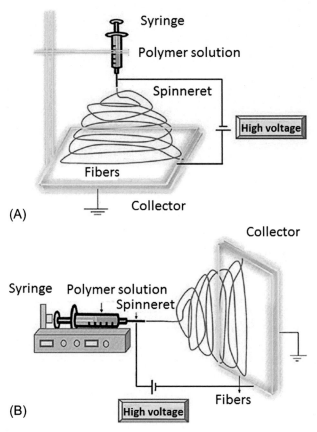

**FIG. 3.16** Schematic diagram of two possible set-ups of the electrospinning apparatus: (A) typical vertical set-up and (B) horizontal set-up. *From Bhardwaj, N., et al., 2010. Biotechnol. Adv.*

helical or wavy fibers, twisted fiber yarns, patterned fiber web, and 3D fibrous stacks) can be also obtained via modified electrospinning process or collectors.

A variety of parameters influence the morphology and diameter of the electrospun fibers, including the intrinsic properties of the solution (e.g., type of polymer, conformation of the polymer chains, viscosity, electrical conductivity, the polarity and surface tension of the solvent), the operating conditions (e.g., strength of applied electric field, distance between spinneret and collector, feeding rate of the polymer solution), as well as the humidity and temperature of the environment (Sun et al., 2014).

## 3.3 METALLIC MATERIALS

The manufacturing processes of metallic materials vary considerably from class to class of material, also depending on the object or component to be obtained.

Most commercial metals ae extracted from their ores by reduction processes in which the ores are reduced to the metallic state and any impurities present are separated and removed as slag. The extracted metal still contains appreciable quantities of dissolved impurities (e.g., oxides, sulphides, and silicates) that have to be removed by refining processes. Metals are subjected to a series of fabricating processes

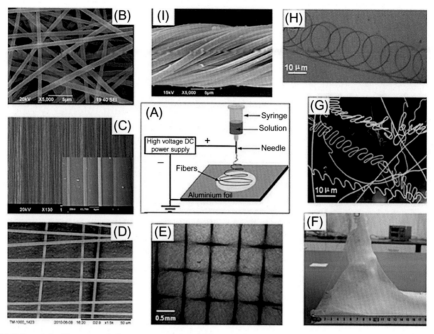

**FIG. 3.17** (A) Schematic diagram of electrospinning method, (B) electrospun polymer nanofiber mesh without orientation, (C) parallel and (D) crossed fiber array, (E) patterned fiber web, (F) 3D fibrous stack, (G) wavy and (H) helical fibers, and (I) twisted fiber yarns. *From Sun, B., et al., 2014. Prog. Polym. Sci.*

that involve fundamental steps such as casting and plastic deformation. Lastly, the products are subjected to various heat treatment processes and finishing operations such as cutting, polishing, burnishing, and coating.

The manufacturing process to obtain a specific object or component can be carried out in different ways and sequences from case to case. Some common lines can however be schematically identified, taking into account the scheme reported in Fig. 3.2:

- casting and powder metallurgy (creation of cohesion),
- processing by hot or cold plastic deformation (maintaining cohesion),
- machining with machine tools (separation of cohesion),
- heat treatments, chemical, and thermochemical treatments (modification of cohesion),

- junction operations between different components,
- surface finishing operations.

It should however be stressed that the choice of the most appropriate processing sequence is related both to the type of material (e.g., hardness, metallurgical structure) and to the type of application (e.g., shape, mechanical characteristics).

Often the selection of the optimal processing technique, or rather of the sequence of techniques, is more important in determining the final cost of the piece than that of the type of metal or metal alloy to be used.

Likewise, the choice of the metallic material with which to obtain an object or a component cannot disregard considerations regarding its workability, often decisive for a correct choice, and the final application.

## 3.3.1 Casting and Powder Metallurgy

Among the processes that allow the achievement of the cohesion of the material (Fig. 3.2), casting is certainly the main one. Powder metallurgy is by now of minor importance in metal processing, for the minor quantities of possible metals or metal alloys to be treated, even though an improvement of its use and development can be highlighted in previous years with the development of additive manufacturing techniques appropriate for metals (see Section 3.7).

The purpose of the casting process is to directly obtain, through a solidification process, pieces of complex geometry, called *castings*, reducing to a minimum the machining processes that would be necessary to obtain the same piece starting from semifinished products of standardized shape. For this purpose, after having melted the chosen metal with or without alloying elements, this is poured into appropriate molds with a shape very similar to that of the final components to be obtained.

The planning of the casting process must take into account the fact that the metal in the liquid state can flow easily throughout the mold, avoiding, for example, that the solidification starts in some areas with a reduced section before the whole mold has been filled. Therefore sections that are too small must be avoided, and the casting channels must be sized appropriately; from this point of view, it has to be taken into account that the different metals are able to flow more or less easily, so the design of the process has to be reconsidered for each material to be casted.

Equally, strong variations in thickness and sharp edges must be avoided to minimize the presence of residual stresses generated by the different cooling speeds of the different parts of the casting or the formation of cracks in correspondence with geometries with acute profiles. Such design precautions are all the more valid the more the material has brittle characteristics, a high thermal expansion coefficient, and lower thermal conductivity (i.e., metals present fewer problems than the other materials).

In the sizing of the molds, it is necessary to oversize them, taking into account the volumetric shrinkage that the casting will undergo during solidification and cooling. Moreover, if the piece must contain internal cavities, removable cores can be inserted in the mold and removed at the end of the solidification, preventing the filling of these cavities.

When obtaining the castings, considerable precautions must be taken to eliminate or at least reduce the formation of internal defects (e.g., micro- and macroholes, concentration of inclusions in particular areas), which can greatly reduce the mechanical characteristics of the casting or even make it useless.

In many cases, accurate nondestructive tests (e.g., X-rays, ultrasounds) must be performed to identify such defects and, if possible, proceed with their elimination, by removing and redepositing material with techniques similar to those of welding.

The castings are then subjected to mechanical surface finishing, of greater or lesser extent depending on the type of casting; further machining operations must be taken into account during the design of the process to give the casting the appropriate dimensions.

Depending on the type of mold and the casting methods, it is possible to classify casting processes as sand castings, shell mold castings, and die castings.

### 3.3.1.1 Sand Casting

The sand casting consists of pouring the metal by gravity into an outer casing made from casting sand that has a very high melting point.

The molds are made by compacting the casting sand around shapes and cores that are removed after casting (Fig. 3.18), in wax forms (lost wax casting), or in expanded polymeric material that is transformed into gaseous products from the liquid metal during the casting.

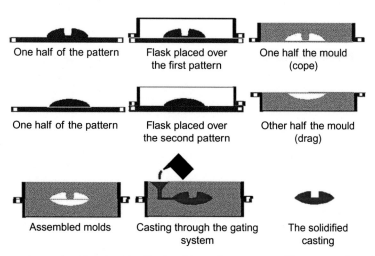

**FIG. 3.18** Basic process for sand casting. A pattern is placed in sand to create a mold; pattern and sand are incorporated in a gating system; the pattern is removed; the mold cavity is filled in with molten metal and the metal is allowed to cool; finally, the sand mold is broken away to remove the casting (https://en.wikipedia.org/wiki/Sand_casting).

Metals with a high melting point (e.g., steel, cast iron, bronze) and also aluminum are processed in this way.

### 3.3.1.2 Shell Mold Casting

Shell mold castings are made by gravity casting the liquid metal, which must have a low melting point (e.g., aluminum) in metal molds made of material with a higher melting point (Fig. 3.19). The separation of the casting from the mold is obtained by the use of appropriate release agents applied on the surface of the mold. Better surface accuracy and finishing are achieved with this process than with sand castings.

### 3.3.1.3 Die Casting

In die castings, the liquid metal is inserted into the molds with a certain overpressure; in this way, it is possible to obtain pieces with less internal voids, lower wall thickness compared to the other casting techniques, and characterized by an excellent surface finish. Only some low-melting alloys (primarily based on aluminum, zinc, and magnesium) can be considered for this technique.

### 3.3.1.4 Lost Wax Casting

This process can be summarized in the following steps:

– a wax model of the object to be manufactured is prepared,
– the mold is built around the wax model using a plaster coating,
– through a thermal cycle, the wax is eliminated while the shape of the object remains,
– the molten metal is then poured into the casting mold.

A rough object is obtained, which is then machined with appropriate machine tools to obtain final dimensions and surface roughness. The processing cycle is illustrated schematically in Fig. 3.20.

## 3.3.2 Powder Metallurgy

In powder metallurgy, the metal or metal alloy is reduced to particles of controlled size by physical, chemical, or electrochemical processes. The process called "atomization" is among the most widespread: the liquid metal, crushed by gas

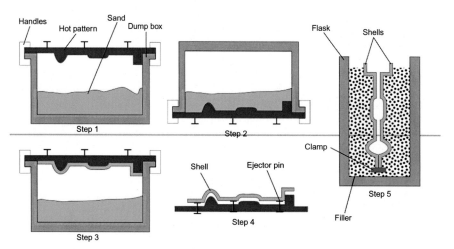

**FIG. 3.19** Steps for shell mold casting: 1. A heated model of the finished cast product is mounted on a heat-resistant box filled with a special resin-bonded sand; 2. The box is turned upside down so that the sand falling on the model due to the result of the heating, sticks and form a sandy shell with the same pattern as the model; 3. The box is turned back up so the excess sand will drain; 4. The hardened shell is then stripped from the pattern; 5. Two or more shells are then combined, via clamping or gluing using a thermoset adhesive, to form a mold. This finished mold can then be used immediately or stored almost indefinitely; 6. (not shown) For casting, the shell mold is placed inside a flask and surrounded with pellets, sand, or gravel to reinforce the shell (https://en.wikipedia.org/wiki/Shell_molding).

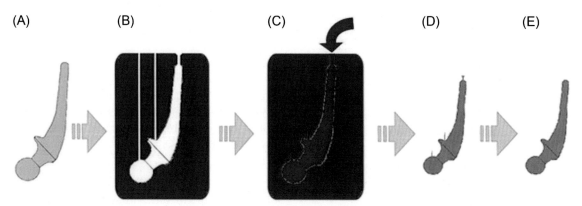

**FIG. 3.20** Processing steps for lost wax casting: (A) wax model of the object to be fabricated, (B) the mold (i.e., a shell) is built around this model by use of a plaster coating, then the wax is eliminated by a thermal cycle, (C) the molten metal is poured into the empty mold, (D) a rough object is obtained, that (E) is finished by machining with appropriate tools.

and/or steam insufflation from a nozzle, is reduced into droplets that solidify to form the particles with different size distribution depending on the process parameters. The powders, after having been suitably screened and after possible addition of binders, can be compacted by pressure in the desired shape in molds to obtain the piece in the so-called "*green*" state. In this state, because the cohesion of the material has not yet been obtained, the piece is larger than the final shape and can be worked easily but must be handled with care. With a high temperature treatment, the "green" is subsequently sintered to obtain the cohesion of the material.

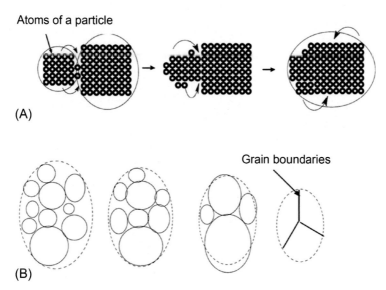

**FIG. 3.21** Steps of a sintering process: (A) diffusion of atoms leading to the bond between two particles; (B) formation of larger grains with volume reduction of the whole piece.

The mechanism of the sintering process consists of the migration of atoms from the surface of one or more particles to the contact zone between the particles with the formation of bridges that bind the particles together (Fig. 3.21A).

The phenomenon is a process of diffusion in the solid state that occurs well below the melting temperature of the compound; the smaller particles (greater global surface energy) increase forming the larger ones (lower overall surface energy), and the particle-particle interstices are gradually filled until the formation of grains having an equilibrium dimension (Fig. 3.21B). This equilibrium is unstable because it depends not only on the temperature but also on the time of the sintering process. The process is promoted by an overall decrease of surface energy and occurs with a decrease in the piece volume.

The powder metallurgy processes produce pieces with complex geometry, with high degrees of surface finish (i.e., very smooth surfaces), and with fewer defects compared to traditional castings. Typical sintered products are metal filters, bearings operating without lubrication (the porosities are preimpregnated with lubricant), gear

wheels, gears, etc. Sintering processes are also used to obtain objects and components made of ceramic materials (advanced ceramics, Section 3.4).

Along with the advantages of casting machining, powder metallurgy has the potential to be less prone to defect formation and to be used with metals and alloys that are hardly castable through traditional techniques. Although currently limited by costs higher than those of traditional techniques, it is a sector of unquestionable future expansion, as reported in Section 3.7.

### 3.3.3 Hot and Cold Plastic Deformation

In most cases, the metallic materials produced in ingots of various shapes must be transformed into semifinished products as close as possible to that of final use.

This is achieved by various plastic deformation[2] techniques, which can be performed under hot or cold conditions.

[2] Plastic deformation changes the dimensions of an object without causing failure.

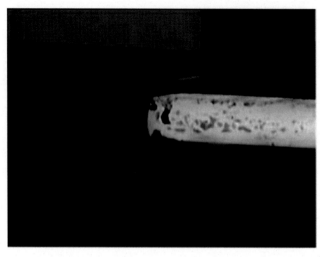

**FIG. 3.22**    Example of oxide on the surface of a component processed by hot plastic deformation.

*Hot plastic deformation* procedures allow greater variations in shape, as the material is more deformable at high temperature. Furthermore, they do not cause work hardening, because the recrystallization processes cancel the work hardening effect. Obviously, however, the surface of the material after hot deformation is strongly oxidized and requires further treatment before final use (Fig. 3.22).

*Cold plastic deformation* procedures, carried out at room temperature, require higher specific powers of the plants and allow minor variations in the shape of the material due to the hardening that is determined. The work hardening leads to an increase in the mechanical characteristics. The surface finish of the cold parts can also be directly the final one, without any further finishing operations.

Among the various plastic deformation techniques, rolling, forging, drawing, and extrusion can be mentioned (Fig. 3.23).

### 3.3.3.1 *Rolling*

In rolling, the parts to be deformed are introduced between a series of two suitably spaced rolls (Fig. 3.21A). A single rolling stand can be used when the piece to be laminated is subject to reciprocating motion or a series of rolling stands within which the piece to be laminated moves in one direction to obtain the desired thickness. In addition to sheets, pipes or bars with a large diameter can also be produced by use of shaped rollers and suitable punches.

### 3.3.3.2 *Forging*

In forging, the material is forced to assume a given shape by the action of a hammer (which causes a dynamic impact) or a press (which determines progressive compression). In this way, pieces of complex shape are obtained, with mechanical properties often clearly better than pieces of the same shape obtained by casting.

Forming operations can be classified by *open mold* forging, to make large components by flat molds (Fig. 3.23B) or with simple shapes (round or shaped), and *closed mold* forging (Fig. 3.23C), performed on pieces of small dimensions between two molds that reproduce the shape of the piece, also using a sequence of molds gradually more similar to the final shape.

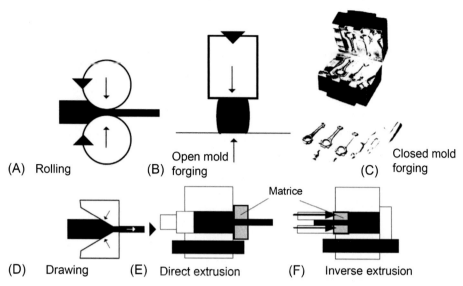

(A) Rolling   (B) Open mold forging   (C) Closed mold forging

Matrice

(D) Drawing   (E) Direct extrusion   (F) Inverse extrusion

**FIG. 3.23** Diagram of plastic deformation operations: (A) rolling, (B) open die forging, (C) closed die forging, (D) drawing, (E) direct extrusion, (F) inverse extrusion.

### 3.3.3.3 *Drawing*

In drawing, bars or metal wires are pulled through one or more drawing dies, made of a very hard ceramic material (for example, tungsten carbide), as shown in Fig. 3.23D. Dies with a progressively smaller diameter are used, until the bars and wires have reached the required dimensions.

Products of small size can be obtained, which, in the case of cold drawing, can result highly work hardened products (due to dislocation movement).

### 3.3.3.4 *Extrusion*

By extrusion, the material compressed with high pressure is passed through a die of suitable shape; it is possible to obtain bars and tubes, and also profiles with a complex cross-section with the most deformable metals such as aluminum.

In *direct extrusion* (Fig. 3.23E), a piston pushes the metal directly into the die, exerting strong pressures. In *inverse extrusion*, instead, the piston holds the matrix (Fig. 3.23F), while an extruder head is closed by a plate. The extruded material therefore exits the extruder through the piston. In this way, there are fewer frictions, but it is possible to exert lower pressures, as the piston is hollow.

### 3.3.4 Machining

Machining can be defined as the process of removing material from a workpiece in the form of chips. Machining is necessary when tight tolerances on dimensions and finishes are required.

In general, the piece is designed with a shape as close as possible to the shape of semifinished products available on the market, or easily obtainable to avoid the additional costs of subsequent processing. The semifinished products can reach the final shape and size by separation of the material cohesion by one or more subsequent removal operations. In many cases, to obtain the desired surface finish, part of the

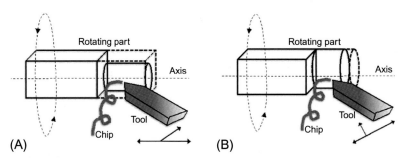

**FIG. 3.24** Tool and workpiece movements in turning machining: (A) movements of the tool are able to reduce the dimensions of the bar along the two directions (longitudinal and radial) and (B) movement of the tool can reduce the length of the bar.

thickness will be appropriately removed or, on the contrary, an extra thickness will be added as a coating. The relative movement between the workpiece and an appropriate tool, which acts as a cutting contact with the piece, allows removal of part of the material (e.g., *chip, particle*). The used tool is equipped with a cutting edge characterized by its shape that can be defined (e.g., drill bit, saw) or undefined (e.g., abrasive powders).

The machining operations for chip removal can be divided into two classes: traditional and nontraditional.

The *traditional ones* include: *turning* (Fig. 3.24), *drilling* (Fig. 3.25), *milling* (Fig. 3.26), *grinding*

(Fig. 3.27), and others such as broaching, planing, filing, choking, reaming, grinding, honing, lapping, blasting, and blanking (Grzesik, 2016). Among the latter ones, lapping and sandblasting are the most used processes.

### 3.3.4.1 Lapping

The abrasive powders are dispersed in a liquid that forms a more or less fluid mixture; the abrasive paste is placed between the tool and the surface to be worked. The relative tool-piece motion allows the abrasive cutting edges to locally deform the material, inducing a strong superficial hardening that allows the break and removal of the work-hardened

**FIG. 3.25** Obtainable shapes (A) and tool movements (B) in drilling machining.

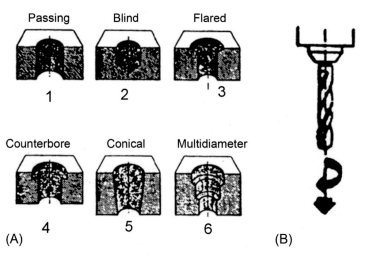

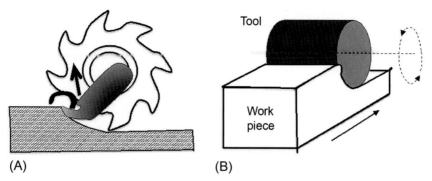

**FIG. 3.26**   (A) Tool and (B) tool and workpiece movements in milling machining.

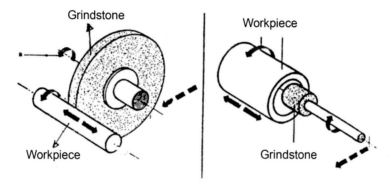

**FIG. 3.27**   Movements of the tool (grindstone) and the workpiece in grinding machining.

material itself. The optimal tool-piece movement must allow the abrasive to act in all directions without any accumulation areas of the abrasive itself (e.g., hollow areas, holes, grooves).

### 3.3.4.2 Sandblasting

This is one of the main processes with sharp edges of an indefinite shape. Abrasive powders are projected at high pressure, by means of a jet of gas (air) or liquid (water) under pressure, on the surface of the piece to be treated with a suitable angle allowing the removal of the surface impurities (dirt, paints, corrosion products, etc.) and the deformation removal of part of the material that constituted the piece. The part of the material involved is that of the outermost

layer, so the surface assumes an opaque and satin-like appearance due to the microrough profile of the surface that reflects light in all directions. In addition to the removal of dirt, the microgeometry of the obtained surface presents an excellent finishing for a subsequent application of coatings because the irregularities of the profile are good mechanical anchors for the coating material. The obtainable roughness depends on the granulometry (i.e., size of the abrasive particles), the surface hardness of the casting, the impact angle of particles with the surface, the type of abrasive, and the treated material.

The *nontraditional processes*, in continuous evolution, include electroerosion, laser cutting, water-jet cutting, and plasma-jet cutting.

### 3.3.4.3 Electroerosion (Electrodischarge, EDM)

EDM (Fig. 3.28) allows the electrical erosion of conductive metallic materials, even if very hard, by sequences of high-frequency electric discharges (20,000–500,000 Hz). These are let loose between the workpiece and the "tool" acting electrode, kept at very close distance from the work surface. The tool is usually constructed with good conductors (e.g., copper, graphite) and has a complementary shape to the cavity to be obtained in the metallic piece. The discharges take place in a circulating dielectric; the part of removed material is vaporized under the action of the discharges. The electrode consumption is very limited. This technology is widely used in the processing of molds for the economic advantages due to its complete automaticity.

### 3.3.4.4 Laser Cutting

Laser processing (i.e., laser is the acronym of amplification of light by stimulated emission of radiation) is mainly used for cutting and welding. This is an innovative technology, because extremely high amounts of thermal energy are used (there are thermal fluxes of the order of $10^8$–$10^9$ W/cm$^2$). This is a very complex process and is not described in this chapter.

### 3.3.4.5 Water-Jet Cutting

When working with a liquid jet, a jet of water is sent to the working area under high pressure, which, by impacting the material, causes its erosion. The disadvantage of this processing is that it allows working only with composite materials for small thicknesses. Therefore water-jet processing is combined with the abrasive one, which can be also used in the case of metallic materials. In this case, the water acts as a carrier for the abrasive particles. This type of processing is able to work up to 100 mm in thickness of aluminum, determining a practically invaluable surface roughness (it is almost a superficial lapping). This technology is interesting because, although expensive, produces finished products that do not require further processing.

### 3.3.4.6 Plasma-Jet Cutting

For plasma processing, the machine consists of a container of a gas (helium or nitrogen) in which a very high voltage is created so that the gas is ionized (as in neon lamps). Because there is a pressurized gas inlet, the newly formed plasma (high temperature ions) is pushed out through the nozzle and carried to the workpiece where it discharges the energy that has just been supplied, raising the temperature of the piece.

## 3.3.5 Junction Operations

Often an object is an assembly of several components joined together by different techniques and systems. The various components may consist of different materials (e.g., because each of them must respond to different functions and properties) or even of the same material. In the latter case, the object is not produced in a single piece due to technical difficulties in obtaining and/or for economic reasons that make it more advantageous to produce the object in individual components despite the additional costs of joining.

Joints, riveting, bolting, welding, and bonding are among the most common joining techniques (Messler, 2004). In particular, it is

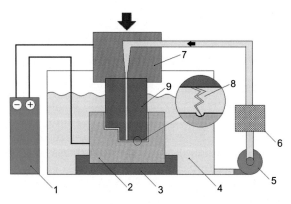

**FIG. 3.28** Schematic of an electric discharge machining system. 1. Pulse generator (DC); 2. Workpiece; 3. Fixture; 4. dielectric fluid; 5. Pump; 6. Filter; 7. Tool holder; 8. Spark; 9. Tool (https://en.wikipedia.org/wiki/Electrical_dis charge_machining).

worth noting that the relatively recent development of polymeric-based adhesives provides the designer with a comfortable and versatile tool for making junctions by gluing.

## 3.3.6 Surface Finishing Operations

Before realizing the surface finish of a piece, it is necessary to carry out appropriate preparation operations of the surface. Among the most used techniques, we can mention:

- sandblasting,
- pickling,
- degreasing,
- electropolishing.

The main proper finishing operations are:

- electrochemical or chemical passivation treatments,
- preparation of metallic coatings by immersion, spraying of molten metal, or galvanizing techniques,
- painting,
- production of ceramic coatings by spraying (plasma spray),
- production of coatings of oxides, carbides or nitrides by physical vapor deposition (PVD) or chemical vapor deposition (CVD),
- increasing the surface content of particular metallic or nonmetallic ions (nitrogen, oxygen) by ion implantation.

The following processes are among the most used in the production of medical devices:

### 3.3.6.1 Plasma Spray

This technique allows the production of thick, dense, or porous metal and/or ceramic coatings using powders as starting materials. The metal or ceramic powders are sprayed by high temperature ionized gas in the plasma phase. The powders in the plasma partially melt, are accelerated, and impact at high speed against the target (i.e., material to be treated). There are two different types of plasma spray: air plasma spray (APS) and vacuum plasma spray (VPS). In particular,

using the VPS technique, the deposition takes place in a controlled atmosphere and is indicated for powders that have a high reactivity with the air, avoiding possible oxidation of the powder particles. VPS is used for the deposition of hydroxyapatite on components of joint prostheses to promote/improve osseointegration.

### 3.3.6.2 Physical Vapor Deposition

Physical vapor deposition (PVD) obtains a thin coating (of a few μm) on a substrate, by condensing a vaporized element or compound. Particularly, titanium nitride deposits are important in the orthopedic and dental fields; examples of possible applications are:

- joint prostheses (reduction of friction and wear), acetabular stems and cups,
- screws (increased biocompatibility),
- surgical instruments (increased durability and cutting efficiency, e.g., of drill bits used in orthopedics).

### 3.3.6.3 Chemical Vapor Deposition

By chemical vapor deposition (CVD), thin films can be deposited on heated substrates starting from compounds in the gaseous phase. Variants of the CVD technique retain the substrate to be treated at a low temperature. The main applications of CVD coatings are:

- reduction of friction coefficient, increase of wear and oxidation resistance, corrosion protection;
- increased biocompatibility.

### 3.3.6.4 Ion Implantation

This technique consists of bombarding the surface of metallic and polymeric materials with ions of different species, suitably accelerated, creating a surface diffusion layer, characterized by high hardness or other specific properties. The technique allows the substrate to be kept at a low temperature during the surface treatment. In the biomedical field, it is used for the following purposes:

- increasing the wear resistance of metals and polymers,
- decreasing the bacterial adhesion, for example, on catheters or nails of external fixators (i.e., Ag ions implantation),
- modifying the superficial properties of polymers (e.g., antithrombogenicity),
- accurately and directly placing radiopaque markers on implantable medical devices.

## 3.3.7 Nondestructive Tests

Nondestructive tests (NDT), also named non-destructive examination (NDE), nondestructive inspection (NDI), and nondestructive evaluation (NDE), are a set of physical tests designed to identify the possible presence of defects in a piece without making it unusable. NDT have assumed great importance in modern industry for various reasons; the primary ones are examined in the following text.

*Prevention of accidents and misfortunes.* The identification by NDT of defects present in a structure due to the initial quality of the materials, or generated during the construction of the structure, or finally created during the exercise, allows us to avoid the defects, if they are of critical dimensions, to determine the sudden collapse of the structure.

*Product quality guarantee.* This aspect concerns the commercial value of the piece and the duration (i.e., lifetime) that a mechanical construction, in a general sense, must have. In fact, the user requires a certain number of years of trouble-free operation. This can only be guaranteed by perfectly knowing the state of the parts that make up the structure. For example, if the designer sizes a machine organ obtained by casting based on the mechanical properties of the compact material, but there is an occult casting defect (e.g., an inclusion of ground), the cross-section area may not be sufficient. In this case, an appropriate NDT identifies and evaluates the defect and decides, with good

reason, the acceptance of the piece or if it has to be repaired, by removing the inclusions and filling in with welding, or it has to be discarded.

*Design improvement.* The NDTs help the designer, because they reduce the safety coefficients, decreasing the thicknesses and therefore the weight of the pieces without decreasing the performances. Moreover, when NDTs are carried out on test pieces, for example, welding or jets, it is possible to systematically identify defective zones and therefore to modify the design so to eliminate them.

*Uniformity of production quality.* This issue is of great importance, particularly in mass production, because similar parts need to perform similar performances. When the level of quality that a certain piece must have is fixed, based on considerations of importance and safety, the NDTs may help to obtain and maintain this standard within the established limits.

*Periodic monitoring in exercise.* On equipment, even if initially free of defects, failure may arise due to factors such as fatigue, mechanical damage, generalized or localized corrosion, or erosion that can lead to breakage, shutdown, or in any case necessary intervention services or replacement. Only a careful and periodic NDT campaign in the zones most affected by these problems can allow intervention in time, without resorting to premature substitutions, but, above all, without taking the risk of failures in exercise, which often involve indirect costs greatly higher than those of a replacement of the component.

*Reduction of manufacturing costs.* NDTs identify the defective piece at the first stages of manufacture, before carrying out long and expensive mechanical machining operations. For example, in forging, it is very important to know the size and distribution of defects contained in the starting ingots, because based on these evaluations, it is possible to choose the ingot suitable for obtaining the different forgings.

Furthermore, it has to be noted that when high-risk components are involved, such as medical devices and prostheses where sudden failures are not acceptable, the risks associated with failure during operation make it necessary to carry out careful controls, even if expensive.

The most widespread NDTs of greater application importance are:

- fluorescent penetrant inspection
- liquid penetrant inspection
- magnetic particle inspection
- ultrasounds,
- X-ray and film processing,
- online and offline hydrostatic testing
- positive material identification
- laser scanning for dimensional compliance inspection.

## 3.3.8 Manufacturing Steps of a Metallic Prosthetic Implant

Understanding the structure and properties of metal materials requires the assessment of the metallurgical significance of the material process history. Because each metal device is different in the details of its manufacture, the general stages of the process are shown in Fig. 3.29.

*Step 1: From the ore containing the metal to the raw metallic product.*

With the exception of noble metals (which do not represent the majority of metals used for medical devices), metals exist on the Earth's crust in the form of minerals and are chemically combined with other elements, as in the case of metal oxides. The mineral deposits (i.e., ore) must be identified, extracted, then separated and enriched to provide a suitable ore to obtain a pure metal through further processes.

Depending on the degree of purity desired in the final product of the specific metal, it is still necessary to refine ore using vacuum furnaces, recasting, and additional stages. All of these processes can be critical in the production of metallic material with appropriate properties.

The resulting metallic crude product is obtained in some types of forms such as ingots for supplying the manufacturers.

In the case of metal alloys for prosthetic implants, the raw metal product must be further processed. Process steps include remelting, adding alloying elements, and solidifying to produce an alloy that meets certain chemical specifications. As an example, to obtain 316L stainless steel in accordance with the ASTM (American Society for Testing and Materials) F138 standard practice, iron is alloyed with certain amounts of carbon, silicon, nickel, and chromium. To obtain an alloy specified in ASTM F75 or F90 standards, cobalt is alloyed with defined amounts of chromium, molybdenum, carbon, nickel, and other elements.

*Step 2: From the metallic raw product to metal forms for stock.*

The manufacturer then processes the raw product (metal or alloy) to obtain "stock" forms such as bars, wires, plates, rods, sheets, tubes, and powders. These forms are then sold to specialized companies (for example, manufacturers of prosthetic implants) that require metal in stock that is as similar as possible to the final shape of the implant.

Bulk forms are transformed into stock forms through a variety of processes, including continuous casting, hot rolling, forging, and cold drawing. Depending on the metal, there may also be heat treatment stages (heating and cooling cycle) designed to facilitate subsequent processing or forming of the stock, to reduce the effects of previous plastic deformations (e.g., by annealing) or to produce a specific microstructure and some properties of the material in stock. Due to the chemical reactivity of some metals at elevated temperatures, high temperature processes may require vacuum conditions or inert atmospheres to prevent unwanted oxygen uptake by the metal. For example, in the production of fine powders of the CoCrMo alloy as prescribed by ASTM F75 standard practice, the molten metal is ejected through a small injector to produce a

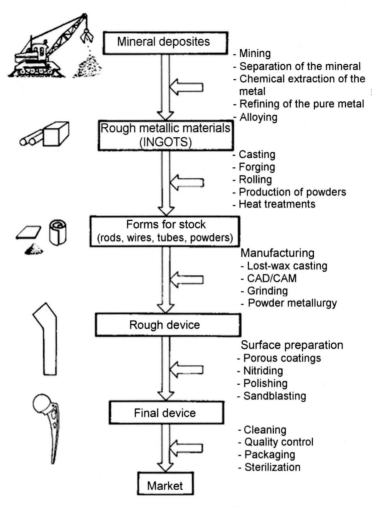

**FIG. 3.29**  Operating steps for the manufacture of a joint prosthesis (femoral stem).

fine spray of atomized droplets that solidify while cooling in an inert atmosphere of argon.

In the case of metals for medical implants, in general, the stock shapes are chemically and metallurgically tested to ensure that the chemical composition and the metal microstructure meet the industrial standards for surgical implants (ASTM standards).

*Step 3: From metallic forms in stock to preliminary and final prosthetic metal implants.*

Typically, a prosthetic implant manufacturer will buy the material in stock and then turn it into preliminary or final shapes. The specific steps depend on a number of factors, which includes the final implant geometry, the forming and machining properties of the metal, the costs of alternative manufacturing methods, and the company that has to perform the processing.

Manufacturing methods include lost wax industrial casting (the so-called "lost wax process"), conventional mechanical processing or computer-based (CAD/CAM), forging, powder metallurgy processes (e.g., hot isostatic pressing, HIP), and different grinding and polishing stages. A variety of manufacturing methods are required because not all alloys for medical implants can be processed in the same way, both in terms of feasibility and in economic terms. For example, cobalt-based alloys are extremely difficult to machine with machine tools in complicated shapes and are often formed in the shape of the medical device by the lost wax process or by powder metallurgy. On the other hand, titanium is relatively difficult to melt and therefore is frequently worked with machine tools even if it is not considered an easily machinable metal.

Another aspect of manufacturing is to provide a surface treatment on the finished product, such as the application of macro- and microporous coatings on the implants. This treatment has become very important in recent years as a means of facilitating implant fixation to the bone (i.e., osseointegration). Porous coatings can take different shapes and require different manufacturing technologies.

Finally, the metallic device is subjected to finishing stages. These vary with the material and processing but typically include chemical cleaning and passivation in an appropriate acid, or controlled electrolyte treatments to remove machining chips or impurities due to mechanical processing. Usually, these steps are performed according to ASTM specifications for the cleaning and finishing of medical devices. Furthermore, these stages can be extremely important for the overall behavior of the implant in the biological environment.

# 3.4 CERAMIC MATERIALS (ADVANCED)

The processing techniques used for ceramic materials are different from those used for metallic and polymeric materials. The main problem is due to the high melting point of ceramic materials (for advanced ones, the melting temperature is around 2000–4000°C, compared to that of steel, which is around 1000°C and that of polymeric materials that is rarely above 300°C).

Furthermore, due to their high hardness and fragility, the workability at room temperature of ceramic materials is limited (e.g., they are not plastically deformable at room temperature; therefore molding cannot be performed).

Processing is possible by machining (e.g., turning, milling, drilling) but only for surface finishing operations, because extremely expensive diamond wheels must be used.

Fig. 3.30 provides a schematic picture for a generic processing path for ceramic products, whereas Fig. 3.31 shows the different steps for advanced ceramics manufacturing.

## 3.4.1 Ceramic Powders

The production of ceramic components requires the use of powders of high purity and uniform particle size. They can be obtained from natural raw materials through a series of chemical purification treatments or through the synthesis of chemical precursors.

An ideal powder should have strictly controlled chemical composition and purity, micrometric or submicrometric particle size, absence of agglomerates, narrow distribution, and centered on the mean value with spherical morphology.

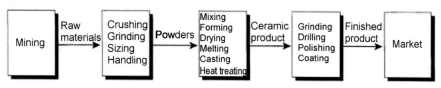

**FIG. 3.30** Steps in a generic processing of ceramic materials.

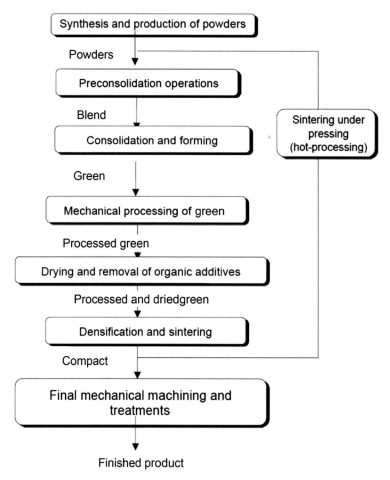

FIG. 3.31   Advanced ceramics manufacturing.

There are various methods of production of powders, through mechanical (i.e., grinding) or chemical processes (e.g., sol-gel, coprecipitation, reaction in plasma).

Because the quality of the ceramic powder is a critical factor in the production of ceramics, some information on their synthesis and production is provided in the following text.

### 3.4.1.1 Production by Solid-State Reaction

This method is widely used: raw materials are mixed and treated at high temperature. During this process, the new phases are formed by reaction in the solid state, during which the transport of matter takes place by diffusion. The product is ground, and it is not unusual for the cooking and grinding to be repeated several times to ensure a complete and homogeneous reaction.

Among the substances produced by this method, mixed oxides and silicon carbide (SiC) can be mentioned.

A variant of the described system may be that of the melting of the reacting powders. In this case, they are exposed to a high-intensity electric arc, which reaches 6000°C. Following cooling, new crystalline phases are formed, which are treated with a method very similar to the previous one. In practice, this technology is applied to electrofused refractories, with which it is possible to obtain oxides (e.g., $Al_2O_3$, $ZrO_2$) and their combinations, and carbides (e.g., SiC, TiC, and $B_4C$).

### 3.4.1.2 Production by Thermal Decomposition

Ceramic oxides can be produced by heat treatment at temperatures equal to or slightly higher than the thermal decomposition of the respective carbonates, nitrites, acetates, oxalates, etc. according to the scheme:

$$A\,(s) = B\,(s) + C\,(g)$$

where A, B, and C represent compounds in the solid (s) or gaseous (g) state. In this way, it is possible to produce $Al_2O_3$, MgO, SiC, etc.

### 3.4.1.3 Production of Powders in Vapor Phase

This technique is becoming increasingly important as it allows the production of nanometric-sized powders, free of agglomerates, often of spherical shape. The main drawbacks are related to the difficulty of producing multicomponent powders with a defined composition and to the fact that they are accompanied by considerable volumes of gas, which require the use of complicated and not always quantitative separation systems, such as mechanical and electrostatic filters.

The reagents can be gases, liquids, or solids that must in any case be brought into the vapor phase; the reaction temperatures can even reach 1300°C. The technique can be divided into three groups: gas-gas, gas-solid reactions, and decomposition in vapor phase. With this technique, powders of MgO, $Al_2O_3$, and $Cr_2O3$, and their combinations are produced industrially.

### 3.4.1.4 Methods in Solution

Normally, these methods offer the advantage of simplicity of preparation even in the case of complex compositions. The main objective is to produce the homogeneity achieved at the atomic or molecular level in the solid state at the time of the solution. Therefore the most delicate step is that of concentration and solvent removal. A general classification of these methods can be made considering the technique used to eliminate the solvent: precipitation-filtration, evaporation of the solvent (spray-drying), freeze-drying, hydrothermal synthesis (i.e., treatment at high temperature and pressure of the reagents, represented by salts, oxides, metal powders in solution or in suspension), or sol-gel process.

Sol-gel chemistry offers some particular advantages, centered on the ability to produce a solid-state material from a chemically homogeneous precursor. It consists of the preparation of inorganic polymers or ceramics from solution through a transformation from liquid precursors to a sol, and finally to a network structure called a "gel".

Furthermore, the sol-gel process offers the possibility to obtain powders of high homogeneity and purity, and to actually synthesize new ceramics. Its versatility prepares dense solids, fibers, thick and thin layers, abrasives, and coatings, in addition to the powders.

A generic sol–gel process can be summarized as follows:

Precursors → Sol → Gel → Xero – gel → Product

Fig. 3.32 gives schematic insight of the sol-gel process to produce dense ceramic products.

## 3.4.2 Forming

In the production process, forming is the stage during which a certain quantity of powder is molded into the desired shape, which after the sintering treatment will be more or less close to the final one. The final object will be refined with finishing operations.

Based on the required performance, the manufacturer chooses the most convenient forming technology.

Because ceramic powders show low cohesion, to obtain a correct forming, they must be

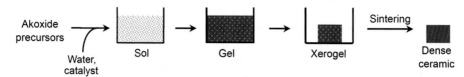

**FIG. 3.32** The sol-gel process leading to dense ceramic products. Starting from precursors, the process can be summarized in the following key steps: (1) synthesis of the *"sol"* from hydrolysis and partial condensation of alkoxides; (2) formation of the *gel* via polycondensation to form metal-oxo-metal or metal-hydroxy-metal bonds; (3) "aging" and drying the gel to form a dense *xerogel* via collapse of the porous network (or an *aerogel* through supercritical drying); (4) sintering to entirely remove porosity.

joined to one or more additives to obtain a piece (i.e., *green*) with mechanical characteristics to allow its handling. This necessity represents a complication and sometimes, to avoid additives that can be a source of defects in the finished piece, a more expensive and challenging forming method is chosen.

The methods of forming advanced ceramics can be divided into *pressing, plastic forming, casting,* and *depositing* on a conveyor belt.

### 3.4.2.1 Forming by Pressing

Ceramic particulate raw materials can be pressed in the dry, plastic, or wet condition into a die to form shaped products.

The pressing of a ceramic powder can be carried out in different ways: cold or hot, by uniaxial or isostatic pressing, or by wet pressing.

### 3.4.2.2 Dry Uniaxial Pressing

Dry pressing may be defined as the simultaneous uniaxial compaction and shaping of a granular powder along with small quantities of water and/or organic binder in a die. Dry pressing is used extensively because it can promptly form a wide variety of shapes with uniform and close tolerances.

The process takes place in three stages: filling, compaction, and extraction (Fig. 3.33).

Common unilateral pressing defects are the lamination and the formation of internal and localized fractures that can cause spontaneous breaking of the piece after extraction and adhesion of the powder to the walls of the mold.

### 3.4.2.3 Cold Isostatic Pressing

The ceramic powder is loaded into an airtight container with flexible walls (e.g., rubber) placed in a chamber filled in with a hydraulic fluid to which pressure is applied. The force of the applied pressure compacts the ceramic particles uniformly in all directions, and the product takes the shape of the flexible container (Fig. 3.34).

Despite the greater workability of the green piece, the defect of the system consists in the fact that the surfaces of the object remain poorly finished, and the production rates are rather limited.

### 3.4.2.4 Hot Uniaxial Pressing

This technique allows densification to be carried out with simultaneous application of pressure and heat. The particular advantages of hot pressing consist in the fact that it is possible to obtain a microstructure with grains even <1 μm. The disadvantages lie in the very high cost and in the possibility of realizing only very simple forms resulting in low productivity.

### 3.4.2.5 Hot Isostatic Pressing

The powder is subjected simultaneously to heat and isostatic pressure obtained by an inert gas (argon, nitrogen). With this process, products that are practically free of porosity are obtained at temperatures lower than those used in free sintering. With this method, it is possible to produce practically finished pieces of even

**Powder pressing**

**Die fill stage**

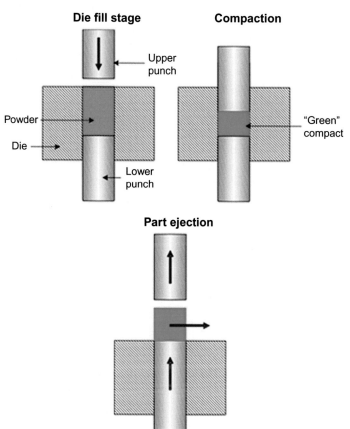

FIG. 3.33   Scheme of the dry (die) uniaxial pressing method. *Courtesy of Substech.com.*

**Compaction**

Upper punch

Powder

Die

Lower punch

"Green" compact

**Part ejection**

complex shape; this reduces the machining costs if surface grinding and finishing, which greatly affect the cost of the other forming techniques, are avoided.

### 3.4.2.6 Wet Pressing

This technique is achieved by pressing a powder suspension and simultaneously removing the liquid. A characteristic of the process is the possibility of obtaining a compact, homogeneously dense green, if the initial powder suspension in the liquid (*slip*) remains constant during the process; in this way, there will be no significant density variations in the final compact material.

The disadvantages of the process are represented by the duration of the cycle, which also includes a drying period, not without risks for the integrity of the piece that, in the absence of an appropriate binder, can be rather brittle.

### 3.4.2.7 Forming of Plastic Material

To use this technique, it is necessary to use powders mixed with relatively large amounts of binders so as to provide the dough with a high

**Cold isostatic pressing**

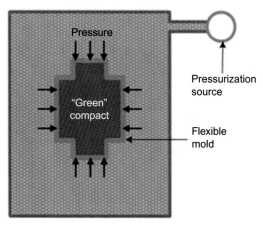

**FIG. 3.34** Scheme of the cold isostatic pressing method. *Courtesy of Substech.com.*

**Injection molding**

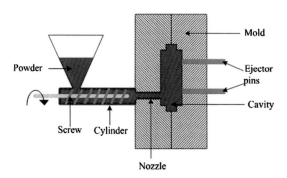

**FIG. 3.35** The fundamental scheme of the injection molding method. Main steps are: mixing the ceramic powder with 30%–40% of a binder (e.g., low melt polymer), injection of the warm powder with molten binder into the mold by means of the screw, removal of the part (green) from the mold after cooling down of the mixture, and debinding (removal of the binder). *Courtesy of Substech.com.*

degree of plasticity. It can be carried out in two ways: by *extrusion* and by *injection*.

Both methods have been already described in Sections 3.2 and 3.3. An example of injection molding for ceramic powders is provided in Fig. 3.35.

### 3.4.2.8 *Forming by Casting and Deposition*

Also in this technique, the powder is mixed with binders and additives to give the mixture specific rheological characteristics: the obtained mass must be much more fluid than in the case of "plastic" molding, and it is very similar to the one used for traditional ceramics. The process is simple, does not require expensive equipment, and allows the creation of complex shapes, even of considerable size.

The deposition techniques can use powders in the form of more or less fluid suspensions or in the dry state. The following procedures can be mentioned: deposition on a continuous conveyor belt (*tape casting*), serigraphy, and spraying.

*Tape casting* produces thick films or ceramic sheets starting from a suspension of the powder in a solvent, binder, plasticizer, and dispersing fluid. The suspension is temporarily deposited on a support (e.g., Teflon, cellulose acetate, glass), dried, cut, and sintered.

*Spray deposition* is often used to cover an object, generally metallic: the system is based on techniques by which a ceramic powder (or a wire) is melted and projected onto a surface. Casting is made with different techniques: by flame, electric arc, or plasma.

### 3.4.2.9 *Green*

The ceramic piece obtained after the forming process is called "green". The material consists of a weakly bound but sufficiently compacted powder.

Density should be constant, without density gradients that produce internal stresses during sintering.

The processing of the green is much cheaper than the processing of the ceramic in its final densified state: it can be machined by machine tools to approximate the final geometry.

### 3.4.3 Sintering

The term "sintering" refers to the process of densification of a powder compact, which includes the removal of porosity between the

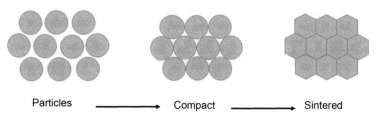

Particles ——————→ Compact ——————→ Sintered

**FIG. 3.36** Schematic representation of solid phase sintering.

starting particles, and the coalescence and formation of strong bonds between adjacent particles (Fig. 3.36).

This process constitutes the central phase of the production of ceramic objects, including traditional ones, and is also used in the metallurgy sector, progressively replacing the casting techniques for the fabrication of complex and small-sized objects.

During sintering, small particles of the ceramic material are consolidated through solid-state diffusion phenomena and, in the manufacture of ceramic materials, this thermal treatment leads to the transformation of a porous product (porosity $\cong$ 50%) into a dense and coherent one.

The particles are agglomerated by diffusion in the solid state at very high temperatures but lower than the melting point of the compound to be sintered. The atomic diffusion occurs between the contact surfaces of the particles that become chemically linked to each other through the formation of "necks" (Fig. 3.37).

At the end of the process, a "grain size equilibrium" is obtained. The motor work of the process is the lowering of the system energy. The high surface energy associated with the individual, original, small particles is replaced by the low total energy of the grain edge surfaces of the sintered product.

The packaging of the grains, which involves an approaching of the grain centers with consequent volumetric shrinkage, occurs in the last step.

To achieve a good densification, it is possible to operate on five variables:

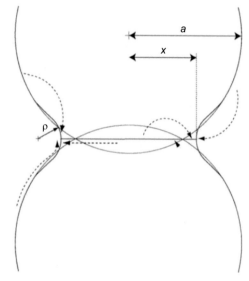

**FIG. 3.37** Formation of a neck between two ceramic particles. The arrows indicate the different sintering mechanisms considered by M.F. Ashby in Acta Metall., 22, 275–289 (1974). *From Goodall, R., et al., 2006. J. Eur. Ceramic Soc. 26, 3487–3497.*

1. temperature;
2. time;
3. initial particle size;
4. chemical composition;
5. pressure.

It can be mentioned that the grain boundaries are discontinuities, so they are stress centers rising (i.e., stress sources), therefore it is important to use powders with a fine particle size.

In summary, the phases of the sintering process can be identified as: (a) creation of necks between the particles, (b) disappearance of the open porosity, (c) disappearance of the closed porosity.

### 3.4.4 Example of Fabrication of a Medical Device: Production of a Femoral Head in Alumina

The processing stages for the fabrication of a femoral head in a joint prosthesis are illustrated in Fig. 3.38. High quality Alumina ($Al_2O_3$) particles are transformed into a green intermediate by cold isostatic pressing, presintered at 800°C, machined for surface finishing, sintered at 1500°C, and finally polished.

### 3.5 MANUFACTURING OF CARBON AND GRAPHITE MATERIALS

As illustrated in Table 3.1, there are a wide variety of materials for technical uses based on carbon (C).

**TABLE 3.1**   Types of Industrially Produced Carbons and Graphite

Commercial carbons and graphites

Pyrolytic carbons and graphites

Carbons deposited in the vapor phase

Glassy carbons

Laminated carbons and graphites

Carbon fibers source: Rayon, Polyacrylonitrile (PAN), pitch

Carbons and graphites in the form of threads, ribbons, braids, fabrics, felts

Composites based on carbon and graphite:

    C/polymer matrix

    C/metal matrix

    C/ceramic matrix

    C/C

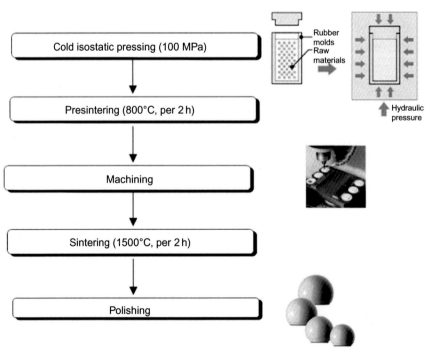

**FIG. 3.38**   Essential processing steps for the production of a femoral head (in $Al_2O_3$) in a joint prosthesis by cold isostatic pressing and sintering.

The different structures of carbon (i.e., graphite, glassy and turbostratic carbon) were already described in Chapter 1, Section 1.4.4. This chapter describes the methods of preparation.

Due to the high temperature of carbon sublimation (>3700°C) and the consequent low atomic diffusion values, the normal methods of powder processing cannot be used for the preparation of pure carbons with useful engineering properties. Processing is predominantly based on techniques that exploit the decomposition of precursors rich in carbon and susceptible to give a carbonaceous residue with the desired properties.

### 3.5.1.1 Pyrolytic Graphite and Isotropic Carbon

They are obtained by pyrolytic decomposition of hydrocarbons, usually methane ($CH_4$) or natural gas, which are continuously flowed at low pressure through the deposition plant on a substrate. The substrate temperature ranges from 800°C to 2800°C. The structure of the deposited layer depends on deposition temperature, flow rate, and hydrocarbon concentration. Impurities can be intentionally introduced into the reaction chamber to improve the characteristics of carbon material.

If the process conditions are such that the deposition occurs molecule on molecule, the maximum anisotropy of the deposit (i.e., *graphitic structure*) is achieved. Very compact materials are obtained that exhibit suitable thermomechanical characteristics, especially for aerospace applications. If, on the other hand, there is supersaturation in the gaseous phase, and deposition occurs almost exclusively by deposition of nucleated and increased crystallites in the vapor phase, dense and almost perfectly isotropic deposits can be obtained, having a *turbostratic* structure, that is, the so-called LTI carbons (*low temperature isotropic*).

The deposition temperature of LTI carbons is usually lower than 1500°C; to increase its mechanical strength, LTI carbon can be codeposited with silicon.

The process of LTI carbon production is illustrated in Fig. 3.39.

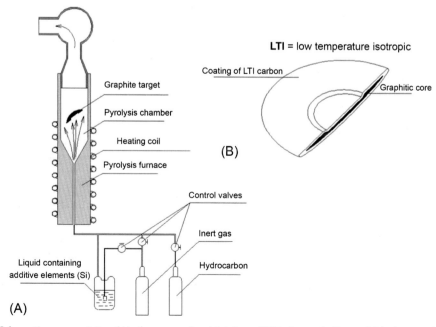

**FIG. 3.39** Schematic representation (A) of a process for obtaining a LTI turbostratic (isotropic) carbon coating onto a thermostable substrate (e.g., graphite). The specific example (B) concerns the production of a mechanical valve cardiac prosthesis disk.

### 3.5.1.2 *Isotropic Carbons Deposited in the Vapor Phase*

There are two types of deposition: physical vapor deposition (PVD) or chemical vapor deposition (CVD), and different production methods are available:

- sublimation or sputtering from a graphitic source;
- evaporation of carbon heated electrically or by electron beam;
- deposition by ionic beam and ionic plating;
- dissociation of hydrocarbons in glow discharge.

In particular, ULTI (ultra low temperature isotropic) carbon is an isotropic form of turbostratic carbon deposited by PVD that can be obtained at low pressure and room temperature with the use of a catalyst (Fig. 3.40) (Haubold et al., 1981). ULTI carbon was developed by Sorin Group Spa with the registered trademark of Carbofilm (http://www.livanova.sorin.com).

This form of carbon is waterproof and has remarkable qualities of elasticity. It is particularly suitable for coating thermolabile substrata, such as polymers and fabrics, and porous metals.

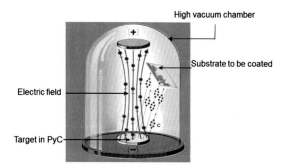

**FIG. 3.40** Representation of the PVD process to create ULTI carbon coatings on a generic thermolabile substrate. During the process, an ionic flow eradicates ions from a pyrolytic turbostratic carbon target, directing them onto the substrate to be covered.

### 3.5.2 Carbon Fibers

The raw material used to make carbon fibers is a precursor represented by polyacrylonitrile (PAN, nearly 90% of the carbon fiber's production), whereas the remaining 10% are made from rayon or petroleum pitch.

An example of carbon fiber production from PAN is shown in Figs. 3.41 and 3.42.

In general, the manufacture of PAN-based carbon fibers takes place in three major steps. The first step is to spin PAN polymer solution to create PAN precursor fibers. The second step is the stabilization (cyclization and oxidation) of PAN precursor fibers through heat treatment ($<300°C$). The third and final step carbonizes fibers (up to 1500°C), removing all noncarbon atoms and creating the final product, that is, the carbon fiber.

With appropriate manufacturing techniques, it is possible to produce a combination of high mechanical strength fibers and high or low modulus (Fig. 3.42). Fibers with mechanical strength up to 5 GPa and an elastic modulus ranging from 200 to 500 GPa are obtainable.

Because several techniques are available to assemble the fibers to form products and composites in a wide variety of configurations and arrangements, the technology based on carbon fibers (monoliths, fibers, coatings, composites) is very versatile and offers interesting opportunities to match the needs of orthopedic implants subjected to complex mechanical stresses with the surrounding bone (Park, 2015).

The field of carbon structures was revitalized by the discovery of *fullerenes* in 1985 (Kroto et al., 1985), carbon *nanotubes* in 1991 (Iijima, 1991), and *graphene* in 2004 (Novoselov et al., 2004) (source: www.cmu.edu/maty/materials/Nanostructured-materials/carbon-nanostructures.html).

Although these advanced structures have acquired increasing interest, they are not described in this book.

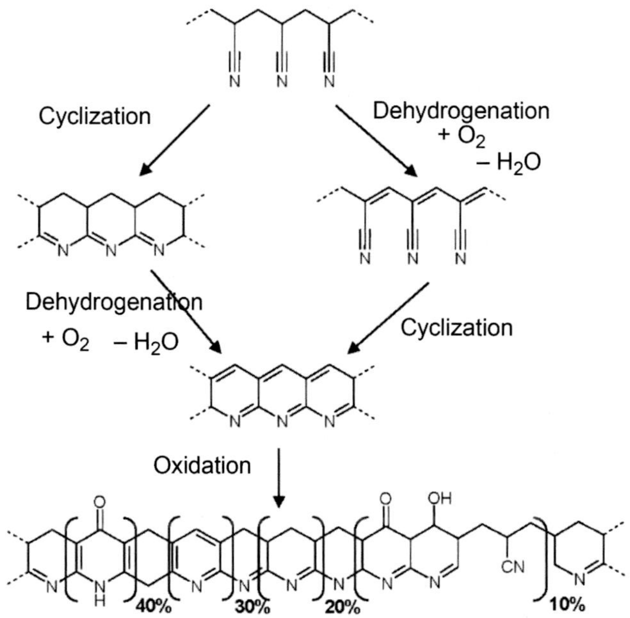

**FIG. 3.41** Chemical transformations of polyacrylonitrile before its transformation into carbon fiber: cyclization, dehydrogenation, and oxidation. *From Bajaj, P., Roopanwal, A.K., 1997. Thermal stabilization of acrylic precursors for the production of carbon fibers: an overview. J. Macromol. Sci. C Polym. Rev. 37 (1), 97–147. doi:10.1080/15321799708014734.*

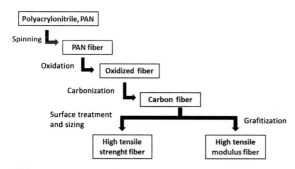

**FIG. 3.42** Manufacturing process of PAN type carbon fibers. *From http://www.carbonfiber.gr.jp. See also: https://www.compositesworld.com/articles/the-making-of-carbon-fiber.*

## 3.6 MANUFACTURING OF COMPOSITE MATERIALS

There are several methods for fabricating composite components. Some methods have been borrowed (e.g., injection molding), but many others were developed to meet specific design or manufacturing challenges. Selection of a fabrication method for a particular part, therefore, will depend on the materials, the part design, and end-use or application (https://www.compositesworld.com/blog/post/fabrication-methods).

According to the American Composite Manufacturers Association (https://acmanet.org/), there are three main types of composite manufacturing processes: open molding, closed molding, and cast polymer molding; there are a variety of processing methods within these molding categories, each with its own benefits and drawbacks.

In *open molding*, composite materials (i.e., resin and fibers) are placed in an open mold, where they cure or harden while exposed to the air. Tooling cost for open molds is often inexpensive, making it possible to use this technique for prototype and short production runs.

In *closed molding*, composite materials are processed and cured inside a vacuum bag or a two-sided mold, closed to the atmosphere. Closed molding may be considered for two cases: first, if a two-sided finish is needed, and second, if high production volumes are required.

In *cast polymer molding*, a mixture of resin and fillers are poured into a mold and left to cure or harden. These molding methods sometimes use open molding and sometimes closed molding techniques (http://compositeslab.com/composites-manufacturing-processes/).

## 3.6.1 Open Mold Processing

### 3.6.1.1 *Process by Hand Lay-Up*

This is the simplest method for producing fiber-reinforced composite parts. To make pieces with this process using, for example, glass fibers (i.e., the filler) and polyester resin (i.e., the matrix), first a coat of gel is applied in the open mold (Fig. 3.43). The reinforcing fibers, which are normally in the form of fabric or mat, are manually placed in the mold. The base resin, mixed with catalysts and accelerators, is then applied by casting, with a brush or by spraying. Rollers are used to impregnate the reinforcement material with resin and to remove trapped air. To increase the wall thickness of the piece

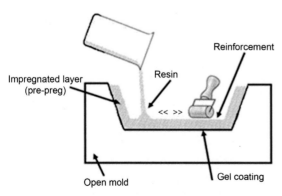

FIG. 3.43  Schematic representation of a hand lay-up manufacturing process for composites.

being manufactured, layers of fabric or mat of fibers and resin are added until the desired thickness is acquired. Applications of this method include the construction of boat keels, tanks, and building panels.

### 3.6.1.2 *Spray Lay-Up Application Process*

The spray application method for manufacturing fiber-reinforced plastic casings is similar to the hand lay-up method and can be used to produce boat hulls, bathtubs, and other medium to large objects.

In this process (Fig. 3.44), if glass fibers are used as filler, a continuous thread is passed through a gun that cuts and sprays at the same time, simultaneously depositing the shredded fibers and the catalyzed resin inside the mold. The deposited layer is then thickened with a roller to remove the entrapped air and to ensure that the resin impregnates the reinforcing fibers. Multiple layers can be applied to obtain the desired thickness. The hardening usually takes place at room temperature or can be accelerated by moderately increasing the temperature.

### 3.6.1.3 *Filament Winding Process*

Another important open mold process used for the production of high mechanical strength pipes is the filament winding process, already mentioned in Section 3.2.2.3. To fabricate composites with this process, the reinforcing fibers are impregnated with a catalyzed resin and wound around a suitable rotating mandrel according to a pattern determined by the desired characteristics of the final product (Fig. 3.45). When a sufficient number of layers have been applied, the composite is cured at room temperature or at high temperature, then detached from the mandrel. The high degree of orientation of the fibers and the high fiber charge obtainable with this method gives extremely high mechanical strength to the obtained products. Applications of this process include pipes, conduits, tanks for chemicals and fuels, pressure tanks, and containers for missile engines.

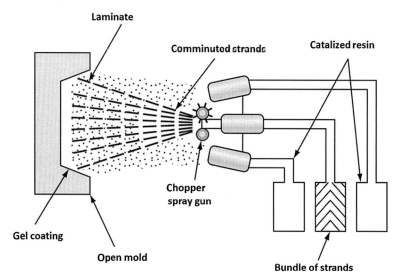

**FIG. 3.44**  Spray lay-up application process in open-mold manufacturing of composites.

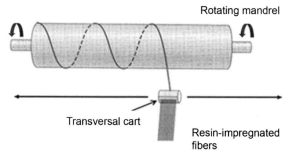

**FIG. 3.45**  Equipment for the filament winding process.

Filament winding is a continuous fabrication method that can be highly automated and repeatable, with relatively low material costs. Computer-controlled filament-winding machines are available, equipped with 2 up to 12 axes of motion.

## 3.6.2 Closed Mold Processing

Demand for faster production rates has pressed the industry to replace hand lay-up methods with alternative fabrication processes and has encouraged fabricators to automate those processes wherever possible.

### 3.6.2.1 *Resin Transfer Molding process*

Resin transfer molding (RTM), also referred to as liquid molding, uses a two-part, matched, closed mold that is made of metal or composite material. Dry reinforcement (typically a preform) is placed into the mold, and the mold is closed. Resin and catalyst are mixed in dispensing equipment, then pumped into the mold under low to moderate pressure through injection ports, following predesigned paths through the preform. Both mold and resin can be heated, as necessary, for particular applications.

In light RTM, low injection pressure coupled with vacuum allows the use of less-expensive, lightweight two-part molds or a very lightweight, flexible upper mold.

### 3.6.2.2 *Reaction Injection Molding Process*

Contrary to reaction injection molding (RTM), where resin and catalyst are premixed prior to injection under pressure into the mold, RIM injects a rapid-cure resin and a catalyst into the mold in two separate streams. Mixing and the resulting chemical reaction occur in the mold instead of in a dispensing head.

### 3.6.2.3 Vacuum-Assisted Resin Transfer Molding Process

In vacuum-assisted resin transfer molding (VARTM), resin is drawn into a preform only by use of vacuum, instead of being forced in under pressure. Because VARTM does not require high heat or pressure values, it operates with low-cost tooling, making it possible to inexpensively produce large and complex parts in one shot.

In this process, fiber reinforcements are placed in a one-sided mold, and a cover (typically a plastic film) is placed over the top to form a vacuum-tight seal. The resin is drawn by vacuum through the reinforcement by a series of designed-in channels that facilitate the impregnation of the fibers. Fiber content in the finished part can be as high as 70%. Current applications include marine, ground transportation, and infrastructure parts.

### 3.6.2.4 Compression Molding and Injection Molding

These processes are essentially the same as those used for polymeric materials (see Section 3.2), except that the reinforcing material is mixed with the resin before molding.

### 3.6.2.5 Sheet-Molding Compound Process

The sheet-molding compound (SMC) process is one of the most widely used closed-mold processes to produce fiber-reinforced polymer matrix parts, especially for automotive industry applications, enabling good mechanical strength properties to be achieved. The advantages of the SMC process compared to the hand lay-up process are the higher production efficiency with higher volumes, better product surface quality, and production uniformity. This technology uses only nonwoven mats, preimpregnated with a resin (usually a polyester one). The method consists in preparing the material inside a mold, which, having a counter-mold (negative mold), compresses the composite material by means of high pressure to allow its compaction. In addition to the pressure, the molds are heated, and the composite material is brought to temperatures of about 120–150°C, to which the resin, under pressure, will flow together with the reinforcement in all the empty spaces contained in the air space of the mold and counter-mold. Production times are quite fast, and the surface quality of the laminate is very high.

### 3.6.2.6 Pultrusion

Pultrusion, like other processing methods for composites, has been used for decades with glass fiber and polyester resins, but in the last 10 years the process has also found application in advanced composites applications. This continuous process is relatively simple and low-cost; the reinforcing fibers (usually roving, tow, or continuous mat) are typically pulled through a heated resin bath and then formed into specific shapes as they pass through one or more forming guides (Fig. 3.46). Then the material moves through a heated die, where it takes its final shape and cures.

After cooling, the resulting profile is cut to the desired length. Pultrusion yields smooth

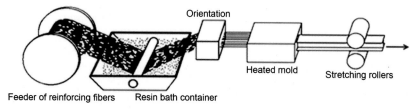

**FIG. 3.46** Schematic of a process of continuous pultrusion.

finished parts that typically do not require post-processing. A wide range of continuous, consistent, solid, and hollow profiles are pultruded, and the process can be custom-tailored to fit specific applications.

Furthermore, pultrusion is the process normally used for manufacturing products in plastic reinforced with fibers, with a constant cross-section such as profiles, channels, tubes, and ducts. The high concentration of fiber reinforcement and the parallel orientation of the fibers with respect to the length of the finished piece guarantee the high mechanical resistance of the product.

## 3.7 ADVANCED TECHNOLOGIES

In the last decades, advanced technologies focus to improve the quality of the objects obtained with the most traditional processing described in the previous sections and to allow a custom-desired production. In fact, the manufacturing landscape is ever-changing, and one of the most important drivers of this change is the emergence of advanced manufacturing technologies that enable more cost- and resource-efficient small-scale production. Among the advanced technologies available, the world concentrates more attention toward *rapid prototyping* (RP) techniques, which were previously used only to obtain prototypes in different industrial areas.

*Additive manufacturing* (AM) is the formalized term for what used to be called RP and what is popularly called *3D printing*. The term rapid prototyping is used in different industries to describe a process for rapidly obtaining physical prototypes directly from digital model data. Moreover, the emphasis is on obtaining something quickly, and the output is a prototype or basis model from which further models and eventually the final product will be derived. These technologies, first developed for prototyping, are now used for many more purposes.

Moreover, users of RP technology have come to realize that this term is becoming inadequate and in particular does not effectively describe more recent applications of the technology. In fact, many parts are now directly manufactured in these machines, so it is not possible to label them as "prototypes," because final objects and components are now fabricated using an additive approach. A recently formed technical committee within ASTM International agreed that new terminology should be adopted. Hence, the ASTM standard F2792-12a asserts the definition for additive manufacturing as follows: "a process of joining materials to make objects from 3D model data, usually layer upon layer, as opposed to subtractive manufacturing methodologies"; synonyms are also highlighted in the standard: additive fabrication, additive processes, additive techniques, additive layer manufacturing, layer manufacturing, 3D printing, rapid prototyping, and freeform fabrication. In addition to those synonyms, presently AM is known as 3D printing technology. We can say that, today, additive manufacturing is the most common term in industry markets and research, whereas 3D printing is more commonly used in the consumer market.

Additive manufacturing is the opposite of subtractive manufacturing (e.g., milling, folding, molding). In fact, with subtractive manufacturing techniques, the idea is to work on a volume and remove material to obtain the desired shape (Fig. 3.47A). On the contrary, with AM, the principle is to add more material to obtain a 3D object (Fig. 3.47B). Additive technologies differ from conventional manufacturing technologies in several aspects and have huge potential if deliberately used with respect to their specific features. Their most important benefit is their high design flexibility. Because the material is built up layer-by-layer until the object is produced, there is no need for molds, which are both time-consuming and costly to make. This means that it is theoretically possible to produce any shape.

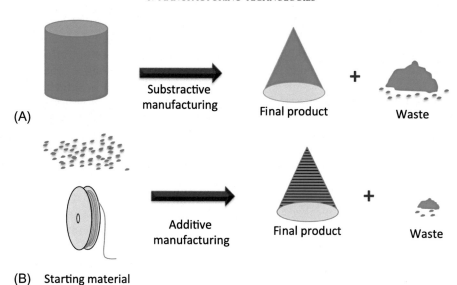

FIG. 3.47   (A) Subtractive manufacturing technique: material is removed to obtain the final object; (B) additive manufacturing technique: material is added to obtain the desired object.

Development in AM technologies is an on-going process as reported in the timeline of Table 3.2, which demonstrates this sector's fast growth.

Different classification of the AM techniques are proposed considering, for example, the starting form of the material to be processed or the principles of functioning of the techniques. Here we propose a classification of the main AM approaches based on the form of the starting material (i.e., liquid, raw materials, e.g., pellets, powder, and semifinished products) (Fig. 3.48). It is important to highlight that not all these techniques can be used for all materials; in fact, the characteristics of different material classes (i.e., polymeric, ceramic, and metallic materials) together with the specific properties of each material have to be considered in the selection of the appropriate technique.

In the next sections, the common basis of the different AM approaches and details on the

**TABLE 3.2**   Additive Manufacturing Applications Timeline: Past, Present, and Possible Future Applications of AM (Royal Academy of Engineering, 2013)

| | |
|---|---|
| 1988–94 | Rapid prototyping |
| 1994 | Rapid casting |
| 1995 | Rapid tooling |
| 2001 | AM for automotive |
| 2004 | Aerospace (polymers) |
| 2005 | Medical (polymers) |
| 2009 | Medical implants (metals) |
| 2011 | Aerospace (metals) |
| 2013–16 | Nanomanufacturing |
| 2013–17 | Architecture |
| 2013–18 | Biomedical implants |
| 2013–22 | In situ biomanufacturing |
| 2013–32 | Full body organs |

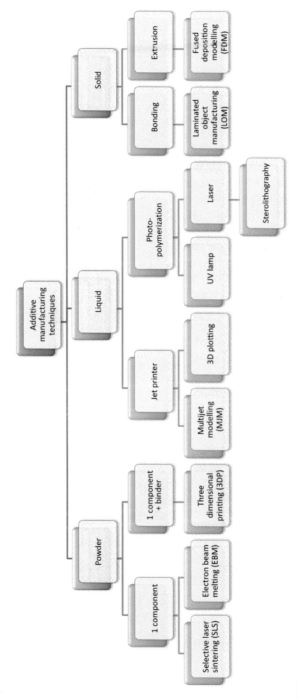

**FIG. 3.48** Classification of AM techniques depending on the nature of the starting materials.

main AM techniques will be described. For further information, suggested readings are indicated at the end of the chapter.

## 3.7.1 The AM Process

### 3.7.1.1 General AM Process Steps

AM involves a number of steps that move from the virtual CAD design to the physical produced part. Different products will involve AM at different degrees in industrial field. Small, relatively simple products may make use of AM for the production of models (e.g., prototypes), whereas larger, more complex products with greater engineering content (i.e., final product) may involve AM during the development process of the fabrication of a new product. Furthermore, early stages of the product development process may only require rough parts, with AM being used because of the speed at which they can be fabricated. At later stages of the process, parts may require careful cleaning and postprocessing (e.g., sanding, surface preparation, and painting) before they are used, with AM being useful here because of the complexity of shapes that can be made without having to consider tooling. To summarize, in its simplest form, most AM processes involve the following eight steps (Fig. 3.49):

– conceptualization and computer aided design (CAD);
– conversion to STereoLithography (STL)/ additive manufacturing file (AMF);
– transfer to AM machine and file manipulation (i.e., slicing);
– machine set-up;
– build;
– removal and clean-up;
– possible postprocessing;
– application.

In the following sections, each step will be briefly described.

Step 1: Conceptualization and computer-aided design (CAD)

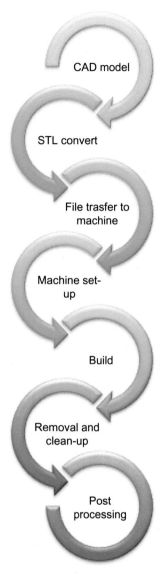

FIG. 3.49   Steps necessary to reach the final production of a product by AM technologies.

The first step in any product development process is to come up with an idea for how the product will look and function. All AM parts must start from a software model that fully describes the external geometry. This can involve the use of almost any professional

CAD solid modeling software, and the output should be a 3D solid or a surface representation. In the biomedical field, geometries from 3D scanners (e.g., MRI, CT) can also be used.

Step 2: Conversion to .STL file

When 3D CAD data are used, the second step is to convert them to .STL format, which has become established as the de facto standard for this process step. Every AM machine accepts the .STL file format (Standard Triangulation Language, STereoLithography or Surface Tessellation Language), and today every CAD system can output such a file format. It works by removing any construction data, modeling history, etc., and approximating the surfaces of the model with a series of triangular facets. Hence, a mesh of triangles is created by CAD or, alternatively, higher-order curves can be computed by the approximation of series of points. This file describes the external closed surfaces of the original CAD model and forms the basis for calculation of the slices. However, the conversion process suffers from a number of serious problems. The .STL format only describes the surface geometry of the original three-dimensional object. This is approximated using a large number of triangles. Other information from the CAD system is lost, for example, material data, curvature radii, and dimensional tolerances.

Step 3: Transfer to AM machine and STL file manipulation.

The STL file describing the part to be obtained must be directly transferred to the AM machine. Ideally, it should be possible to press a "print" button, and the machine should build the part straight away. Unfortunately, there are some many actions required prior to printing the object, such as general manipulation of the file so that it is the correct size, position, and orientation for building.

Step 4: Machine set-up

The AM machine must be properly set up prior to the build process start. Such settings would relate to the build parameters (e.g., material constraints, energy source, layer thickness, timings). In fact, for example, some machines are only designed to run a few specific materials and give the user few options to vary layer thickness or other build parameters. Normally, an incorrect set-up procedure will still result in a part with unacceptable final quality.

Step 5: Build

Building the part is mainly an automated process, and the machine can largely carry on without supervision. However, the first few stages of the AM process are semiautomated tasks that may require considerable manual control, interaction, and decision-making. Once these steps are completed, the process switches to the computer-controlled building phase. All AM machines have an analogous sequence of layering (e.g., height adjustable platform, material deposition/spreading mechanisms). Some machines combine the material deposition and layer formation simultaneously whereas others separate them.

Step 6: Removal and clean-up

Once the AM machine has completed the build, the parts must be removed. Ideally, the part should be ready for use. Often, parts will require a significant amount of postprocessing before they are ready for use. The part must be either separated from a build platform on which the object was produced or removed from excess build material surrounding it. Different AM processes have different clean-up requirements. The clean-up stage may also be considered as the initial part of the postprocessing stage.

Step 7: Postprocessing

Postprocessing refers to the step of finishing the AM parts for the final application use. This may involve abrasive finishing (e.g., polishing and sandpapering), or application of coatings. This step is very application-specific.

In fact, some applications require only a minimum of postprocessing; on the contrary, other applications may require careful handling of the parts to maintain good accuracy and finish.

Step 8: Application

After the postprocessing step, parts are ready to be used. However, they may also require additional treatment before they are acceptable for use. For example, they may require priming and painting to give an acceptable surface texture and finish. Treatments may be laborious and lengthy if the finishing requirements are very demanding. They may also require to be assembled together with other mechanical or electronic components to form a final model (e.g., prototype) or product. It has to be taken into account that AM-created parts behave differently than parts made using a more conventional manufacturing approach (e.g., molding and casting). This behavior may be better or worse for a particular application, and thus a designer should be aware of these differences and take them into account during the design stage.

### 3.7.1.2 Classification of AM Techniques

AM techniques can be classified in different manners as previously described (Fig. 3.48). Here, common features of power bed fusion, bonder jet, and materials jetting techniques are briefly reported (Fig. 3.50).

### POWDER BED FUSION

Powder bed fusion (PBF) methods deposit layers of powder, which are sequentially fused together by an energy source, resulting in solid parts residing in a powder bed. The prominent methods are selective laser sintering (SLS) and electron beam melting (EBM), and these two are well compared. EBM processes benefit from the flexibility and high energy of their heat source, which can move nearly instantaneously and with split beams; EBM builds typically maintain the bed at high temperature and can create parts with a cast, low porosity, and low residual stress microstructure. However their requirement for a vacuum chamber and conductive target material restricts material capabilities, whereas SLS machines can process materials such as polymers and ceramics in gaseous atmospheres.

PBF is a well-commercialized technology, but due to the lapsing of major patents, there are some open-source machines available, which have led inventors to create nonengineering applications of the technology as well.

### BINDER JETTING

Binder jetting (BJ) methods were primarily developed at Massachusetts Institute of Technology (MIT), United States, during the early 1990s and feature a powder bed similar to that of the PBF process (see previous section). However, instead of using an energy source to fuse

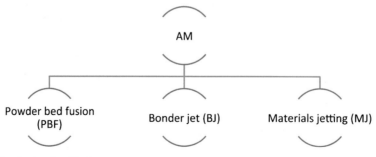

FIG. 3.50   AM technologies classification.

the material together, an inkjet head deposits a binder onto each layer to form part cross-sections. The binder forms agglomerates with the powder particles and provides bonding with the layer below. Different commercial materials are available for BJ approach, but many of them require postprocessing to increase strength, for example, by reinforcement with another material. Materials include polymers (e.g., PMMA), ceramics, casting sands, and metals (e.g., stainless steel or nickel-chromium alloy). This technology can be scaled up quite easily, as demonstrated by machines with an incredible $4 \times 2 \times 1 \, \text{m}^3$ build volume.

## MATERIALS JETTING

The first generation of materials jetting (MJ) machines was commercialized in the 1980s. They relied on heated waxy thermoplastics deposited by inkjet print heads, lending themselves to modeling and investment casting manufacture; however, the more recent focus has been on deposition of acrylate photopolymers, wherein droplets of liquid monomer are formed and then exposed to UV light to initiate polymerization. Machines with build spaces as big as $1000 \times 800 \times 800 \, \text{mm}^3$ are available, as well as multimaterial capability, which can print many materials by varying the composition of several photopolymers. Research groups around the world are working on other materials such as nonphotopolymers, ceramic suspensions, and low melting point metals (including aluminum). The challenge of forming molten droplets, then controlling deposition and solidification characteristics, has kept these material systems in the research arena.

## 3.7.2 Additive Manufacturing for Metals

The use of AM with metal powders is a new and growing industry sector, and it became a suitable process to produce complex metal net shape parts, and not only prototypes, as before. In fact, AM now enables both a design and

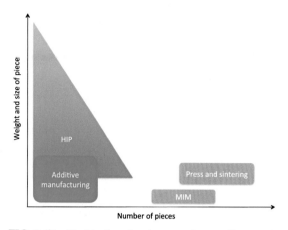

FIG. 3.51 Positioning of various powder metallurgy technologies according to part weight or size and production series compared to additive manufacturing.

industrial revolution, in various industrial sectors such as aerospace, energy, automotive, biomedical, tooling, and consumer goods.

Additive manufacturing is complementing to other powder metallurgy (PM) technologies (Fig. 3.51). As with hot isostatic pressing (HIP), AM is more suitable for the production of small or medium series of parts. Although the HIP process is generally used for the manufacturing of massive near-net shape parts of several hundred kilograms, the AM process is more suitable for smaller parts of a few kilos, and it offers an improved capacity to produce complex metal parts due to a greater design freedom. Moreover, metal injection molding (MIM) and press and sintering technologies also offer the possibility to produce net shape parts, but they are recommended for large series of small parts.

Metal AM technologies offer many key benefits and some drawbacks; some of them have been previously discussed, other ones are strongly related to the metal metallurgy technology and are here reported. Among the advantages of AM compared to traditional PM, the main ones are as follows:

- lightweight structures, either by the use of the appropriate design or by designing parts where material is only present where it needs to be, without other constraints;
- new functions such as complex internal channels or several parts builtin one;
- net shape process, meaning less raw material consumption, up to 25 times less versus machining, important in the case of expensive or difficult to machine alloys; the net shape capability helps create complex parts in one step only, thus reducing the number of assembly operations (e.g., welding and brazing);
- no tools needed, unlike other conventional metallurgy processes that require molds and metal-forming or removal tools;
- short production cycle time: complex parts can be produced layer-by-layer in a few hours in additive machines; the total cycle time including postprocessing usually amounts to a few days or weeks, and it is usually much shorter than conventional metallurgy processes that often require production cycles of several months.

To take full advantage of AM technologies, it is important to be aware of some possible drawbacks and limitations:

- part size: in the case of powder bed technology, the part size is limited to powder bed size; however, part sizes can be greater with direct energy deposition (or laser metal deposition) processes. Unfortunately, due to the low thickness of powder layers, it can be very slow and costly building high parts or massive parts;
- production series: AM processes are generally suitable for unitary or small series; for small-sized parts, series up to 25,000 parts/year are already possible;
- part design: in case of powder bed technology, removable support structures are

needed when the overhang angle is below 45 degrees;
- material choice: though many alloys are available, nonweldable metals cannot be processed by AM, and difficult-to-weld alloys require specific approaches;
- material properties: parts made by AM tend to show anisotropy in the z axis (i.e., layer-by-layer direction);
- although densities of 99.9% can be reached, some residual internal porosities can be present.

### 3.7.2.1 AM Process

Metal additive manufacturing processes can be classified as resumed in Fig. 3.52; in particular, the powder bed technologies will be here briefly described, as they represent the most used ones. Among them, the 3D printing technology does not allow a final piece at the end of the process; in fact, binder binds green parts and then sintering creates the final solidification of the part.

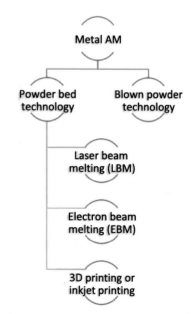

FIG. 3.52 Scheme of the main metal powder AM technologies.

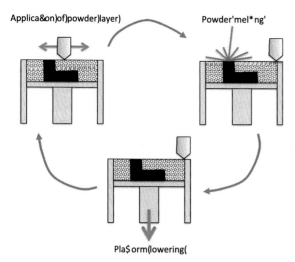

Applica&on)of)powder)layer)

Powder'mel*ng'

Pla$orm(lowering(

**FIG. 3.53** The powder bed manufacturing cycle: (1) a layer of metal powder is homogeneously spread on the platform; (2) the powder is melted where needed (i.e., accordingly to the STL file); (3) platform is lowered and the cycle from (1) can restart.

**TABLE 3.3** Comparison Between SLM and EBM Processes

|  | SLM | EBM |
|---|---|---|
| Power source | Laser beam (up to 1 kW) | Electron beam (3 kW) |
| Operating environment | – Inert gas environment<br>– Plate preheat up to 200°C | – High vacuum<br>– Chamber temperature at 600–1000°C |
| Powder material | Metals, polymers, ceramics, composites | Metals and metallic-based composites |
| Powder layer thickness | 20–100 μm | 50 μm |
| Powder particle size | About 20–63 μm | About 45–105 μm |
| Melting method | – Contour melting<br>– Hatch melting | – Powder bed preheating<br>– Contour melting<br>– Hatch melting |
| Build rate | Slow | Fast |

The powder bed process is generally reported in Fig. 3.53. In beam-based powder bed systems (LBM or EBM), a powder layer is first applied on a building platform. Then a laser or electron beam selectively melts the upper layer of powder. After melting, the platform is lowered and the cycle is repeated until the part is fully built, embedded in the powder bed.

A comparison between EBM and LBM is reported in Table 3.3; in particular, compared to SLM, EBM is faster in producing fully dense builds with a high energy electron beam, allowing full melting of powder particles and a faster scanning rate.

### 3.7.2.1.1 LASER BEAM MELTING (OR SELECTIVE LASER MELTING, SLM)

In the laser beam melting (LBM) process (Fig. 3.54A), a layer of metal powder is first applied on a building platform with a recoater (blade or roller), and a laser beam (or multiple lasers) selectively melts the layer of powder. Then, the building platform is lowered by 20 up to 100 μm, and a new metal powder layer is applied. The laser beam melting operation is repeated until the piece is completely built up layer-by-layer in the powder bed. After a few thousand cycles (depending on height of the part), the built part is removed from the powder bed. The full metal melting enables the production of solid, dense metal parts in a single process (i.e., not using binders or postprocess furnace operations). A variety of metals can be used, including stainless steels, cobalt-chrome, and titanium (Fig. 3.54B).

### ELECTRON BEAM MELTING

The electron beam melting (EBM) process is based on a high power electron beam that generates the energy needed for high melting capacity and high productivity. The electron

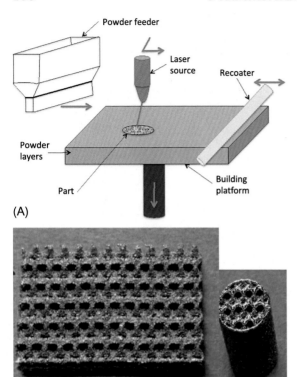

**(A)**

**(B)**

FIG. 3.54 (A) LBM process; (B) Porous implants obtained by LBM using commercially pure titanium (Grade 2) with a pore size about 600 micron: (i) plate; (ii) cylinder. *Adapted from Taniguchi, N., Fujibayashi, S., Takemoto, M., Sasaki, K., Otsuki, B., Nakamura, T., Matsushita, T., Kokubo, T., Matsuda, S., 2016. Effect of pore size on bone ingrowth into porous titanium implants fabricated by additive manufacturing: an in vivo experiment. Mater. Sci. Eng. C Mater. Biol. Appl. 59, 690–701.*

beam is achieved by electromagnetic coils providing extremely fast and accurate beam control. The EBM process takes place in vacuum (e.g., $P = 1 \times 10^{-5}$ mbar) and at high temperature, resulting in stress-relieved components. For each layer, the electron beam heats the entire powder bed to an optimal temperature, specific for the used material. As a result, the parts produced with the EBM process are almost free from residual stresses and have a microstructure free from martensitic structures.

## 3D PRINTING

The 3D printing process is an indirect process composed of two steps. After applying a powder layer on the build platform, the powder is agglomerated thanks to a binder fed through the printer nozzle. The operation is repeated until parts are produced, which are then carefully removed from the powder bed, as they are in a "green" stage. The metal part solidification takes place in a second step, during a debinding and sintering operation.

The 3D printing technology is more productive than laser beam melting and requires no support structure. Besides, it provides a good surface quality by using a postprocessing technique, as, for example:

- peening, blasting, and tumbling for $R_a$ about 3.0 μm;
- superfinishing for $R_a$ about 1.0 μm or lower.

Unfortunately, the range of available materials is limited and mechanical properties achieved can be lower than those obtained by laser and electron beam melting.

## DIRECT ENERGY DEPOSITION

With the direct energy deposition or laser metal deposition (LMD) process, a nozzle mounted on a multiaxis arm deposits melted material onto the desired surface, where it solidifies. This technology offers a higher productivity than selective laser melting and also the ability to produce larger parts, but the freedom in design is much more limited; for example, lattice structures and internal channels are not possible.

## 3.7.3 Additive Manufacturing for Polymeric Materials

Different AM processes are nowadays used for polymers; here the main ones are briefly described (for further information suggested readings can be found at the end of the chapter). As polymeric materials are very different in terms of structures, transition temperatures, and molecular weights, the appropriate AM process has to be selected, so as to achieve the optimal results on the parts (e.g., finish, mechanical properties).

### 3.7.3.1 Stereolithography

Stereolithography (SLA, Fig. 3.55) is one of the most important additive manufacturing technologies currently available. The resin is a liquid photosensitive polymer that cures or hardens when exposed to ultraviolet radiation. This technique involves the curing or solidification of a liquid photosensitive polymer through the use of the irradiation light source that induces a chemical reaction (curing reaction), bonding small polymeric molecules and forming a highly cross-linked polymer. The UV light comes from a laser beam, which is controlled to scan across the surface by a computer-guided mirror according to the cross-section of the part that corresponds to the layer. The laser penetrates into the resin for a short distance that corresponds to the layer thickness. The first layer is bonded to a platform, which is placed just below the surface of the resin container. The platform lowers by one layer of thickness, and the scanning is performed for the next

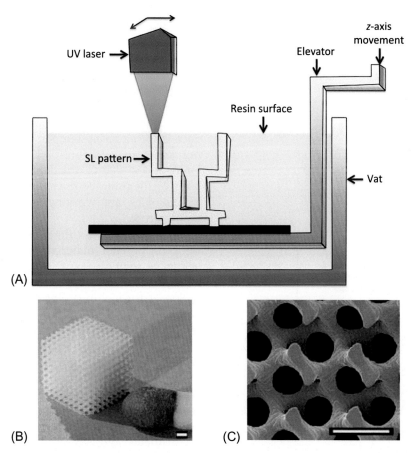

**FIG. 3.55** (A) Scheme of stereolithography process and (B, C) images of 3D object fabricated from stereolithography. Scale bar: 500 μm (Melchels et al., 2009).

layer. This process continues until the part has been completed. SLA is the technique that produces the finest surface finish and dimensional accuracy. It requires a support structure under production, which has to be removed postproduction.

The two-photon approaches (Fig. 3.56) have been developed in the last decade, essentially providing further enhancement of SLA resolution. In fact, in that case, photopolymerization occurs at the intersection of two scanning laser beams and the effect is to greatly increase the resolution of photopolymerization processes. In detail, two-photon absorption is the simultaneous absorption of two photons of identical or different frequencies in order to excite a molecule from one state (usually the ground state) to a higher energy electronic state; the energy difference between the involved lower and upper states of the molecule is equal to the

sum of the energies of the two photons. Two photons of lesser energy together can cause an excitation "normally" produced by the absorption of a single photon of higher energy in the process called two-photon excitation. In this process, special initiator molecules in the monomer start the polymerization reactions if activated by two photons simultaneously, then the laser intensity field can be tuned so that this event only happens in a very small region near the focus, resulting in an extremely local polymerization, with resolutions in the tens of nanometers range.

### 3.7.3.2 Selective Laser Sintering

Selective laser sintering (SLS, Fig. 3.57) is a similar powder-based process where a $CO_2$ laser fuses fine polymeric powders in layers, directed by a computer-guided mirror, realizing parts by sintering the powder particles. Similar to the SLM technologies for metals, the plastic part will be produced layer-by-layer, but the major difference is the sintering temperature. For example, polyamide (e.g., Nylon, PA12) needs to be sintered at 160–200°C, thanks to a high-wattage laser; on the contrary, the metal laser 3D printing technologies are working at a much higher temperature.

During the process, the SLS machine preheats the bulk powder material in the powder bed below its melting point, to make it easier for the laser to raise the temperature of the selected regions to the melting point. A roller will apply a layer of fresh polymer powder, then the laser will sinter the powder according to the 3D file (direction $x$-$y$), and the build platform will get down before applying a new layer of powder over the build area. The process will be repeated until the desired part is created. Once it is finished, the plastic objects need to cool down before being extracted. The nonsintered powder encapsulates and supports the sintered part. This eliminates the need for support material being printed along with the part.

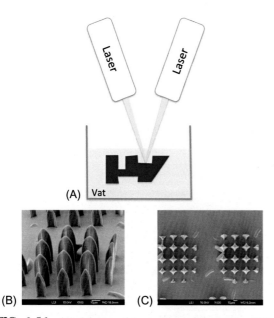

FIG. 3.56  (A) Scheme of two photon lithography; 3D microstructures obtained by two-photon lithography: (A) side view and (B) top view. Scale bar: 10 μm (Hsieh et al., 2010).

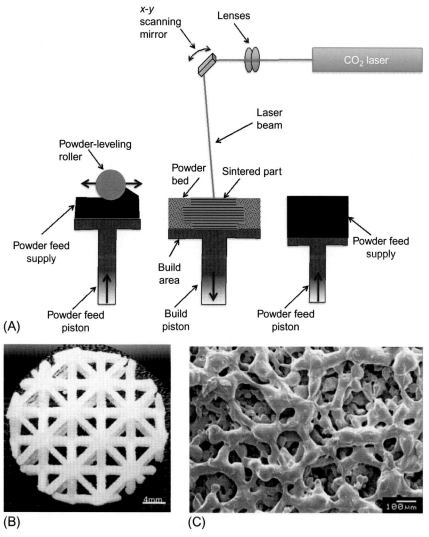

**FIG. 3.57** Selective laser sintering: (A) scheme of the process; (B) macrostructure of the sintered object; and (C) SEM image of sintered polymeric powder. Scale bar: (B) 4mm; (C) 100μm (Yeong et al., 2010).

### 3.7.3.3 *Extrusion-Based Systems*

Extrusion-based technology is currently the most popular on the market. Although there are other techniques for polymer extrusion, heat is normally used to melt bulk material in a small chamber. The material is pushed through by a feed system, which creates the pressure to extrude, creating the object by AM approach. The material extruded must be in a semisolid state when it comes out of the nozzle. This material must fully solidify while remaining in that shape. Furthermore, the material must bond to material that has already been extruded so that a solid structure can result. Because material is extruded, the AM machine must be capable of scanning in a horizontal plane as well as starting and stopping the flow of material while scanning. Once a layer is completed, the machine must index upward, or move the part downward, so that a further layer can be produced.

There are two primary approaches when using an extrusion process. The most commonly used approach is to use temperature as a way of controlling the material state. Molten material is liquefied inside a reservoir so that it can flow out through the nozzle and bond with adjacent material before solidifying. An alternative approach is to use a chemical change to cause solidification. In that case, a curing agent allows bonding to occur. Parts may therefore cure or dry out to become fully stable.

## FUSED DEPOSITION MODELING

Fused deposition modeling (FDM), well known as 3D printing, is the most widely used 3D printing technology; it represents the largest installed base of 3D printers globally, and it is often the first technology people are exposed to. FDM is a process mainly composed of several steps as reported in Fig. 3.58.

In FDM, the materials mostly used are thermoplastic polymers (e.g., PLA, ABS, polyurethane) and come in a preextruded filament shape. In the process (Fig. 3.59), a coil of thermoplastic filament is first loaded into the machine. Once the nozzle has reached the desired temperature (e.g., close to the melting temperature of the polymer), the filament is fed to the extrusion head, then in the nozzle where it melts. The extrusion head is attached to a two-axis system that allows it to move in x-y directions. The melted material is extruded in thin strands and deposited layer-by-layer in predetermined locations, as described in the STL file. Sometimes the cooling of the material on the platform is accelerated through the use of cooling fans attached on the extrusion head. When a layer is finished, the build platform moves down in z direction, and a new layer is deposited. This process is repeated until the part is complete.

### 3.7.3.4 Powder and Binder-Based 3D Printing

Unlike FDM, powder-binder printing uses two separate materials that come together to form each printed layer: a fine dry powder and a binder. A 3D powder-binding printer is composed of a powder dispenser, a binder inkjet nozzle, and a platform where the printer prints the object. The printing of the object begins by lowering the platform on which the

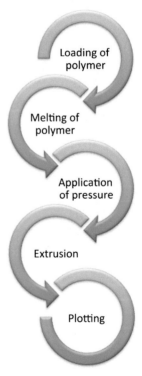

FIG. 3.58 Key features of the extrusion-based AM system.

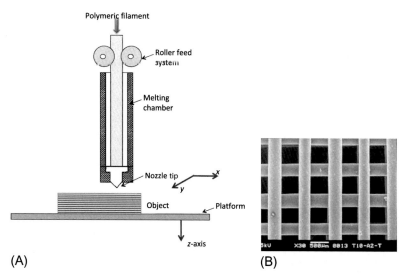

**FIG. 3.59** Fused deposition modeling: (A) scheme of the process; (B) example of PCL structure obtained by FDM, scale bar 500 μm (Zeina et al., 2002).

first layer of material powder is spread by the leveling roller (Fig. 3.60). Drops of a low viscosity binder (e.g., solvent, solvent solution) are dispensed in *x-y* directions, from a nozzle for printing on a thin bed of powder. Hence, the drops, by binding to the powder, compact the powder and form a 2D layer. After printing a layer and drying it, the platform is lowered in *z*-direction by a fixed distance, a new layer of powder will then be spread, and the printing process is repeated until the whole structure is printed. Although there are some advantages, there is a drawback as well. In fact, this process requires a posttreatment to remove the surplus of unsolidified powder present on the printed parts.

## LAMINATED OBJECT MANUFACTURING

Laminated object manufacturing (LOM) is a AM method based on the use of layers of plastic or paper fused together using heat and pressure, then cut into the desired shape with a computer-controlled laser or blade (Fig. 3.61). In particular, the system is mainly composed of a feed mechanism that advances a sheet over a build platform, a heated roller to apply pressure to bond the sheet to the layer below, and a laser to cut the outline of the part in each sheet layer. A laser cuts the outline of the part into each layer, then the platform lowers in *z*-direction by a depth equal to the sheet thickness (typically 0.05–0.5 mm) and another sheet is advanced on top of the previously deposited layers. The platform rises slightly and the heated roller applies pressure to bond the new layer to the previous one. The process is repeated until the part is completed.

The process is much faster than competitive techniques, as the laser does not have to scan the entire area of a cross-section. Instead, it just has to go around its periphery. Layers are glued to the stack quickly, reducing layer formation time. The speed advantage of the process increases with the size of the manufactured part. On the downside, this process has more limited 3D geometries than other AM processes.

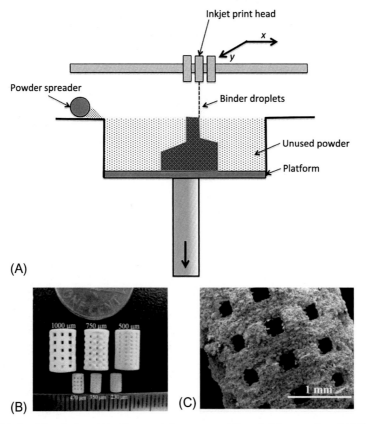

FIG. 3.60   Powder-binder 3D printing: (A) scheme of the process; (B) and (C) examples of TCP structure, Scale bar: (C) 1mm. (Bose et al., 2013).

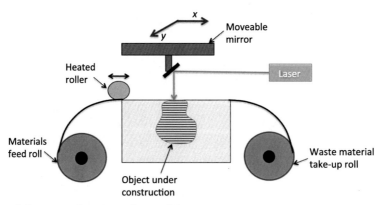

FIG. 3.61   Laminated object manufacturing: scheme of the process.

# References

Bertoldi, S., Farè, S., Denegri, M., Rossi, D., Haugen, H.J., Parolini, O., Tanzi, M.C., 2010. Ability of polyurethane foams to support placenta-derived cell adhesion and osteogenic differentiation: preliminary results. J. Mater. Sci. Mater. Med. 21 (3), 1005–1011.

Bose, S., Vahabzadeh, S., Bandyopadhyay, A., 2013. Bone tissue engineering using 3D printing. Mater. Today 16 (12), 496–504.

Ghanbari, H., et al., 2009. Polymeric heart valves: new materials, emerging hopes. Trends Biotechnol. 27 (6), 359–367.

Grzesik, W., 2016. Advanced Machining Processes of Metallic Materials, second ed. Elsevier Publ., New York City, NY, USA. ISBN: 9789444637116.

Haubold, A.D., Shim, H.S., Bokros, J.C., 1981. Carbon in Medical Devices. In: Williams, D.F. (Ed.), Biocompatibility of Clinical Implant Materials, Vol II. CRC Press, Boca Raton, FL, pp. 3–42.

Hsieh, T.M., Ng, C.W., Narayanan, K., Wan, A.C., Ying, J.Y., 2010. Three-dimensional microstructured tissue scaffolds fabricated by two-photon laser scanning photolithography. Biomaterials 31 (30), 7648–7652.

Iijima, S., 1991. Helical microtubules of graphitic carbon. Nature 354 (6348), 56–58.

Kroto, H.W., Heath, J.R., O'Brien, S.C., Curl, R.F., Smalley, R.E., 1985. C60: buckminsterfullerene. Nature 318, 162–163.

Mackay, T.G., et al., 1996. New polyurethane heart valve prosthesis: design, manufacture and evaluation. Biomaterials 17, 1857–1863.

Melchels, F.P.W., Feijen, J., Grijpma, D.W., 2009. A poly(D,L – lactide) resin for the preparation of tissue engineering scaffolds by stereolithography. Biomaterials 30 (23–24), 3801–3809.

Messler, R., 2004. Joining of Materials and Structures, first ed. Elsevier Publ., New York City, NY, USA. ISBN: 9780750677578

Novoselov, K.S., et al., 2004. Electric field effect in atomically thin carbon films. Science 306 (5696), 666–669.

Park, S.-J., 2015. Carbon Fibers. Springer Series in Materials Science, vol. 210. Springer, Berlin, Germany. https://doi.org/10.1007/978-94-017-9478-7_2.

Sun, B., et al., 2014. Advances in three-dimensional nanofibrous macrostructures via electrospinning. Prog. Polym. Sci. 39, 862–890.

Tanzi, M.C., Farè, S., Petrini, P., Tanini, A., Piscitelli, E., Zecchi Orlandini, S., Brandi, M.L., 2003. Cytocompatibility of polyurethane foams as biointegrable matrices for the preparation of scaffolds for bone reconstruction. J. Appl. Biomater. Biomech. 1 (1), 58–66.

Yeong, W.Y., Sudarmadji, N., Yu, H.Y., Chua, C.K., Leong, K.F., Venkatraman, S.S., Boey, Y.C., Tan, L.P., 2010. Porous polycaprolactone scaffold for cardiac tissue engineering fabricated by selective laser sintering. Acta Biomater. 6 (6), 2028–2034.

Zeina, I., Hutmacher, D.W., Tan, K.C., Swee Hin Teoh, S.H., 2002. Fused deposition modeling of novel scaffold architectures for tissue engineering applications. Biomaterials 23, 1169–1185.

# Further Reading

Advani, S., Hsiao, K.T., 2012. Manufacturing Techniques for Polymer Matrix Composites (PMCs). Woodhead Publishing, New York City, NY, USA.

Anon, 1981. ASM Metals Handbook. Vol. 5, Surface Cleaning, Finishing, and Coating. American Society for Metals.

ASTM Standard Practice, F2792—12a, Standard Terminology for Additive Manufacturing Technologies.

Baird, D.G., Collias, D.I., 2014. Polymer Processing: Principles and Design. Wiley, Weinheim, Germany.

Beddoes, J., Bibby, M.J., 1999. Principles of Metal Manufacturing Processes. Elsevier Ltd Publ., Amsterdam, Netherlands.

Bhardwaj, N., Kundu, S.C., 2010. Electrospinning: a fascinating fiber fabrication technique. Biotechnol. Adv. 28 (3), 325–347.

Burtrand, L., Sridhar, K., 2005. Chemical Processing of Ceramics. CRC Press, Boca Raton, FL, USA.

Callister, W.D., Rethwisch, D.G., 2007. Materials Science and Engineering: An Introduction. John Wiley & Sons, Weinheim, Germany.

Campbell, J., 2015. Metal Casting Processes, Metallurgy, Techniques and Design. In: Complete Casting Handbook. Butterworth-Heinemann Publ., Amsterdam, Netherlands.

Chanda, M., Roy, S.K., 2009. Plastics Fabrication and Recycling. CRC Press, Boca Raton, FL.

Chanda, M., Roy, S.K., 2018. Plastics Technology Handbook, fifth ed. CRC Press, Boca Raton, FL.

Coradin, T., Livage, J., 2011. Sol-gel synthesis of solids. In: Encyclopedia of Inorganic and Bioinorganic Chemistry. 1, Whiley online libraryhttps://doi.org/10.1002/9781119951438.eibc0207.

Danks, E., Hallb, S.R., Schnepp, Z., 2016. The evolution of 'sol-gel' chemistry as a technique for materials synthesis. Mater. Horiz. 3, 91–112. https://doi.org/10.1039/C5MH00260E.

European Powder Metallurgy Association, Introduction to Additive Manufacturing Technology. www.epma.com/am.

Ford, S., Despeisse, M., 2016. Additive manufacturing and sustainability: an exploratory study of the advantages and challenges. J. Clean. Prod. 137, 1573–1587.

Gibson, I., Rosen, D., Stucker, B., 2015. Additive Manufacturing Technologies: 3D Printing, Rapid Prototyping, and Direct Digital Manufacturing. Springer Science, New York City, NY, USA. ISBN: 978-1-4939-2112-6.

Groza, J.R., Shackelford, J.F., 2007. Materials Processing Handbook. CRC Press, Boca Raton, FL, USA.

Harbison, W.C., 2011. Casting. In: Encyclopedia of Polymer Science and Technology. Wiley, Berlin, Germany.

Hopkinson, N., Hague, R., Dickens, P., 2005. Rapid Manufacturing: An Industrial Revolution for the Digital Age. Wiley, Berlin, Germany. ISBN: 978-0-470-01613-8.

http://compositeslab.com/composites-manufacturing-processes/.

https://acmanet.org/.

https://www.compositesworld.com/blog/post/fabrication-methods.

https://www.metaltek.com/capabilities/processes/metal-processing/metal-testing-analysis/non-destructive-testing.

Huang, Z.M., Zhangb, Y.Z., Kotakic, M., Ramakrishna, S., 2003. A review on polymer nanofibers by electrospinning and their applications in nanocomposites. Compos. Sci. Technol. 63, 2223–2253.

King, A.G., 2001. Ceramic Technology and Processing. A Practical Working Guide. William Andrew Publ., Amsterdam, Netherlands.

Mallick, P.K., 2007. Fiber-Reinforced Composites. Materials, Manufacturing, and Design. CRC Press, Boca Raton, FL, USA.

Mazumdar, S., 2001. Composites Manufacturing. Materials, Product, and Process Engineering. CRC Press, Boca Raton, FL, USA.

Narayanan, R.G., Dixit, U.S., 2014. Advances in Material Forming and Joining. Springer, New York City, NY, USA.

Ramakrishna, S., Fujihara, K., Teo, W.E., Lim, T.C., Ma, Z., 2005. An Introduction to Electrospinning and Nanofibers. World Scientific Publishing Co., Singapore.

Randall, M.G., 1996. Sintering Theory and Practice. Wiley-VCH, Berlin, Germany.

Rawlings, R.D., 2009. Materials Science and Engineering. Encyclopedia of Life Support Systems, vol. III. EOLSS Publ. Co. Ltd, Oxford.

Smith, W.F., Hashemi, J., 2006. Foundations of Materials Science and Engineering. McGraw-Hill Series in Materials Science, McGraw-Hill, New York.

Tan, X.P., Tan, Y.J., Chow, C.S.L., Tor, S.B., Yeong, W.Y., 2017. Metallic powder-bed based 3D printing of cellular scaffolds for orthopaedic implants: a state-of-the-art review on manufacturing, topological design, mechanical properties and biocompatibility. Mater. Sci. Eng. C 76, 1328–1343.

Taniguchi, N., Fujibayashi, S., Takemoto, M., Sasaki, K., Otsuki, B., Nakamura, T., Matsushita, T., Kokubo, T., Matsuda, S., 2016. Effect of pore size on bone ingrowth into porous titanium implants fabricated by additive manufacturing: an in vivo experiment. Mater. Sci. Eng. C Mater. Biol. Appl. 59, 690–701.

Youssef, H.A., El-Hofy, H.A., Ahmed, M.H., 2017. Manufacturing Technology: Materials, Processes, and Equipment. CRC Press, Boca Raton, FL, USA.

Zhang, L., Allanore, A., Wang, C., Yurko, J.A., Crapps, J., 2013. Materials Processing Fundamentals. The Minerals, Metals & Materials Society, TMS, Pittsburgh, PA, USA.

# BIOMATERIALS AND BIOCOMPATIBILITY

# 4

# Biomaterials and Applications

## 4.1 BIOMATERIALS AND BIOCOMPATIBILITY

This chapter intends to give an overview of the main (bio)materials used in the medical field, in particular for the manufacture of devices and prostheses.

Before starting with the different types of biomaterials, it is appropriate to provide the definition of biomaterial and related concepts.

### 4.1.1 Biomaterial

It is difficult to give a concise and unambiguous definition of biomaterial, as the elements to be taken into account are manifold. Over the years, different definitions of biomaterial have been given, and presently the main organizations (e.g., the American National Institute of Health (NIH)) and associations of the sector are inclined toward some specific definitions, reported in the following sections.

According to the *Miller-Keane Encyclopedia & Dictionary of Medicine, Nursing, and Allied Health*, 7th Edition (© 2003 by Saunders, Elsevier Inc.), a biomaterial is: *"any substance (other than a drug), synthetic or natural, that can be used as a system or part of a system that treats, augments,* or replaces any tissue, organ, or function of the body; especially, material suitable for use in prostheses that will be in contact with living tissue."* This definition is substantially similar to the one employed by NIH.

According to the *Segen's Medical Dictionary* (© 2012 by Farlex, Inc.), the definition covers two aspects: *"(1) Any synthetic material or device—e.g., implant or prosthesis—designed to treat, enhance or replace an aging, malfunctioning, or cosmetically unacceptable native tissue, organ, or function in the body; (2) A native material used for its structural, not biological, properties—e.g., collagen in cosmetics, carbohydrates modified for biomedical applications, or as bulking agents in food manufacture."* This definition is almost similar to that reported by the McGraw-Hill *Concise Dictionary of Modern Medicine* (© 2002 by The McGraw-Hill Companies, Inc.).

Another definition of biomaterial endorsed by a large consensus of experts in the field (European Society of Biomaterials, Satellite Consensus Conference, Sorrento, 2005) is: *"A material intended to interface with biological systems as an integral part of a process designed to evaluate, monitor, or treat tissues of the body, to replace or augment tissues, or to facilitate the regeneration of tissues."*

*Foundations of Biomaterials Engineering*
https://doi.org/10.1016/B978-0-08-101034-1.00004-9

As a science, "biomaterials" is about 50 years old. In the modern era, the study of biomaterials, called "biomaterials science" or "biomaterials engineering," is marked by a high degree of interdisciplinarity, encompassing elements of medicine, biology, chemistry, tissue engineering, and materials science.

Although the introduction of materials into the human body took place throughout history, the word *biomaterial* was not known, and just 60 years ago biomaterials, as we think of them today, did not exist.

Ancient Egyptians used linen threads to close wounds as long as 4000 years ago, and Europeans used sutures made from catgut in the Middle Ages. However, the use of materials to repair the body prior to the era of modern medicine was not limited to sutures. Incas regularly repaired cranial fractures with gold plates; Mayans used seashells to create artificial teeth; and early Europeans shaped artificial teeth out of iron 2300 years ago. Nevertheless, early attempts at using materials in the body were rather hit-and-miss. Because of a poor understanding of biocompatibility and sterilization, even most implants before the 1950s had a low probability of success, and only in recent times scientists have begun to systematically examine interactions between the body and materials.

## 4.1.2 Biocompatibility

According to *The Williams Dictionary of Biomaterials* (Williams, 1999), "*Biocompatibility refers to the ability of a material to perform with an appropriate host response in a specific application.*"

Although this definition seems ambiguous at a first glance, it represents a huge step forward from the prevailing opinion at the time of its introduction that successful materials should play inert roles in the human body.

More specifically, biocompatibility can be defined as the "*ability of a biomaterial to perform its desired function with respect to a medical therapy,*

*without eliciting any undesirable local or systemic effects in the recipient or beneficiary of that therapy, but generating the most appropriate beneficial cellular or tissue response to that specific situation, and optimizing the clinically relevant performance of that therapy.*"

The biological response is the local and systemic response of the host organism to the implanted biomaterial or device. Examples of appropriate host responses include resistance to blood clotting, resistance to bacterial colonization, and normal, uncomplicated healing. In this context, the evaluation of biological responses is a measure of the magnitude and duration of the adverse alterations in homeostatic mechanisms that lead to the host response. Practically speaking, the evaluation of biological responses is carried out to determine that the biomaterial performs as intended and presents no significant harm to the patient or user, with the aim of predicting whether a biomaterial presents potential harm to the patient or user by evaluating conditions that simulate clinical use (Anderson, 2001).

This definition offers no insights the mechanisms of biocompatibility, how to test the biocompatibility of a material, or how to enhance or optimize the biocompatibility of a material.

Another important aspect to be considered is the difference in biocompatibility requests between a long-term implantable device and a structure (scaffold or matrix) to be used for tissue regeneration:

*Biocompatibility for long-term implantable medical devices.* "*Ability of the device to perform its intended function, with the desired degree of incorporation in the host, without eliciting any undesirable local or systemic effects in that host*" (ESB, Satellite Consensus Conference, Sorrento, 2005);

*Biocompatibility for scaffolds or matrices for regenerative medicine.* "*Ability to perform as a substrate that will support the appropriate cellular activity, including the facilitation of molecular and mechanical signalling systems, in order to optimise tissue regeneration, without eliciting any undesirable effects in those cells, or inducing any undesirable local or systemic responses in the eventual host"* (ESB, Satellite Consensus Conference, Sorrento, 2005).

The statements that follow explicitly define the concepts of repair, replacement, and regeneration (from: D.F. Williams, Symposium on Tissue Engineered Product Regulations, Society for Biomaterials, Memphis 2005):

*Repair*: Nonfunctional restoration of continuity (natural process augmented by products regulated as drugs, devices, or biologics);

*Replacement*: Partly functional replacement of diseased or damaged tissues or organs, usually without replication of natural structure (prostheses, artificial organs, transplants, or grafts);

*Regeneration*: Generation of new tissue functionally and structurally analogous to the diseased or damaged tissues or organs.

Based on the reaction of the tissue with the biomaterial, biomaterials can be classified into three possible categories:

1. *Bioinert (Biotolerant)*. No chemical reactions occur at the interface between implant and living tissue. Bioinert materials are usually separated from bone tissue by a layer of fibrous tissue, although in some cases, they may establish a direct contact with the adjacent tissue.
2. *Bioactive*. Bioactive materials have the property of establishing chemical bonds with tissues (see Section 6.3 in Chapter 6). When related to bone tissue, the interaction is known as *osseointegration*, consisting of the deposition of collagen and mineral phase directly on the implant surface.
3. *Biomimetic*. Biomimetic materials are synthetic or nonnatural materials that mimic natural materials or that follow a design motif derived from nature (Zhang et al., 2014). Current regenerative medicine strategies are addressed at the design of biomimetic materials by the combination of biocompatible synthetic polymer, providing

structural and mechanical properties, and natural polymers or peptides, able to direct cell response (see Section 8 in Chapter 8).

To conclude this first part, it is appropriate to point out that the choice of material or combination of materials has followed a historical evolution, as shown in Fig. 4.1 considering the bone tissue. However, in any case the selection of the strategy depends on the type of device or on the specific application.

## 4.2 POLYMERIC BIOMATERIALS

Thanks to their versatility, polymeric materials find various uses as biomaterials for the construction of devices and prostheses.

Applications of polymers in the biomedical field can be divided into three main categories, depending on whether there is direct contact with the human body:

1. *Use within the human body*. Typical examples are: high density polyethylene (HDPE), ultra-high molecular weight polyethylene (UHMWPE), polytetrafluoroethylene (PTFE), polypropylene (PP), polymethylmethacrylate (PMMA), hydrogels (e.g., poly-hydroxyethyl-methacrylate (PHEMA)), polyamides (Nylon) and polyimides, polyesters (e.g., polyethylene terephthalate (PET)), biodegradable polyesters (PGA, PLA), polyurethanes, silicones, and polyetheretherketone (PEEK).
2. *Use in contact with the human body*. Typical examples are: low-density polyethylene (LDPE), polyvinylchloride (PVC), polyacrylonitrile (PAN), and polysulfones.
3. *Use not in contact with the human body* (biomedical devices). Typical examples are: polystyrene (PS, PST), copolymers (SAN, ABS), and polycarbonate (PC).

In this chapter, the use within the human body will be given greater importance, because

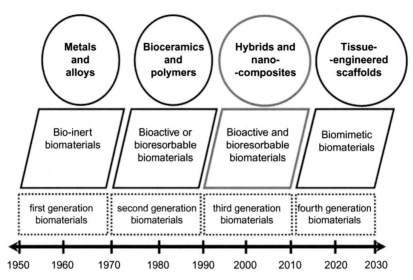

FIG. 4.1   Evolution of biomaterials in bone repair and regeneration. *Modified from Allo B.A., et al., 2012. J. Funct. Biomater. 3, 432–463; with permission.*

it requires the most restricted requirements from the materials.

The main polymeric biomaterials and their significant properties and applications will be illustrated in the following sections.

## 4.2.1 Vinyl Polymers

Vinyl polymers owe their name to the presence in the main chain of the $CH_2$-CHR group, which derives from a monomer containing a C=C double bond. The main vinyl monomers and the polymers obtained from them are shown in Fig. 4.2.

### 4.2.1.1 Polyethylene

Polyethylene (PE) (Fig. 4.3) is obtained by radical (chain) polymerization of ethylene monomer, $CH_2$=$CH_2$, gaseous. Given its chemical stability, PE is produced without additives or a stabilizer; in addition, due to the chemical inertness and absence of toxicity, it can be used for food and medical applications.

Depending on the method of production, the structure of polyethylene macromolecules can be prevalently ramified or linear (see later).

Polyethylene can be characterized in terms of *density*. Commercial linear polyethylenes have density values ranging from 0.91 to $0.96 g/cm^3$ and corresponding molecular weights (weight = average molecular weight) ranging from 30,000 to 300,000.

Polyethylene is semicrystalline. The degree of crystallinity of the material is directly proportional to the density and is influenced by the molecular weight, its polydispersity index (PDI), and the number and length of the branches. The formation of more or less extensive ramifications along the main chain prevents the packing of the structure in crystallites; the longer the ramifications are, the greater the effect. Depending on the chain configuration and length, there are three main types of PE as described in the following:

**1.** *Low-density polyethylene (LDPE)*

The structure is branched and is obtained by high-pressure polymerization (1200 atm) at 130–350°C. Density ranges from 0.91 to $0.93 g/cm^3$, and the crystalline regions may represent 50%–60% of the whole structure, showing a $T_m = 102–112°C$.

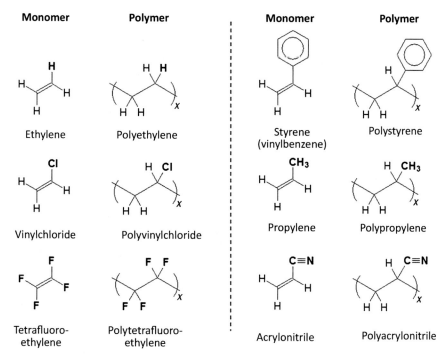

| Monomer | Polymer | | Monomer | Polymer |
|---|---|---|---|---|
| Ethylene | Polyethylene | | Styrene (vinylbenzene) | Polystyrene |
| Vinylchloride | Polyvinylchloride | | Propylene | Polypropylene |
| Tetrafluoro-ethylene | Polytetrafluoro-ethylene | | Acrylonitrile | Polyacrylonitrile |

**FIG. 4.2** Structural formulas representing the 3D shape of the main vinyl polymers and their respective monomers.

| Condensed formula | Structural formula |
|---|---|
| -[CH₂-CH₂]- | H H structure with x |

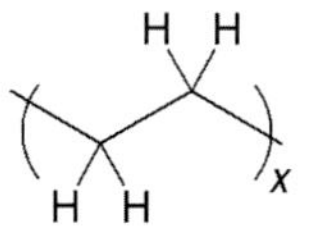

**FIG. 4.3** Chemical structure of polyethylene.

*Properties*: Tenacity, high impact resistance, flexibility, processability, transparency, chemical resistance, and low water permeability. LDPE has good heat stability in the absence of oxygen. In the presence of oxygen, degradation proceeds even at room temperature.

*Applications*: Cling films, packaging, flexible containers and bags.

**2.** *High-density polyethylene (HDPE)*

Polyethylene with density of $0.94–0.96 \, g/cm^3$, substantially linear, with crystallinity even higher than 90%, and $T_m = 128–135°C$. HDPE is synthesized using "stereospecific" catalysts (process called *Ziegler-Natta*), at low pressures (1–10 atm) and low temperatures of about 40°C.

*Properties*: HDPE films are translucent and much less transparent than LDPE film (i.e., the higher crystallinity lowers transparency). It is less flexible than LDPE and relatively stable when heated but undergoes thermo-oxidation in the presence of oxygen and photo-oxidation (see Chapter 5) even at room temperature.

*Applications*: Food and drug packaging, containers, and catheters.

**3.** *Ultra-high molecular weight polyethylene (UHMWPE)*

Linear, with a molecular weight higher than 2,000,000 Da. It is obtained with a process similar to that of HDPE. UHMWPE is substantially a very high molecular weight HDPE and is generally available in the form of powder, bars obtained by injection molding or sintering, and sheets produced by thermoforming. Although UHMWPE is a polymer with a substantially linear structure, it has a relatively low density $(0.935 \, g/cm^3)$ and is less crystalline than other types of polyethylene (crystallinity = 45%–60%). This is due to the high viscosity of the material in the fluid phase, which considerably limits the normal crystallization mechanisms. These structural characteristics make the material ductile and tenacious but also more subject to creep and absorption of liquids. For biomedical applications, UHMWPE must meet specific requirements established by standards that fix its mechanical properties (DIN 58836) and the residual content after combustion (DIN 58834).

The high molecular weight makes it difficult to work with using conventional methods, and in general, it is transformed by mechanical processing of sheets via compression-molded or piston-extruded profiles; high-speed and high-pressure injection machinery, equipped with advanced control systems, are presently available.

*Properties*: High abrasion resistance, higher impact resistance than any other plastic material, low surface friction coefficient, and high resistance to radiation.

In general, polyethylene is a tough material; in the presence of detergents, however, it is sensitive to the phenomenon of stress-cracking.

UHMWPE is very resistant to chemical agents, so much so that there are solvents able to dissolve it only at high temperatures; it is also very resistant to acids. It has a tendency to oxidation and photodegradation; exposure to photons promotes the breaking of the chains, with a reduction of the molecular weight and worsening of the mechanical characteristics (see also Chapter 5.2 and Chapter 6.3).

UHMWPE *fibers* with very high mechanical strength, which are also used for the preparation of composite materials, are produced by gel spinning (spinning in the gel state). These fibers have a highly crystalline-oriented structure and are defined as 10 times more resistant than steel.

The structure of LD, HD, and UHMWPE is depicted and compared in Fig. 4.4.

*Application of UHMWPE as a biomaterial*: UHMWPE is mainly used for components of articular prostheses (acetabular cup in hip prosthesis, tibial insert in knee prosthesis), and also in internal and external osteosynthesis devices. Fig. 4.5 shows some examples.

The use of UHMWPE as a biomaterial represents a very small percentage of the total polyethylene market. Despite being a limited market in terms of the quantity of UHMWPE employed, it is still an interesting market from an economic point of view.

*Disadvantages of UHMWPE*: Histological studies of periprosthetic tissues have shown the presence of polyethylene particles of varying sizes, released by wear phenomena of the acetabular cup or tibial insert in UHMWPE. Some,

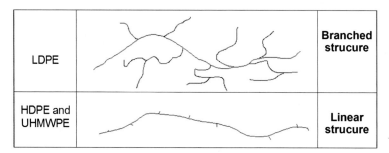

FIG. 4.4  Comparative illustration of low-density (LD), high-density (HD), and ultra-high molecular weight (UHMW) polyethylene.

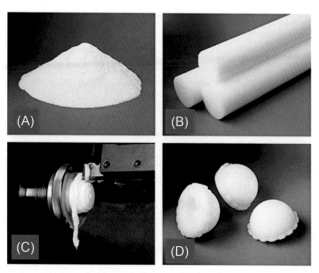

**FIG. 4.5** (A)–(D) Example of processing steps for manufacturing UHMWPE components of joint prostheses: (A) UHMWPE powder, (B) semifinished rods obtained by consolidating the UHMWPE powder, (C) machining of the rods, (D) finished acetabular components. *Adapted from Kurtz, Handbook of UHMWPE, third ed. (Chapter 2).*

appreciable macroscopically, reach dimensions of 3–4 mm, are surrounded by a fibrous capsule, and behave like inert foreign bodies; others, of smaller size, activate an intense inflammatory reaction and, in particular, recall a large number of macrophages (see Chapter 6, Section 6.3.3).

The most damaging effects are therefore caused by those wear patterns that generate smaller particles, because they can lead to the necrosis of the bone surrounding the prosthesis, that is, the greater the proliferation of the cells involved in the inflammation caused by the wear debris, the greater the tendency of invasion and replacement of bone tissue with granulomatous tissues. Because the stability of the implant depends on a strong bone support, phenomena of mobilization of the device can occur.

There is a minimum wear rate threshold below which the action of phagocytic cells can remove debris faster than these can be generated. The minimization of wear and its rate therefore project specifications necessary for the proper functioning of the articular prosthetic implant.

*Effects of sterilization on the properties of UHMWPE:* The prosthetic components in UHMWPE were sterilized in the past in almost all cases by a dosage of 25 kGy rays in a noninert atmosphere (air). Under these conditions, degradation phenomena can be triggered with consequences on mechanical characteristics and wear resistance. Presently, sterilization with ethylene oxide or gas plasma is preferred, but γ-ray irradiation in inert atmosphere and stabilization with antioxidants (i.e., vitamin E, a group of eight compounds that include four tocopherols and four tocotrienols) is used to purposely cross-link UHMWPE and render it stable to oxidation and degradation (Kurtz, 2016).

### 4.2.1.2 Polypropylene

PP is a thermoplastic polymer, obtained by chain polymerization of the propylene gaseous monomer (Fig. 4.6). Its structure is partially crystalline, with a $T_m = 174°C$ and a $T_g = -10°C$.

By stereospecific polymerization (i.e., by use of metal-organic catalysts like Ziegler-Natta

| Condensed formula | Structural formula |
|---|---|
| –[CH₂–CH(CH₃)–]– | |

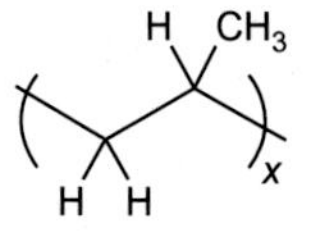

FIG. 4.6    Chemical structure of polypropylene.

ones), a highly regular material is obtained, with a high degree of crystallinity (*isotactic* and *syndiotactic*) with $T_m > 170°C$. The different types of PP are shown in Fig. 4.7.

*Properties*: PP has greater rigidity, hardness, and mechanical resistance than PE but lower impact resistance. When isotactic or syndiotactic, PP has greater crystallinity and then higher mechanical strength.

The carbon atom that carries the $CH_3$ group is tertiary, thus more subject to oxidation and degradation; therefore PP degrades more than PE (i.e., turns yellow) and must be stabilized to resist external agents.

Films and fibers of PP can be produced; copolymers are advantageously prepared. Copolymers with polyethylene (PE/PP) are mainly intended to replace PVC. The advantages of PVC are the absence of plasticizers, but the cost is higher. The improved transparency of these materials allows autoclave sterilization because there is no excessive development of opalescence.

*Applications and problematic aspects*: PP has practically all the same applications of HD polyethylene, with respect to which it has the advantages of lower density, higher mechanical

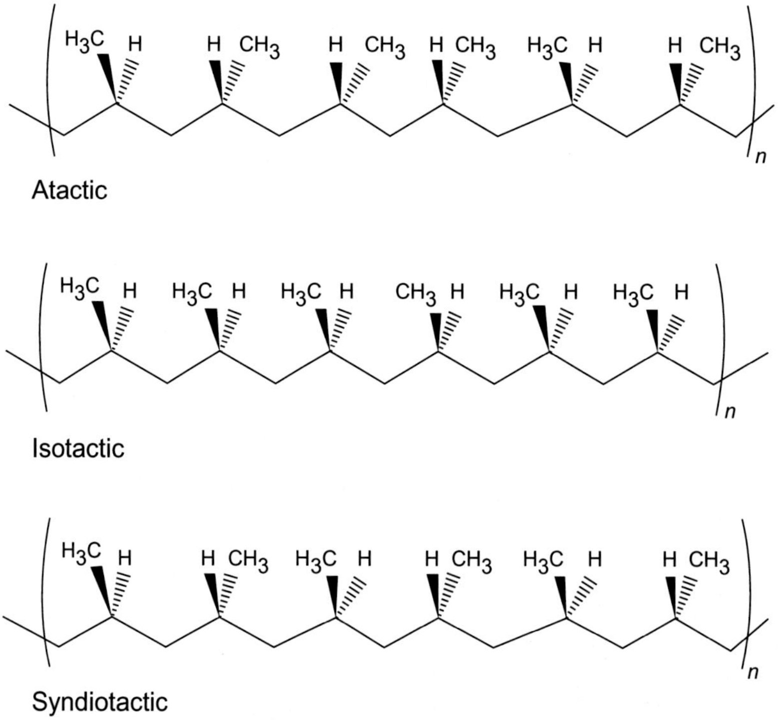

Atactic

Isotactic

Syndiotactic

FIG. 4.7    Structural formula of the three possible types of polypropylene. In the atactic, the methyl group is placed randomly above and below the plane of the main chain; whereas in the isotactic, it is always on the same side; and in the syndiotactic, it alternates in a regular way above and below the plane.

strength, greater surface hardness, and superior heat resistance.

Together with PVC, it is the most commonly chosen material for sterile packaging, syringes, and tubes. Both polymers are brittle and tend to change color; they do not tolerate ionizing radiation and, for this reason, stabilizers are used. Gamma ray sterilization, in particular, causes chain breaking and cross-linking, with degradation of the properties of the material. It is possible to prepare special degrees of radiation-resistant PP.

PP is also used in the medical field for connectors, bottles, membranes for oxygenators, finger joints, nonabsorbable sutures (longer lasting than polyamide ones), and meshes for hernia repair. For example, some commercial meshes for surgical repair are shown in Fig. 4.8.

### 4.2.1.3 Polyvinylchloride

PVC is a substantially amorphous material, with the possibility of exhibiting 7%–20% crystallinity. It is obtained by chain polymerization from the vinyl chloride monomer (Fig. 4.9).

Different types of PVC can be obtained, according to different production processes: from rigid (not plasticized) to semirigid, to flexible.

*Properties*: The intrinsic characteristics of PVC are toughness and translucency. PVC is unstable

| Condensed formula | Structural formula |
|---|---|
| -[CH₂-CH(Cl)-]- |  |

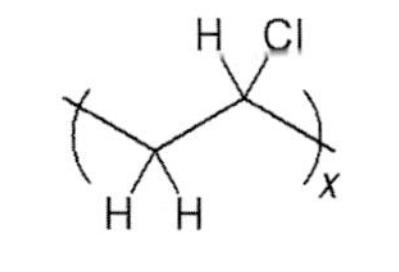

FIG. 4.9  Chemical structure of polyvinylchloride.

at processing temperatures (near 150°C) and tends to decompose with the development of hydrochloric acid, assuming a yellowish or brownish color and worsening in many of its physical-mechanical characteristics. PVC has a tendency to discolor, turning yellow to brown with loss of properties even after irradiation (including gamma radiation). Therefore stabilizers are added.

It cannot be sterilized in an autoclave, and resterilization must be avoided. The development of HCl can corrode any metal parts in contact.

*Applications and problematic aspects*: PVC is still the most used material in medicine, with 75% in the form of material plasticized with dioctyl (or 2-ethylhexyl) phthalate (DOP or DEHP). DOP, however, presents problems of toxicity, has a low molecular weight, and can therefore be released into the surrounding environment.

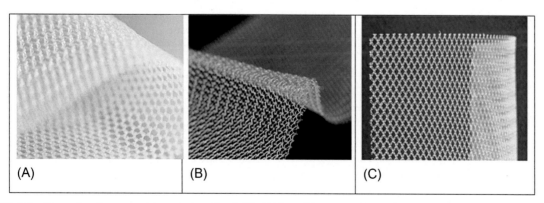

FIG. 4.8  Examples of commercial surgical meshes in PP. (A) http://mysurgicalspecialist.uk/chronic-pain-mesh-after-hernia-surgery/; (B) http://penangsurgeon.com/laparascopic-surgery/laparoscopic-hernia-repair/; (C) http://proxybiomedical.com/vitamesh-macroporous-pp-surgical-mesh/.

This is a problem, especially for dialysis patients who undergo dialytic treatment several times a week. However, there is a regulation on the maximum quantity that can be released from medical grade PVC.

Plasticizers that can replace it are often more expensive, because they are still produced in a smaller scale. Alternative plasticizers are compounds based on citrate, trimellitate, adipate, and terephthalate esters (Van Vliet et al., 2011; Eljezi et al., 2017; Nielsen et al., 2014). Alternatively, polymeric blends based on TPU mixtures (polyurethane thermoplastic elastomers) and PVC, or PE, EVA (ethylene-vinylacetate) copolymers, and copolymers of PP (Dypro Z) have been proposed.

The greatest use in medicine of PVC is for single-use items. Other uses are listed in Table 4.1.

### 4.2.1.4 Polystyrene

Polystyrene is obtained by radical (chain) polymerization of the styrene monomer (Fig. 4.10).

*Properties*: PS homopolymer is amorphous, with $T_g \cong 100°C$. It is rigid, hard, transparent, and brittle (also called improperly "crystal polystyrene"), with excellent insulating properties, low cost, good moldability, low water absorption, and good dimensional stability.

From a physiological point of view, it is well tolerated.

| Condensed formula | Structural formula |
|---|---|
| $-[CH_2-CH(C_6H_5)-]-$ | 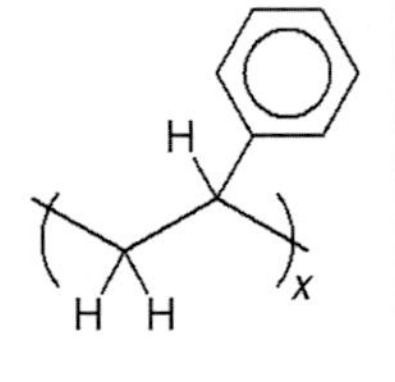 |

FIG. 4.10　Chemical structure of polystyrene.

*Applications and problematic aspects*: Polystyrene in the medical field finds application for the production of syringes, disposable material, and rigid packaging (kit for analysis). Properly surface-modified, it represents the material of choice of plates (wells) for cell cultures (tissue culture PS or TCPS). In the expanded form, it is used for thermal and acoustic insulation and for packaging.

One of the drawbacks presented by polystyrene is the presence in the final articles of residual monomer (toxic to auditory organs, i.e., ototoxic, by repeated exposure and toxic for reproduction of category 2, i.e., suspected of damaging the fetus) that manufacturers are committed to reduce. The manufacturer has an obligation to verify that the monomer content is below a certain limit value.

One of the advantages of PS is the resistance to gamma radiation, which offers the possibility of using this type of sterilization.

Following reaction with other monomers, such as butadiene, acrylonitrile, acrylates, it is possible to produce plastic materials with a wide spectrum of characteristics (Fig. 4.11), ranging from brittle polystyrene, to materials with greater toughness and impact resistance.

Major copolymers and blends are high impact polystyrene (HIPS), styrene-acrylonitrile (SAN), acrylonitrile-butadiene-styrene (ABS), styrene-butadiene block copolymers (SBS), and styrene-acrylate copolymers (SMMA).

*High-impact polystyrene (HIPS).* HIPS is manufactured by dissolving unsaturated rubber in a

TABLE 4.1　Biomedical Applications of PVC

| Flexible PVC | Rigid PVC |
|---|---|
| Blood bags | Laboratory instruments |
| Tubes | Hospital equipment, operating theater equipment |
| Disposable gloves | Medical packaging |
| Catheters (nonprolonged use) | Containers |
| Intravenous probes | Disposable accessories |

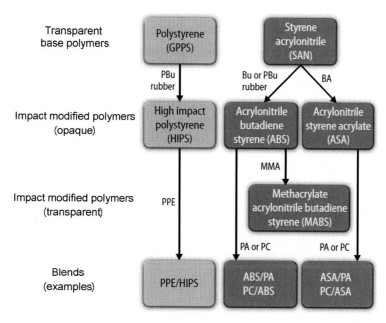

**FIG. 4.11** Some significant examples of different materials obtainable from styrene copolymers and blends. *BA*, butyl acrylate; *Bu*, butadiene; *GPPS*, general purpose polystyrene; *PA*, polyamide; *PBu*, polybutadiene; *PC*, polycarbonate; *PPE*, polyphenylene ether. *Modified from Niessner, N., et al., 2014. Kunststoffe Int. 10.*

styrene monomer and polymerizing the monomer in a solution or mass-suspension process. The rubber is generally polybutadiene. In this process, the resultant blend will contain not only rubber and polystyrene but also a graft polymer where short styrene side chains have been attached to the rubber molecules, which enhances the impact strength. HIPS obtained by this process has an impact strength seven times greater than general purpose PS (GPPS).

*Butadiene-styrene copolymers (SBS).* SBS is transparent, unbreakable, and sterilizable with ethylene oxide and gamma radiation. Butadiene gives impact resistance to the material. These copolymers are used for the production of disposable items and containers for liquid collection, impact resistant packaging, tubes, surgical instruments, and other operating room items.

*Acrylonitrile-butadiene-styrene (ABS).* These materials are complex blends and copolymers. In most types, acrylonitrile and styrene are grafted onto a polybutadiene backbone. The product also contains unreacted polybutadiene and some acrylonitrile-styrene copolymers. The reasons for its widespread acceptance are high impact resistance, good stiffness, excellent surface quality, and high dimensional stability at elevated temperatures. Its main disadvantages are lack of transparency, poor weathering resistance, and poor flame resistance.

*Styrene acrylonitrile copolymer (SAN).* These copolymers show excellent dimensional stability, very good tensile and flexural strength, good abrasion resistance and impact strength, high chemical resistance, and better resistance to stress cracking and crazing. They have a low cost (e.g., about half compared to PC), can be sterilized with radiation, are rigid and resistant, and have good chemical resistance to alcohols, drugs, and disinfectants. They are used for hospital instrumentation components (nozzles, blood aspirators) and for external dialyzer support.

| Condensed formula | Structural formula |
|---|---|
| -[CF₂-CF₂-]- | |

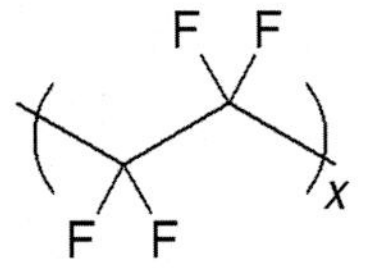

FIG. 4.12   Chemical structure of polytetrafluoroethylene.

### 4.2.1.5 Polytetrafluoroethylene

PTFE is obtained by chain polymerization from the gaseous monomer tetrafluoroethylene (TFE) (Fig. 4.12).

The polymer is translucent white, highly crystalline (the TFE molecule is symmetric and therefore the ordered conformation is favored), with high thermal stability and very high molecular weight ($10^6$–$10^7$ Da). PTFE has a $T_g$ of $-120°$ C and different $T_m s$. The main melting temperature is $+327°C$.

*Properties*: Chemical inertia, expensive, heat resistant, excellent electrical insulation.

It cannot be transformed with the usual techniques; due to its chemical inertia and high molecular weight, it flows with difficulty. It is called a *thermoelastomer* because it undergoes a thermoelastic transition via heating, but after the crystalline zone has melted, the plastic material does not become sufficiently fluid to be worked as a normal thermoplastic. Therefore PTFE is mechanically worked.

Fluorinated polymers are nonflammable and physiologically inert. The nonadhesiveness is due to the fact that fluorine atoms prefer their similar, attracting each other, rejecting any other type of molecule (e.g., water).

PTFE is sold in the form of granular resins, fine powder, and aqueous dispersion.

*Applications*: PTFE in its compact form is used for accessories and parts of biomedical instrumentation. It can not be used in artificial joints because it releases microscopic wear debris, which provokes an intense inflammatory reaction.

One of its commercial names is Teflon (Du Pont).

In its expanded form (patented process), it is waterproof, allowing the exchange of gas and transpiration. As a biomaterial, it is used for the production of medium-caliber vascular prostheses (Fig. 4.13), sutures, cardiac patches, tissues for hernia repair, and membranes for guided bone regeneration in

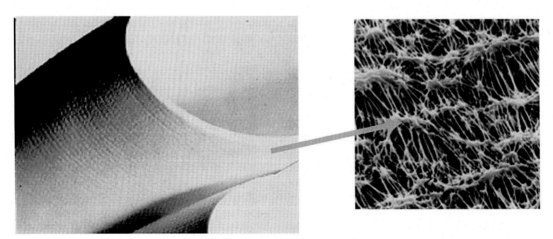

FIG. 4.13   Vascular prosthesis in expanded PTFE, ePTFE (on the right: electron microscope image of the expanded wall structure).

| Condensed formula | Structural formula |
|---|---|
| -[CH₂-CH(CN) ]- | 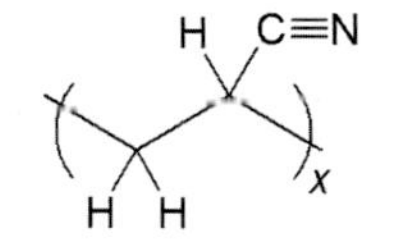 |

FIG. 4.14  Chemical structure of polyacrylonitrile.

| Condensed formula | Structural formula |
|---|---|
| -[CH₂-CH(OH)-]- | 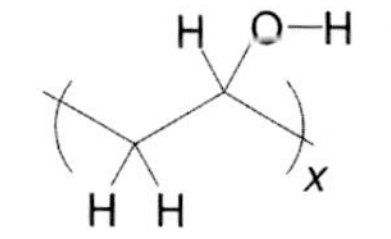 |

FIG. 4.15  Chemical structure of polyvinylalcohol.

dentistry. One of its well known commercial names is *Goretex*.

### 4.2.1.6 Polyacrylonitrile

PAN is obtained by chain polymerization of acrylonitrile monomer (Fig. 4.14). It is highly crystalline, with $T_g = +87°C$ and $T_m > 200°C$.

*Properties*: PAN decomposes above the melting temperature and therefore it is not possible to process it as a thermoplastic. Copolymers, for example, with styrene (SAN), have increased heat its resistance. PAN has a limited permeability to gases (oxygen, $CO_2$), for which it represents a "barrier" material, yet it shows water permeability.

Copolymers can be used in the food or medical field, provided they meet the requirement to contain <10 mg/kg (10 ppm) of monomer.

*Applications*: Fibers produced by dry or wet spinning. Thanks to its water permeability, PAN is used for the production of dialysis membranes. It is also the precursor for carbon fibers (see Section 1.5 in Chapter 1).

### 4.2.1.7 Polyvinylalcohol

Polyvinylalcohol (PVA) is a linear water-soluble homopolymer, with 350–2500 repeating units. Unlike many vinyl polymers, PVA is not prepared by polymerization of the corresponding monomer but is obtained by hydrolysis (i.e., by partial or total de-esterification with methanol in an alkaline environment) of polyvinylacetate (Fig. 4.15).

A copolymer, poly(vinyl alcohol-*co*-vinyl acetate) is obtained in case of partial hydrolysis (namely when about 20% of the acetate groups are left on the polymer, as roughly shown in the following formula):

$$- [-CH_2CH(OCH_3)]_n + CH_3OH\,(NaOH) \longrightarrow$$
$$-[-CH_2CH(OH)-]_{n-x}$$
$$-[-CH_2CH(OCH_3)-]_x$$

*Properties*: PVA has a highly polar structure and tends to align the chains forming crystalline-oriented structures. The melting temperature varies, depending on the degree of esterification, from 180°C (highly esterified structure) to 240°C (lower content of ester groups), with a $T_g$ varying between 40°C and 80°C.

As in the case of PP, isotactic, syndiotactic, and isotactic configurations are possible. The solubilization of PVA becomes more difficult with increasing the content of OH groups and the molecular weight. PVA has excellent emulsifying, adhesive, and film-forming properties. As a film, it has high tensile strength, flexibility, and excellent barrier properties against oxygen and aromas. PVA is completely biodegradable.

*Applications*: As a film, it is used for packaging, and as a $CO_2$ barrier in PET bottles. In medicine, it is used as an embolizing agent; in ophthalmology, in eye drops; and as a lubricant in contact lenses.

It can be cross-linked with different chemical reactions and secondary bond formation, to obtain "slimes" or hydrogels with adequate biocompatibility and sufficient mechanical properties, suitable for use as wound dressing materials and other medical applications such as drug delivery (Kamoun et al., 2015, 2017; Maitra and Shukla, 2014).

## 4.2.2 Acrylic Resins

Acrylic resins are divided into two main groups: *polyacrylates* and *polymethacrylates*. Although part of the vinyl polymers, they are described separately because they are brought together by the presence of an ester group linked to the vinyl group.

Fig. 4.16 shows the structural formulas of these polymers and their respective monomers.

### 4.2.2.1 Polyacrylates

$R=CH_3$:    polymethyl-acrylate;    $R=C_2H_5$ polyethyl-acrylate.

*Properties*: They are soft, tough, and rubbery. Polyethyl-acrylate is considerably softer and more extensible than polymethyl-acrylate. Both acrylates are often copolymerized to achieve the desired hardness, flexibility, and strength; two important acrylic comonomers are butyl and 2-ethylhexyl acrylate (http://polymerdatabase.com).

*Applications*: Major applications are coatings, paints, textiles, tape adhesives, and emulsion paints.

### 4.2.2.2 Polymethacrylates

*Properties*: They are transparent, rigid, chemically stable, and resistant to aging. They are obtained by chain polymerization of the corresponding monomers and are generally *amorphous*. The value of $T_g$ varies from +20°C to +140°C, and is influenced by the type of group R present in the polymer formula.

In general, acrylic resins after polymerization are well tolerated by the human organism and are used in orthopedics and dentistry in various forms (self-curing, curing by application of heat or ultraviolet light, with a linear or branched structure, or cross-linked; pure or composite form with mineral fillers in particles).

### 4.2.2.3 Polymethylmethacrylate

This polymer is thermoplastic, linear, and obtained by chain polymerization of methyl methacrylate, a liquid monomer (Fig. 4.17). PMMA is completely amorphous, with $T_g$ in the range of +100°C to +125°C.

*Properties*: PMMA has excellent optical transparency and exceptional stability to UV rays and atmospheric agents. The transmission of light is 92%, the highest among those presented by all other plastics, and higher than that of common inorganic glass.

*Applications as biomaterial*: Rigid contact lenses, intraocular lenses (IOL), bone cement, orthopedic and dental prostheses.

**APPLICATION EXAMPLE OF PMMA: BONE (ACRYLIC) CEMENT**

*Preamble*: The considerable difference in mechanical properties, and in particular of

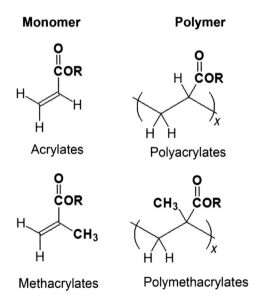

**Monomer**          **Polymer**

Acrylates          Polyacrylates

Methacrylates          Polymethacrylates

**FIG. 4.16**  Structural formulas of acrylates, methacrylates, and their corresponding polymers.

| Condensed formula | Structural formula |
|---|---|
| -[CH₂-C(CH₃) (COOCH₃)-]- | CH₃, COCH₃ ... |

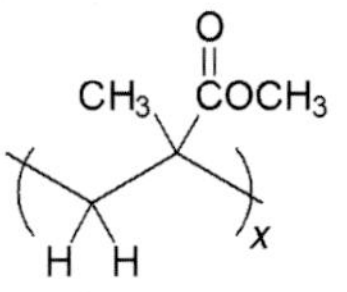

**FIG. 4.17**  Chemical structure of polymethylmethacrylate.

elastic behavior, between materials used for the construction of prosthetic devices on one side and bone tissue on the other hand causes a high concentration of loads at the bone-prosthesis interface. The use of *acrylic cement* is an advantageous method also for achieving better distribution of loads to this interface.

The acrylic cement, or bone cement, has the main function of fixing joint prostheses in the established bone site.

Its use is linked to the fact that, during the preparation phase, it presents plasticity, a property that allows it to adapt to the shape of the bone in which it is introduced and to the prosthetic element inserted in it; whereas, when complete polymerization is achieved, it hardens and blocks the prosthetic element in the bone.

It is classified among the self-curing acrylic resins based on PMMA.

The bone cement is prepared in the operating room at the time of surgery by manually, or by use of appropriate tools, mixing two components (a "powder" and a "liquid") present in the commercial packaging and appropriately predosed by the manufacturer.

The "powder" consists of prepolymerized PMMA, in the form of spherical particles of a few tens of μm in diameter, and contains the polymerization initiator (i.e., benzoyl peroxide, PB). In place of or in addition to pure PMMA (homopolymer), copolymers with styrene or other acrylic monomers, such as methyl acrylate (MA) or butyl methacrylate (BMA), are also used. BMA, in particular, reduces fragility by imparting a more "rubbery" character to the cement.

The "liquid" consists of an MMA monomer, plus a polymerization accelerator (an aromatic amine, dimethyl-para-toluidine, or DMPT) and other activators of the reaction, in addition to traces of an inhibitor (hydroquinone, HQ) added with the aim of avoiding premature polymerization of the monomer.

The chemical structure of the polymeric material obtained is therefore predetermined and depends on its commercial formulation, whereas its physical and mechanical characteristics are greatly influenced by the mixing modalities and by the surgical implantation techniques, as well as by the design of the prosthetic device used.

At the time of mixing, the solid polymer particles are swollen by the monomer and, due to the amine/peroxide system, the polymerization starts in and around the swollen particles.

By varying the particle size, the initiator and accelerator concentration and using different acrylic monomers, the properties of the cement can be modified. However, the type of chemical formulation allows hardening of the cement (indicative of the occurrence of the polymerization reaction) in a short time (within 15 min).

Depending on the viscosity of the mixture, the cements can be inserted into the implant site manually or with a syringe.

From a practical point of view, two stages can be distinguished:

- *Dough time*: From the beginning of the mixing to the moment in which the insertion of the cement can begin. There is a mixing time recommended by the manufacturer, above which it is good not to proceed because of the increase in viscosity due to the advancement of the polymerization reaction.
- *Setting time*: From the beginning of mixing to the one in which the hardening due to polymerization begins. It includes the dough time and the time of workability of the dough (*working time*).

These stages are followed by an increase in temperature, due to the polymerization reaction, with the attainment of a maximum temperature, or *exothermic peak* (see Fig. 4.18).

*Variables and involved problems*: Problems associated with the use of acrylic cement in the orthopedic field are related to the type of chemical formulation, the type of reaction at stake (polymerization), and the preparation technique. These variables have great influence on

**FIG. 4.18**  Time/temperature ($t/T$) diagram of acrylic cement.

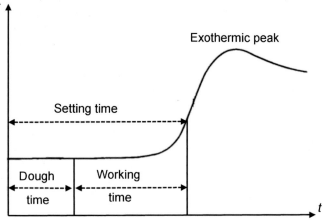

the chemical-physical structure of the material and involve serious aspects of toxicity to the human body.

1. *Toxicity*: Polymerization inhibitors, aromatic amines and peroxides, as well as the radicals created during the reaction are toxic substances for the human body. In particular, methyl methacrylate monomer (MMA), which is toxic just as a monomer (highly reactive) and used in much higher quantities than the other components, is critical. It is accused of causing allergy and to have cytotoxic effects, not to mention systemic effects, the most relevant of which is the pressure drop that occurs when the cement is inserted.

2. *Rise in temperature*[1]: The polymerization reaction is exothermic (130 cal/g of monomer), and the heat produced results in an increase of temperature (exothermic peak), to the extent of which the bone necrosis that can be found at the interface between bone

and cement has been attributed. Between 48°C and 60°C, cell necrosis depends on the exposure time, which at 50°C can be between 30″ and 400″; in addition, protein denaturation begins at 56°C. Experimental measures have shown that the temperature of the exothermic peak at the cement/bone interface seldom reaches such critical values. It is however very important for this purpose to keep the thickness of the cement mantle low.

3. *Volumetric shrinkage and porosity*: PMMA is normally a compact and transparent polymer, whereas acrylic cement is, roughly speaking, an opaque and porous material, and the nontransparency is due to the porosity of the material obtained.

The porosity of the cement comes from the preparation technique that involves, on one hand, the incorporation of air during mixing, and on the other, the evaporation of monomer due to the polymerization temperature (which reaches sufficiently high values inside the mass). The formation of porosity causes a decay of the PMMA mechanical properties. To reduce porosity, slow and nonprolonged mixing, as well as pressurization, centrifugation, vacuum mixing, and reduced thickness of the cement mantle, are important factors.

[1] *Important note*: The exothermicity of the reaction is linked solely to the polymerization reaction of the monomer and should not be confused with the term: *thermosetting*. Cement, like commercial PMMA, is linear and therefore not cross-linked, therefore thermoplastic.

On the other hand, porosity is opposed to an undesired phenomenon but intrinsic to the type of reaction: the volumetric shrinkage. In fact, the amount of possible shrinkage is reduced by the formation of porosity to values equal to 2%–4% of the initial volume.

Although there is no reliable assessment of the optimum porosity of the cement and the best technique for obtaining it, it is however reasonable to conclude that the porosity of the cement can and should be controlled.

### 4.2.2.4 Poly-Hydroxyethyl-Methacrylate

From a chemical point of view, PHEMA is part of the acrylic polymers, but it is mainly used in a weakly cross-linked form, where it has hydrogel properties (Fig. 4.19).

Usually, PHEMA is obtained by chain polymerization of the corresponding monomer (HEMA) in the presence of a divalent monomer, the *ethylene-dimethacrylate*, which acts as a cross-linking agent. The resulting polymer, which is brittle and glassy, and with properties similar to those of PMMA, is able to absorb about 40% by weight of water, taking on a rubbery consistency. The addition of more than 50% by weight of water leads to spongy, opaque materials.

*Applications*: Today, PHEMA is still the most used material for the manufacture of soft contact lenses, being moderately permeable to oxygen, unlike PMMA. However, this permeability in some cases is not sufficient to ensure proper oxygenation of the cornea, for which HEMA copolymers have been studied or proposed with other materials.

### 4.2.2.5 Acrylic Hydrogels as Biomaterials

Hydrogels are described in Chapter 1, Section 1.2. Here, the use of acrylic hydrogels as biocompatible materials is concisely presented.

Table 4.2 provides some important information on this category of polymeric materials, considering their ability to absorb water and their end use as biocompatible materials.

Important hydrogels that find a biomedical use are:

- PHEMA and copolymers
- PVP (polyvinyl-pyrrolidone)
- poly-acrylamide (PAM)
- poly-vinyl alcohol (PVA)
- polyacrylic acid and polymethacrylic acid
- poly-$N,N'$-dimethylaminopropyl-acrylamide (poly-NIPAAm)

and combinations thereof with hydrophobic and hydrophilic comonomers, cross-linking agents, and other modifying agents.

**TABLE 4.2** Classification of Hydrogels According Water Swelling and End Use

| Property | Classification |
|---|---|
| Water content (% weight at the equilibrium) | Low swelling (20%–50%) |
| | Medium swelling (50%–90%) |
| | High swelling (90%–99.5%) |
| | Super-absorbent forms (>99.5%) |
| End use | *Structural hydrogels* For applications requiring a defined shape, mechanical integrity, shape memory (*contact lens, membranes, coatings, drug-releasing matrices, sponges, tubes, implants*) |
| | *Absorbent additives* For applications that do not require a specific shape (*urine-adsorbents, soil conditioners, emulsifiers, rheological additives, lubricants, fillers, etc.*) |

| Condensed formula | Structural formula |
|---|---|
| -[CH₂-C(CH₃) (COOCH₂CH₂OH)-]- | CH₃ COCH₂CH₂OH |

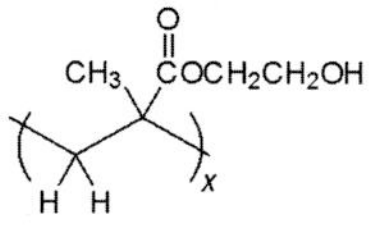

**FIG. 4.19** Chemical structure of poly-hydroxyethyl-methacrylate.

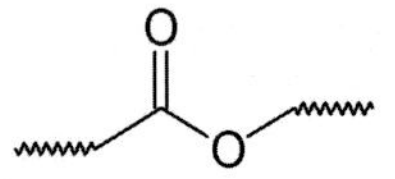

FIG. 4.20 Chemical structure of the ester group inserted in the polymer backbone.

## 4.2.3 Polyesters

These polymers contain the chemical group *ester* in the main chain (Fig. 4.20). They are obtained by step polymerization of monomers containing hydroxyl and/or carboxylic groups.

What distinguishes one polyester from another are the chemical groups linked to the ester bonds. Polyesters can be aliphatic or aromatic, saturated or unsaturated. They can also be linear or cross-linked, biodegradable or biostable. Homopolymers and copolymers, in particular block copolymers, can be produced with very different physical properties.

### 4.2.3.1 Polyethylene Terephthalate

PET is a saturated polyester, thermoplastic, aromatic, with the structural formula reported in Fig. 4.21.

PET is obtained by step polymerization of terephthalic acid with ethylene glycol. It is highly crystalline, with $T_g = 70°C$ and $T_m = 265°C$.

*Properties*: Excellent surface hardness, good chemical resistance, and dimensional stability. Excellent transparency, wear resistance, and good antifriction properties, low creep.

Increasing the degree of crystallinity increases resistance to chemical agents and degradation.

*Limitations*: Strict molding conditions.

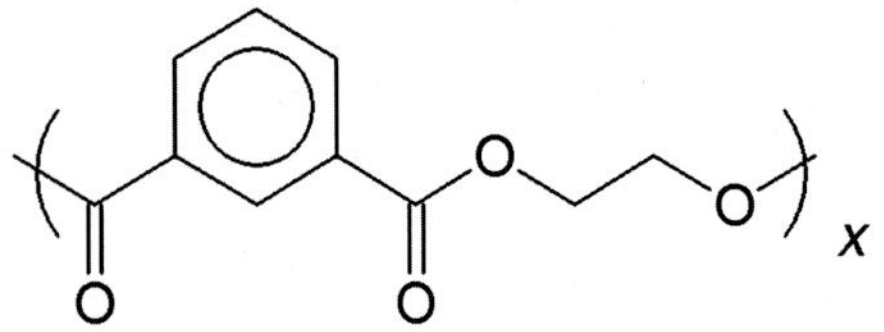

FIG. 4.21 Chemical structure of polyethylene terephthalate.

*Applications*: Bi-oriented films for electrical insulation and food packaging, containers, blow-molded bottles.

**PET FIBERS**

The most widely used fibers are those obtained from linear PET. The process of obtaining these fibers allows the production of a large number of products for the textile industry: from clothing to furniture, from the automotive industry to that of carpets.

High modulus fibers with high mechanical strength, low volumetric shrinkage, heat, light, and chemical agents can be obtained. The properties are influenced by the fiber structure, which depends above all on the parameters of the fiber formation process, described in Chapter 3 (Section 3.2.2.2).

*Applications*: In the textile industry for fabrics and ropes. In medicine, with the names of Dacron, Mylar, etc., mainly for vascular prostheses of large caliber ($\varnothing \geq 8mm$), patches and gauzes, suture rings, nonbiodegradable sutures.

**DACRON**

Depending on the type of weaving, two types of textile structures can be obtained: *woven* or *knitted*.

A *woven* graft is manufactured by interlacing two sets of yarn (warp and weft) oriented at 90 degrees to each other. These grafts are currently available in different types of weave designs namely, plain, twill, and satin (Fig. 4.22).

*Knitted* grafts have a looped filament construction in which a continuous interconnecting chain of yarn loops spirals around the graft circumference. Knitted structures are softer, more flexible, compliant, and have better handling characteristics than woven structures. The most common types of knits, which are used for graft design, are the *weft* (machining carried out in the radial direction) and *warp* (machining carried out in the longitudinal direction) knit constructions (Fig. 4.23).

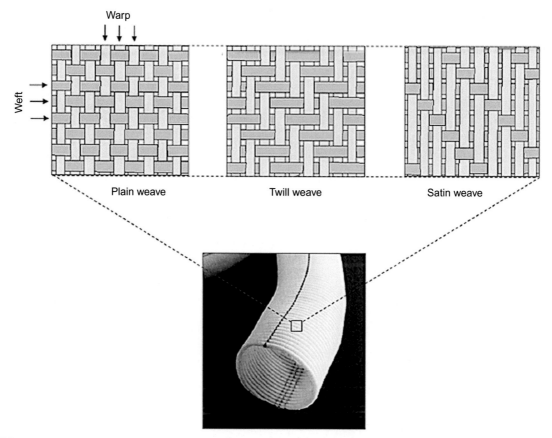

FIG. 4.22  Textile patterns of a woven Dacron vascular prosthesis. *From Fig. 4 of Singh, C., et al., 2015. J. Funct. Biomater.*

The warp is denser than the weft and resists fraying, whereas the weft is more flexible and stretchable, and its use in vascular prostheses requires precoagulation and is more prone to dilatation.

In the production of vascular grafts, additional technologies can also be used, such as the *velour* represented by many filaments anchored to the woven or knitted structure. These filaments are exposed on the surface of the prosthesis (the surface looks very rough and irregular), and can be added both on the outer and on the inside surface (in contact with blood) of the prosthesis.

The vascular graft, therefore, can be flat, without velour, or present internal velour, external, or both. The more in-depth treatment of the problem concerning vascular prosthesis, however, goes beyond the scope of this book and therefore it is referred to other bibliographic sources.

### 4.2.3.2 Biodegradable (or Bioabsorbable) Polyesters

The most known and used bioabsorbable polyesters are poly-α-hydroxy acids (Fig. 4.24).

Both polymers are obtained by ring opening polymerization of the respective monomers, glycolide and lactide (Fig. 4.25).

Therefore although the final product is a polyester, theoretically obtainable from the reaction between alcoholic groups (OH) and acid groups (COOH), in the case of PGA and PLA,

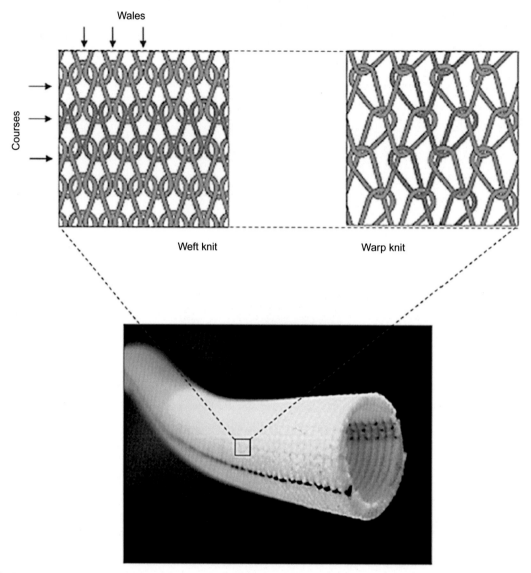

**FIG. 4.23** Textile patterns of a knitted Dacron vascular prosthesis. *From Fig. 6 of Singh, C., et al., 2015. J. Funct. Biomater.*

| Poly-$\alpha$-hydroxy acid | Polyglicolide (PGA) | Polylactide (PLA) |
|---|---|---|
| | R = H | R = CH$_3$ |
| | | |

**FIG. 4.24** Chemical structure of poly-$\alpha$-hydroxy acids, PGA, and PLA.

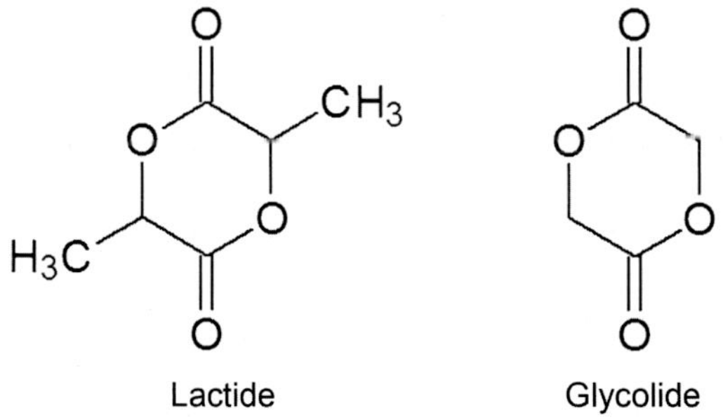

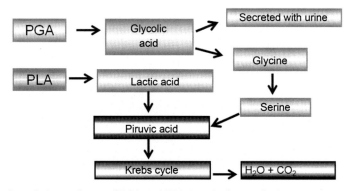

FIG. 4.25  Chemical structure of lactide and glycolide.

differently oriented in space. Actually, the forms with which PLA is marketed are the racemic (PD, L-LA) or the L form (PL-LA).

PGA, due to its simplicity of structure, easily crystallizes (50% of crystallinity, $T_m \cong 220°C$), whereas only PL-LA is semicrystalline ($T_m \cong 170°C$).

The amorphous zones of PGA exhibit a $T_g = 36°C$, therefore at the temperature of the human body they are straddling this transition (i.e., lower rigidity of the material). The $T_g$ of PLA is near 60°C.

To change the properties and modulate the degradation kinetics, copolymers can be prepared with PGA and PLA, PL-LA and PD, L-LA, and with other monomers.

*In vivo degradation*: The in vivo degradation mechanism involves the hydrolysis of ester bonds, the release of lactic and/or glycolic acid, their metabolism according to the pathway shown in Fig. 4.27, and the final elimination from the human body in the form of water and $CO_2$.

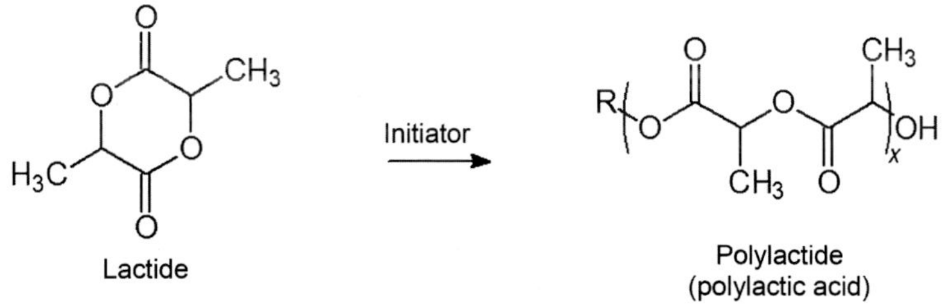

it is more correct to speak about *polyglycolide* and *polylactide*. An example of such polymerization reaction for polylactide is shown in Fig. 4.26.

*Configurational stereoisomerism and morphology:* Unlike polyglycolide, polylactide has a chiral carbon atom (see Annex 1), with the possibility of forming optical isomers. Theoretically, therefore, for PLA it is possible to obtain the laevorotatory (L) and the dextrorotatory (D) forms, in which the methyl group is found

FIG. 4.26  Polymerization of lactide to produce polylactide (PLA).

FIG. 4.27  Schematic degradation pathway of PGA and PLA into the human body.

B. BIOMATERIALS AND BIOCOMPATIBILITY

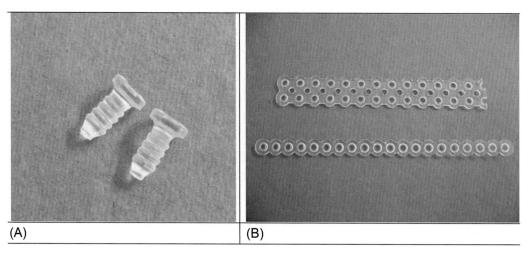

**FIG. 4.28** Devices used in maxillofacial surgery and cranioplasty manufactured in PL-LA/PD, L-LA (70:30); (A) pins; (B) resorbable fixation plates.

Depending on the type of polyester (PGA, PL-LA, PD, L-LA, copolymers) and their physical properties, the degradation time varies from hours to years, with lowering of the molecular weight and loss of mass. For a more in-depth analysis of the degradation mechanisms, refer to Chapter 8, Section 8.1.

*Applications*: PLA and PGA are among the few degradable polymers to have the approval of the Food and Drug Administration (FDA) for clinical use in humans. Applications include bioabsorbable sutures, osteosynthesis plates, pins and screws, drug delivery systems, bioresorbable stents, and supports for cell growth in tissue engineering. Fig. 4.28 shows some devices made of biodegradable polyesters.

### 4.2.3.3 Polycaprolactone

Polycaprolactone (PCL) is semicrystalline, with $T_g = -60°C$ and $T_m$ of $\cong 64°C$. PCL is prepared by ring-opening polymerization of the cyclic ε-caprolactone monomer (Fig. 4.29).

PCL mechanical properties are rather low (tensile strength $\cong 23 MPa$, deformation at break $\cong 80\%$); degradation is slower than PLA and useful in controlled drug delivery systems (degradation time of 2–3 years). PCL is easy to process and compatible with many polymers in forming blends.

Copolymers with other lactones (e.g., *valerolactone*) degrade more rapidly.

### 4.2.4 Polyamides

Polyamides are a class of polymeric materials that contain *amide* groups as an integral part of the main chain. They are obtained by step polymerization of monomers containing amino and/ or carboxylic groups. If the amide groups are bound at both ends to chemical groups of the aliphatic type, they generally take the name of Nylon,[2] the DuPont trademark (Fig. 4.30, left side).

[2] *Nomenclature*: The name of each type of Nylon is always combined with one or two numbers. These numbers refer to the starting products (monomers) used in the polymerization and correspond to the number of carbon atoms; if the numbers are two, it means that two starting monomers were used. In the case of a single number, the polymerization reaction started from a single monomer. For example, Nylon 6,6 is obtained from hexamethylenediamine (six C atoms) + adipic acid (six C atoms); Nylon 6 is obtained from caprolactam (six C atoms).

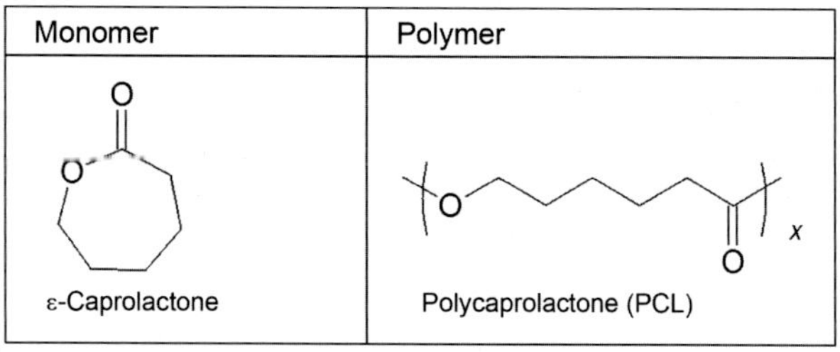

| Monomer | Polymer |
| --- | --- |
| ε-Caprolactone | Polycaprolactone (PCL) |

FIG. 4.29 Chemical structure of the cyclic monomer and polycaprolactone (PCL).

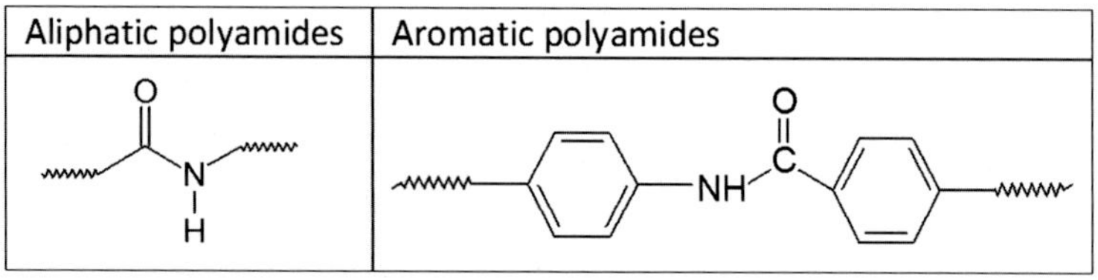

| Aliphatic polyamides | Aromatic polyamides |
| --- | --- |

FIG. 4.30 Schematic chemical structure of aliphatic and aromatic polyamides.

If the amide groups are bound at both ends with aromatic rings, we speak of Aramides. Aramides are defined as all polyamides having at least 85% amide bonds attached directly to two aromatic rings (Fig. 4.30, right side).

*Morphology*: Linear polyamides, with a high content of hydrogen bonds, are generally highly crystalline ($T_g = 50–80°C$, $T_m = 200–280°C$), and their excellent mechanical properties are also due to this property.

*Degradability*: No synthetic biodegradable polyamides are known. Polyamides are stable to hydrolysis in water, but they undergo its plasticizing and swelling action (with modification of the properties).

*Applications*: Nylon are used for the production of synthetic fibers, waterproof coatings, packaging, and a number of other industrial purposes. One of the applications in the medical field is that of suture threads (not bioabsorbable).

## 4.2.5 Polyimides

These polymers contain the chemical group *imide* in the main chain (Fig. 4.31) and may be completely aromatic (i.e., resulting from diamines and dianhydrides, both of an aromatic nature) or partially aromatic (e.g., obtained from aromatic dianhydrides and from aliphatic diamines).

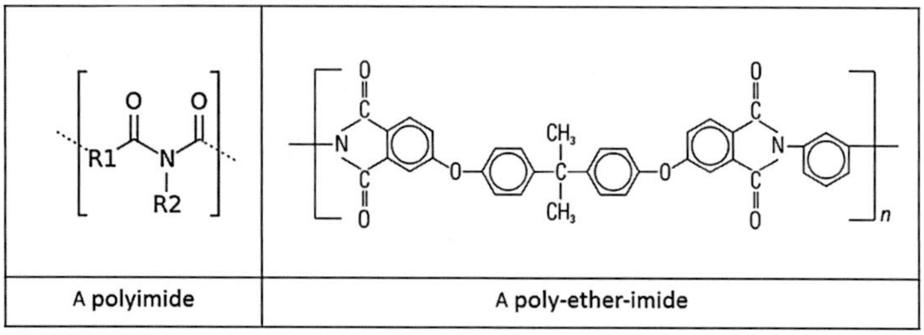

| A polyimide | A poly-ether-imide |
| --- | --- |

FIG. 4.31 Schematic structure of a polyimide showing the imide bond (*left side*) and of a poly-ether-imide copolymer (*right side*).

Their structure can be linear or cross-linked. Copolymers with interesting properties can be prepared as well (e.g., poly-ether-imides, poly-amide-imides, poly-anhydrides-imides).

*Properties*: Favorable wear and friction characteristics even at high temperatures. Good mechanical strength and rigidity, good hydrolytic and dimensional stability, with a wide thermal range of suitability for use (from $-43°C$ to $+480°C$).

The exceptional thermal stability, chemical resistance, and tensile strength make polyimides an ideal material for high-performance medical applications (see later).

*Typical applications*: Cardiovascular catheters, urological retrieval devices, neurovascular applications, fiber optics, intravascular drug delivery, balloon angioplasty, and stent delivery systems. Tubing are lightweight, flexible, and resistant to heat and chemical interaction. In membrane technology, polyimides are excellent polymers because of their outstanding heat resistance and good mechanical strength, as well as chemical resistance to many solvents.

## 4.2.6 Polyurethanes

All the polymeric materials that have the *urethane* group (Fig. 4.32A) in the main chain are classified with this term; urethane groups can also be joined by *urea* groups (Fig. 4.32B), and then the resulting material is a *poly-urea-urethane*.

Polyurethanes (PU), depending on their structure and composition, may exist in a substantially *linear* form (thermoplastic polyurethanes that are, typically, segmented copolymers), or *cross-linked*. They can be also produced in an expanded form (rigid or flexible PU *foams*).

Fig. 4.33 shows, schematically, the reactions between difunctional monomers, which lead to the formation of linear polyurethanes and linear polyureas.

– *Linear polyurethanes* present the advantages of thermoplastics (solubility, processability from the molten state), low hygroscopicity, and good mechanical characteristics.
– *Cross-linked polyurethanes* possess electrical, mechanical, and thermal properties superior to those of other thermosetting resins.
– *Expanded polyurethanes*, or polyurethane *foams*, generally with a cross-linked structure, may have a rigid or flexible consistency, depending on the degree of cross-linking and the pore size.

*Manufacturing*: In general, PUs are obtained by step reaction of difunctional or polyfunctional alcohols and/or amines with difunctional or plurifunctional isocyanates.

*Properties*: Wide spectrum of mechanical and chemical-physical properties, high resistance to fatigue, bio- and hemocompatibility.

Depending on the starting products, physical properties and technological characteristics of the various types of polyurethanes differ significantly.

*Applications*: Linear polyurethanes are used in devices in contact with blood and soft tissues: cardiovascular prostheses (artificial ventricles, small-diameter vascular prostheses, biomorphic cardiac valves, catheters, pacemakers), membranes, cannulas, and endotracheal tubes.

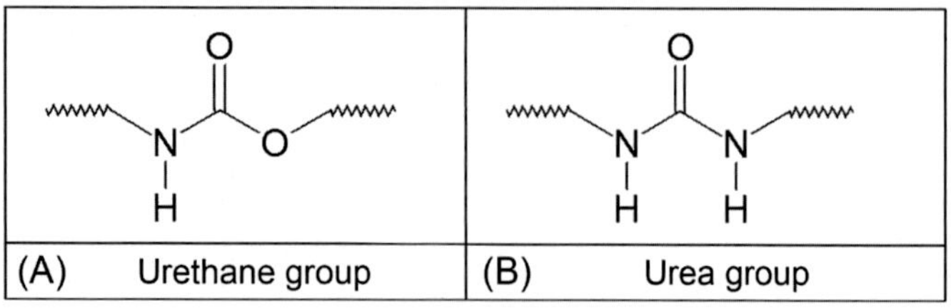

**FIG. 4.32** Structure of (A) the urethane and (B) the urea groups inserted in a polymer chain.

| (A) Urethane group | (B) Urea group |

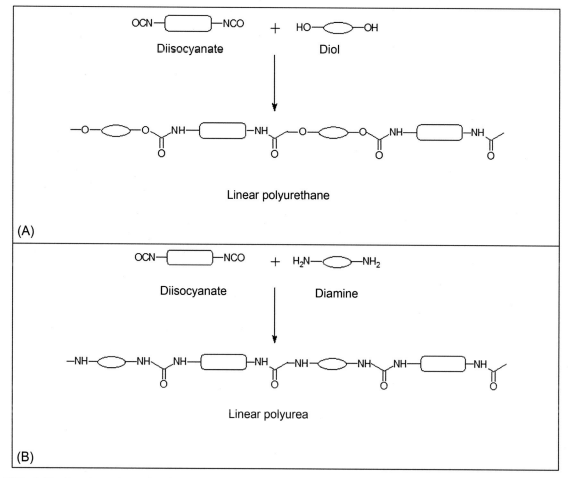

**FIG. 4.33** Reaction scheme for obtaining linear polyurethanes (A) and linear polyureas (B) starting from difunctional monomers. A diisocyanate is reacted with (A) a diol or (B) a diamine.

Cross-linked polyurethanes are used for parts of medical instrumentation or accessories where a high mechanical resistance is required, whereas foams have potential applications as scaffolds for tissue regeneration in tissue engineering.

- *Linear segmented polyurethanes (thermoplastic polyurethanes, TPU, or segmented polyurethanes, SPU)*

Segmented linear polyurethanes are produced by chemical reaction of three different molecules: diisocyanate, macroglycol (or macrodiol), and chain extender (a low molecular weight dialcol or diamine), as shown in Fig. 4.34. TPU are block copolymers of the type (AnBm) consisting of alternating blocks of long and flexible segments (soft segment, An) connected at the ends with rigid segments (hard segment, Bm) through covalent bonds.

### 4.2.6.1 Typology

There are hundreds of different types of isocyanates, macroglycols, and chain extenders used in the plastics industry, and there are

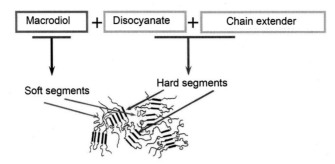

FIG. 4.34    Schematic of the synthesis of a linear segmented polyurethane and of the resulting structure.

consequently thousands of possible syntheses of polyurethanes with different chemical and physical properties.

In the majority of medical grade commercial polyurethanes (in the case of segmented poly-urethanes), the most commonly used chain extenders are *butanediol* [HO-CH$_2$-CH$_2$-CH$_2$-CH$_2$-OH] and *ethylene diamine* [H$_2$N-CH$_2$-CH$_2$-NH$_2$], the latter for poly-urea-urethanes. Di- and polyisocyanates can be grouped as aromatic or aliphatic. There are different lines of thought on the use of aromatic isocyanates, as they are more reactive and able to impart better mechanical properties to the polyurethane but suspected of carcinogenicity, compared to the use of less toxic aliphatic isocyanates.

There are various types of macroglycols (or macrodiols). The majority of polyurethanes currently in use are composed of polyester-like macrodiols [HO-(OC-R-OCO-R-COO)R-OH], and are also called poly-ester-urethanes, which are however subject to hydrolytic degradation.

However, for the implant in the human body, macrodiols of the polyether type [HO-(ORORO-)R-OH], less subject to hydrolytic degradation (but easily oxidizable), which give rise to poly-ether-urethanes, and macrodiols of the PC type [HO-(COO-R-OCO-R-OCOO)R-OH], which impart better structural stability to the corresponding poly-carbonate-urethanes, are used.

In the last years, biodegradable polyurethanes have been investigated, in which the soft segment comes from macrodiols with a biodegradable polyester structure (such as polycaprolactone-

diol), and the diisocyanate is chosen so as not to give rise to toxic degradation products.

### 4.2.6.2 Phase Separation

In linear polyurethanes, the chains actively interact with each other through physical interactions (intermolecular bonds). For example, hard urethane segments tend to attract each other, until they are ordered in aggregates (crystalline or semicrystalline "domains") within the polymer matrix.

This tendency is favored by intermolecular forces exerted by the hydrogen bonds established between NH groups of the urethane units, acting as a donor, and acceptor groups, for example, C=O groups of the hard urethane segment. The result of these interconnections and interactions on the different polymer chains is a two-phase characteristic structure where the aggregates, the semicrystalline or glassy hard domains at body temperature, are arranged in an amorphous and flexible matrix composed of soft segments (Fig. 4.34).

### 4.2.6.3 Mechanical Properties

In general, the presence of hard/soft phase separation gives the segmented polyurethanes elastomeric properties, similar to those of vulcanized rubbers.

A high content of hard segments increases the modulus and the resistance, reducing the elongation at break, whereas an increase of the molecular weight of the soft segments produces the opposite effect, as simplified in Fig. 4.35.

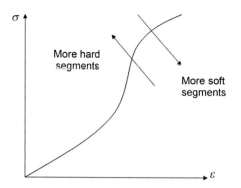

**FIG. 4.35** A high content of hard segments in TPUs increases modulus and strength, reducing the elongation at break whereas a high content and an increase of the molecular weight of the soft segments produces the opposite effect.

The mechanical response of a polyurethane therefore depends on the size and concentration of the hard domains, on the aggregation force of the segments themselves, and on the ability of the segments to orientate themselves in the direction of the stress, but also on the ability of the soft segments to crystallize partially under deformation.

### 4.2.6.4 TPU as Biomaterials

Segmented polyurethanes are used for advanced applications, instead of vulcanized rubbers, because they have superior resistance to abrasion, aging, solvents, and heat. They have excellent resistance to fatigue in flexion, and good bio- and hemocompatibility. For this reason, they are particularly suitable for cardiovascular applications and all those that have soft tissue interfaces. Segmented thermoplastic polyurethanes are used in the manufacture of catheters, cannulas, endotracheal tubes, intra-aortic balloons, dialysis membranes, prosthetic hearth valves (see Fig. 3.15 in Chapter 3), ventricular chambers and artificial ventricles, pacemaker leads, stent coatings, wound dressings, breast implants, and maxillofacial prostheses. Some examples are shown in Fig. 4.36.

The most important and striking application can be the fabrication and clinical use of circulatory assistance devices in which the polyurethane is used for manufacturing the ventricular chambers. After successful experiments on animals, in 1982, the first artificial heart intended to sustain life indefinitely was implanted in Barney B. Clark. The achievement of 112 days of patient survival has to be remembered as one of the greatest achievements of modern technologies applied to medicine.

### 4.2.6.5 Biodegradation Phenomena

Although poly-ether-urethanes (PEUs) have been used in the biomedical field because they are more stable than poly-ester-urethanes, they have shown susceptibility to degradation in the

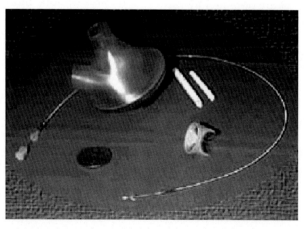

**FIG. 4.36** Examples of devices made of segmented polyurethane. From right: catheter, cardiac ventricular chamber, small-caliber vascular prostheses, biomorphic cardiac valve (prototype). The coin is a demonstration of the real dimensions.

long-term in vivo implant. There are three main mechanisms believed to be responsible for the in vivo degradation of poly-ether-urethanes, namely hydrolysis, oxidation, and environmental stress cracking (ESC) (see Chapter 5.2).

To minimize the degradation phenomena that drastically limits the time of use of the devices, new materials were developed, such as poly-carbonate-urethanes, which, not containing ether groups, are less, but not completely susceptible to hydrolytic and oxidative biodegradation phenomena (Pinchuk et al., 2008).

## 4.2.7 Silicones (Polysiloxanes)

These polymers contain the (-Si-O)- group in the main chain; silicon atoms are also linked to organic groups, typically methyl groups. Therefore they are partially inorganic polymers.

The basic repeating unit is known as "siloxane" (Fig. 4.37A), and the most common repeating unit is dimethyl-siloxane (Polydimethylsiloxane (PDMS), Fig. 4.37B).

Many other groups such as vinyl, phenyl, and halogenated groups can represent the R groups.

The simultaneous presence of organic groups attached to an "inorganic" backbone (i.e., Si atoms alternating to oxygen atoms) gives silicones peculiar properties that make possible their use in a number of applications and fields.

*Production*: Polysiloxanes are obtained by use of organochlorosilanes ($Cl_2$-Si-$R_2$, e.g., dichloro-dimethylsilane, or $CH_3Cl$-Si-$R_2$, e.g., chloro-methylsilane) following different steps (hydrolysis or methanolysis to cyclic and/or linear oligodimethylsiloxanes); the different polymers then are produced in a second step by

polymerization of the cyclic oligomers or polycondensation of the linear oligomers (schematic in Fig. 4.38 for PDMS).

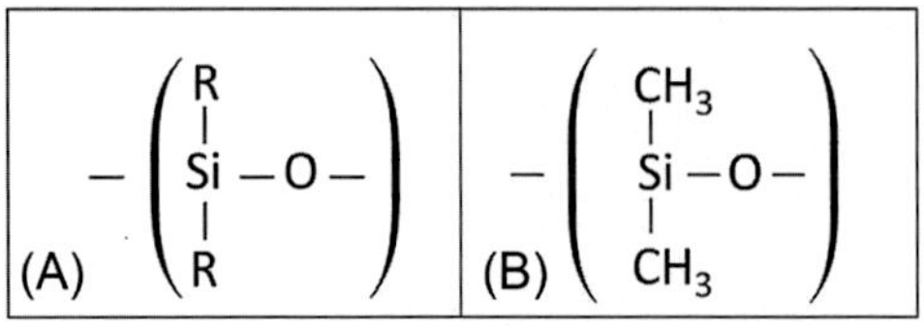

*Distinctive Properties*: Hydrophobicity, stability to hydrolysis and oxidation, excellent electrical properties (insulation). Silicones have constant properties over a wide range of temperatures ($-75°C$ to $+260°C$) and a rather high cost. The mechanical characteristics at room temperature are not very high, but silicones have exceptional characteristics of structural stability. They are chemically inert and highly biocompatible. Mechanical properties and dimensional stability can be increased by cross-linking.

### 4.2.7.1 Silicone Elastomers

Polysiloxanes are easily transformed into a 3D network by cross-linking reactions. There are three possible types of cross-linking reaction:

(1) By *action of radicals* ($R^•$) generated by organic peroxides. High-consistency silicone rubbers (HCSRs) are obtained;
(2) By *condensation* of ad hoc prepared reactive polysiloxanes with moisture (e.g., the humidity in the atmosphere) usually in the presence of a catalyst. When the condensation reaction is applied to a two-part system, cross-linking takes place upon mixing the two components and, in this case, no atmospheric moisture is needed. Adhesives and sealants are examples of products obtained in this way;
(3) By *addition* using an addition-curing reaction between Si-H groups and vinyl-terminated polymers (Fig. 4.39). Organometallic compounds catalyze this reaction.

*Medical applications of silicones:* General and aesthetic surgery (breast implants, penile and testicular prostheses, ear lobes, etc.); small prostheses (chin, nose, fingers), tracheal prosthesis; blood lines and catheters; membranes of oxygenators; artificial skin.

In ophthalmology, silicones were initially used in the form of oils, particularly in retinal

$$- \left( \begin{array}{c} R \\ | \\ Si - O - \\ | \\ R \end{array} \right) \qquad - \left( \begin{array}{c} CH_3 \\ | \\ Si - O - \\ | \\ CH_3 \end{array} \right)$$

(A)                    (B)

**FIG. 4.37** Chemical structure of (A) siloxane repeating unit and (B) dimethyl-siloxane repeating unit in silicones.

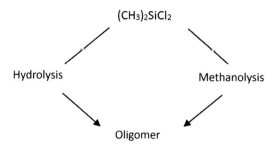

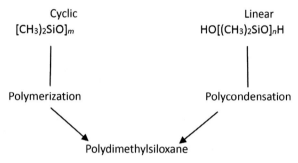

**FIG. 4.38** Schematic for the production of PDMS starting from dimethyldichlorosilane.

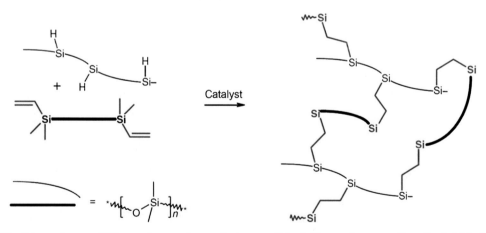

**FIG. 4.39** Scheme for an addition-curing reaction to obtain cross-linked silicone elastomers. *From Brook, M.A., 2006. Biomaterials 27, 3274–3286, Scheme 1.*

detachment surgery and then for contact lenses, corneal implants, and foams for retinal detachment surgery.

In dentistry, silicones are used as impression material.

### 4.2.7.2 Polydimethylsiloxane

PDMS is an amorphous-crystalline polymer that has a $T_g = -120°C$ and a melting temperature of the crystalline component at about $-40°C$. This implies that, at room temperature,

PDMS is a viscous liquid and has lower mechanical properties than those requested for the fabrication of medical devices. To make PDMS a material with a suitable degree of cohesion and usable for the fabrication of devices, a more or less dense cross-linking is therefore necessary, according to the desired structural characteristics, as previously described.

PDMS is the most largely used silicone for the fabrication of medical implants, particularly for breast implants. Although there are different types of breast prostheses, the design of most of them consists of an outer shell, with a smooth or textured (rough) surface, filled inside with a gel or saline solution. Fig. 4.40 presents two images of breast prostheses.

### 4.2.7.3 Silicone-Polyurethane Copolymers

More recently, silicone-polyurethane copolymers have been developed, which appear to have improved properties: greater biostability than polyurethanes but retaining the excellent biocompatibility and elastomeric properties of the polyurethane family.

## 4.2.8 Polycarbonate

Although this class comprises all polymers containing the carbonate group (O-CO-O) in the main chain, actually the name "polycarbonate" identifies the polymeric material obtained from bisphenol A and fosgene (or from bisphenol A and diphenyl carbonate) having the chemical structure presented in Fig. 4.41B.

Other PCs are not used as thermoplastic materials, because they do not have a sufficiently high molecular weight and can at most be used as intermediates in the preparation of other materials, such as block copolymers.

The main characteristics of polycarbonate of bisphenol A (PC) are: very high toughness in thin layers, excellent thermal stability, transparency equal to that of glass (light transmission = 87%), good processability, resistance to radiation (it does not yellow), high durability, and low flammability.

PC is totally amorphous, with $T_g = 150°C$.

*Applications*: 25% of the industrial consumption of PC is addressed to glazing.

Medical applications concern transparent containers, reservoirs and containers for cardiotomy, connectors, housings, membranes for oxygenators, and plastic surgery of the skull.

The use of PC for disposable products, drug delivery systems, and fiber optics is increasing.

## 4.2.9 Acetals Resins

Family of polymers derived from formaldehyde (H-CHO) by step polymerization. They present the -CH$_2$-O- group in the main chain.

FIG. 4.40   Images of breast implants: (A) with a smooth surface and (B) with a textured surface of the shell. *From Brook, M.A., 2006. Biomaterials 27, 3274–3286, Fig. 2.*

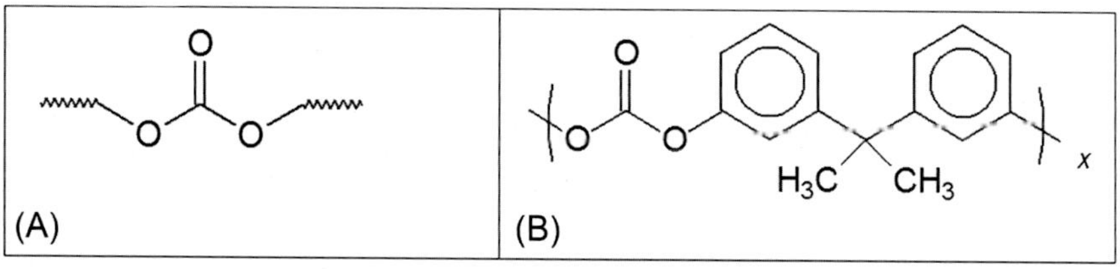

**FIG. 4.41** (A) Structure of a generic polycarbonate showing the presence of the carbonate group in the main chain; (B) Polycarbonate of bisphenol A.

The simplest homopolymer presents only this group as a repeating unit and takes the name of *Polyoxymethylene*, POM, (including the Delrin, DuPont).

Copolymers have structure:

$$-\left[-(CH_2-O)_x-Y-\right]_n-$$

Acetal resins must be stabilized to be processed with conventional techniques in the molten state (i.e., the end groups of the chains has to be chemically modified to prevent depolymerization).

These polymers have excellent resistance to many chemicals, even to strong mineral acids. They have remarkable dimensional stability, hardness, rigidity, toughness, and favorable antifriction and wear behavior.

POM is a linear, crystalline polymer with a degree of crystallinity higher than 70% and is white-opaque. In commercial products, there can be significant differences in mechanical characteristics depending on the degree of crystallinity and the average size of crystallites.

*Applications:* POM can be used in many applications where most other plastics are not suitable. Acetal copolymers are used in the medical field for components of equipment and instruments (precision parts, springs, gears, connectors, pump parts, bottles, etc.).

## 4.2.10 Polysulfones

Polysulfones contain the chemical group - $SO_2$- and aromatic rings in the main chain.

They are rigid, tough thermoplastics, with a $T_g$ between +180°C and +250°C. The stiffness derives from the difficulty of movement of the aromatic and $SO_2$ groups, the toughness from the oxygen (-O-) connection bridges.

*Advantages:* Excellent thermal stability ant to oxidation and hydrolytic degradation. They are chemically inert.

*Disadvantages:* Sensitivity to notch impact, which is the reason why, in manufactured products, it is necessary to avoid reduced bending radii and sharp section changes. They undergo stress cracking in the presence of organic solvents.

Fig. 4.42 shows the structural formula of two typical polysulfones.

*Applications:* Polysulfones are used in case of continuous exposure to steam and/or hot water: in medical-surgical instruments, and in food-use equipment subjected to repeated cycles of washing and/or sterilization in temperature. Some examples are: centrifuge covers, outer casings of anesthesia masks, dental instrument tips, and endoscopy instruments.

*Applications in devices and prostheses:* Membranes for oxygenators and dialyzers. They are also investigated as matrices in advanced composites for orthopedic applications.

## 4.2.11 Polyaryl-Ether-Ketones

Thermoplastic polymers with molecular structure in which aromatic rings alternate with ketone groups (C=O), and ether -O-.

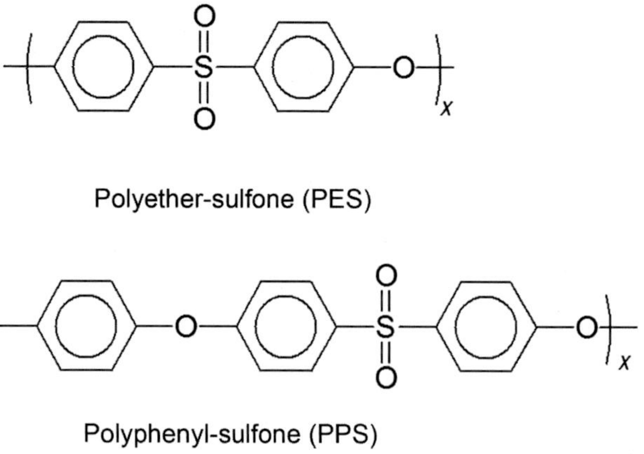

Polyether-sulfone (PES)

Polyphenyl-sulfone (PPS)

FIG. 4.42   Chemical structure of two polysulfones: PES and PPS.

Two main routes of step polymerization have been exploited in the production of polyaryl-ether-ketones (PAEKs): a first method consists of linking aromatic ether species through ketone groups, whereas a second method involves linking aromatic ketones by an ether bond.

The best known is the *poly-ether-ether-ketone* (PEEK), synthesized for the first time by ICI in 1978 and marketed in 1981 (Fig. 4.43).

PEEK can reach a 40% crystallinity, with $T_g = +143°C$, and $T_m$ of +334°C. It is chemically inert and insoluble in all conventional solvents at room temperature, with the exception of 98% sulfuric acid.

By the late 1990s, PEEK emerged as the leading high-performance thermoplastic candidate for replacing metal implant components, especially in orthopedics.

*Advantages*: Crystallinity of PEEK is the reason for its high performance. The aromatic structure imparts excellent resistance and stability to γ and β radiation, flame, and heat.

*Disadvantages*: Difficult solubility and processability, high production prices.

*Applications*: PEEK is a technopolymer for special applications in composites resistant to high temperatures and is currently also used as a resin as such (Kurtz and Devine, 2007). In particular, it is used for the production of advanced composites (see Section 4.5). A cross-linked PEEK has a modulus of elasticity similar to cortical bone and is now broadly accepted as a radiolucent alternative to metallic biomaterials in spinal implants (Fig. 4.44).

## 4.3 NATURAL POLYMERS AS BIOMATERIALS

The challenge of using natural polymers or developing new biomaterials is not only to understand their mode of action in nature but also to correctly coordinate the complex interplay between chemistry, biology, physics, and

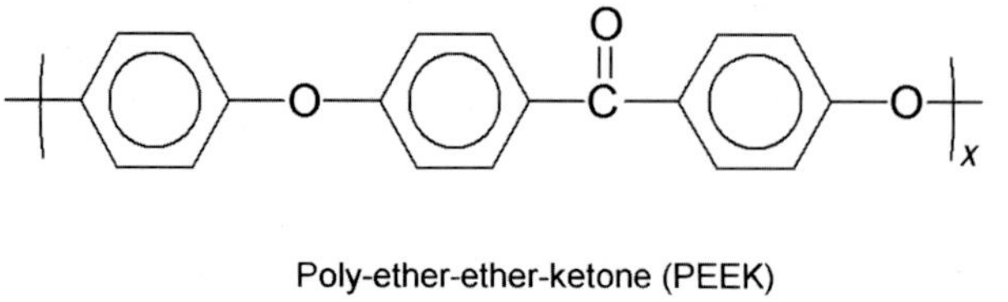

Poly-ether-ether-ketone (PEEK)

FIG. 4.43   Chemical structure of PEEK.

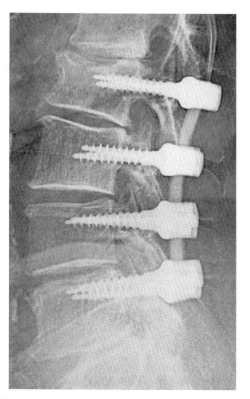

**FIG. 4.44** Lateral view of PEEK rods stabilization at four levels without fusion in a 71-year-old patient. *From Benezech, J., Garlenq, B., Larroque, G., 2016. Advances in orthopedics. J. Biomater. Appl. 2016, Article ID 7369409, 7 pages, Hindawi Publ. Co., https://www.hindawi.com/journals/aorth/2016/7369409/.*

engineering. The number of possible parameters seems to be endless because not only are pure polymers considered on their own but also different blends thereof, with the possibility to even include chemical steps to further functionalize and tailor the final material properties. However, the main challenges that are currently faced are the low-cost manufacturing and the scale-up of surface modifications (e.g., micro- and nanopatterned) to mimic nature's choice of materials (e.g., promising adhesion mechanism of geckos or frogs) and to use them in a different (e.g., wet) environment to provide new solutions in biomedicine and tissue engineering.

## 4.3.1 Proteins

Proteins are synthetized by cells throughout a process that involves the genetic information stored in the DNA to be converted into functional sequences of amino acids.

Proteins can be produced by and extracted from different natural sources, such as animals, bacteria, and plants. As the result of the great variability/diversity of protein types and sources, different extraction and purification methods have been envisioned so far, all strongly dependent on the end-use. In this context, the complex protein three-dimensional folding can be strongly influenced by several factors, including temperature, pH, and the presence of organic solvents, which can ultimately affect the final protein conformation. The changes in protein structure that take place when they are exposed to one of these factors might result in protein denaturation, which is mostly an irreversible process. Because protein function relies on protein conformation, in order to avoid denaturation, several efforts have been devoted to find innovative strategies for protein extraction and processing.

Silk, elastin, collagen and gelatin, keratin, and resilin are some of the more common structural proteins used to shape protein-based biomaterials (Fig. 4.45). These proteins are all characterized by highly repetitive amino acid motifs (i.e., the primary sequence).

### 4.3.1.1 Collagen (see Chapter 1 - Proteins, Fig. 1.64)

Collagen is the most abundant protein in human body, accounting for 20%–30% of total body proteins (Harkness, 1961). Among the 28 different types of collagen, type I (found in skin, tendon, and bone), type II (in cartilage), and type III (in skin and blood vessels) collagens belong to fibrillar structures that form an essential part of the architecture and integrity of the tissues, essentially conferring some mechanical functions. For instance, type I collagen is found in tendons and ligaments, to ensure the interconnectivity

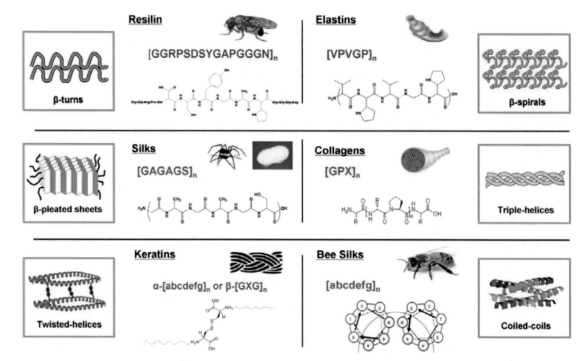

**FIG. 4.45** Examples of natural proteins, including resilins, elastins, spider and silkworm silks, collagens, hair keratins, and bee silks. *From Hu, X., Cebe, P., Weiss, A.S., Omenetto, F., Kaplan, D.L., 2012. Protein-based composite materials. Mater. Today 15 (5), 208–215.*

between bone and muscles, and consequently skeleton movements. Besides, collagen is also present in mineralized tissues such as teeth or bones.

Because of its abundance, excellent biocompatibility, biodegradability, and low antigenicity, collagen has been largely investigated as a biomaterial for biomedical and pharmaceutical applications, as well as for tissue engineering purposes. One of the main benefits of collagen is in the ease of shaping in hydrogels, sheets, and sponges. However, although collagen can be extracted from rat tails and bovine tendons, calf skin, human placenta, and bovine cartilage, among others, the properties of collagen-derived materials are strongly dependent on the collagen source and the extraction method. For these reasons, substantial lot-to-lot collagen variation in terms of different cross-linking density, fiber size, and presence of impurities are still main issues to be solved in realizing collagen-based materials.

On the other hand, collagen-based materials do possess very poor mechanical strength, which limit their applications in specific tissues such as hard tissues. In this context, chemical and/or physical cross-linking methods or blending with other materials (natural or synthetic materials) is frequently used to enhance the mechanical strength of collagen (Dong and Lv, 2016).

Gelatin, which is a mixture of peptides and proteins produced by partial hydrolysis of collagen extracted from the skin, bones, and connective tissues of animals such as domesticated cattle, chicken, pigs, and fish, has been widely used as a biomaterial. Gelatin is of two types, type A gelatin obtained through acid hydrolysis and type B obtained through basic (alkaline) hydrolysis. During hydrolysis, the natural molecular bonds between individual collagen strands are broken down into a form that rearranges more easily. Its chemical composition is, in many aspects, closely similar to that of its parent collagen. Gelatin readily dissolves in hot water and sets to a gel upon cooling. When added directly to cold water, it does not dissolve well, however. Gelatin also is soluble in most polar solvents. Gelatin solutions show viscoelastic flow. Solubility is determined by the method of manufacture.

*Applications*: The use of collagen in biomedical applications has been largely explored, displaying suitable properties for the development of biocomposites, films/substrates for cell attachment, proliferation and differentiation, and scaffolds for regenerative medicine.

Collagen has been used as drug delivery system, for example, collagen shields in ophthalmology, sponges for burns/wounds, minipellets and tablets for protein delivery, gel formulation in combination with liposomes for sustained drug delivery, as controlling material for transdermal delivery, and nanoparticles for gene delivery. It has also been used for tissue engineering purposes, including skin replacement, bone substitutes, and artificial blood vessels and valves.

Collagen also served as: (i) dressing for the treatment of wounds and burns (e.g., OrCel® from Ortec International Inc. (United States), which is composed of donor neonatal human keratinocytes and fibroblasts cultured in separate compartments of a bilayered collagen sponge; CellerateRX® from WNDM Medical Inc. (United States) (Fig. 4.46), which is collagen granules or gels); (ii) sponge graft for dura mater repair (e.g., DuraMatrix-Onlay PLUS from Collagen Matrix Inc., United States).

### 4.3.1.2 Silk

Silks are fibrous proteins produced within specialized glands where the proteins are stored prior to spinning into fibers. Silks are a variety of

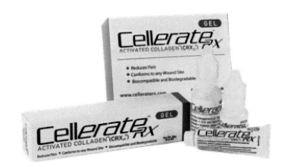

**FIG. 4.46** An example of commercial collagen-based material. CellerateRX from WNDM Medical Inc. (United States) (http://www.wndm.com/).

proteins that differ in composition (i.e., different amino acid composition), structure, and properties depending on the source. The most extensively characterized silks are produced by silkworm (*Bombyx mori, B. mori*) and spiders (*Nephila clavipes* and *Araneus diadematus*).

Silk from *B. mori* is basically composed of silk fibroin protein coated with glue-like, adhesive proteins named sericins. When the sericin component is removed from the silk filament core protein, the fibroin fibers can be dissolved into aqueous solutions or organic solvents (Fig. 4.47) and next processed into gels, sponges, fibers, films, tubes, and microspheres (Rockwood et al., 2011).

Very similarly to collagen, silks are characterized by a highly repetitive sequence (for *B. mori*, [GAGAGS]$_n$), which results in a very homogenous secondary structure (i.e., β-sheets). In particular, silk fibroin is a block copolymer rich in hydrophobic β-sheets-forming blocks linked with small hydrophilic spacer (Vepari and Kaplan, 2007 #262). Because of the abundance of β-sheets within the fibroin structure, silk-based materials usually display impressive mechanical properties, in terms of high mechanical strength and toughness, even greater than synthetic materials such as Kevlar. Therefore silk fibroin is a worthy candidate polymer for a wide range of biomedical application,

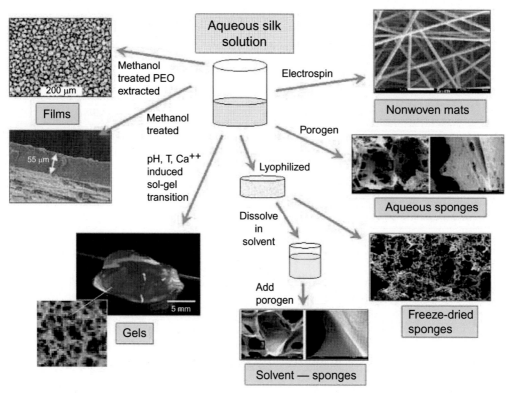

**FIG. 4.47**　Examples of silk material formats from aqueous silk fibroin solution into nonwoven silk fibers, sponges, gels, and films. *From Vepari, C., Kaplan, D.L., 2007. Silk as a biomaterial. Prog. Polym. Sci. 32 (8–9), 991–1007.*

including scaffolds for tissue engineering purposes, as wound dressings, and for drug delivery/controlled release applications.

Although the degradation rate in vivo is highly dependent on the implantation site, mechanical and chemical environment as well as the mode of processing used to produce silk-based materials, silk fibroin generally displays a slow degradation rate.

*Applications*: Biomedical applications of silk fibroin include scaffolds, in the form of sponges, electrospun fibers, films, and hydrogels for tissue engineering applications, including bone, cartilage, ligaments, tendons, vascular tissue, and skin. Silk fibroin in the form of microspheres has been investigated for the delivery of drugs, grow factors, and small molecules.

Silk fibroin-based materials have been used for soft tissue support and repair. For instance, NAD from Leonardino SRL (Italy) (Fig. 4.48) has been thought for treating chronic wounds such as those happening in burns.

### 4.3.1.3 Elastin (see Chapter 1 - Proteins, Fig. 1.66)

Elastin is a protein insoluble in water composed by cross-linked chains of its precursor called *tropoelastin*. The latter is, instead, a soluble, nonglycosylated protein, composed by alternating hydrophilic and hydrophobic stretches.

In animals, elastin confers rubber-like elasticity to those connective tissues undergoing repetitive distensions, such as ligaments, tendons, blood vessels, skin, and even distensible

**FIG. 4.48** An example of silk material used in medical device manufacturing. NAD nanodressing from Leonardino SRL (Italy) (http://leonardino.eu/).

organs like lungs. However, due to its chemical composition and cross-linked nature, elastin is difficult to isolate. Although attractive, for such reasons there are fewer reports about elastin used as a biomaterial compared to other proteins.

Elastin shows features such as elasticity, tenacity, stability, and self-assembly behavior. Altogether, these characteristics make elastin a suitable substrate for biomaterials of different forms, such as hydrogels, sheets, sponges, and (nano)composite materials for biomedical applications.

Recently, recombinant polypeptides with peculiar properties resembling those of the native protein because inspired by elastin have recently been obtained (e.g., HELP = human elastin-like protein).

*Applications*: Due to its elasticity, elastin has been used to fabricate hybrid scaffolds (with collagen or PCL, as a few examples) for engineered vascular tissue substitutes (Fig. 4.49). Elastin has been also used to engineer skin substitutes, cartilage, and, more recently, for the development of bioactuators, implantable sensors, and drug delivery vehicles.

### 4.3.1.4 Keratin (see Chapter 1 - Proteins)

Keratins belong to the broad family of fibrous structural proteins that display different morphologies and features. They are the main constituent of hair, skin, horn, and fingernails.

Keratins are cysteine-rich proteins endowed by nature with high mechanical strength owing to the large number of disulfide bonds. Keratins can be isolated from a variety of body parts, such as human and animal hair, feathers, and hooves. Because of the presence of such disulfide bonds, this protein is basically insoluble in any solvent such as pure water, aqueous acidic or basic solutions, and organic solvents. Therefore the dissolution and the extraction of keratin is difficult compared to other natural polymers, such as collagen or starch.

Keratin-based biomaterials have been widely used in various biomedical applications. Indeed, the presence of cell-binding motifs in keratin, as well as its ability to self-assemble, make it an ideal natural polymer to be used for tissue regeneration. Moreover, hydrophobic and hydrophilic moieties have been also tethered to keratin to modulate the degradation rate of such protein matrix.

As with other natural proteins, keratin unfortunately suffers from poor mechanical stability and processing properties. In this light, keratin has been often combined with plasticizers, cross-linkers, and other synthetic or natural polymers to overcome these issues.

Keratin-based materials for biomedical applications have been produced in the form of films, hydrogels, scaffolds, and composites fibers (Shavandi et al., 2017).

**FIG. 4.49** Examples of elastin-based vascular scaffold (Wise et al., 2009).

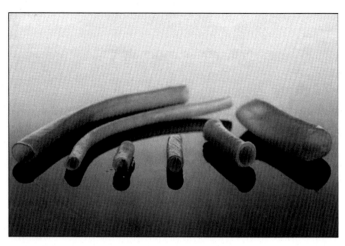

*Applications:* Due to its inherent biocompatibility and biodegradability behaviors, keratin was found very useful for a wide range of biomedical applications, such as bone and cartilage tissue engineering, ocular regeneration, skin replacement, controlled drug delivery, and nerve regeneration. Keratins and keratin-based materials are particularly suitable for wound healing purposes, and some products for the treatment of burns and acute/chronic wound, such as Keramatrix (Fig. 4.50), are already on the market.

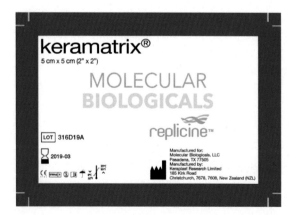

**FIG. 4.50** Example of commercial keratin-based matrix scaffold Keramatrix® for use in the healing of burns, acute and chronic wounds. *From Molecular Biologicals LLC., United States. http://www.molecularbiologicals.com/.*

## 4.3.2 Polysaccharides

Differences in the monosaccharide composition, linkage types and patterns, chain shapes, and molecular weight dictate their physical properties, such as solubility, viscosity, gelling potential, and/or surface and interfacial properties. Often, polymers of natural origin have limitations in terms of their solubility and industrially acceptable processability factors such as high temperature of melting, which are commonly applied to synthetic polymers. For instance, the majority of polysaccharides are water-soluble and oxidize at elevated temperatures beyond their melting point (Kumbar et al., 2014).

Polysaccharides are derived from renewable resources, like plants, animals, and microorganisms, and are therefore widely distributed in nature. A list of polysaccharides from varying sources is given in the following text:

- examples of polysaccharides from higher plants include starch and cellulose;
- examples of algal polysaccharides are alginates;
- examples of polysaccharides from animals are chitin and chitosan, and hyaluronic acid (HA);

- examples of polysaccharides from microorganisms are dextran, gellan gum, pullulan, xanthan gum, and bacterial cellulose.

### 4.3.2.1 Cellulose (see Chapter 1 - Classification of Polysaccharides, Fig. 1.76)

Cellulose, the most abundant polymer on Earth, holds a big potential for different applications in the biomedical field (Lin and Dufresne, 2014). The degree of linearity and the presence of extensive hydroxyl (-OH) groups throughout the cellulose chain are responsible for formation of inter- and intra-molecular hydrogen bonds throughout the polymer chain. This causes cellulose chains to organize in parallel arrangements into crystallites and crystallite strands, the basic elements of the supramolecular structure of the cellulose fibrils and the cellulose fibers. This arrangement of fibers in the polymer is termed its supramolecular structure, and it in turn influences its physical and chemical properties. Cellulose is known to exist in at least five allomorphic forms. Cellulose I is the form found in nature. Cellulose may occur in other crystal structures denoted celluloses II, III, and IV, with cellulose II as the most stable and relevant structure. This structure can be formed from cellulose I by treatment with an aqueous solution of sodium hydroxide to give raise to the regeneration of native cellulose. This treatment renders the cellulose II more accessible to chemical treatments and hence more reactive. Cellulose behaves as an active chemical due to the three -OH in each glucose unit, in which the -OHs at the second and third positions behave as secondary alcohols, whereas the hydroxyl group at the sixth position acts as a primary alcohol.

Cellulose is nontoxic and has good biocompatibility, therefore it offers several possibilities in medical applications. Unfortunately, cellulose has certain drawbacks. These include poor solubility in common solvents (because of extensive hydrogen bonding), poor crease resistance, poor dimensional stability, lack of thermoplasticity, high hydrophilicity, and lack of antimicrobial properties. To overcome such drawbacks, the controlled physical and/or chemical modification of the cellulose structure is essential (Kumbar et al., 2014). Esters of cellulose with interesting properties such as bioactivity, and thermal and dissolution behavior can be obtained by esterification of cellulose with nitric acid in the presence of sulfuric acid, phosphoric acid, or acetic acid. Commercially important cellulose esters are cellulose acetate, cellulose acetate propionate, and cellulose acetate butyrate. Carboxymethyl cellulose (CMC) is the major cellulose ether. By activating the noncrystalline regions of cellulose, selective regions of alkylating reagents can attack the cellulose. This is termed the concept of reactive structure fractions and is used widely for the production of CMC.

Apart from the conventional plant source, cellulose is also obtained from bacteria, termed bacterial cellulose (BC) or microbial cellulose (MC), which is highly pure cellulose, because it does not have lignin and hemicellulose conventionally produced by static and shaking culture methods. The BC fibers have a high aspect ratio along with a diameter of 20–100 nm. Consequently, BC has a very high surface area per unit mass. These peculiar properties with a very highly hydrophilic nature result in a very high liquid loading capacity. Besides, MC possesses high tensile strength, moldability, and extreme insolubility in most of the solvents, and is 100 times thinner than the cellulose fibrils obtained from plants with good shape retention. The typical Young modulus of BC is found to be in the range of 15–35 GPa, with the tensile strength in the range of 200–300 MPa. Its water-holding capacity is up to 100 times its own weight. Of note, BC shows excellent biodegradability and biocompatibility (Mohite and Patil, 2014).

Although biodegradation remains a challenge for the absorption of cellulose, treatments that markedly reduced crystallinity led to degradation as well as high biocompatibility of regenerated cellulose (Kumbar et al., 2014).

*Applications*: Cellulose and its derivatives are used as coating materials for drugs, additives of pharmaceutical products, blood coagulant, supports for immobilized enzymes, artificial kidney membranes, stationary phases for optical resolution, in wound care, and as implant material and scaffolds in tissue engineering (Eo et al., 2016). In addition, cellulosic scaffolds can also be used for tissue engineering and other biomedical applications.

BC has already found application as a wound dressing material, as it does provide a moist environment to a wound, resulting in better wound healing. In this regard, XCell from Medline Industries Inc. (United States) and Dermafill™ from Cellulose Solutions LLC (United States) are commercially available BC-based products for topic wound dressing applications that have the benefit of conforming remarkably well to almost any contour and are ideal moisturizing applicants as they maintain proper water balance by either absorbing or releasing fluid according to the behavior of the wound. Besides, BC has been used in drug delivery, vascular grafting, and as scaffold material for in vitro and in vivo tissue engineering. Besides, BC has been widely used in the papermaking, food, pharmaceutical, and cosmetic industries.

Cellulose is primarily used in pharmaceutical coatings because of its good film-forming properties. Additionally, BC exhibits desirable mechanical properties, including high tensile strength, elastic modulus, and high wet strength due to its uniform and ultrafine fibrous network structure. It can be sterilized without any changes to its structure and properties. The previously mentioned properties make BC a suitable candidate for application as a tablet film-coating agent.

Although BC itself has no antimicrobial activity that prevents wound infectivity, it can be attained by imparting silver nanoparticles into BC by immersing BC in silver nitrate solution (Mohite and Patil, 2014).

Concerning cellulose derivatives, cellulose esters have been put to use in many biomedical applications. Hemodialysis membranes used in purification of blood for patients with renal failure have employed melt-spun cellulose diacetate membrane. Besides, MC is used in several drug delivery and tissue engineering purposes.

### 4.3.2.2 Starch (see Chapter 1 - Classification of Polysaccharides, Fig. 1.77)

Starch is a material composed of anhydroglucose units, which form two different polymers, namely amylose and amylopectin. Amylose is a linear polymer with a degree of polymerization ranging from $3 \times 10^2$ to $1 \times 10^4$, whereas amylopectin typically displays a degree of polymerization of about $10^8$. Therefore amylose is typically much smaller than amylopectin. Their content in starch granules varies with the botanical source of starch and is affected by climatic conditions and soil type during growth. For instance, amylose content of potato starch varies from 23% to 31% for different genotypes, whereas the amylose content of rice is specified as waxy, 0%–2%; very low, 5%–12%; low, 12%–20%; intermediate, 20%–25%; and high 25%–33% (Mozafari, 2007).

Pure starch is a white, tasteless, and odorless powder that is insoluble in cold water or alcohol. When heated in the presence of water, starch undergoes a process known as gelatinization, during which the granules swell, leach amylopectin, and lose birefringence (Torres et al., 2013). The unique physicochemical and functional characteristics of starches isolated from different botanical sources, such as corn, potato, rice, and wheat, make them useful for a wide variety of biomedical and pharmaceutical applications. Starch properties such as swelling power, solubility, gelatinization, rheological characteristics, mechanical behavior, and enzymatic digestibility are of utmost importance while selecting starch source for distinctive applications (Mozafari, 2007). Swelling power and solubility provide evidence of the magnitude of interaction between starch chains within the amorphous and crystalline domains. The extent of this interaction is influenced by the amylose/amylopectin ratio, and by the characteristics of amylose and amylopectin in terms of molecular weight/distribution, degree and length of branching, and conformation.

Starch is hydrolyzed to glucose, maltose, and malto-oligosaccharides by α- and β-amylase and related enzymes.

*Applications*: The starch for biomedical applications is available in various forms as shown in Fig. 4.51.

Starches from various plant sources, such as wheat, maize, and rice, have received extensive attention with relation to structural and physicochemical properties. Limitations such as low shear stress resistance, thermal resistance, and thermal decomposition of native starches limited their industrial applications. Earlier native starches were only used as a tablet binder and a disintegrant because of various restrictions like low viscosity, cold water insolubility, and low gelling property. In recent years, physically and chemically modified starches have been suitably modified to meet industrial needs. The modifications were aimed to add some functional groups or alter them during processing and make them suitable for tailor-made use. For instance, the chemically modified starches with more reactive sites to carry biologically active compounds were found suitable biocompatible carriers, easily metabolized in the human body (Mozafari, 2007). The modified starch forms either a hydrophilic or a hydrophobic derivative, which was earlier used in food industries. They have been well explored for controlled and sustained delivery of therapeutic agents, tissue engineering scaffolds, substrates for cell seeding, bone replacement implants, and wound dressings (Torres et al., 2013). For instance, starch-based biodegradable bone cements are highly advantageous because they can provide for immediate structural support and, as they degrade from the site of application, allow the ingrowth of new bone for complete healing of bone fracture.

### 4.3.2.3 Chitin and Chitosan (see Chapter 1 - Classification of Polysaccharides, Fig. 1.78)

After cellulose, chitin is the most far-reaching biopolymer in nature. Chitin and its subsidiaries have awesome monetary worth due to their biological activities and their industrial and biomedical applications. Chitin can be obtained from three sources, specifically shellfish, insects, and microorganisms. However, the principle commercial sources of chitin are shells of shrimps, crabs, lobsters, and krill supplied in huge amounts by the shellfish-handling commercial enterprises. Extraction of chitin includes two stages: demineralization and deproteinization,

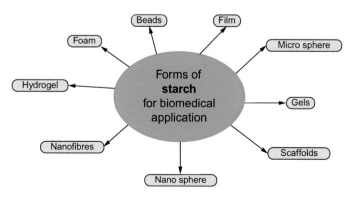

FIG. 4.51 Forms of starch for biomedical applications (Hemamalini and Giri Dev, 2018).

which can be directed by two techniques, chemical or biological. The chemical method obliges the utilization of acids and bases, whereas the biological method includes microorganisms. In industrial processing, chitin is removed from crustaceans by acid treatment to break down calcium carbonate via alkaline extraction to solubilize proteins. Furthermore, a decolorization step is regularly added to evacuate extra pigments to acquire a colorless material. These treatments must be adjusted to every chitin source, owing to contrasts in the ultrastructure of the introductory materials. The subsequent chitin needs to be evaluated as far as purity and color as residual protein and pigment can result in issues for

further use, particularly for biomedical items. Because chitin is highly hydrophobic and insoluble in water and in most organic solvents (Ravi Kumar, 2000), by fractional deacetylation under alkaline conditions and high temperature, one acquires chitosan, which is the most imperative chitin derivative with regard to applications (Fig. 4.52). Because every glucosamine monomer constituting chitin is *N*-acetylated, the *N*-deacetylation of chitin leads to the formation of chitosan. Of note, the degree of conversion of the acetamido group to amine group is never really complete. This is given by the degree of deacetylation (DD) of the chitosan. The DD of chitosan can vary from 30% to 95%. This

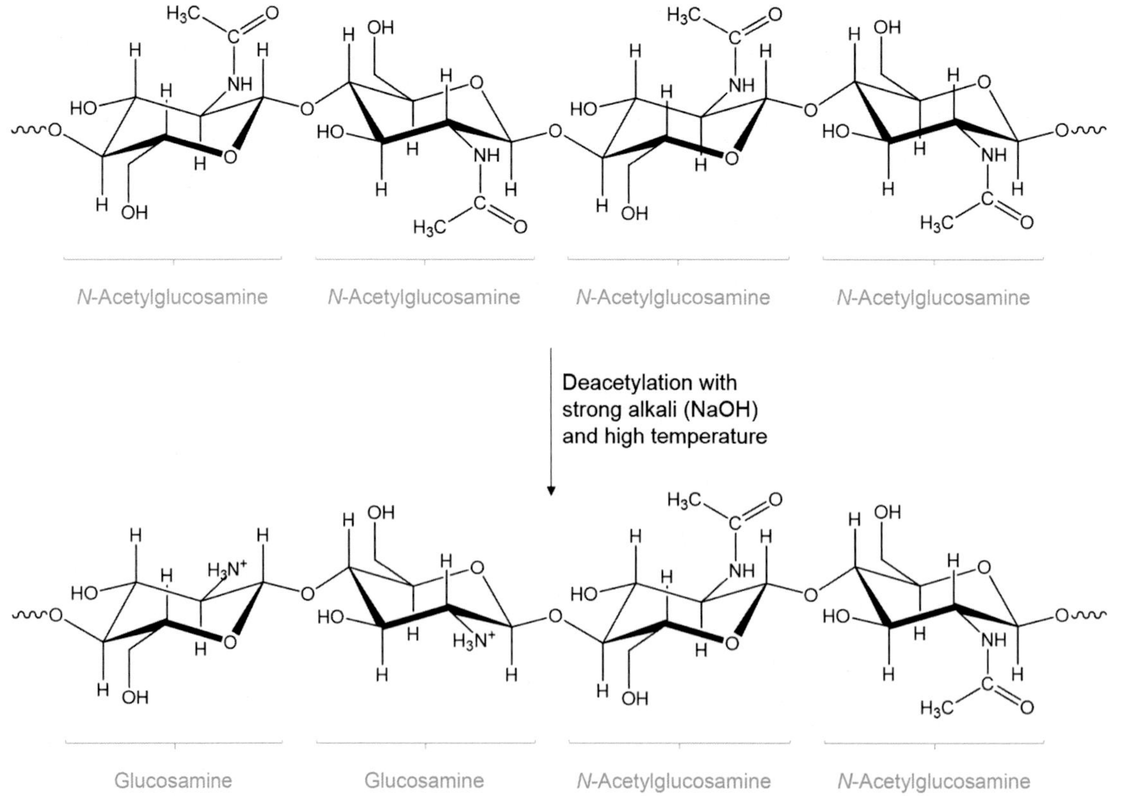

FIG. 4.52 Structure of chitin and chitosan. Chitin, a homopolysaccharide composed of *N*-acetylglucosamine units, is deacetylated with strong alkali (e.g., sodium hydroxide, NaOH) at high temperature (e.g., 120°C) for different time durations (e.g., 1–3h) to give chitosan at variable degree of deacetylation (DD). A chitosan molecule with a DD of 50% is reported in the figure.

conversion of chitin to chitosan renders the material more readily soluble and processable for various applications (Kumbar et al., 2014). Although crystalline chitosan is insoluble in aqueous solution at a pH > 7.0, in dilute acid where pH < 6.0, the positively charged amino group facilitates its solubility. The acetylated residues of chitosan are targeted by lysozyme in vivo, and this seems to be the major mechanism of chitosan degradation. Therefore DD and crystallinity of the polymer are inversely related to degradation. This means that the higher the DD (>85%), the more crystalline the chitosan polymer and the slower its degradation in the body.

On average, the molecular weight of commercially produced chitosan is between 3.8 and 20 kDa.

*Applications*: Bioapplications of chitosan were presumably more advanced in the most recent three decades. This polymer is distinct from other available polysaccharides due to the presence of nitrogen in its molecular structure, its cationicity, and its capacity to form polyelectrolyte complexes (Bhattarai et al., 2010). When two oppositely charged polymers (a polycation and a polyanion), in a solution phase, separate out, a dense polymer phase called coacervate and a supernatant with low polymer content are obtained. This process is termed polyelectrolyte complex formation. Polyelectrolyte complex formation has been used in a number of chitosan drug delivery systems where controlled release of the loaded drug was desired (Bernkop-Schnürch and Dünnhaupt, 2012). Chitin and chitosan have also been successfully applied to fabricate polymer scaffolds in tissue engineering.

Some of the particular characteristics of chitosan such as its optical clarity, mechanical stability, sufficient optical correction, gas permeability (particularly toward oxygen), wettability, and immunological compatibility make it ideal to be used for fabricating contact lenses for use in the eye.

Moreover, chitosan is an orally administered dietary supplement thought to lower low-density cholesterol and considered helpful for weight loss (Mhurchu et al., 2005). Apart from its direct consumption, chitosan is used in drug formulations of different types such as microparticles, granules, and gels for oral, ocular, nasal, vaginal, buccal, parenteral, and intravascular drug delivery. In the same way as other polysaccharides, chitosan is degraded in the colon. By making use of this colon-specific degradation, chitosan was discovered as a useful coating to achieve site-specific delivery. In most applications, chitosan is physically or chemically cross-linked to increase stability. Other examples of covalent modifications of chitosan include acylation and quarternization.

It is worth noting that chitosan is known to have antibacterial properties that are attributed to the interaction between the anionic cell wall components and the cationic chitosan molecules. This would lead to bacterial wall disruption and osmotic lysis of cells (Vaz et al., 2018).

### 4.3.2.4 Alginate (see Chapter 1 - Classification of Polysaccharides, Fig. 1.79)

Alginic acid is an anionic, strictly linear (unbranched) copolymer of mannuronic acid (M block) and guluronic acid (G block) units arranged in an irregular pattern of varying proportions of GG, MG, and MM blocks. It is generally seen that alginates with high G block content are highly suitable for biomedical application due to the ease of processability. Hence, the content of G and M blocks is a crucial factor that determines the properties and applications of the resultant alginate. The molecular weight of alginate can vary widely between 50 and 100,000 kDa, and is a critical factor to influence its viscosity in solution, besides the concentration of the polymer.

Alginates are polysaccharides produced by a wide variety of brown seaweeds (*Laminaria* sp., *Macrocystis* sp., *Lessonia* sp., etc.). Additionally, bacteria also synthesize alginates, and these can be used as tools to tailor alginate production, by understanding the biosynthesis of the polymer in these bacteria. By genetically selecting and engineering *Pseudomonas* strains that contain only a single epimerase, the production of high G-containing alginates has been possible. Using such strategies, alginates with up to 90% G content and extremely long G blocks have been produced. Although such strategies are useful to engineer alginates, most of the alginates extracted for large-scale applications originate from natural sources such as seaweeds. The quality is determined by the species and even the seasonal variations. The alginate could contain from 10% to 70% G. Techniques of separation such as fractionation and precipitation in calcium can help separate the G block- and M block-rich alginates.

The most important property of alginate is its ability to gel in the presence of cations (like $Ca^{2+}$ and $Ba^{2+}$) (Fig. 4.53).

Factors that influence the stiffness of the gel are molecular weight distribution of the alginate polymer (dependent on M/G ratio) and the stoichiometry of alginate with the chelating cation (Kumbar et al., 2014).

It is worth noting that different mechanisms reported herein may contribute to the degradation of alginate in the human body:

- disintegration of the alginate material by exchange of gelling calcium ion with sodium;
- acid and alkaline hydrolysis;
- degradation by reactive oxygen species. Most polymers including alginates undergo "free-radical depolymerization" or "oxidative-reductive depolymerization." Exposure to γ radiation can be used to enhance the rate of polymer degradation as well.

The major drawback of using alginate relies on the lack of enzymatic degradation. This is an inert polymer that is nonadherent for cells.

*Applications*: Alginate by itself is widely used in many industrial sectors. It is especially useful in the kitchen due to its thickening, gelling, and

stabilizing properties. It is also used as a stabilizer and emulsifier in the food industry, as it interacts with proteins, fats, and fibers. Of note, alginate-pectin mixtures are used as gelling agents independent of sugar content in foods. Hence, alginate is used in many low-calorie substitute foods.

Alginate is also largely employed by pharmaceutical industry as drug excipient, dietary supplement, diet aid, dental impression material, in the treatment of heartburn, acid reflux, and gastroesophageal reflux disease (GERD) (e.g., Gaviscon from GlaxoSmithKline plc, United Kingdom), as a wound dressing material (e.g., ALGICELL Ag, a silver calcium alginate dressing, from Derma Sciences, United States; Safe N Simple Simpurity Silver Alginate Wound Dressing from Safe n Simple, United States), and as a thickener and moisturizer in cosmetics.

### 4.3.2.5 Hyaluronic Acid (see Chapter 1 - Classification of Polysaccharides, Fig. 1.80)

Hyaluronic acid (HA), also named hyaluronan, is a high molecular weight, nonsulfated, linear polysaccharide constituted by a repeating disaccharide. HA is a ubiquitous substance abundantly spread in nature (Falcone et al., 2006; Prestwich, 2011). Although the molecular weights may differ depending on the source, the uniqueness of HA relies on its structure that is apparently identical when synthesized by bacteria, birds, and mammals. HA is an essential GAG component of connective tissue and synovial fluid (i.e., the fluid that lubricates joints). Because of its hydration properties, HA has the ability to bear compressive loads in vivo and provide lubrication at the same time. It is also found at higher concentrations in the vitreous humor of the eye and in the hyaline cartilage. HA does mediate cellular signaling, wound repair, morphogenesis, and matrix organization (Toole, 2004).

HA has been largely purified from cartilage and bacteria to be used as a biomaterial in medical devices. Its rheological properties as a viscoelastic polymer become apparent after

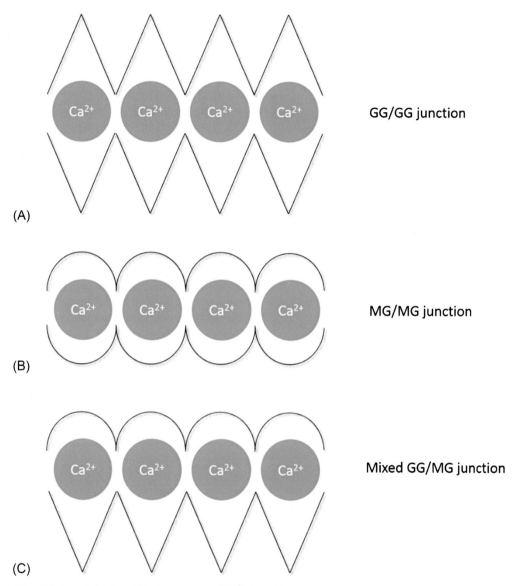

(A)

(B)

(C)

GG/GG junction

MG/MG junction

Mixed GG/MG junction

**FIG. 4.53** Gelation of alginic acid in the presence of $Ca^{2+}$. Possible junctions: (A) GG/GG junctions, (B) MG/MG junctions, and (C) mixed GG/MG, with $Ca^{2+}$ (Kumbar et al., 2014).

purification. HA solutions (especially those of high molecular weight) exhibit viscoelastic properties that make them excellent biological absorbers and lubricants, with rheological properties that depend on the polysaccharide concentration, ionic strength, pH, and molecular weight.

*Applications:* HA and its derivatives have been clinically used as medical products for more than three decades. In the past decade, HA has become recognized as an important building block for the design of new biomaterials for use in cell therapy, three-dimensional cell cultures, and tissue engineering applications. In this sense, HA-based materials have been processed in different forms, such as hydrogels, films, fibers, and nonwoven pads. Because they are viscous, cohesive, lubricious, and hydrophilic, HA solutions have been used as viscoelastic adjuncts in ophthalmic surgery (e.g., iCross from OFFitalia srl, Italy, in Fig. 4.54), and as covalently bound lubricious hydrophilic coatings for medical devices, such as stents and catheters. More recently, HA has been proposed as a biomaterial for cartilage repair and dermis augmentation/wound healing. Hyalofast® and Hyalomatrix® from Anika Therapeutics Inc. (United States), HA-based 3D scaffolds for cartilage and skin tissue engineering purposes, respectively, are already on sale.

HA is also used as a basic component for several make-up items, for eye and skin care, and anti-aging action (e.g., ELEVESS® from Anika Therapeutics Inc., an injectable dermal filler used to correct facial lines and wrinkles).

For medical applications that require extended residence time *in situ*, HA has also been cross-linked by means of a variety of chemical methods.

FIG. 4.54 A commercial HA-based medical device. iCross from OFFitalia srl (Italy) (http://www.offitalia.it/en/).

### 4.3.3 Nucleic Acids (see Chapter 1 - Nucleic Acids, Fig. 1.85)

Nucleic acids (NAs) are involved in the most important processes occurring in living cells and have an enormous potential for specific interactions with other different biopolymers (Vlassov et al., 1997). Specificity of recognition of DNA and RNA species by complementary NAs is unique in the world of biopolymers. A great number of the proteins involved in processes of replication, transcription, and translation are also capable of highly specific recognition of certain nucleotide sequences. Therefore NAs have always attracted the attention of pharmacologists dreaming of highly specific therapeutics, that is, the ability to target other biopolymers. Unfortunately, their use as biomaterials is very limited at the present.

Recent studies have pointed out that oligonucleotides can interact with a broad spectrum of proteins capable of binding polyanions. Some pioneering studies have shed light on the possibility of using DNA as nanocages for protein encapsulation. DNA cages are basically nanometer-scale polyhedrons formed by self-assembly from synthetic oligonucleotides (Erben et al., 2006). The cages are shaped so that they display a central cavity larger than the wall openings (Fig. 4.55) and thus they may hold and seclude proteins and other biomolecules.

*Applications*: Potential applications of such engineered DNA nanocages include in vivo imaging and the targeted delivery of macromolecules into living cells (Walsh et al., 2011), paving a new way for controlled delivery of theranostics into target cells.

## 4.4 METALLIC BIOMATERIALS

Because numerous biomedical applications rely on implants with one or more metallic components (e.g., hip joint, mechanical heart valves), metals have a highly noteworthy importance in

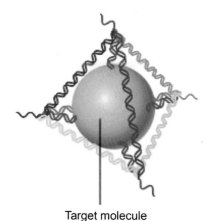

Target molecule

FIG. 4.55 Schematic representation of a molecule-encapsulated DNA cage (Griggs, 2009).

the biomaterials field. Some applications are specific to the replacement of damaged or dysfunctional tissues to restore function, as, for example, in orthopedic applications in which all or a part of bone or joint (e.g., hip, knee, shoulder) are replaced or reinforced by metal alloy devices. In dental applications, metals were used as filler material for decayed teeth and are now used as screws for dental implants and orthodontic wires. In cardiovascular applications, metals are used for the repair of damaged arterial vessels or vein (e.g., stent), or as a component of artificial heart valves (e.g., housing of heart valves).

In general, metals have specific characteristics that make them suitable for application in the human body, and they should have the following properties:

- biocompatibility, nontoxicity, not allergenicity;
- corrosion resistance (in particular, pitting and crevice corrosion);
- adequate mechanical characteristics;
- high wear resistance.

The internal environment of the body is highly corrosive and degrades the implant materials, resulting in the release of harmful metal ions. Thus a stable oxide on the surface of the metallic biomaterials is needed to enhance their corrosion resistance, and at the same time, should allow a good biocompatibility, in terms, for example, of noncarcinogenicity and nontoxicity. The mechanical properties of a metal are important and have to match the requirements of a specific application in the human body. In fact, the properties necessary for a hip replacement should be different from the ones required for a vascular stent, as the condition and stress pattern the metal is subjected to is completely different. Moreover, an important characteristic is related to the endurance to large and variable, cyclic stresses in the highly corrosive environment. In fact, the importance of the load-bearing capacity of the metal may be realizing that a person may average experience 1–2.5 million cycles of stress on the hip per year (depending on the daily activity). Hence, in the life of the person, it means 50–100 million cycles of stress over a 50-year period. Therefore the metallic biomaterial must be strong enough, and fatigue- and wear-resistant, in a highly corrosive environment.

Several metals are used for a great variety of clinical applications (Fig. 4.56), with titanium and its alloys, stainless steel and cobalt-chromium alloy, being the most commonly used metals for different implantable applications.

In the following, the main aspects related to the composition design, structure, and properties of metallic biomaterials are provided. In particular, this section of the chapter reports the mechanical properties of metals used in implant devices, as well other properties, such as surface properties and wear properties that have to be considered, because in some applications they can be more critical to control.

## 4.4.1 Stainless Steels

Stainless steel (SS) is the term used to describe an extremely versatile family of engineering materials, which are selected primarily for their corrosion and heat resistant properties.

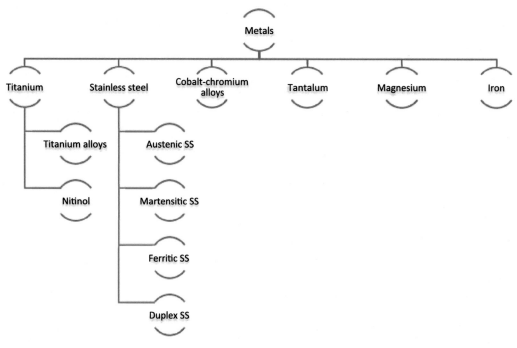

**FIG. 4.56**    Most used metals for biomedical applications.

SS are iron-based alloys containing at least 10.5% w/w chromium. The corrosion resistance of stainless steel is attributed to the formation of a stable, protective chromium oxide ($Cr_2O_3$) film (i.e., passive film, 2–3nm thickness) on the surface. In fact, chromium reacts with oxygen and moisture in the environment to form a protective, adherent, and coherent oxide film that envelops the entire surface of the material. The passive layer on SS exhibits a truly remarkable property: when damaged (e.g., abraded), it self-repairs as chromium in the steel reacts rapidly with oxygen and moisture in the environment to reform the oxide layer. Moreover, resistance to corrosion phenomena as well as other physical and mechanical properties can be improved by varying the alloying elements. For example, addition of molybdenum increases pitting corrosion resistance, and the addition of nitrogen can increase mechanical strength and pitting corrosion resistance. Table 4.3 summarizes the possible effects of the alloying elements on the structure and properties of stainless steel alloys.

### 4.4.1.1 Phase Diagram

Taking into account the Fe-C phase diagram, it is interesting to notice that some elements can affect the allotropic transformation of steels.

In particular, some elements extend the γ-loop in the iron-carbon equilibrium diagram (see Chapter 1), for example, nickel and manganese (Table 4.3). When sufficient alloying element is added, it is possible to preserve the face-centered cubic austenite at room temperature, either in a stable or metastable condition. Focusing on nickel (Fig. 4.57), for a certain nickel content (>15%–20%), the transformation Fe-α → Fe-γ takes place at a temperature lower than room temperature, so that it is possible that a

**TABLE 4.3** Effect of Alloying on Structure and Properties of Stainless Steels

| Element | Effect on SS |
|---|---|
| Chromium | – The most important alloying element in SS production<br>– Minimum of 10.5% Cr required for the formation of a surface protective layer of chromium oxide<br>– Strength of protective layer increases with increasing chromium content<br>– Prompts the formation of ferrite within the alloy structure and is described as ferrite stabilizer |
| Nickel | – Improves general corrosion resistance and prompts the formation of austenite (i.e., it is an austenite stabilizer)<br>– 8%–9% nickel content gives a fully austenitic structure<br>– Content beyond 8%–9% further improves both corrosion resistance (especially in acid environments) and workability |
| Molybdenum (and tungsten) | – Increases resistance to both local (e.g., pitting, crevice corrosion) and general corrosion<br>– Ferrite stabilizers that, when used in austenitic alloys, must be balanced with austenite stabilizers to maintain the austenitic structure<br>– Added to martensitic stainless steels to improve high temperature strength |
| Nitrogen | – Increases mechanical strength<br>– Enhances resistance to localized corrosion<br>– Austenite former |
| Copper | – Increases general corrosion resistance to acids<br>– Reduces the rate of work-hardening<br>– Austenite stabilizer |
| Carbon | – Enhances strength (especially, in hardenable martensitic stainless steels)<br>– May have an adverse effect on corrosion resistance by the formation of chromium carbides<br>– Austenite stabilizer |
| Titanium (and niobium, zirconium) | – Titanium or niobium may be used to stabilize stainless steel against intergranular corrosion<br>– As titanium (niobium and zirconium) have greater affinity for carbon than chromium, titanium (niobium and zirconium) carbides are formed in preference to chromium carbide and thus localized depletion of chromium is prevented<br>– Ferrite stabilizers |
| Sulfur | – Added to improve the machinability of SS<br>– Sulfur-bearing SSs exhibit reduced corrosion resistance |
| Cerium | – Improves the strength and adhesion of the oxide film at high temperatures |
| Manganese | – Austenite former<br>– Increases the solubility of nitrogen in the steel and may be used to replace nickel in nitrogen-bearing grades |
| Silicon | – Improves resistance to oxidation<br>– Used in special stainless steels exposed to highly concentrated sulfuric and nitric acids<br>– Ferrite stabilizer |

steel containing nickel has an austenitic structure (CFC) at room temperature.

Moreover, chromium added alone to plain carbon steel tends to close the $\gamma$-loop and favor the formation of ferrite (Table 4.3). In fact, beyond 13% Cr, the transformation Fe-$\alpha$ → Fe-$\gamma$ no longer occurs and the Fe-$\alpha$ field joins the one of Fe-$\delta$, so that a Cr-containing steel may have a ferritic structure (CCC) at room temperature. However, when chromium is added to

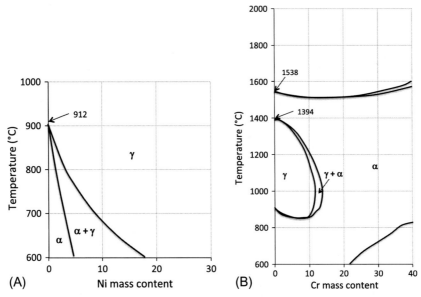

FIG. 4.57    (A) Phase diagram of Fe-Ni; (B) phase diagram of Fe-Cr.

a steel containing nickel, it retards the kinetics of the $\gamma \rightarrow \alpha$ transformation, thus making it easier to retain austenite at room temperature. The presence of chromium greatly improves the corrosion resistance of the steel by forming a very thin stable oxide film on the surface, so that chromium-nickel stainless steels are now the most widely used materials in a wide range of corrosive environments.

### 4.4.1.2 Stainless Steel Classification

Based on their microstructure, SS can be classified in four main classes:

- martensitic SS;
- ferritic SS;
- austenitic SS;
- duplex SS.

Within each of these groups, there are several "grades" of stainless steel defined according to their compositional ranges. These compositional ranges are defined in European (and others, e.g., United States) standards, and within the specified range, the stainless steel grade will exhibit all of the desired properties (e.g., corrosion

resistance and/or heat resistance and/or machinability).

In the next sections, the main characteristics of these classes are reported, together with their possible applications in biomedical fields. It has to be highlighted that all the classes of SS are regulated by international standards, and for application in medical fields, medical (nonimplant) devices are described in terms of both EN 10088 and U.S. designations, but in addition, reference is made to specific standards applicable to medical devices (e.g., ISO 7153-1). Medical implants have specific material specifications (e.g., ISO 5832-1 and ISO 5832-9), which do not have an equivalent EN 10088 grade.

### 4.4.1.3 Schaeffler Diagram

One of the most convenient ways of representing the effect of various elements on the basic structure of chromium-nickel stainless steels is the Schaeffler diagram. It plots the compositional limits at room temperature of austenite, ferrite, and martensite in terms of nickel and chromium equivalents (Fig. 4.58). At its simplest

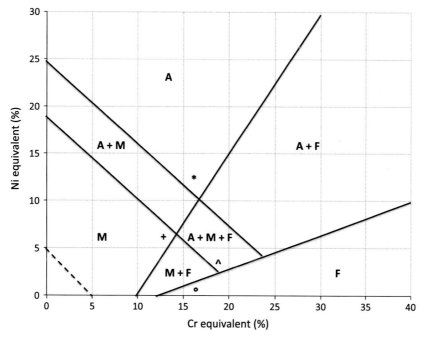

**FIG. 4.58** Schaeffler diagram: effect of alloying elements on the basic structure of Cr-Ni stainless steels.

level, the diagram shows the regions of existence of the three phases for iron-chromium-nickel alloys. However, the diagram becomes of much wider application when the equivalents of chromium and of nickel are used for the other alloying elements. The chromium equivalent has been empirically determined using the most common ferrite-forming elements in a weighted sum, as reported in Eq. (4.1):

$$\mathrm{Cr_{equivalent}} = (\mathrm{Cr}) + 2(\mathrm{Si}) + 1.5(\mathrm{Mo}) + 5(\mathrm{V})$$
$$+ 5.5(\mathrm{Al}) + 1.75(\mathrm{Nb}) + 1.5(\mathrm{Ti}) + 0.75(\mathrm{W}) \quad (4.1)$$

whereas the nickel equivalent has likewise been determined as the weighted sum of the familiar austenite-forming elements, as in Eq. (4.2):

$$\mathrm{Ni_{equivalent}} = (\mathrm{Ni}) + (\mathrm{Co}) + 0.5(\mathrm{Mn}) + 0.3(\mathrm{Cu})$$
$$+ 25(\mathrm{N}) + 30(\mathrm{C}) \quad (4.2)$$

All concentrations are expressed in weight percentages.

From this diagram, it is possible to appropriately select the alloy elements to obtain steels with an austenitic (FCC), ferritic (BCC), or martensitic structure (TCC).

For example, the three most typical compositions (i.e., with the minimum possible number of alloy elements) corresponding to the three structures are:

- austenitic structure (point *, Fig. 4.58): 18% chromium, 8%–10% nickel, 0.06% carbon;
- ferritic structure (point °, Fig. 4.58): 17% chromium, 0.1% carbon;
- martensitic structure (point +, Fig. 4.58): 13% chromium, 0.15% carbon.

The duplex SS, partially austenitic and partially ferritic, are also detectable in the diagram (point ˆ, Fig. 4.58): 23% chromium, 4% nickel, 0.03% carbon.

### 4.4.1.4 Martensitic Stainless Steel

Martensitic SS have a body-centered tetragonal crystal structure. The Cr content in martensitic SS varies from 10.5% to 18%, and the carbon content can be greater than 1.2%. The amount of

Cr and C are adjusted in such a way that a martensitic structure is obtained. Several other elements, for example, tungsten, niobium, and silicon, can be added to alter the toughness of the martensitic SS. The addition of small amounts of nickel enhances the corrosion resistance and toughness, and the addition of sulfur in this alloy improves the machinability. They have good mechanical properties and moderate corrosion resistance, and they are ferromagnetic. These alloys may be heat treated, in a similar manner to conventional steels (e.g., tempering and hardening), to provide a range of mechanical properties but offer higher hardenability and have different heat treatment temperatures. Their corrosion resistance may be described as moderate (i.e., their corrosion performance is poorer than other stainless steels of the same chromium and alloy content).

*Applications.* Martensitic stainless steels (e.g., grades 1.4006 [AISI 410], 1.4021 [AISI 420], 1.4028 [AISI 429], and 1.4125 [AISI 440C]) are used extensively for dental and surgical instruments. These stainless steels can be hardened and tempered by heat treatment. Thus they are capable of developing a wide range of mechanical properties (i.e., high hardness for cutting instruments and lower hardness with increased toughness for load-bearing applications). Martensitic stainless steels used in medical devices usually contain up to 1% nickel. Although there are some martensitic grades with higher nickel contents, these grades are not generally used for medical device applications. For example, applications of martensitic SS (Fig. 4.59) include bone curettes, chisels and gouges, dental burs, dental chisels, curettes, explorers, root elevators, forceps, hemostats, retractors, orthodontic pliers, and scalpels.

### 4.4.1.5 Ferritic Stainless Steel

Ferritic SS have a body-centered cubic, BCC crystal structure, and the Cr content can vary from 11% to 30%. Ferritic stainless steels are essentially nickel-free. Other elements, for example, niobium, silicon, and molybdenum, can be added to obtain specific characteristics. Similar to martensitic SS, sulfur and selenium can be added to improve machinability. Compared to martensitic SS, ferritic SS cannot be strengthened by heat treatment but exhibit superior corrosion resistance to martensitic stainless steels and possess good resistance to oxidation. Additionally, cold working is not commonly performed, because it decreases the ductility of these alloys.

*Applications*: The few applications of ferritic SS in biomedical field include handles for instruments and medical guide pins.

### 4.4.1.6 Austenitic Stainless Steel

Because chromium stabilizes the magnetic ferrite (i.e., BCC) structure, other alloying elements must be added to stabilize the desired austenite phase (i.e., FCC). This role is filled primarily by additions of chromium (content: 15%–20%), nickel (content: 3%–14%), and manganese (content: 1%–7.5%). Moreover, nitrogen additions also increase mechanical strength and corrosion resistance. In particular, corrosion performance may be varied to suit a wide range of service environments by careful alloy adjustment, for example, by varying the carbon or molybdenum content. In fact, molybdenum, silicon, and niobium additions have a beneficial impact on the pitting corrosion and oxidation resistance. Similar to ferritic SS, austenitic SS cannot be hardened by heat treatments, but cold working can be performed to harden them. These alloys have excellent cryogenic properties, high-temperature strength, oxidation resistance, and formability. The increasing amount of Ni influences the formability of these alloys.

Austenite SS make up over 70% of total stainless steel production for a wide range of industrial applications. They are essentially nonmagnetic; this is crucial as patients may be exposed to high intensity magnetic fields during MRI examinations. It should be noted that modest amounts of heating and displacement could occur for these alloys under MRI exposure.

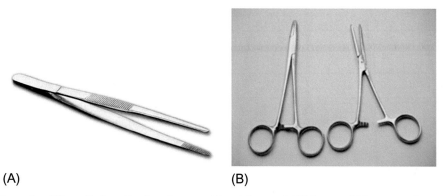

(A)                              (B)

FIG. 4.59   Examples of biomedical application of martensitic stainless steel: (A) forceps; (B) hemostats (hemostatic clamp).

*Applications*: Austenitic SS are extensively used for biomedical implants and devices, and 316L is the most commonly used austenitic SS. For that reason, in the next section, more information on 316L will be provided.

*Austenitic stainless steels for biomedical applications:* It is important to distinguish between stainless steel grades used for implant applications and the commercial grade stainless steels (e.g., 1.4305 [AISI 303], 1.4301 [AISI 304], and 1.4401 [AISI 316]) used for other medical devices (e.g., dental scalers, dental explorers, dental and surgical forceps, kidney dishes, and theater tables).

*ISO standards 5832-1* and *5832-9* specify wrought stainless steel and high-nitrogen stainless steel, respectively, for surgical implants. These materials were originally developed from 1.4401 (AISI 316) stainless steel, but their chemical composition is now enhanced (i.e., higher chromium, nickel, and molybdenum contents). In addition, implant grade SS have specific requirements for resistance to pitting corrosion and for cleanliness that do not apply to commercial stainless steels (e.g., 1.4401 [AISI 316]).

*ISO 7153-1* specifies stainless steel for surgical and dental instruments. It should be stressed that the grades in ISO 7153-1 are generic, although there is a reference to another little used ISO steel standard. This standard also provides an indication of typical applications for each grade. Therefore they are not special steels and certainly not specifically prepared for surgical applications. They are, however, used worldwide by all dental and surgical instrument manufacturers for their nonimplant products.

*Grade 1.4305 (AISI 303)* stainless steel is used in medical devices, where its free-machining properties enhance the ease of manufacture, for example, medical devices with screw threads, with drilled and/or tapped holes. Handles of multipart dental instruments are often manufactured in grade 303. In this application, its lower corrosion resistance is not a disadvantage. In fact, the handle rarely comes into contact with the patient and, if it does, contact is transient. The dentist is not at risk from contact dermatitis, because latex gloves are worn to reduce the risk of cross-contamination. *Grade 304* has applications in medical devices, where good corrosion resistance and moderate strength are required (e.g., dental impression trays, hollowware, retractors, and guide pins). Once again, most of these applications involve minimal and transient contact with patients.

*Grade 316 SS*, as previously assessed, is the most often used austenitic stainless steel for implantable devices. This metal is commonly used because it is relatively inexpensive and

**TABLE 4.4**  Chemical Composition (%) and Mechanical Properties of Grade 316L Stainless Steel

| Cr | Ni | Mo | Mn | Si | Cu | N | C | S | P | Fe |
|---|---|---|---|---|---|---|---|---|---|---|
| 17–20 | 12–14 | 2–4 | 2 | 0.75 | 0.50 | 0.10 | 0.03 | 0.03 | 0.03 | Balance |

| 316L Properties | Annealed | 30% Cold-Worked |
|---|---|---|
| Density (g/cm$^3$) | 7.6 | |
| Elastic modulus (GPa) | 190 | 190 |
| Yield strength (MPa) | 172 | 690 |
| Tensile strength (MPa) | 485 | 860 |

Composition is expressed as maximum percentage content of the element in the alloy.

can be shaped easily with existing metal forming techniques. It consists primarily of iron, chromium, nickel, and smaller amounts of molybdenum, manganese, copper carbon, nitrogen, phosphorus, silicon, and sulfur (18Cr-14Ni-2.5Mo, ASTM F138, Table 4.4). Shown in Table 4.4 are the main mechanical properties of annealed and 30% cold-worked material. The alloy is characterized by a low carbon content (>0.03%, Table 4.4) that imparts an excellent corrosion resistance (the "L" in 316L designates low carbon content). In fact, when the C content exceeds 0.03%, it may precipitate as carbide (Cr$_{23}$C$_6$) at grain boundaries, reducing the amount of Cr in the region close to the grain boundaries, hence the formation of the protective Cr oxide layer. Other commonly used alloying elements are molybdenum and silicon that stabilize the ferritic phase; nickel is added to strengthen the austenitic phase. The density of stainless steel is 7.9 g/cm$^3$, almost twice that of Ti and its alloys. For temporary bone fracture treatments, where today austenitic stainless steel is the leading material, this is not so important. However, for fixed larger implants (e.g., joint replacements), this characteristic is important due to the possibility of stress shielding.

*Corrosion behavior in human body:* Austenitic stainless steels contain nickel as alloying, so that allergic reactions can occur in many patients. In addition, they are susceptible in the human body to crevice corrosion phenomena, especially in the grades with low molybdenum and nitrogen content.

The occurrence of this local corrosion form, which occurs in interstices, such as those between the heads of the screws and their places in the osteosynthesis plates, or in the areas of contact between intramedullary nails, causes a significant increase in the metal ion release in tissues close to the implant. Such metal ions can cause local irritations, systemic effects, and above all, in particular in presence of nickel in the alloy, allergic sensitization phenomena in a nonnegligible percentage of patients (especially female). The onset of such phenomena involves the retrieval of the metallic device and the implantation of a new one.

*Applications:* The most effective applications (Fig. 4.60) are in bone screws, pins, plates, intramedullary bone nails, and other temporary fixation devices.

### 4.4.1.7 Duplex Stainless Steel

Duplex SS are two-phase alloys containing equal proportion of ferritic and austenitic phases in their microstructure, providing a combination of the corrosion resistance of austenitic stainless steels with greater strength. Hence, they have a mixed microstructure of austenite and ferrite, the aim usually being to produce a

(A)

(B)

**FIG. 4.60** Example of biomedical application of austenitic stainless steel: (A) plate and screws for treatment of bone fracture; (B) possible use of stainless steel (and other biomaterials) in articular hip joint prosthesis.

50/50 mix, although in commercial alloys the ratio may be 40/60. With carbon content lower than 0.03%, the amount of Cr and Ni content can be varied from 20% to 30% and from 5% to 8%, respectively. Minor alloying elements contained in duplex SS include molybdenum, nitrogen, tungsten, and copper. Duplex stainless steels have roughly twice the strength compared

to austenitic stainless steels and also improved resistance to localized corrosion, particularly pitting, crevice corrosion, and stress corrosion cracking. Moreover, they have improved toughness and ductility compared to ferritic SS. Duplex stainless steels are weldable, but care must be exercised to maintain the correct balance of austenite and ferrite. They are ferromagnetic and their formability is reasonable, but higher forces than those used for austenitic stainless steels are required.

#### 4.4.1.8 Other Stainless Steel Alloys Under Evaluation for Biomedical Devices

Recently, nitrogen-strengthened SS have been developed with improved mechanical properties and corrosion resistance compared to 316L stainless steel. Nitrogen-strengthened stainless steels are used for fracture fixation devices and other medical implants, and are classified under ASTM F1314, F1586, and F2229. The nitrogen contents can vary from 0.20% to 0.40%, 0.25% to 0.50%, and 1.0%, respectively for ASTM F1314, F1586, and F2229. These alloys are typically ferrite-free and can be cold worked to improve mechanical tensile strength, fatigue strength, and crevice and pitting corrosion resistance, compared to grade 316L. In particular, ASTM F2229 is particularly attractive for biomedical applications because it does not contain nickel and patients' nickel allergies can be avoided.

### 4.4.2 Cobalt Alloys

Cobalt alloys are classified into two different classes: casting alloys and wrought by (hot) forging alloys. In particular, four cobalt-chromium alloys are commonly used for biomedical applications:

- ASTM F75: Co-28Cr-6Mo, casting alloy;
- ASTM F799: Co-28Cr-6Mo, thermodynamically processed alloy;
- ASTM F90: Co-20Cr-15W-10Ni, wrought alloy;
- ASTM F562: Co-35Ni-20Cr-10Mo, wrought alloy.

The chemical composition and the mechanical properties of these alloys are listed in Table 4.5. It has to be highlighted that, although ASTM75 and ASTM799 alloys possess similar compositions, the different processing methods result in different mechanical properties

(Table 4.5). Moreover, the cobalt content in ASTM F90 and ASTM F562 is lower than that in ASTM75 and ASTM799. On the contrary, ASTM F562 contains more nickel and ASMT contains more tungsten (Table 4.5).

Co-based alloys can be described as nonmagnetic, wear and corrosion resistant, and stable at elevated temperatures. The corrosion resistance of these alloys is, similar to stainless steel, based on the formation of a passivation thin layer of $Cr_2O_3$. The two basic elements of the Co-Cr alloys form a solid solution of up to 65% Co. The molybdenum is added to produce finer grains, which results in higher strengths after casting or forging. The chromium enhances corrosion resistance as well as solid solution strengthening of the alloy.

In the next sections, the properties of the four Co-Cr alloys are briefly described together with their use in biomedical field.

#### 4.4.2.1 Cast Cobalt-Chromium Alloys

*ASTM F75 alloy:* The ASTM F75 is a casting Co-28Cr-6Mo alloy; it is commercially available with different trade names, for example, Vitallium (Howmedica Inc.), Haynes Stellite 21 (Cabot Co.), Protasul-2 (Sulzer AG), and Zimaloy (Zimmer Co.). F75 alloy has a long history in both the aerospace and biomedical implant industries. The main attribute of this alloy is its excellent corrosion resistance, even in chloride environments, which is related to its bulk composition (i.e., chromium content) and surface oxide. When F75 is cast into shape by investment casting (e.g., the "lost wax" process), the alloy is first melted at 1350–1450°C, then poured or pressurized into ceramic molds of the desired shape (e.g., femoral stems for artificial hips, oral implants, dental partial bridgework). Different microstructural features can be observed depending on casting conditions of the alloy and can adversely affect the mechanical properties of the F75 alloy. In particular, the grain sizes are large, decreasing the tensile

**TABLE 4.5** Chemical Composition (% Max) and Mechanical Properties of the Four Main Co-Cr Alloys Used in the Biomedical Field

| Elements | ASTM F75 | ASTM F799 | ASTM F90 | ASTM F562 |
|---|---|---|---|---|
| Co | 58.9–69.5 | 58–59 | 45.5–56.2 | 29.0–38.8 |
| Cr | 27.0–30.0 | 26.0–30.0 | 19.0–21.0 | 19.0–21.0 |
| Mo | 5.0–7.0 | 5.0–7.0 | – | 9.0–10.5 |
| W | 0.2 | – | 14.0–16.0 | – |
| Ni | 2.5 | 1.0 | 9.0–11.0 | 33.0–37.0 |
| Mn | 1.0 | 1.0 | 2.00 | 0.15 |
| Si | 1.0 | 1.0 | 0.40 | 0.15 |
| Fe | 0.75 | 1.5 | 3.00 | 1.0 |
| C | 0.35 | 0.35 | 0.15 | 0.025 |
| N | 0.25 | 0.25 | – | – |
| P | 0.02 | – | 0.04 | 0.015 |
| Ti | 0.10 | – | – | 1.0 |
| S | 0.01 | – | 0.03 | 0.010 |

| Property | ASTM F75 | | ASTM F799 | ASTM F90 | | ASTM F562 | |
|---|---|---|---|---|---|---|---|
| | As-Cast/ Annealed | Powder Metallurgy/ Hot Isostatic Pressing | Hot Forged | Annealed | 44% Cold Worked | Hot Forged | Cold Worked, Aged |
| Elastic modulus (GPa) | 210 | 253 | 210 | 210 | 210 | 232 | 232 |
| Yield strength (MPa) | 450 | 841 | 896–1200 | 448–648 | 1606 | 965–1000 | 1500 |
| Tensile strength (MPa) | 655 | 1277 | 1399–1586 | 951–1220 | 1896 | 1206 | 1795 |

strength of the alloy. Moreover, casting defects are present, causing possible fatigue fracture of the implant under in vivo conditions.

*ASTM F799 alloy:* The ASTM F799 is known as heat-treated, cast Co-28Cr-6Mo alloy because it is similar in composition to the F75 alloy, but it is processed by hot forging at about 800°C after casting, in a series of steps. In most cases, high temperature is used in the early forging step to allow for a significant flow (i.e., increase in deformability); later, in the final stages, a low temperature is used to induce cold working, that is, strengthening. Hence, the final shape of the device can be achieved with yield and tensile stress twice as high as the ones of as-cast F75 alloy.

#### 4.4.2.2 *Wrought Cobalt-Chromium Alloys*

*ASTM F90 alloy:* ASTM F90 alloy is a wrought Co-20Cr-15W-10Ni alloy; it is commercially available as Haynes Stellite 25 (Cabot Co.). The relevant level of Ni and W in the alloy allows an improvement in machinability, processing, and fabrication characteristics. In particular, in the annealed state, its properties matches the ones of F75 alloy, whereas in the 44% cold-worked state (Table 4.5), the yield, fatigue, and tensile stress are twice the ones of F75. It is important, in the latter case, to be careful in achieving uniform properties across the thickness of the component, otherwise it will be prone to unexpected failures.

*ASTM F562 alloy:* ASTM F562 is a wrought Co-35Ni-20Cr-10Mo alloy. This alloy has the most effective combination of strength, ductility, and corrosion resistance. Its excellent strength is mainly due to the variety of techniques such as cold working, involved in its processing. In particular, aged ASTM F562 has a tensile strength of 1795 MPa, which is the highest among the metals used for biomedical applications, maintaining a ductility of about 8%.

*Wear behavior of Co-Cr alloys:* The abrasive wear properties of the wrought Co-Ni-Cr-Mo alloy are similar to the cast Co-Cr-Mo alloy (about 0.14 mm/year in joint simulation tests with UHMWPE acetabular cup); however, the former is not recommended for the bearing surfaces of joint prostheses because of its poor frictional properties with itself or other materials. The superior fatigue and ultimate tensile strength of the wrought Co-Ni-Cr-Mo alloy make it suitable for the applications, which require long service life without fracture or stress fatigue. Such is the case for the stems of the hip joint prostheses. This advantage is better appreciated when the implant has to be replaced, because it is quite difficult to remove the failed piece of implant embedded deep in the femoral medullary canal. Furthermore, the revision arthroplasty is usually inferior to the primary surgery in terms of its function due to poorer fixation of the implant.

*Applications:* Among the four alloys, at present only two are used extensively in implant fabrications, the castable Co-28Cr-6Mo and the wrought Co-35Ni-20Cr-10Mo alloy. The castable Co-28Cr-6Mo alloy has been used for many decades in dentistry and, relatively recently, in making artificial joints (Fig. 4.61). In total hip joint replacement, the damaged femoral head and cup on which it articulates (i.e., acetabular cup) are replaced with artificial prostheses. The artificial joint consists of a metallic base (metal back, Fig. 4.63D) and a cup insert on which femoral articulation tales place. The metal-back, in Ti alloys, and the cup, generally made of UHMWPE, are fixed to the pelvic bone. The femoral head can be made of Co-Cr alloy (Fig. 4.61B); unfortunately, the combination metal-UHMWPE can result in the wear of the UHMWPE surface and the loosening of the prostheses.

The wrought Co-35Ni-20Cr-10Mo alloy is used for making permanent fixation devices and joint replacement components (e.g., stems of prostheses for heavily loaded joints such as the knee and hip), due to the combination of long-term corrosion resistance and strength. In addition, Co-Cr alloys are used for the realization of the housing in mechanical heart valves, either tilting disc and bileaflet models (Fig. 4.61D).

### 4.4.3 Titanium and Titanium Alloys

Titanium is well known for its lightweight, excellent corrosion resistance, and enhanced biocompatibility. In particular, the density of Ti ($4.5 \, g/cm^3$) is significantly lower than the one of other metals used for biomaterials as stainless steel (density $= 7.9 \, g/cm^3$) and Co-Cr alloys (density $= 8.3 \, g/cm^3$). Even if its density is lower, Ti and Ti alloys mechanical and chemical properties are competitive with the ones of commonly used stainless steel and Co alloys.

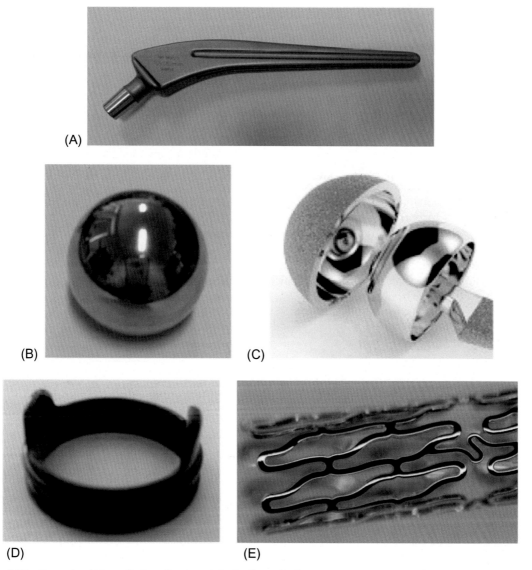

**FIG. 4.61** Example of biomedical application of Co-Cr alloys: (A) hip stem; (B) hip joint head; (C) metal/metal articular joint; (D) housing of a mechanical heart valve; (E) vascular stent.

The commercially pure (CP) titanium is an allotropic material and exhibits two crystallographic forms. At high temperature, beta-phase has a body-centered cubic crystalline structure; a phase transition occurs at temperature above 882.5°C, with a transformation of CP-Ti to an alpha-phase, having a hexagonal crystalline structure. Ti alloys, as described in the following, can stabilize both the crystal structure at room temperature, by selectively alloying it with appropriate elements. Based on the stabilized structure, alpha, beta, and alpha+beta alloys can be obtained, each of them with specific properties.

Titanium and titanium alloys, including alpha, beta, and alpha+beta, have mechanical and forming characteristics that are attractive for different implantable applications. The most important feature of these alloys is their outstanding corrosion resistance even in some aggressive environments as human body. Apart for fretting corrosion, the corrosion resistance of Ti alloys is higher than that of both stainless steel and Co-Cr alloys. In particular, their corrosion resistance is related to the ability to form a protective, stable oxide layer of $TiO_2$. From a biomedical point of view, Ti's excellent biocompatibility, high corrosion resistance, and low elastic modulus are highly desirable.

*Commercially pure titanium:* CP-titanium (CP-F67) is classified by ASTM in four different grades, Grade-1, Grade-2, Grade-3, and Grade-4, depending on the quantity of impurities (e.g., oxygen, nitrogen, iron) present in the metal. In particular, classification is based on the amount of oxygen in Ti that can influence its mechanical properties (Table 4.6). When the O percentage increases from 0.18 (Grade-1) to 0.40 (Grade-4), the tensile strength increases, as well as the yield strength. In general, the presence of impurities or alloying elements leads to increased strength, reducing ductility. Moreover, fatigue strength is increased by increasing the amount of oxygen.

*Alpha Ti alloys:* Alpha alloys contain, as a stabilizer, aluminum, gallium, tin, and/or zirconium. The phase diagram of Ti alloys with alpha stabilizer is reported in Fig. 4.62A. These

**TABLE 4.6** Chemical Composition (% Max) and Mechanical Properties of the Main Ti Alloys Used in the Biomedical Field

| Elements | CP-Ti | | | | Ti-6Al-4V |
|---|---|---|---|---|---|
| | Grade-1 | Grade-2 | Grade-3 | Grade-4 | |
| C | 0.10 | 0.10 | 0.10 | 0.10 | 0.08 |
| O | 0.18 | 0.25 | 0.35 | 0.40 | 0.13 |
| N | 0.03 | 0.03 | 0.05 | 0.05 | 0.05 |
| H | 0.015 | 0.015 | 0.015 | 0.015 | 0.0125 |
| Fe | 0.20 | 0.30 | 0.30 | 0.50 | 0.25 |
| Al | – | – | – | – | 6 |
| V | – | – | – | – | 4 |
| Ti | Balance | Balance | Balance | Balance | Balance |

| Property | CP-Ti | | | | Ti-6Al-4V |
|---|---|---|---|---|---|
| | Grade-1 | Grade-2 | Grade-3 | Grade-4 | |
| Elastic modulus (GPa) | 103 | 103 | 103 | 104 | 110 |
| Yield strength (MPa) | 170 | 275 | 380 | 485 | 795 |
| Tensile strength (MPa) | 240 | 345 | 450 | 550 | 860 |

Content of each element is expressed as maximum percentage in the alloy.

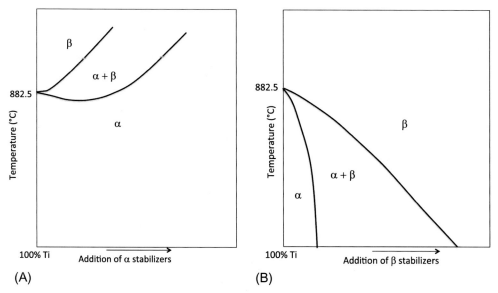

**FIG. 4.62** (A) Phase diagram of Ti-alpha stabilizers; (B) phase diagram of Ti-beta stabilizers.

alloys cannot be significantly strengthened by heat treatment and, as a result, they do not exhibit any significant improvement compared to CP alloys; they are weldable. Alpha alloys have limited biomedical applications due to their low strength at room temperature compared to beta and alpha+beta alloys.

*Beta Ti alloys:* Beta alloys containing alloying elements (e.g., molybdenum, iron, vanadium, tantalum, niobium, nickel, chromium, cobalt, copper, tungsten, manganese) that allow stabilizing at room temperature the body-centered cubic crystal structure (Fig. 4.62B). Beta alloys show an excellent forgeability, as they can be heat-treatable and cold-formable. However, they can be solution-treated and aged to high strength levels above those of alpha-beta alloys. The beta alloys exhibit a lower elastic modulus among the titanium alloys used for biomedical implants. In addition, an advantage of beta alloys compared to alpha Ti alloys and alpha-beta Ti alloys, is related to the presence, as beta stabilizers, of elements such as tantalum and

molybdenum, more biocompatible than the ones used to form alpha Ti alloys (e.g., tin and aluminum) and alpha-beta alloys (e.g., vanadium).

*Alpha-beta Ti alloys:* Alpha-beta alloys contain alpha stabilizers (e.g., aluminum) and beta stabilizers (e.g., vanadium, molybdenum). As a result, at room temperature, a mixture of alpha and beta phases coexists. They exhibit good formability, and solution treatment can be used to increase the strength of these alloys by 30%–50% when compared to the annealed state. Alpha-beta alloys are characterized by high tensile strength, and they are not weldable if the beta phase is higher than 20%. Examples of alpha-beta alloys used for biomedical applications are Ti-6Al-4V (F1472), Ti-6Al-7Nb, and Ti-5Al-2.5Fe. In particular, Ti-6Al-4V is commonly used for biomedical applications, but vanadium is reported to cause adverse reactions in some cases.

*Applications:* CP Ti is used mainly in orthopedic applications that do not require high strength such as screws and staples for spinal

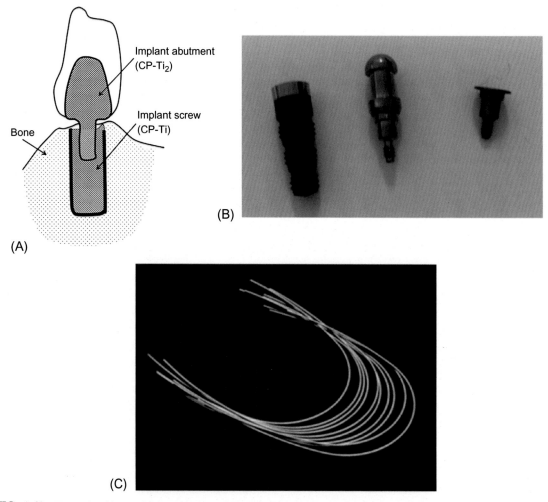

**FIG. 4.63**  Example of biomedical application of CP-Ti and Ti alloys: (A) sketch of a dental implant; (B) components of a dental implant (components on the left and centered in CP-Ti, screw on the right in Ti alloy); (C) orthodontic wire in beta-Ti alloy;

*continued*

surgery, maxillofacial, and craniofacial applications (e.g., bone plate and screws) and dental applications (e.g., osteointegrated implants, Fig. 4.63).

Ti alloys are used for a wide range of applications (Fig. 4.63), including total joint replacements, dental implants, pacemaker cases, and housing for ventricular assist devices. Among the alpha-beta alloys, the F1472 alloy is the most common used for implantable devices such as total joint replacement. The other alpha-beta alloys (i.e., Ti-6Al-7Nb and Ti-5Al-2.5Fe) are used in femoral hip stems, plates, screws, rods, and nails.

*Disadvantages*: The main disadvantages of titanium alloys in medical implantable applications are the following:

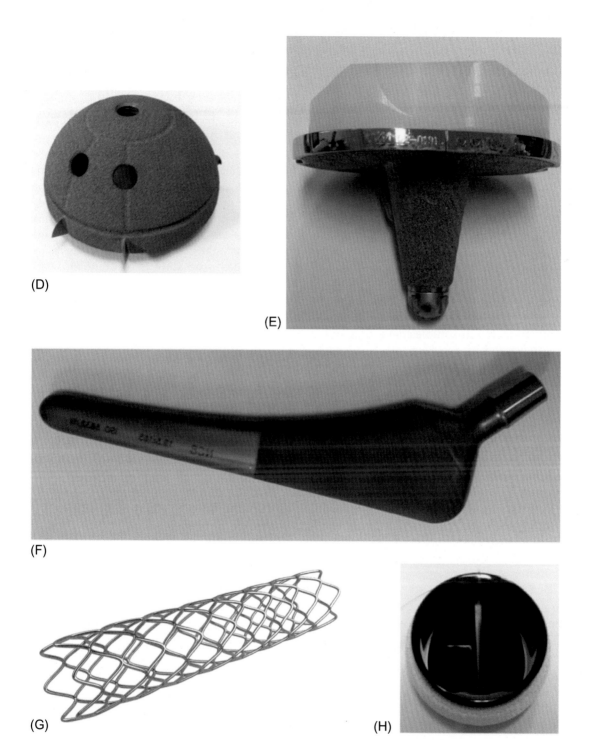

**FIG. 4.63—cont'd** (D) metal back, surface roughness is aimed to improve bone ingrowth; (E) tibial component of a knee joint prosthesis; (F) femoral hip in Ti alloy, the *gray part* of the prosthesis is a osteoinductive coating to improve bone ingrowth; (G) vascular stent in Nitinol; (H) housing of a bileaflet mechanical valve.

B. BIOMATERIALS AND BIOCOMPATIBILITY

– poor wear resistance;
– high notch sensitivity, that is, the presence of a scratch or notch reduces the fatigue life of the implant.

For their low wear resistance, titanium alloys should not be used in load-bearing surfaces, such as hip and knee joint, unless adequate surface treatments (e.g., ion implantation, coating) are performed to improve wear resistance.

Issues regarding the possible problems related to the wide use of Ti-6Al-4V as implantable material, have been expressed concerning the presence in long-term implants of elements such as vanadium, which are toxic both in the elemental state and as oxides, and which are present at the surface in the form of $V_2O_5$, which is toxic. In vitro studies have shown that debris generated by the wear of Ti-6Al-4V as compared to the debris of alloy Ti-6Al-6Nb stimulates phagocytic cells to a greater extent than Ti-6Al-6Nb or Ti.

## 4.4.4 Advanced Alloys

Particular alloys or metals are used in biomedical applications, in general but not always with a minor impact in terms of production volume than the alloys previously described. In this section, some alloys already used in the biomedical field are described together with other ones that are now under evaluation.

### 4.4.4.1 Ni-Ti Alloy

Ni-Ti alloy (also known as Nitinol) is an alloy with a near-equiatomic composition (i.e., 49%–51%) of nickel and titanium. Ni-Ti belongs to the class of shape memory alloys that can be deformed at a low temperature and are able to recover their original, permanent shape when exposed to a high temperature. The shape memory effect of Nitinol is related to its martensitic transformation. In fact, at high temperature, the atoms are arranged in an orderly manner, in a BCC crystal structure (i.e., austenitic phase). In this phase, the alloy is resistant to twisting and bending. At low temperature, Ni-Ti alloy exhibits a monoclinic distorted crystal structure, that is, a martensitic phase. At this stage, the alloy can be easily deformed. The temperature at which the austenitic phase transforms into a martensitic one is called the transformation temperature. The atomic percentage composition (Table 4.7) is important in determining the shape memory effect at body temperature; in fact, the presence of a higher content of nickel (even by 0.1%) can induce a decrease in the transformation temperature and an increase in yield strength of the alloy. Moreover, contaminants, such as carbon and oxygen, can vary the transformation temperature, affecting the mechanical properties. Hence, attention should be taken to minimize the concentration of the contaminants.

TABLE 4.7  Mechanical Properties of the Ni-Ti Alloy and Tantalum Used in the Biomedical Field

| Property | Ni-Ti Alloy | Tantalum | Magnesium | Iron |
|---|---|---|---|---|
| Elastic modulus (GPa) | 83 (austenitic phase) 28–41 (martensitic phase) | 185 | 41–45 | 200 |
| Yield strength (MPa) | 195–690 (austenitic phase) 70–140 (martensitic phase) | 138 | 65–100 | 150 |
| Tensile strength (MPa) | 895 | 207 | 207 | 210 |
| Fracture toughness (MPa $\sqrt{m}$) | 895 | 207 | 207 | N/A |

*Applications*: The shape memory effect can be used in different biomedical devices; among them, self-expanding vascular stents represent the main example. Nitinol stents have a small diameter at room temperature so they can be crimped onto the delivery system and inserted into the human body via catheter, using a minimally invasive approach. Once it reaches the correct site, it can recover its permanent shape (i.e., larger diameter) at body temperature, as the stent can remember the shape in its austenitic phase at elevated temperature.

### 4.4.4.2 Tantalum

Tantalum has been used for biomedical devices as a commercially pure element (i.e., 99.9%), and as an alloying element in titanium alloys. Ta exhibits excellent corrosion resistance and biocompatibility, caused by the stable surface oxide layer, as well as a high density (16.6 g/cm$^3$). Mechanical properties of Ta are interesting for biomedical applications (Table 4.7). In fact, elastic modulus is close to the one of 316L stainless steel, but yield and tensile strengths are lower compared to those of Ti, 316L and Co-Cr alloys.

*Applications*: Ta can be used, due to its high density, as a radiographic marker for diagnostic investigations. Moreover, Ta has also used for coronary stents, vascular clips, cerebral covering for cranial effects, fracture fixation, and dental implants.

### 4.4.4.3 Biodegradable Alloys

Magnesium (Mg) alloys and pure iron (Fe) are metallic materials that are now under evaluation for possible applications in biomedical field. The main characteristic of these metals is the fact that they are degradable, so that an in vivo absorption mechanism occurs by metal corrosion due to the aggressive body fluid, and the corrosion products are tolerated and metabolized by the human body. An example of possible degradation of metal in vivo is reported in Fig. 4.64, in the case of a vascular biodegradable stent. The use of these materials for orthopedic and vascular applications dates back nearly half a century. The mechanical properties of a bioresorbable metallic material are similar to stainless steel ones. In addition to that, corrosion product are well tolerated by the human body, as, for example, Mg is present in bone tissue, as well as Fe is in human diet.

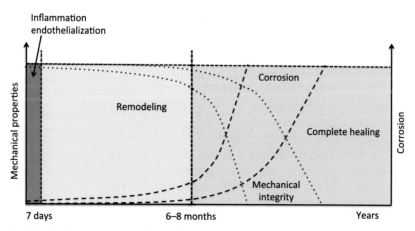

**FIG. 4.64** Sketch of the timeline between mechanical properties and corrosion of biodegradable metals in the case of coronary stent. Mechanical stability of the structure is needed high for the first 6–8 months to allow vessel remodeling, as well as corrosion rate has to start low. In the case of bone plate, mechanical stability is needed for the first 3–6 months.

*Magnesium alloys.* Even if the mechanical properties are very interesting, the rapid corrosion (i.e., few weeks) of pure Mg and its alloys under physiological conditions limits its use in load-bearing application. Under physiological conditions, the main corrosion products are magnesium hydroxide, $Mg(OH)_2$, and hydrogen, $H_2$, gas. Although $Mg(OH)_2$ is not very soluble in water, it reacts immediately with chloride ions in the physiological environment to produce magnesium chloride, $MgCl_2$, which is readily soluble in water. The reactions involved in the corrosion of Mg are as follows:

$$Mg + 2H_2O \rightarrow Mg(OH)_2 + H_2$$

$$Mg(OH)_2 + 2Cl^- \rightarrow MgCl_2$$

The $H_2$ gas involved during the degradation of Mg is a significant concern, because it could accumulate as gas bubbles under the skin; in this case, patients are treated by drawing off the gas using a subcutaneous needle. To improve corrosion resistance, alloying elements are used (e.g., aluminum, zinc, manganese, zirconium, yttrium, rare earth elements). Even if the Mg alloys exhibit higher corrosion resistance and mechanical properties, the degradation products should be carefully investigated as the release of metal ions may have toxic effects in human body. Among the most studied Mg alloys, AE21 (2% Al, 1% rare earth metals, remaining Mg) and WE43 (4% Yt, 3.4% rare earth metals, 0.6% Zr, remaining Mg) represent examples of possible alloys under investigation for biomedical applications.

*Applications*: Mg alloys have been recently used for making biodegradable coronary stents, plates, and screws for bone fracture repair.

*Pure iron and alloys:* Pure Fe (>99.5%) is another biodegradable metal considered for biomedical applications; in particular, it is under investigation for vascular stents. Its mechanical properties (Table 4.7) are much closer to 316L stainless steel than Mg alloys, with high ductility and mechanical strength.

The high strength is helpful in making stents with thinner struts, whereas the ductility is very important for plastic deformation during in vivo expansion. It does not generate toxic products, and the biological performances seem to be favorable; in fact, the corrosion of Fe produces ferrous and ferric ions, which dissolve in biological fluids. A serious limitation is that its ferromagnetism could negatively impact the compatibility with certain imaging devices, such as magnetic resonance imaging (MRI). Another important drawback is the slow degradation rate in vivo. In fact, large portions of the pure Fe stent were found intact in the blood vessels 12 months postsurgery, which was considered to cause reactions similar to those found in bare metal stents (Peuster et al., 2006). To reduce corrosion rate of iron, different aspects are investigated to reach this goal (Zheng et al., 2014):

(1) alloying: new Fe-based alloys using Mn, Co, Al, W, Sn, B, C, and S as alloy elements;
(2) surface modifications: thin films of Fe-O could be formed on pure iron surface by plasma ion implantation or deposition;
(3) new manufacturing techniques: among them, electroforming technique is under investigation.

*Applications*: Iron and its alloys have been under investigation for biodegradable metal stents.

## 4.4.5 Possible Choice of Metal for Orthopedic Applications

In the application of orthopedic implants, mechanical properties are critical, such as yield strength, to resist plastic deformation under load; fatigue strength, to resist cyclic loads; hardness, to resist wear when joint articulation is involved; and adequate elastic modulus, to achieve bone-metal load-bearing proportionality. To clearly understand this, consider that, prior to the fracture, all the acting forces are balanced. After the fracture, this balance is lost and the implant of prosthesis is needed. If the fracture is perfectly reconstructed, the bone will still carry a significant portion of the load, and the implant act mainly as the structure around which the fractured bone is reconstructed. On the other hand, in many situations, due to a complex fracture or inadequate fixation, the implant will carry a disproportionate amount of the load and could also undergo elastic twisting and bending. All of these could give rise to fatigue-related failure of the implant. For these reasons, the yield, tensile, and fatigue strength of metals are critical and must be most favorable.

### 4.4.5.1 Fatigue Behavior

When a joint prosthesis is implanted in a patient, as already described in the previous section, the prosthesis is subject to forces that are 3.3–4.3 the human body weight, and considering a walking patient, the force is applied about 700 times per kilometer. Even for an aged patient, each leg is subjected to the force related to their own weight $5–10 \times 10^3$ times/day, hence about $10^6$ times/year. These forces have to be considered together with the low cross-section of the implant, because fatigue failure can occur. In addition, fatigue failure could be promoted by the design of the implant that exhibits discontinuity (e.g., change in cross-section, holes), defects introduced during fabrication (e.g., inclusions), application (e.g., cracks due to surgery), or interaction with human environment (e.g., localized corrosion). Ex vivo studies on different implants, both orthopedic and cardiac, detected that, in the majority of cases, failure is caused by fatigue break.

### 4.4.5.2 Stiffness and Elastic Modulus

The elastic modulus of the metal is a different issue. Metal stem, as well as other components of orthopedic joint components, could be cemented or cementless, and load would have a different distribution, depending on the elastic modulus of the selected metallic material. The elastic modulus of bone, in the load-bearing direction, is close to 17 GPa. In comparison to this, the Young modulus of Ti alloys, stainless steel, Co-based alloys are, respectively 110, 190, and 240 GPa. The stiffness of the material is inversely proportional to the elastic modulus, which for stainless steel is approximately 80% greater than for Ti. This consequently means that, for an implant of the same size, stainless steel implants are significantly stiffer than Ti implants.

Let's consider a situation in which the hip joint is damaged (e.g., femur is broken); to fix the fracture, a metal prosthesis used replacing the natural joint.

In the case of *cementless stem*, if the used metal has a significantly higher elastic modulus than bone, it carries a disproportionate portion of the load. Hence, the metal implant will shield the bone from carrying the load that it would support under normal conditions, a phenomenon called stress shielding. Although by an engineering point of view, that sounds reasonable, but by a biological point of view, it is undesirable. In fact, in normal conditions, bone tissue responds to stress by remodeling itself to the applied level of stress. If the elastic modulus is high (e.g., stainless steel, Co-alloys), due to stress shielding, bone remodels itself to the lower load level, and its quality deteriorates. Hence, in the case of cementless implant, titanium alloys, with the lowest elastic modulus among the three major alloys, is the most desirable in lowest applications.

In the case of *cemented stem*, PMMA is used as bone cement. If stress is concentrated in the proximal part of the stem, a high force is transferred to PMMA cement that could undergo to fragmentation. This effect can induce local inflammation or/and wear mechanism in head/acetabular cup (i.e., three-body wear), causing a possible failure of the implant. On the contrary, if the force is transferred even in the distal zone of the implant, there is a decrease of the local force on the bone cement. Hence, for that reason, a stiff stem is needed, so that austenitic SS or Co-Cr stem is used for the cemented implant. In this case, failure of the prosthesis can occur at a longer time from the implant.

## 4.5 CERAMIC BIOMATERIALS

Advanced ceramics (see Chapter 1) are used in the biomedical field, in particular for orthopedic joint implants and dental applications. In particular, ceramics used in medical applications are named bioceramics. The reasons that make ceramic biomaterials an excellent candidate for biomedical applications are biocompatibility, corrosion resistance, high stiffness, resistance to wear in applications in surface articulation (e.g., hip and knee implants), and low friction. At present, different ceramic and glass materials are used for structural functions or tissue replacement, as well as coatings to improve the biocompatibility of metal and/or polymeric implants. In addition to mechanical integrity requirements and physicochemical compatibility in biological environments, implantable bioceramics should have the following properties:

- nontoxicity;
- noncarcinogenicity;
- does not induce allergic reactions;
- does not induce inflammatory response;
- induces tissue regeneration or tissue integration, depending on the specific application.

Bioceramics can be classified according to their reactivity in human body environment (Fig. 4.65). They can be grouped referring to the main material/tissue physiological response, as follows:

- nearly inert bioceramics: they can be divided into two groups: fully dense, relatively inert crystalline ceramics that attach to tissue by either a press fit or tissue ingrowth onto a roughened surface; and porous, relatively inert ceramics into which tissue ingrowth occurs, creating a mechanical attachment;
- resorbable bioceramics: resorbable ceramics that integrate with tissue and eventually are replaced by host tissue;
- surface reactive bioceramics: fully dense, surface-active ceramics that attach to tissue via a chemical bond.

Bioceramics may therefore be classified by their macroscopic surface characteristics (smooth, fully dense, roughened, or porous) or their chemical stability (inert, surface reactive, or bulk reactive/resorbable). The integration of biological (i.e., inductive) agents with ceramics further expands the clinical potential of these materials. As better described in the following sections, relatively inert ceramics elicit minimal tissue response and lead to a thin layer of fibrous tissue immediately adjacent to the surface; surface-active ceramics are partially soluble, resulting in ion exchange and the potential to lead to a direct chemical bond with bone; bulk bioactive ceramics are fully resorbable, have much greater solubility than surface-active ceramics, and may ultimately be replaced by an equivalent volume of regenerated tissue. The relative level of bioactivity mediates the thickness of the interfacial zone between the biomaterial surface and host tissue (Fig. 4.65). There are, however, no standardized measures of reactivity, but the most common are pH changes, ion solubility, tissue reaction, and any number of assays that assess some parameter of cell function.

In the following sections, the applications of ceramics in biomedical devices will be exposed. In particular, some of the bioceramics commonly used today in biomedical applications are described in the following sections, taking into account their different interaction with human body.

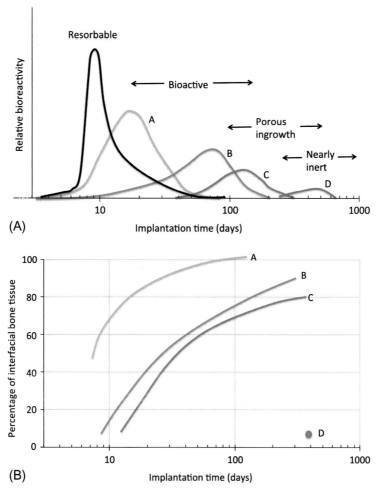

FIG. 4.65    Bioactivity spectra for selected bioceramics: (A) relative magnitudes and rates of bioactivity; (B) time dependence of formation of bone at bioceramic surface and ceramic-bone bonding. A: 45S5 Bioglass; B: 55S4.3 Bioglass; C: hydroxyapatite; D: alumina. *From Ratner, B.D., Hoffman, A.S., Schoen, F.J., Lemons, J.E., (Eds.), 2012. Biomaterials Science: An Introduction to Materials in Medicine. Academic Press, San Diego, CA, pp. 73–84.*

## 4.5.1 Nearly Inert Bioceramics

The nearly inert bioceramics are not subject, or if they are, this occurs in a negligible extent, to chemical variations when implanted for long-term in human body. In particular, nearly inert bioceramics are chemically stable, that is, they do not corrode, wear, or react to the host environment. Little or no chemical change occurs during their long-term exposure to the physiological environment.

This group includes the "advanced ceramics," made up of pure or almost pure compounds; among them, the most used as bioceramics are aluminum oxide ($Al_2O_3$), zirconia oxide ($ZrO_2$), different structures of carbon (e.g., LTI and ULTI carbon), silicon carbide (SiC), and silicon nitride ($Si_3N_4$).

Relatively inert ceramics elicit minimal tissue response and lead to a thin layer of fibrous tissue immediately adjacent to the surface. In fact, the host environment (i.e., human body) recognizes the implant bioceramics material as a foreign body and tends to isolate it with a very thin (about 10 μm) fibrotic membrane around the ceramic implant. The major disadvantage related to the formation of the fibrous membrane around the implant is the isolation with respect to biological tissues, which can lead to detachment and to the loosening of the implants, with consequent mobilization or nonintegration with the surrounding tissues.

### 4.5.1.1 Alumina (Al₂O₃)

High-density aluminum oxide, also named alumina, $Al_2O_3$ is relatively biological inert and biocompatible; it has excellent corrosion resistance, high wear resistance, and high mechanical strength. The outstanding properties of alumina depend on its grain size and purity.

In the alumina crystalline structure, oxygen ions are in the reticular positions of a compact hexagonal cell, as shown in Fig. 4.66. In the EC structure as in the CFC structure, there are as

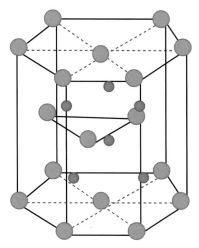

FIG. 4.66  Alumina structure. Oxygen ions occupy the positions of the EC. $Al^{3+}$ ions *(green spots)* occupy interstitial positions to maintain charge neutrality.

many octahedral interstitial positions as there are the atoms in the cell. However, because aluminum has a valence value of +3 and oxygen is −2, there may be only two $Al^{3+}$ ions for every three $O^{2-}$ ions to maintain electrical neutrality. Therefore aluminum ions can occupy only two-thirds of the octahedral positions of the EC $Al_2O_3$ lattice, and this leads to a certain distortion of the structure.

Among the different possible grades of $Al_2O_3$, high $Al_2O_3$ grades have at least 99% purity. The amount of impurities and alloying elements accounts for the differences between the alumina grades. ASTM F603-78 details that $Al_2O_3$ biomedical implants should contain greater than 99.5% (99.8% is recommended) $Al_2O_3$ and less than 0.1% combined $SiO_2$ and alkali oxides.

High purity alumina powder, prepared by calcining alumina tri-hydrate, is typically isostatically compacted and shaped. Subsequent sintering at temperatures in the range 1600–1800°C transforms a preform into a dense polycrystalline solid having a grain size of less than 5 μm. Fluxes such as CaO and magnesia, MgO, can be added to $Al_2O_3$ to decrease sintering temperature and to limit grain growth. In particular, doping of dense $Al_2O_3$ with <0.5% MgO inhibits abnormal grain growth, whereas the presence of a small amount of CaO results in the production of liquid phases or segregation at the grain boundaries, thereby inducing abnormal grain growth and a possible loss of strength over time. $SiO_2$ can also be added during sintering to promote grain growth. If processing is kept below 2050°C, α-$Al_2O_3$, which is the most stable phase, forms. Alternatively, single crystals (sapphire) may be grown by feeding powder onto a seed and allowing buildup.

One of the most important properties of $Al_2O_3$, related to its wear behavior, is the very low surface roughness and high surface wettability, leading to a very low coefficient of friction when sliding against itself or UHMWPE. The high wettability is related to the nonsaturation of oxygen ions on the outer surface of the implant. Hence, when implanted in human body, the nonsaturation of oxygen ions allows the formation of an absorbed layer containing water and biological molecules, minimizing direct contact between the two bearing surfaces (Fig. 4.67A).

The major limitation of alumina is that it possesses relatively low tensile and bending strengths and fracture toughness and, as a consequence, is sensitive to stress concentrations and overloading (Table 4.8). Clinically retrieved alumina total hip replacements exhibit damage caused by fatigue, impact, or overload. In

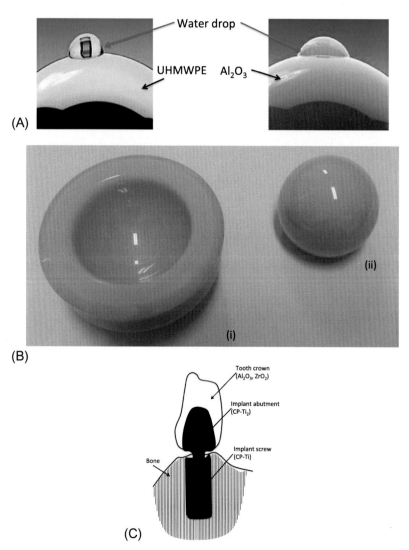

**FIG. 4.67** Example of biomedical application of alumina: (A) wettability of a UHMWPE (on the left) acetabular cup and an alumina (on the right) one; (B) artificial articulation made of alumina: (i) femoral head and (ii) acetabular cup; (C) sketch of a dental implant.

**TABLE 4.8**  Mechanical Properties of Alumina and PSZ Used in the Biomedical Field

| Property | Alumina | PSZ |
|---|---|---|
| Density (g/cm$^3$) | 3.94–3.95 | 6 |
| Grain size (μm) | 2–4 | <1 |
| Hardness, Vickers (HV) | 2300 | 1300 |
| Flexural strength (MPa) | 420–500 | 200 |
| Yield strength (MPa) | 65–100 | 1000 |
| Elastic modulus (MPa) | 380–420 | 220 |
| Fracture toughness (MPa $\sqrt{m}$) | 5–6 | <10 |

general, a large number of ceramic failures can be attributed to material processing or design deficiencies, and can be minimized by an accurate material choice and quality control. Moreover, based on clinical experience and retrieval analyses, implant geometries should have a sphericity of less than 1 μm and a radius tolerance between components of 7–10 μm (Boutin et al., 1988). These specifications are based on the rationale that too small a gap between the components does not provide sufficient room for necessary lubrication or an escape route for alumina particles, whereas too large a gap increases contact pressures.

*Applications:* Alumina is commonly used in orthopedic and dentistry applications. Mainly, alumina is used in joint prostheses (e.g., hip and knee joints) and for the fabrication of crowns (Fig. 4.67C).

To avoid the formation of wear debris (see Section 4.4.2), alumina can be used as femoral head and acetabular cup (Fig. 4.67B). Due to the excellent tribological properties, low surface roughness, hardness, and wettability, the coefficient of friction of both alumina-alumina and alumina-UHMWPE is lower than that of metal-UHMWPE.

Moreover, the articulating surfaces (i.e., ball and cup) must have a high degree of symmetry and tight tolerance; grinding and polishing the

paired cup and femoral head achieve this. The low friction coefficient of the ceramic-on-ceramic coupling is similar to the physiological joint and results in wear debris generation that is 10 times lower than metal-on-polymer combinations.

### 4.5.1.2 Zirconia (ZrO$_2$)

Zirconia, or zirconium oxide, $ZrO_2$, has three polymorphic forms (Fig. 4.68A). At room temperature, pure zirconia has a monoclinic crystal structure; on heating at about 1000–1100°C up to approximately 2000°C, it transforms to a tetragonal phase. At a temperature above 2000°C, a face-centered cubic (FCC) phase is formed. A volumetric shrinkage of about 3%–10% occurs during the monoclinic-tetragonal transformation of pure zirconia leading to residual stresses and cracking. In addition, the cubic-tetragonal transformation induces volume expansion, large stresses, and cracks upon cooling from high temperature. Furthermore, due to the large volume reduction, pure zirconia cannot be sintered.

To prevent volume expansion during phase transformations, several stabilizing oxides, such as yttria or yttrium oxide, $Y_2O_3$, and magnesium oxide, $MgO$, are added to zirconia to stabilize the tetragonal and/or cubic phases by producing multiphase ceramic material known as partially stabilized zirconia (PSZ). Depending on the amount of the stabilizer, cubic $ZrO_2$ is generally the major phase in PSZ at room temperature, and monoclinic and metastable tetragonal $ZrO_2$ represent the minor phases. If an adequate quantity of metastable tetragonal phase is present, an applied stress, magnified by the stress concentration at a crack tip, can cause the tetragonal phase to convert to monoclinic, with the associate volume expansion. This is caused by the elevated energy that induces the metastable tetragonal grains to transform into monoclinic grains in this part of the microstructures (Fig. 4.68B). This phase transformation can put the crack into compression, retarding its growth, because more energy is needed to advance the crack, and

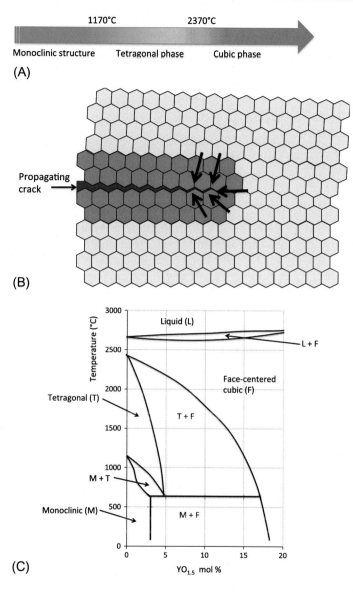

(A)

(B)

(C)

FIG. 4.68 (A) Zirconia polymorphism: monoclinic crystal structure, tetragonal phase, and cubic phase; (B) sketch of the microstructure in PSZ undergoing transformation toughening at a crack tip: *light blue* = tetragonal grain, *red* = transformed monoclinic grain, *arrows* = compressive stress ahead of the crack; (C) zirconia-yttria phase diagram.

enhancing the fracture toughness and relatively high strength (Table 4.8). This mechanism allows extending the lifetime of PSZ products.

Analyzing the zirconia-yttria phase diagram (Fig. 4.68C), if a 3%–10% $Y_2O_3$ is added, PSZ is obtained; however, adding lower quantity of $Y_2O_3$ (i.e., 2%–3%), it is possible to obtain at room temperature a ceramic with only a metastable tetragonal phase, that is called $ZrO_2$ polycrystal (TZP).

*Applications*: As pure zirconia cannot be used, $ZrO_2$ stabilized with $Y_2O_3$ or MgO is used as a nearly inert bioceramic. Stabilized zirconia can be used in bulk form or as a coating on metal components to improve wear resistance in joint prosthesis. The modulus of PSZ is approximately half that of alumina, whereas the bending strength and fracture toughness are two to three times greater and two times greater, respectively. In vitro experiments showed that the wear rate of PSZ on UHMWPE could be five times lower than the wear rate of alumina on UHMWPE (Derbyshire et al., 1994). Partially stabilized zirconia can be used for femoral head and for acetabular cup (Fig. 4.69) in orthopedic applications. In dental applications, PSZ can be used for tooth crowns (Fig. 4.67C).

### 4.5.1.3 Carbon

Carbon has a hexagonal crystal structure formed by strong covalent bonds. It can exist in different allotropic forms such as graphite, and amorphous and turbostratic carbon. Graphite has a planar hexagonal array structure where the carbon-carbon bond energy within the planes is large, whereas the bond between the planes is weak. Therefore carbon derives its strength from the strong in-plane bonds, whereas the weak bonding between the planes results in a low modulus, near that of bone. Turbostratic isotropic carbon (see Chapter 1), on the other hand, has no preferred crystal orientation and hence possesses isotropic material properties. For that reason, turbostratic carbon, either obtained by pyrolysis (i.e., low temperature isotropic, LTI) and physical vapor deposition (i.e., ultra-low temperature isotropic, ULTI) is more used in biomedical applications. Properties that make carbon suitable for biomedical applications are:

– excellent electrical and thermal conductivity;
– low density;
– sufficient corrosion resistance;
– low elasticity;
– low thermal expansion.

LTI and ULTI turbostratic carbons are widely used for implant production and surface coatings for their longer stability and better biocompatibility (i.e., nearly inertness) with blood. In fact, PTI exhibits optimal compatibility with blood and soft tissue and have also excellent durability, high mechanical strength, excellent wear, and fatigue resistance.

*Applications*: The benign biological reaction elicited by carbon-based materials, along with the similarity in stiffness and strength between carbon and bone make carbon a candidate material for musculoskeletal reconstruction. The turbostratic carbons are particularly suitable for applications in mineralized tissue subjected to high loads because they can be produced to have an elastic modulus similar to that of bone. Hence, due to the similarity in elastic modulus, the risk for bone reabsorption followed by necrosis is reduced. Moreover, LTI can also be doped with up to 20% silicon to improve stiffness, hardness, and wear resistance.

Due to its excellent strength, wear resistance and durability, hemocompatibility, and thromboresistance, turbostratic carbon is the preferred materials for construction or surface coating of artificial mechanical heart valves (Fig. 4.70); in particular, discs are coated with LTI turbostratic carbon, whereas metal housing and suture ring are coated by ULTI turbostratic carbon. In addition, ULTI carbon can also be used as vascular prosthesis coating, mainly made in PET (Fig. 4.70C).

FIG. 4.69   Acetabular cup made of $ZrO_2$.

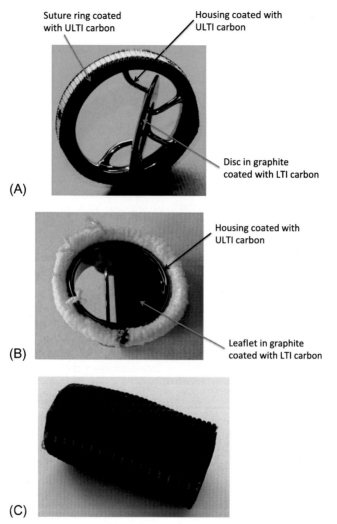

Suture ring coated
with ULTI carbon

Housing coated with
ULTI carbon

Disc in graphite
coated with LTI carbon

(A)

Housing coated with
ULTI carbon

Leaflet in graphite
coated with LTI carbon

(B)

(C)

FIG. 4.70    Example of biomedical application of turbostratic carbon: (A) tilting disc mechanical heart valve; (B) bileaflet mechanical heart valve; (C) PET vascular graft coated with ULTI carbon.

## 4.5.2 Surface Active Bioceramics and Reabsorbable Bioceramics

The concept of bioactivity was introduced with respect to bioactive glasses via the following hypothesis: the biocompatibility of an implant material is optimal if the material elicits the formation of normal tissues at its surface and in addition if it establishes a contiguous interface capable of supporting the loads that normally occur at the site of implantation (Hench et al., 1971).

These materials have the advantage compared to relative inert bioceramics in that they bond to bone. In fact, resorbable and surface reactive ceramics react to the host or physiological environment resulting in surface or bulk chemical changes. Bioactive ceramics promote and facilitate the formation of a bond between the implant material and the surrounding tissue. Moreover, they develop an adherent interface with surrounding bone tissue that is very strong and is able to support load. Hence, the purpose of the bioabsorbable is to be recognized as self by the bone cells, which can metabolize them to reconstruct the physiological mineral phase, whereas the active materials are designed to be chemically stable but able, by affinity, to establish chemical bonds with bone, allowing an adequate anchorage of the peri-implant bone tissue.

The nature of the biomaterial-tissue interface and the reactions that occur at the ceramic surface and in the tissues dictate the resulting mechanical, chemical, physical, and biological processes that occur. In general, four factors determine the long-term effect of bioactive ceramic implants:

- site of implantation;
- tissue trauma;
- bulk and surface properties of the material;
- relative motion at the implant-tissue interface.

For resorbable materials, additional design requirements include the following:

- strength/stability of the material-tissue interface needs to be maintained during the period of degradation and replacement by host tissue;
- material reabsorption and tissue repair rate should be matched;
- reabsorbable material should consist only of metabolically acceptable species.

Under appropriate conditions, three classes of ceramics may fulfill these requirements: bioactive glasses and glass ceramics, calcium phosphate ceramics, and composites of these glasses and ceramics (see Section 4.5).

The main problems related to this class of bioceramics are the difficulty of mimicking both the bone composition and its natural amorphous/crystalline conformation. In general, an amorphous structure is quickly reabsorbed, whereas the presence of crystallinity tends to confer biostability: the greater the crystallinity, the lower the rate of reabsorption.

### 4.5.2.1 Calcium Phosphate-Based Bioceramics

Calcium phosphate (CaPs) ceramics are of special interest in the biomaterials field, because they are present in normal, physiological bone. Bone, in fact, is made of 33% organic matrix and 67% minerals. Several CaPs form can be found in human bone, among them, the most important are summarized in Table 4.9, together with the natural Ca/P ratio. CaPs can also be synthetically produced mimicking composition, biodegradation, bioactivity, and osteoconductivity of the biological calcium phosphate. Unfortunately, the Ca/P ratio in the synthetic CaPs can be different from the natural one (Table 4.9) due to the presence of other constituents, added during the synthesis. A factor that has to be taken into account is related to the fact that synthetic bioactive ceramics should mimic the mineral phase of the bone, with its complex structure. In addition to the dependence of degradation rate from amorphous phase/crystalline phase, the crystallite size is another important parameter. In fact, fine crystals or smaller crystallite-sized CaPs tend to degrade faster compared to larger crystals.

Moreover, the degradation of each CaP is dependent on the phase (Table 4.9) and the biological environmental conditions, such as the pH.

TABLE 4.9 Different Calcium Phosphate Components of the Inorganic Phase of Bone

| $Ca_{8.3} (HPO_4, CO_3)_{1.7} (PO_4)_{4.3} (CO_3, OH)_{0.3}$ | | |
|---|---|---|
| | | Ca/P Ratio |
| Octacalcium phosphate (OCP) | $Ca_8H_2(PO_4)_6$ $5H_2O$ | 1.33 |
| Tricalcium phosphate (TCP) | $Ca_3(PO_4)_6$ | 1.5 |
| Hydroxyapatite (HA) | $Ca_5(PO_4)_3(OH)$ | 1.67 |
| Tetracalcium phosphate | $Ca_4(PO_4)_2O$ | 2 |

*Hydroxyapatite (HA):* The most commonly known CaP bioceramic is hydroxyapatite, HA, which is present in bone and teeth. HA is one of the many crystalline forms of CaPs and is classified as a resorbable bioceramic. Synthetic HA can be used as an osteoconductive material and can be produced mimicking the natural HA using different techniques, such as sol-gel synthesis, coprecipitation, and solid-state reactions. HA is very brittle, but it is strong in compression; its strength depends on grain size, with finer grain size associated with higher strength. The compressive strength of HA is also dependent on its density and porosity, which in turn are dependent on the sintering temperature and time. Aside from chemicophysical properties, HA solubility also depends on the biological environment, such as powder weight to liquid volume ratio, pH, and ionic concentration, as well as the protonation state of the phosphate ions when immersed in a solution. As an example, calcium ions in HA lattice are held in place by trivalent phosphate ions ($PO_4^{3-}$). If a decrease in pH occurs, it changes the protonation state of some trivalent phosphate ions, converting them to divalent phosphate ions ($HPO_4^{2-}$). This conversion weakens the Ca—P bond, allowing some calcium to be released into the physiological solution. Moreover, divalent phosphate ($HPO_4^{2-}$) released into the solution binds with hydrogen ions to form monovalent phosphate ions ($HPO_4^-$). This binding of $HPO_4^-$ with hydrogel ions in solution acts as a buffer, altering the pH of the solution and slowing down the HA degradation.

*Tricalcium phosphate (TCP):* Another investigated and used CaPs is tricalcium phosphate (TCP). TCP can have two forms, called α-TCP and β-TCP, even if the Ca/P ration is always 1.5. The two forms differ in the solubility rate; α-TCP is more rapid compared to β-TCP. Hence, α-TCP degrades quicker in human body; this information could be useful in tailoring a bioceramic for a particular application. It's been noticed that TCP has a higher solubility rate compared to that of HA. To take advantage of the benefits of both HA and TCP, biphasic CaPs have been used, and the properties depend on the TCP/HA ratio. They take advantage of the osteoconductivity of HA and the absorbability of the TCP. In fact, TCP dissolves, supplying the localized environment with a concentrated source of calcium and phosphorus. HA slows the degradation of the structure, giving more support to the healing tissue, allowing cell growth.

### 4.5.2.2 Bioactive Glasses

Glasses that contain $SiO_2$, $Na_2O$, $CaO$, and $P_2O_5$ are materials that exhibit bioactivity. These glasses differ from conventional soda lime glasses in their compositional ratios; in fact, bioactive glasses contain less that 60% silica, high $Na_2O$ and $CaO$ content, and a high $CaO$ to $P_2O_5$ ratio. These specific compositions allow high reactivity of the implant surface, thereby bonding the bioactive glass to the bone in the physiological environment (Fig. 4.71A). Originally suggested by Larry Hench in 1970s, bioactive glass involves a silicate glass-based system. Examples of bioactive glasses are reported in Table 4.10.

In the absence of contact with an aqueous environment, bioactive glasses are inert. In the presence of a physiological solution, rapid surface reaction of the bioactive glass occurs, forming a silica-rich gel layer within an hour. There is a rapid ionic exchange of $Na^+$ with $H^+$ or $H_3O^+$ from the solution. Loss of soluble silica in the form of $Si(OH)_4$ to the solution occurs, resulting in breaking of Si-O-Si bonds and the formation of Si-OH at the bioactive glass-solution interface. The silica-rich gel layer is favorable for the rapid nucleation, formation, and growth of a CaP-rich amorphous apatite layer. The amorphous CaP layer eventually crystallizes, and $CO_3^{2-}$ substitutes for $OH^-$ in the apatite lattice, leading to the formation of a carbonate apatite layer. Moreover, incorporation of $OH^-$ allows the formation of different forms of CaPs (Fig. 4.71B). The reactions occurring in vivo depend on local pH and

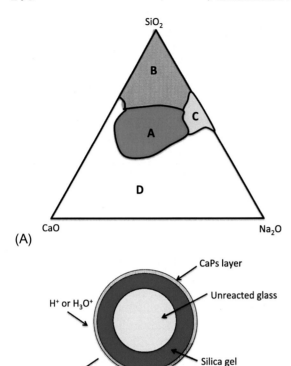

(A)

(B)

**FIG. 4.71** (A) Ternary diagram ($SiO_2$-$Na_2O$-$CaO$, at fixed 6% $P_2O_5$): compositional dependence of bone bonding and fibrous tissue bonding to the surfaces of bioactive glasses and glass ceramics: zone A=bioactive bone bonding ceramics (bonding within 31 days); zone B=nearly inert ceramics (fibrous tissue formation, no bonding; $SiO_2$ content is too high and reactivity is too low; these glasses form a surface hydration layer or too dense silica-rich layer to enable further reactivity and ion exchange); zone C=resorbable glasses (too high reactivity, no bonding; a thick and porous, unprotected $SiO_2$ rich film is formed but it dissociates at a high rate); zone D=no glass formation (no bonding). (B) Scheme of the reactions that occur for bioactive glass in a physiological environment. *(A) From Hench, L.L., 1996. In: Ratner, B.D., Hoffman, A.S., Schoen, F.J., Lemons, J.E., (Eds.), Biomaterials Science: An Introduction to Materials in Medicine. Academic Press, San Diego, CA, pp. 73–84.*

**TABLE 4.10** Composition of Two of the Most Used and Investigated Bioactive Glasses

|  | 45S5 | 52S4.6 |
|---|---|---|
| $SiO_2$ (%) | 45 | 52 |
| $P_2O_5$ (%) | 6 | 6 |
| $CaO$ (%) | 24.5 | 21 |
| $Na_2O$ (%) | 24.5 | 21 |

device is implanted. In parallel with these physicochemical mediated reactions, in an in vivo environment, proteins adsorb/desorb from the silica gel and carbonated layers. The bioactive surface and subsequent preferential protein adsorption can enhance osteoblast adhesion, differentiation, proliferation, and secretion of extracellular matrix.

### 4.5.2.3 Applications

The different classes of materials/biological constituents are used clinically and are under investigation for a wide variety of applications, including bulk implants (surface-active), coatings on metallic or ceramic implants (surface-active), permanent bone augmentation devices/scaffold materials (surface-active), temporary tissue engineering devices (surface- or bulk-active), fillers such as in cements (surface- or bulk-active), and drug-delivery vehicles (bulk-active).

The main applications of bioactive surface bioceramics dwell in the maxillofacial and dental fields and in the orthopedic field (mainly femoral stem coatings). In particular, the bioabsorbable ceramics degrade in physiological fluids and are mainly used for temporary bone fixation systems, with lifetime balanced to the time necessary for the mechanisms of regeneration of bone tissue, allowing for the gradual reabsorption of the bioceramic and, at the same time, the replacement of the implant with neotissue.

HA is used to enhance bone healing and to establish high interfacial bone-implant strength, in low load applications. Owing to its brittleness,

reactive cell constituents; hence, as explained in the ternary phase diagram (Fig. 4.71A), these reactions can be either biologically beneficial or adverse, therefore they should be well controlled to avoid unexpected results once the

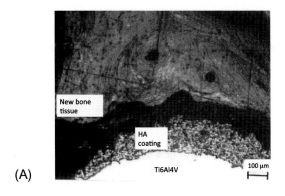

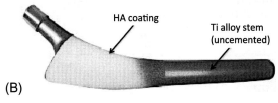

FIG. 4.72 (A) Bone tissue formation (animal model in vivo test, data unpublished) in contact with HA coating on Ti6Al4V implant; (B) HA coating on a femoral stem made of Ti alloy.

HA is used as a coating on metal implant surfaces in dental (e.g., dental implant) and orthopedic (e.g., femoral stem, Fig. 4.72B) applications, as filler for bone ingrowth, as implants in the middle ear (as bulk implants), or as bioactive phase in composite materials (see Section 4.5).

Because of their low mechanical strength characteristics, bioactive glasses should be used as coatings on metal or ceramic materials with high mechanical performances, for example, for devices subjected to high loads such as artificial joints.

## 4.6 COMPOSITE BIOMATERIALS

### 4.6.1 Overview

The design flexibility of composite materials appears particularly interesting for the creation of prostheses capable of imitating the properties of the tissues they replace (Murphy et al., 2016).

Indeed, the complex organization of the organic tissues and the vast range of properties they possess often make it impossible to use traditional materials for fabricating prostheses.

Dental resins, cardiac patches, and hybrid vascular prostheses are examples of application of composite materials in which the function of the reinforcement phase (particles, fibers, meshes) inside the matrix is to improve their mechanical resistance. In these applications, the design of the material can be obtained, through the concepts and technology of composite materials, with the desired mechanical and physical properties.

This flexibility of design appears not replaceable, today, in the case of many biomedical applications to make devices able to mimic, at best, the properties of damaged native tissues.

*Matrices* of composites for biomedical use are usually polymeric. Superpolymers and technopolymers (e.g., PEEK, polysulfones, polyimides) can be used as matrixes in advanced composites for application in orthopedics; more flexible and biocompatible matrices, such as silicones, hydrogels, and polyurethanes, have good possibilities and can be used in composites for the reconstruction of soft tissues (e.g., artificial skin); biodegradable polymeric matrices are used in the preparation of biodegradable composite materials.

In particular, composites with a biodegradable matrix (e.g., PLA, PGA) are used when the mechanical properties are to be time-dependent, determining the complete reabsorption of the implant and eliminating all the problems of long-term biocompatibility. The great advantage is that there is no need for a second surgical operation to remove the implant.

The *reinforcing materials/systems* for composite biomaterials can be selected among particulate and fibrous typologies (e.g., carbon fibers, polymer fibers, ceramic particles, glass fibers, and particles), depending on the specific application. Matrices loaded with nanofibers or nanoparticles are presently investigated, and the resulting systems are referred as *nanocomposites*.

*Fibers* are more effective on mechanical properties than particles, and polymer-fiber composites can reach stiffness and strengths per unit area comparable to metals and even higher. In addition, *particle*-reinforced systems provide composites with isotropic properties, whereas fiber-reinforced composites are anisotropic. The opportunity to design anisotropic properties is undoubtedly one of the most important advantages offered by composites with respect to monolithic materials. Most of the living tissues are composite systems with anisotropic properties, and, when possible, matching the properties of the device with the properties of the replaced tissue is the first design criterion.

Appropriate arrangement of fibers in a soft polymer matrix can create materials that mimic the structure and mechanical properties of soft tissues, such as cartilage, ligaments, blood vessels, muscle, and skin. On the other hand, the addition of inorganic particles (e.g., calcium phosphates, see Section 4.5.2) into a stiff biodegradable polymer will improve compression properties of the matrix while improving osteoblast proliferation and bone mineralization.

Prostheses made of carbon-fiber-reinforced composites can be adapted to meet the specific requirements of amputee athletes who still want to compete (a well-known example is shown in Fig. 4.73). Most successful applications of carbon fiber-reinforced composites are for prostheses exhibiting unique properties in term of lightness, stiffness, and strength.

In all cases, *sterilization* methods must be accurately selected and tested, as they play an important role due to the potential sensitivity of the matrix and sometimes of the reinforcement, as well as their interface, to chemical agents, temperature, humidity, and radiation (see Chapter 5.1).

## 4.6.2 Composites in Dentistry

Polymeric matrix composites are widely used in dentistry, and their use has considerably grown in the last few years, substituting in some cases for materials that are more traditional. Polymer matrix composites are used in restorative dentistry to fill in cavities, to restore fractured teeth, and to replace missing teeth. High

FIG. 4.73   Oscar Pistorius's transtibial prostheses, made of carbon-fiber reinforced polymer, replacing legs and feet amputated below the knee. They were developed by medical engineer Van Phillips in 1984 and are designed to store kinetic energy like a spring, allowing the wearer to jump and run effectively (http://edition.cnn.com/2011/SPORT/03/24/athletics.oscar.pistorius.olympics/index.html).

dimensional stability, wear resistance, and mechanical properties are the most important requirements for these materials.

### 4.6.2.1 Restorative Dentistry

Restorative composites have evolved significantly since they were first introduced in the early 1960s, with most of the development concentrating on filler technology.

In most applications, dental composite consists of a polymeric acrylic or methacrylic matrix reinforced with ceramic particles. The commercial formulations of matrices are mainly based on bisphenol-A-glycidyldimethacrylate (Bis-GMA). Other components, such as triethylenglycoldimetachrylate (TEGDMA) are added to reduce viscosity (Fig. 4.74); other monomers, such as urethane dimethacrylate (UDMA, Fig. 4.74) and ethylene-glycol dimethacrylate (EDMA) are often used. The adhesion between filler and matrix can be improved by treating the particles with a functional silane agent as a coupling agent. The composite formulation may be of the self- (such as in PMMA bone cement) or light-curing type.

The most critical aspects of these composites are the dimensional stability, wear, and mechanical properties. The addition of inorganic filler particles reduces matrix shrinkage; wear and mechanical properties strongly depend on the filler-matrix adhesion.

Because adhesion between matrix and filler is a critical issue for the composite performance, coupling agents (e.g., silanes) are used to improve the bonding force between the constituents.

Wear is still a cause of clinical failure, particularly in high stress areas such as molar restorations. The monomer-polymer conversion greatly affects the mechanical and biological properties of dental composites and can be varied and controlled by changing the curing conditions. However, there are individual factors, because a composite material may perform satisfactorily in one patient whereas in another it may erode, degrade, or fracture.

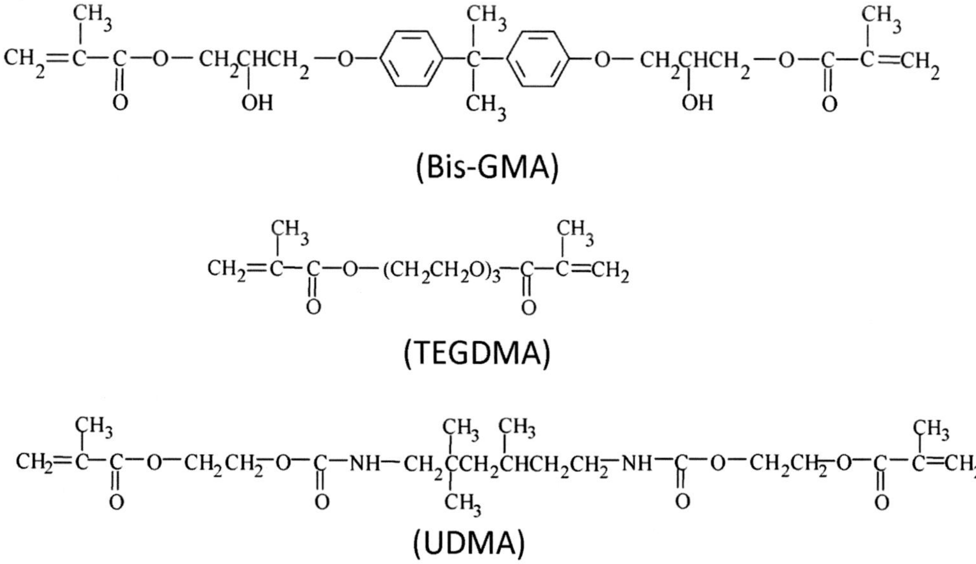

(Bis-GMA)

(TEGDMA)

(UDMA)

**FIG. 4.74** Chemical structure of monomers most used in dental restorative composites.

Another important aspect is linked to micro-cavities resulting from material detachment that could play a decisive role in bacterial adhesion and bacterial plaque formation.

### 4.6.2.2 Dentures, Bridges, and Dental Implants

Fixed bridges and removable dental prostheses are used to replace one or more teeth and to restore masticatory, phonetic, and aesthetic functions (Fig. 4.75).

Composite materials that can be used as bridges are carbon, Kevlar, UHMWPE, or glass fiber-reinforced PMMA and, as removable prostheses, glass fiber-reinforced PC, polyamides, PP, or PMMA.

However, there are several disadvantages: fixed bridges interfere with healthy adjacent teeth used to attach and support the bridge, and removable dentures are uncomfortable and do not transfer the occlusal stresses to the bone. The natural bone under a removable denture may deteriorate over time, permanently changing the appearance of personal smile and face.

Dental implants were developed to overcome these problems; the damaged or missing tooth is replaced by an artificial permanent implant. Current dental implants are composed of three components: a part implanted in the jawbone (known as fixture), a connector (known as an abutment), and a customized crown (Fig. 4.76).

As already described in Section 4.4, fixture and abutment are made of commercially pure titanium (CP-Ti), although some composite materials (i.e., SiC and carbon, or carbon fiber-reinforced carbon) and graded composites combining Ti and hydroxyapatite have been proposed.

To improve osseointegration, various types of fixture finishing are possible: simple mechanical processing, sandblasting, chemical etching, sandblasting with HA, bioactive coatings (cathodic depositions), and innovative anodic spark deposition (ASD) treatments (in particular, high thickness anodization in acid solutions). Biomimetic treatments are aimed at modifying the composition of the superficial oxide film of titanium (but

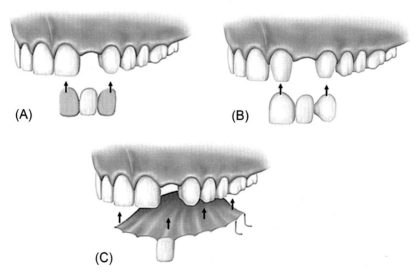

(A)

(B)

(C)

**FIG. 4.75**   Drawings representing fixed and removable dental prostheses. (A) Resin-bonded bridge; (B) Tooth-supported fixed bridge; (C) Removable partial denture.

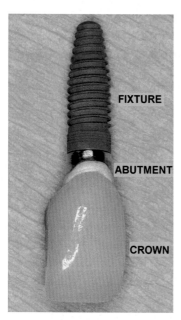

**FIG. 4.76** A dental implant with a crown attached used for a single tooth replacement. *From Wikipedia: https://en. wikipedia.org/wiki/Dental_implant.*

also tantalum and others) to create chemical conditions that facilitate the precipitation of calcium and phosphorus with the formation of apatite, before the surface is "shielded" by proteins (Sandrini et al., 2003, 2005).

## 4.6.3 Composite Materials for Orthopedic Applications

The fracture or failure of bones due to various traumas, pathologies, or natural aging frequently requires an operative treatment such as the implantation of a temporary or a permanent prosthesis, which still is a challenge for orthopedic surgeons, especially in the cases of large bone defects.

Composite materials can reproduce both the macroscopic and microscopic structure, and the most important mechanical properties of bone tissue.

### 4.6.3.1 Bone Grafts

When bone is injured, it regenerates naturally over time. However, when the injury is severe, healing will be incomplete, and there will be the need for bone grafts to restore mechanical functionality. In addition to bone grafts belonging to the same individual (i.e., autografts) or from a donor (i.e., allografts), researchers have developed new materials and composites to replace natural bone.

Great attention is given to the use of bioceramics because of their ability to bind to the natural bone tissue. Most studies deal with the combination of HDPE and hydroxyapatite (HA) particles as bone replacement materials. In this composite, HA (20%–40% $v/v$; by weight, more than 70% HA), used as a filler, gives the materials bioactivity, whereas HDPE gives the fracture toughness. Successful results have allowed its commercialization with the trade name of HAPEX (40% HA in volume, more than 70% by weight) (1995, Smith and Nephew Richards, TN, United States) which was pioneered at Queen Mary and Westfield College, London, United Kingdom (Bonfield et al., 1981; Wang and Bonfield, 2001). However, due to inadequate mechanical properties, this material is generally used for nonload-bearing applications (e.g., middle ear implants).

Coupling agents, for example, 3-trimethoxysiyl propylmethacrylate for HA and acrylic acid for HDPE, can be used to improve bonding (by both chemical adhesion and mechanical coupling) between HA and HDPE.

Biodegradable composites made of bioabsorbable polyesters (PGA, PLA, PLGA copolymers) reinforced with Ca phosphates (e.g., HA short fibers) have been tested for use in bone regeneration.

### 4.6.3.2 Fracture Fixation Devices

Internal fixation devices are temporarily implanted inside the human body to hold together bone fragments and promote healing.

Fracture-fixation devices should have approximately the same stiffness as bone to avoid stress shielding, and be strong enough to avoid fracture. Thermoplastic composites such as carbon fiber-reinforced PMMA, polybutyleneterephthalate (PBT), and PEEK can be used to produce fracture-fixation devices with higher flexibility and adequate strength.

In particular, PEEK matrix composites with carbon fibers and/or bioactive particles have been developed and commercialized by INVIBIO (Fig. 4.77).

Composite plates made of biodegradable polymeric matrix (e.g., PLLA) are already used. To improve bone regeneration, these devices can be reinforced with HA particles (Fig. 4.78); in addition, these composites exhibit very high mechanical strength and elastic modulus close to that of cortical bone. The composite devices show an adequate degradation rate, resulting in a gradual transmission of the load to the healing fracture site.

### 4.6.3.3 Joint Prostheses

Composite materials are also under investigation for the fabrication of joint prosthesis. In fact, new material combinations as the bearing surfaces of hip prostheses are aimed to prolong their life by overcoming the problems of failure due to wear-particle-induced osteolysis. This would reduce the need for revision surgery. UHMWPE is widely used in joint prosthesis, but it is a weak component of the prosthesis due to its low wear behavior. Research is underway for a possible substitute, evidencing in PEEK reinforced with carbon fibers an improvement in wear resistance (Latif et al., 2008). Efforts are also underway to use PEEK-based composite to produce femoral stem for hip implants (Bandoh et al., 2007).

### 4.6.3.4 Cardiovascular Applications

Several biomaterials are used in the repair and replacement of impaired cardiovascular tissues. These biomaterials fall into two main

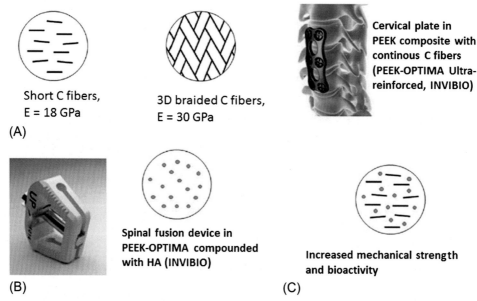

**FIG. 4.77** PEEK-matrix composites: the reinforcement with C fibers enhance the mechanical properties, whereas bioactive particles (HA, Sr-HA, nano-HA) increase bioactivity. (A) PEEK reinforced with carbon fibers, (B) PEEK loaded with bioactive particles, and (C) PEEK loaded with bioactive particles and reinforcing fibers. *Modified from Monich, P.R., et al., 2016. Mater. Lett., with permission of Elsevier.*

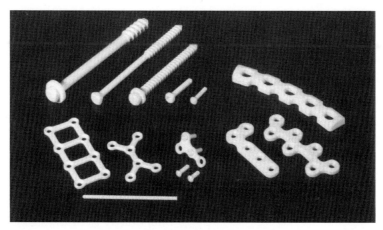

FIG. 4.78    Internal fixation devices (screws, washers, pin, plates) made of μ-HA/PLLA composites. *From Shikinami, Y., Okuno, M., 1999. Biomaterials 20, 863, Fig. 1, with permission of Elsevier.*

categories: synthetic and natural. As we already know, natural biomaterials are superior in functionality and biocompatibility but lack robustness; synthetic biomaterials hold the advantages of strength and durability but are deficient in functional capabilities.

In addition, so far, no single foreign material is able to counteract unwanted effects, primarily blood coagulation, which occurs when a foreign material is placed in contact with the blood (see Chapter 6), especially in the case of long-term use.

Combining materials into composites has created better options that benefit from the strengths of different materials and minimizes the weaknesses. Composite materials are obtained via the methods of blending (physically combined without forming chemical bonds), coating (by submersion or spraying), multilayering, or interpenetrating different combinations and materials design.

Different solutions are now available on the market. Some examples are:

– Coated stents for the prevention of restenosis (Babapulle and Eisenberg, 2002),
– Collagen or gelatin-impregnated vascular grafts (Singh et al., 2015),
– Bioprosthetic valves reinforced with polymeric material to abate the innate lack of strength of the natural tissue (Wang et al., 2010).

### 4.6.3.5  Tendons and Ligament Prostheses

Tendons and ligaments can withstand high stresses due to their particular structure of collagen fibers aligned in the direction of the applied stress. Following injury, tendons and ligaments have a low intrinsic healing capacity due to limited vascularization and thus surgery is required to repair or replace the injured tissue.

Various kinds of artificial ligaments were developed and brought to clinics; each had its own specifications for the design pattern, weaving methods, and fabric materials. However, after trials in clinics for 20 years, most of these prostheses were no longer used because of high complication and failure rates.

Dacron and carbon-fiber ligaments have been abandoned in early clinical usage because of the occurrence of complications such as severe synovitis after the implantation of those prostheses. Expanded PTFE (Gore-Tex) prosthesis used for ACL reconstruction, with extended follow-up, showed an occurrence of mechanical failure and infections.

However, owing to the advancements in surgical techniques, materials processing, and weaving methods, clinical application of some artificial ligaments so far has demonstrated good outcomes and will become a trend in the future.

In this context, the design and fabrication of composite tendon/ligament prostheses reproducing the structure, resistance, and flexibility of natural tissue is very challenging. Today, no composite devices are available for clinical practice and, although the field of tissue engineering has attempted to create a regenerated functional tissue, there are currently no approved tissue engineered tendon and ligament replacements.

# References

Anderson, J.M., 2001. Biological responses to materials. Annu. Rev. Mater. Res. 31, 81–110.

Babapulle, M.N., Eisenberg, M.J., 2002. Coated stents for the prevention of restenosis: Part I, special review. Circulation 106, 2734–2740.

Bandoh, S., Uchida, T., Ohkawa, H., Hibino, S., Zako, M., Yoshikawa, H., Sugano, N., Horikawa, T., 2007. In: The development of composite stem for hip joint, an application of composite materials for medical implant device. Proceedings of the 16th International Conference on Composite Materials, Kyoto, Japan, July 8–13http://www.iccm-central.org/Conferences.html.

Bernkop-Schnürch, A., Dünnhaupt, S., 2012. Chitosan-based drug delivery systems. Eur. J. Pharm. Biopharm. 81 (3), 463–469.

Bhattarai, N., Gunn, J., Zhang, M., 2010. Chitosan-based hydrogels for controlled, localized drug delivery. Adv. Drug Deliv. Rev. 62 (1), 83–99.

Bonfield, W., Grynpas, M.D., Tully, A.E., Bowman, J., Abram, J., 1981. Hydroxyapatite reinforced polyethylene—a mechanically compatible implant material for bone replacement. Biomaterials 2, 185–189.

Boutin, P., Christel, P., Dorlot, J.M., Meunier, A., de Roquancourt, A., Blanquaert, D., Herman, S., Sedel, L., Witvoet, J., 1988. The use of dense alumina-alumina ceramic combination in total hip replacement. J. Biomed. Mater. Res. 22 (12), 1203–1232.

Derbyshire, B., Fisher, J., Dowson, D., Hardaker, C., Brummitt, K., 1994. Comparative study of the wear of UHMWPE with zirconia ceramic and stainless steel femoral heads in artificial hip joints. Med. Eng. Phys. 16, 229–236.

Dong, C., Lv, Y., 2016. Application of collagen scaffold in tissue engineering: recent advances and new perspectives. Polymers 8(2).

Eljezi, T., Pinta, P., Richard, D., Pinguet, J., Chezal, J.-M., Chagnon, M.-C., Sautou, V., Grimandi, G., Moreau, E., 2017. In vitro cytotoxic effects of DEHP-alternative plasticizers and their primary metabolites on a L929 cell line. Chemosphere 173, 452–459.

Eo, M.Y., Fan, H., Cho, Y.J., Kim, S.M., Lee, S.K., 2016. Cellulose membrane as a biomaterial: from hydrolysis to depolymerization with electron beam. Biomater. Res. 20(1).

Erben, C.M., Goodman, R.P., Turberfield, A.J., 2006. Single-Molecule Protein Encapsulation in a Rigid DNA Cage. Angew. Chem. Int. Ed. 45 (44), 7414–7417.

Falcone, S.J., Palmeri, D., Berg, R.A., 2006. Biomedical applications of hyaluronic acid. In: Polysaccharides for Drug Delivery and Pharmaceutical Applications, ACS Symposium Series. Vol. 934. American Chemical Society, pp. 155–174. https://doi.org/10.1021/bk-2006-0934.

Griggs, J., 2009. DNA cages hint at the secret of nanoparticle self-assembly. New Sci. 201(2699).

Harkness, R.D., 1961. Biological functions of collagen. Biol. Rev. 36 (4), 399–455.

Hemamalini, T., Giri Dev, V.R., 2018. Comprehensive review on electrospinning of starch polymer for biomedical applications. Int. J. Biol. Macromol. 106, 712–718.

Hench, L.L., Splinter, R.J., Allen, W.C., Greenlee Jr, T.K., 1971. Bonding mechanisms at the interface of ceramic prosthetic materials. J. Biomed. Res. Symp. 5 (6), 117–141. Interscience, New York.

Kamoun, E.A., Chen, X., Mohy Eldin, M.S., Kenawy, E.-R.S., 2015. Crosslinked poly(vinyl alcohol) hydrogels for wound dressing applications: a review of remarkably blended polymers. Arab. J. Chem. 8, 1–14.

Kamoun, E.A., Kenawy, E.-R.S., Chen, X., 2017. A review on polymeric hydrogel membranes for wound dressing applications: PVA-based hydrogel dressings. J. Adv. Res. 8, 217–233.

Kumbar, S.G., Laurencin, C.T., Deng, M., 2014. Natural and Synthetic Biomedical Polymers. Elsevier, Burlington and San Diego, US.

Kurtz, S., 2016. UHMWPE Biomaterials Handbook, third ed. William Andrew, Oxford. https://doi.org/10.1016/B978-0-323-35401-1.00043-0.

Kurtz, S.M., Devine, J.N., 2007. PEEK biomaterials in trauma, orthopedic, and spinal implants. Biomaterials 28, 4845–4869.

Latif, A.M.H., Mehats, A., Elcocks, M., Rushton, N., Field, R.E., Jones, E., 2008. Pre-clinical studies to validate the MITCH PCRTM Cup: a flexible and anatomically shaped acetabular component with novel bearing characteristics. J. Mater. Sci. Mater. Med. 19, 1729–1736.

Lin, N., Dufresne, A., 2014. Nanocellulose in biomedicine: current status and future prospect. Eur. Polym. J. 59, 302–325.

Maitra, J., Shukla, V.K., 2014. Cross-linking in hydrogels—a review. Am. J. Polym. Sci. 4 (2), 25–31.

Mhurchu, C.N., Dunshea-Mooij, C., Bennett, D., Rodgers, A., 2005. Effect of chitosan on weight loss in overweight and obese individuals: a systematic review of randomized controlled trials. Obes. Rev. 6 (1), 35–42.

Mohite, B.V., Patil, S.V., 2014. A novel biomaterial: bacterial cellulose and its new era applications. Biotechnol. Appl. Biochem. 61 (2), 101–110.

Mozafari, M.R., 2007. Nanomaterials and Nanosystems for Biomedical Applications. Springer, Dordrecht, The Netherlands.

Murphy, W., Black, J., Hastings, G., 2016. Handbook of Biomaterials Properties, second ed. Springer Science +Business Media LLC, New York, ISBN: 978-1-4939-3303-7.

B. S. Nielsen, D. N. Andersen, E. Giovalle, M. Bjergstrøm, P. Bo Larsen, Alternatives to Classified Phthalates in Medical Devices, The Danish Environmental Protection Agency, 2014 (www.mst.dk, 978-87-93178-27-4).

Peuster, M., Hesse, C., Schloo, T., Fink, C., Beerbaum, P., von Schnakenburg, C., 2006. Long-term biocompatibility of a corrodible peripheral iron stent in the porcine descending aorta. Biomaterials 27 (28), 4955–4962.

Pinchuk, L., Wilsonb, G.J., Barryc, J.J., Schoephoersterd, R.T., Parele, J.-M., Kennedy, J.P., 2008. Medical applications of poly(styrene-block-isobutylene-block-styrene) ("SIBS"). Biomaterials 29 (4), 448–460.

Prestwich, G.D., 2011. Hyaluronic acid-based clinical biomaterials derived for cell and molecule delivery in regenerative medicine. J. Control. Release 155 (2), 193–199.

Ravi Kumar, M.N.V., 2000. A review of chitin and chitosan applications. React. Funct. Polym. 46 (1), 1–27.

Rockwood, D.N., Preda, R.C., Yücel, T., Wang, X., Lovett, M.L., Kaplan, D.L., 2011. Materials fabrication from *Bombyx mori* silk fibroin. Nat. Protoc. 6 (10), 1612–1631.

Sandrini, E., Chiesa, R., Rondelli, G., Santin, M., Cigada, A., 2003. A novel biomimetic treatment for an improved osteointegration of titanium. J. Appl. Biomater. Biomech. 1 (1), 33–42.

Sandrini, E., Morris, C., Chiesa, R., Cigada, A., Santin, M., 2005. In vitro assessment of the osteointegrative potential of a novel multiphase anodic spark deposition coating for orthopaedic and dental implants. J. Biomed. Mater. Res. B Appl. Biomater. 73 (2), 392–399.

Shavandi, A., Silva, T.H., Bekhit, A.A., Bekhit, A.E.-D.A., 2017. Keratin: dissolution, extraction and biomedical application. Biomater. Sci. 5 (9), 1699–1735.

Singh, C., Wong, C.S., Wang, X., 2015. Medical textiles as vascular implants and their success to mimic natural arteries. J. Funct. Biomater. 6, 500–525.

Toole, B.P., 2004. Hyaluronan: from extracellular glue to pericellular cue. Nat. Rev. Cancer 4 (7), 528–539.

Torres, F.G., Commeaux, S., Troncoso, O.P., 2013. Starch-based biomaterials for wound-dressing applications. Starch-Starke 65 (7–8), 543–551.

Van Vliet, E.D.S., Reitano, E.M., Chhabra, J.S., Bergen, G.P., Whyatt, R.M., 2011. A review of alternatives to di (2-ethylhexyl) phthalate-containing medical devices in the neonatal intensive care unit. J. Perinatol. 31 (8), 551–560.

Vaz, J.M., Pezzoli, D., Chevallier, P., Campelo, C.S., Candiani, G., Mantovani, D., 2018. Antibacterial coatings based on chitosan for pharmaceutical and biomedical applications. Curr. Pharm. Des. 24 (8), 866–885.

Vepari, C., Kaplan, D.L., 2007. Silk as a biomaterial. Prog. Polym. Sci. 32 (8–9), 991–1007.

Vlassov, V.V., Pautova, L.V., Vlassova, I.E., 1997. Oligonucleotides and Polynucleotides as Biologically Active Compounds. Prog. Nucleic Acid Res. Mol. Biol. 57, 95–143.

Walsh, A.S., Yin, H., Erben, C.M., Wood, M.J.A., Turberfield, A.J., 2011. DNA cage delivery to mammalian cells. ACS Nano 5 (7), 5427–5432.

Wang, M., Bonfield, W., 2001. Chemically coupled hydroxyapatite-polyethylene composites: structure and properties. Biomaterials 22, 1311–1320.

Wang, Q., McGoron, A.J., Pinchuk, L., Schoephoerster, R.T., 2010. A novel small animal model for biocompatibility assessment of polymeric materials for use in prosthetic heart valves. J. Biomed. Mater. Res. A 93 (2), 442–453.

Williams, D.F., 1999. Williams Dictionary of Biomaterials. University Press, Liverpool, UK.

Wise, S.G., Mithieux, S.M., Weiss, A.S., 2009. Engineered tropoelastin and elastin-based biomaterials. Adv. Protein Chem. Struct. Biol. 78, 1–24.

Zhang, Z., Ortiz, O., Goyal, R., Kohn, J., 2014. Biodegradable polymers. In: Principles of Tissue Engineering, fourth ed, pp. 441–473 (Chapter 23).

Zheng, Y.F., Gu, X.N., Witte, F., 2014. Biodegradable metals. Mater. Sci. Eng. R 77, 1–34.

## Further Reading

Abdel-Hady Gepreela, M., Niinomi, M., 2013. Biocompatibility of Ti-alloys for long-term implantation. J. Mech. Behav. Biomed. Mater. 20, 407–415.

Agrawal, C.M., 2013. Introduction to Biomaterials: Basic Theory With Engineering Applications. Cambridge Texts in Biomedical Engineering. Cambridge University Press, Cambridge, UK.

Allo, B.A., Costa, D.O., Dixon, S.J., Mequanint, K., Rizkalla, A.S., 2012. Bioactive and biodegradable nanocomposites and hybrid biomaterials for bone regeneration. J. Funct. Biomater. 3, 432–463.

Banoriyaa, D., Purohita, R., Dwivedi, R.K., 2017. Advanced application of polymer based biomaterials. Mater. Today Proc. 4, 3534–3541.

Benezech, J., Garlenq, B., Larroque, G., 2016. Advances in orthopedics. J. Biomater. Appl. 2016. 7 pages, Hindawi Publ. Co.

Brook, M.A., 2006. Platinum in silicone breast implants. Biomaterials 27, 3274–3286.

Chen, T., Jiang, J., Chen, S., 2015. Status and headway of the clinical application of artificial ligaments. Asia-Pacific J. Sports Med. Arthrosc. Rehabil. Technol. 2 (1), 15–26.

Colas, A., Curtis, J., 2013. Silicones. In: Ratner, B.D., Hoffman, A.S., Schoen, F.J., Lemons, J.E. (Eds.), Biomaterials Science, An Introduction to Materials in Medicine, third ed. Academic Press, Kindlington, Oxford, UK, pp. 82–91.

Davis, J.R., 2003. Handbook of Materials for Medical Devices. ASM International, Materials Park, OH.

De Santis, R., Gloria, A., Ambrosio, L., 2017. Composite materials for hip joint prostheses. In: Biomedical Composites. second ed. Woodhead Publishing Series in Biomaterials, pp. 237–259 (Chapter 11).

Dickinson, B.L., 1988. UDEL® polysulfone for medical applications. J. Biomater. Appl. 3 (4), 605–634.

Dorozhkin, S.V., 2015. Calcium orthophosphate-containing biocomposites and hybrid biomaterials for biomedical applications (review). J. Funct. Biomater. 6, 708–832.

Eisenbarth, E., Velten, D., Muller, M., Thull, R., Breme, J., 2004. Biocompatibility of β-stabilizing elements of titanium alloys. Biomaterials 25, 5705–5713.

Fugolin, A.P.P., Pfeifer, C.S., 2017. New resins for dental composites, critical reviews in oral biology & medicine. J. Dent. Res. 96 (10), 1085–1109.

Gergely, R.C.R., Toohey, K.S., Jones, M.E., Small, S.R., Berend, M.E., 2016. Towards the optimization of the preparation procedures of PMMA bone cement. J. Orthop. Res. 34 (6), 915–923.

Goldenberg, R.A., Driver, M., 2000. Long-term results with hydroxylapatite middle ear implants. Otolaryngol. Head Neck Surg. 122 (5), 635–642.

Hench, L.L., 1996. In: Ratner, B.D., Hoffman, A.S., Schoen, F.J., Lemons, J.E. (Eds.), Biomaterials Science: An Introduction to Materials in Medicine. Academic Press, San Diego, CA, pp. 73–84.

Hermawan, H., 2018. Updates on the research and development of absorbable metals for biomedical applications. Prog. Biomater. 7 (2), 93–110. https://doi.org/10.1007/s40204-018-0091-4.

Hu, X., Cebe, P., Weiss, A.S., Omenetto, F., Kaplan, D.L., 2012. Protein-based composite materials. Mater. Today 15 (5), 208–215.

International Standard ASTM F138-13a. Standard Specification for Wrought 18Chromium-14Nickel-2.5 Molybdenum Stainless Steel Bar and Wire for Surgical Implants.

Jaganathan, S.K., Supriyanto, E., Murugesan, S., Balaji, A., Asokan, M.K., 2014. Biomaterials in cardiovascular research: applications and clinical implications. Biomed. Res. Int. 2014, Article ID 459465, 11 pages, Review Article.

Koerner, G., Schulze, M., Weis, J., 1991. Silicones: Chemistry and Technology. Vulkan Publ., Essen, ISBN: 3-8027-2161-6

Kurtz, S.M., 2012. PEEK Biomaterials Handbook. William Andrew, Elsevier, Kidlington, Oxford, UK. https://doi.org/10.1016/B978-1-4377-4463-7.10019-3.

Kurtz, S.M., Bracco, P., Costa, L., Oral, E., Muratoglu, O.K., 2016. Vitamin E-blended UHMWPE biomaterials. In: UHMWPE Biomaterials Handbook. third ed. pp. 293–306 (Chapter 17).

Lam, M.T., Wu, J.C., 2012. Biomaterial applications in cardiovascular tissue repair and regeneration. Expert. Rev. Cardiovasc. Ther. 10 (8), 1039–1049.

Lamba, N.M.K., Woodhouse, K.A., Cooper, S.L., 1998. Polyurethanes in Biomedical Applications. CRC Press LLC, Boca Raton, Florida.

Li, H., Zheng, Y., Qin, L., 2014. Progress of biodegradable metals. Prog. Nat. Sci. Mater. Int. 24, 414–422.

Migliaresi, C., 2013. Composites. (Chapter I 2.9)In: Ratner, B.D., Hoffman, A.S., Schoen, F.J., Lemons, J.E. (Eds.), Biomaterials Science. An Introduction to Materials in Medicine. third ed. Academic Press, Kidlington, Oxford, UK, pp. 223–241.

Monich, P.R., Henriques, B., Novaes de Oliveira, A.P., Souza, J.C.M., Fredel, M.C., 2016. Mechanical and biological behaviour of biomedical PEEK matrix composites: a focused review. Mater. Lett. 185, 593–597.

Murugan, R., Ramakrishna, S., 2005. Development of nanocomposites for bone grafting. Compos. Sci. Technol. 65 (15–16), 2385–2406.

Narayan, V., 2016. Alternate antioxidants for orthopedic devices. In: UHMWPE Biomaterials Handbook, third ed. pp. 326–351 (Chapter 19).

Niessner, N., Wagner, D., 2013. Practical Guide to Structures, Properties and Applications of Styrenic Polymers. Smithers Rapra, Akron, OH.

Niessner, N., et al., 2014. Styrene copolymers. Kunststoffe Int. 10, 29–33.

Park, J.B., Bronzino, J.D., 2002. Biomaterials: Principles and Applications. CRC Press, Boca Raton, Florida.

Pivec, T., Smole, M., Gašparič, P., Stana-Kleinschek, K., 2017. Polyurethanes for medical use. Tekstilec 60 (3), 182–197.

Watson, W.D., Wallace, T.C., 1985. Polystyrene and styrene copolymers. In: Applied Polymer Science, second ed.

ACS Symposium Series, vol. 285. Dow Chemical Company, pp. 363–382. (Chapter 17). Copyright © 1985 American Chemical Society, Washington, DC. https://doi.org/10.1021/bk-1985-0285.ch017.

Ratner, B.D., Hoffman, A.S., Schoen, F.J., Lemons, J.E., 2013. Biomaterials Science. An Introduction to Materials in Medicine, third ed. Academic Press, Kidlington, Oxford, UK.

Saleh, K.J., El Othmani, M.M., Tzeng, T.H., Mihalko, W.M., Chambers, M.C., Grupp, T.M., 2016. Acrylic bone cement in total joint arthroplasty: a review. J. Orthop. Res. 34, 737–744.

Salernitano, E., Migliaresi, C., 2003. Composite materials for biomedical applications: a review. J. Appl. Biomater. Biomech. 1, 3–18.

Shikinami, Y., Okuno, M., 1999. Bioresorbable devices made of forged composites of hydroxyapatite (HA) particles and poly-L-lactide (PLLA): Part I. Basic characteristics. Biomaterials 20, 859–877.

Stansbury, J.W., 2012. Dimethacrylate network formation and polymer property evolution as determined by the selection of monomers and curing conditions. Dent. Mater. 28, 13–22.

Vaishya, R., Chauhan, M., Vaish, A., 2013. Bone cement. J. Clin. Orthop. Trauma 4 (4), 157–163.

Vanherck, K., Koeckelberghs, G., Vankelecom, I.F.J., 2013. Crosslinking polyimides for membrane applications: a review. Prog. Polym. Sci. 38 (6), 874–896.

Wang, W., Wang, C., 2012. Polyurethane for biomedical applications: a review of recent developments. In: Paulo Davim, J. (Ed.), The Design and Manufacture of Medical Devices. Woodhead Publishing Reviews: Mechanical Engineering, Woodhead Publ. Ltd, Cambridge, UK, pp. 115–151.

Williams, D.F., 2014. Essential Biomaterials Science. Cambridge Texts in Biomedical Engineering. Cambridge University Press, Cambridge, UK.

# Sterilization and Degradation

## 5.1 STERILIZATION

Effective sterilization of biomaterials used in implants and devices is a critical prerequisite for their successful clinical application. It requires knowledge of sterility concepts and sterilization technologies that render products sterile. Moreover, understanding the effect of sterilization processes on the biomaterials themselves is also increasingly important. These topics are the focus of this chapter.

New surgical devices are constantly being introduced by bioengineering technology. Because of their effectiveness and cost efficiency, steam sterilization units have long been the preferred method for reprocessing surgical devices. The materials used to construct many of the newest surgical instruments, however, become damaged when they are sterilized in an autoclave, also known as a steam sterilizer, and so these devices require a low temperature sterilization process. In fact, the more complicated and technologically advanced the equipment is, the more likely it will be damaged by the high temperature and humidity of steam sterilization. For example, an endoscope will be damaged and eventually ruined by standard autoclaving. Therefore we must find a low

temperature sterilization method that is more delicate on the life of the equipment but still effective enough to fully sterilize all germs on said equipment.

Sterilization is the killing or removal of all microorganisms, including bacterial spores, which are highly resistant. Sterilization refers to the antimicrobial process during which all microorganisms are killed or eliminated in or on a substance or substrate by applying different processes. Microbes react in their own way to the antimicrobial effects of various physical treatments or chemical compounds, and the effectiveness of treatments depends on many other factors as well (e.g., population density, condition of microorganisms, concentration of the active agent, environmental factors). Sterilization procedures involve the use of heat, radiation, chemicals, or "physical removal" of microbes (Table 5.1). The type of sterilization should always be chosen as required, by taking into consideration the quality of materials and tools used and the possible adverse effects of sterilization on them.

Disinfection methods are different from sterilization as they kill many but not all microorganisms. It is a process of reduction of number of contaminating organisms to a level that

**TABLE 5.1**  Main Sterilization Techniques

| | |
|---|---|
| Heat | Flaming |
| | Incineration |
| | Boiling |
| | Moist heat |
| | Dry heat |
| Radiation | UV |
| | Gamma rays |
| | Electron beam |
| | X-rays |
| Chemicals | Ethylene oxide |
| | Ozone |
| Plasma | Ionizing gas |
| Filtration | Membrane |

cannot cause infection, that is, pathogens must be killed. Some organisms and bacterial spores may survive. Disinfection is provided with chemical disinfectants. Disinfectants should be used only on inanimate objects. On the other hand, antiseptics are mild forms of disinfectants used externally on living tissues to kill microorganisms, for example, on the surface of skin and mucous membranes.

The current regulatory expectation of the term "sterile" for blood-contacting medical devices and implants is to produce only one nonsterile device out of one million. The challenge for the biomaterials scientist responsible for defining an appropriate sterilization process is an optimization problem, to determine and define a cost-effective sterilization process window where sterility is achieved, yet deleterious effects on material are minimized. Many traditional medical devices are made with materials that are compatible with multiple sterilization modalities. In this scenario, finding a terminal sterilization solution is straightforward. However, the optimization problem is much more complex for biologics and combination devices, both of which are important and rapidly growing markets.

## 5.1.1 High Temperature Sterilization Methods

### 5.1.1.1 Sterilization by Heat

Heat is the most effective and rapid method of sterilization and disinfection. Excessive heat acts by coagulating cell proteins, whereas less heat interferes with metabolic reactions. Sterilization occurs by heating above 100°C, which ensures lolling of bacterial spores. Sterilization by hot air in a hot air oven (i.e., dry heat) and sterilization by autoclaving (i.e., moist heat) are the two most common methods used and will be described in the following. In particular, when sterilization occurs by heat, two types of techniques can be identified and will be briefly illustrated here:

**a.** sterilization by dry heat;
**b.** sterilization by moist heat.

### 5.1.1.2 Dry Heat

The use of dry heat is based on the removal of the water content of microbes and subsequent oxidation, thus killing or removing all microorganisms, including bacterial spores. Dry heat sterilization technique requires longer exposure time (1.5–3h) and higher temperatures than moist heat sterilization (see next section). Various methods of dry heat sterilization are available, among them: hot air oven, incineration, and flaming (wire loop). Open flame can be used for sterilization if the object is not directly exposed to flame damage. Different laboratory devices (e.g., scalpel, knife, inoculating loop, or needle) can be sterilized quickly and safely by crossing over open flame or by ignition.

Dry heat does most of the damage by oxidizing molecules. The essential cell constituents are destroyed and the organism dies. The temperature is maintained for almost an hour to kill the most resistant spores. Dry heat sterilization (Table 5.2) is performed in a hot air sterilizer (i.e., an oven). It is an electric box with adjustable temperature like an incubator. To achieve

**TABLE 5.2** Advantages and Disadvantages of Dry Heat Oven Sterilization

| Advantages | Disadvantages |
|---|---|
| A dry heat cabinet is easy to install and has relatively low operating costs | Time-consuming method because of slow rate of heat penetration and microbial killing |
| It penetrates materials | High temperatures are not suitable for most materials |
| It is nontoxic and does not harm the environment | |
| It is noncorrosive for metal and sharp instruments | |

uniform chamber temperature, hot air is circulated. The most common time-temperature relationships for sterilization with hot air sterilizers are the following:

- 170°C for 30 min;
- 160°C for 60 min;
- 150°C for 150 min or longer depending on the volume.

Sterilization with dry heat is limited to devices made of metal, glass, ceramic, and other thermostabile materials, like glycerol, soft paraffin, oils, and fats. In the dry heat sterilization system, they have to withstand the temperature needed to kill the spore-forming bacteria (i.e., at 160°C for 45 min; at 180°C for 25 min; at 200°C for 10 min). Dry heat ovens are used to sterilize items that might be damaged by moist heat or that are impenetrable to moist heat (e.g., powders, petroleum products, sharp instruments). Sterilizing by dry heat is accomplished by conduction. The heat is absorbed by the outside surface of the item, then passes toward the center of the item, layer-by-layer. The entire item will eventually reach the temperature required for sterilization to take place.

### 5.1.1.3 Moist Heat (Autoclave)

Due to the fact that the heat conductivity of water is several times higher than that of the air, heat sterilizes more quickly and effectively in the presence of hot water or steam than dry heat. Moist heat acts by denaturation and coagulation of protein, breakage of DNA strands, and loss of functional integrity of cell membrane; in addition, the structural components for microorganism replication are destroyed. Different techniques can be used accounting for the temperature of the sterilization process; in any case, the sterilization facility design should guarantee that all surfaces of the device/material are in contact with moist heat, and the packaging should allow moist penetration. When the sterilization is performed at 100°C, it is possible to boil the devices at 100°C for 30 min in a water bath. Syringes, rubber goods, and surgical instruments may be sterilized by this method. All bacteria and certain spores are killed, so it leads to disinfection. Boiling is the simplest and oldest way of using moist heat. The temperature of boiling water does not exceed 100°C at normal atmospheric pressure. Heat-resistant, endospore-forming bacteria can survive the 10–30-min heat treatment of boiling, so no sterilizing effect can be expected from boiling.

Steam ($T = 100$°C) is ideal for destroying bacteria, fungi, and spores. It is the most cost-effective and simple method available, but steam is not applicable to all materials and instruments. Moreover, it is more effective than dry heat at the same temperature because: (a) bacteria are more susceptible to moist heat, (b) steam has more penetrating power, and (c) steam has more sterilizing power as more heat is given up during condensation. Steam sterilization works at 100°C under normal atmospheric pressure, that is, without extra pressure. It is ideally suitable for sterilizing media that may be damaged at a temperature higher than 100°C. The sterilizer is a metallic vessel having two perforated diaphragms, one above boiling water, and the other about 10 cm above the floor; water is boiled by electricity, gas, or stove; steam passes up; and there is a small opening on the roof of the instrument for the escape of steam.

Sterilization by moist heat can be performed at temperatures above 100°C, mainly using autoclaving, which is one of the most used sterilization techniques because it is simple, low cost, and devoid of any risk for the operators (Table 5.3). Moreover, the use of saturated steam under high pressure is the most effective method to kill microorganisms. In an autoclave, sterilization is carried on by steam under pressure, in particular steaming at temperatures higher than 100°C is used. The temperature of boiling depends on the surrounding atmospheric pressure and a higher temperature of steaming is obtained by using a higher pressure. When the autoclave is closed and made airtight, and water starts boiling, the inside pressure increases and the water boils above 100°C, reaching a temperature of 121°C ($P = 1$ atm), which is kept for 15–30 min to kill spores. To achieve sterilization, generally 15 min of heat treatment at 121°C under $1.1 \, kg/cm^2$ pressure has to be applied. Most microbes are unable to tolerate this environment for more than 10 min.

However, the time used for sterilization depends on the size and content of the load.

Autoclave is a metallic cylindrical vessel (Fig. 5.1); on the lid, there is a:

**(1)** gauge for indicating the pressure,
**(2)** safety valve, which can be set to blow off at any desired pressure,
**(3)** stopcock to release the pressure. It is provided with a perforated diaphragm.

Water is placed below the diaphragm and heated by electricity, gas, or stove. Typical sterilization process is composed by three phases (Fig. 5.2): heating phase, stationary phase ($t = 15$–30 min), and cooling phase.

Autoclaves are widely used mainly for sterilization of aqueous liquids, glassware, and heat-resistant plastic products before their use, and also for contaminated materials prior to disposal as municipal solid waste.

## 5.1.2 Ethylene Oxide

A wide range of chemicals is suitable to inhibit or kill microbes. Some of the antimicrobial agents only inhibit the growth of microorganisms (e.g., bacteriostatic, fungistatic, and virostatic compounds) whereas others kill them (e.g., bactericidal, fungicidal, and virucidal agents). The bacteriostatic or bactericidal effect of a substance depends on the applied concentration and exposure time, in addition to its quality. Only bactericidal effect substances are used for chemical sterilization. These substances have the following requirements: they should have a broad-spectrum effect, they should not be toxic to higher organisms, they should not enter detrimental reactions to the materials being treated with, they should not be biodegradable, and they should be environmentally friendly, easy to apply, and economical.

The materials used in chemical sterilization are liquids or gases. Liquid agents are used primarily for surface sterilization. Among sterilizing gases, those working at low temperature

**TABLE 5.3** Advantages and Disadvantages of Autoclave Sterilization

| Advantages | Disadvantages |
|---|---|
| Nontoxic to patient, staff, environment | Deleterious for heat-sensitive or moist-sensitive instruments/materials |
| Cycle easy to control and monitor | Microsurgical instruments damaged by repeated exposure |
| Rapidly and efficacy microbicidal | May leave instruments wet, causing them to rust (in case of metallic devices) |
| Least affected by organic/ inorganic soils among sterilization processes listed | Potential for burns |
| Rapid cycle time | |
| Penetrates medical packing, device lumens | |

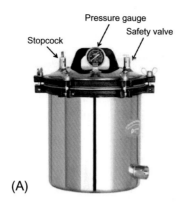

(A)

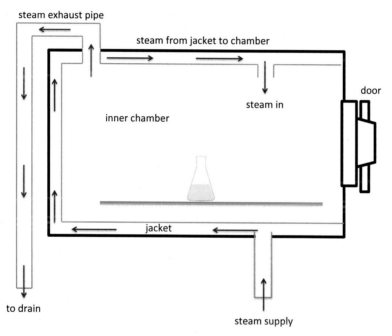

FIG. 5.1 Sterilization by autoclave. (A) Example of a laboratory autoclave equipment; (B) main components of an industrial steam sterilizer; in particular, exhaust valve: to remove steam after sterilization, operating valve: controls steam from jacket to chamber, pressure regulator: for steam supply, automatic ejector valve: thermostatically controlled and closes on contact with pure steam when air is exhausted. *From 2001, Benjamin Cummings, Addison Wesley Longman, Inc.*

function by exposing the materials to be sterilized to high concentrations of very reactive gases (e.g., ethylene oxide (EtO), beta-propiolactone, or formaldehyde). Due to their alkylating effect, these compounds cause the death of microbes by damaging their proteins and nucleic acids. The chemical agents used for sterilization must be chemically compatible with the substances to be sterilized; therefore they have a great importance in sterilization of

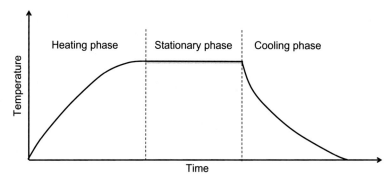

**FIG. 5.2** Scheme of the autoclave sterilization process: three phases can be detected: heating, stationary, and cooling phase.

pharmaceutical and thermoplastic materials. The chemicals used by the gas sterilizers are harmful to humans as well. Therefore the application of gas sterilizers requires compliance with the precautions by the users. Among all the chemical sterilization methods, EtO gas is widely used, as it can be applied to several polymeric materials without causing any degradation effects to them. For that reason, this sterilization technique is here described.

Ethylene oxide (Fig. 5.3) is a common gas used for low temperature sterilization (boiling point at 11°C); it is colorless, explosive, and dangerous to handle. Pure EtO, in concentrations in air above 3%, is highly explosive and flammable; to eliminate the risk of explosion, it can be mixed with 90% carbon dioxide or other inert compounds. EtO is a mutagen (i.e., a substance or agent that causes an increase in the rate that genes change), a carcinogen (i.e., a cancer-causing substance), and a reproductive hazard. A possible problem is related to the fact that EtO odor is pleasant and can be perceived at a minimum concentration of 700 ppm. Because the maximum safety concentration in work areas is estimated to be lower than one part

per million, there is a significant hazard that personnel are unaware of being exposed to hazardous concentrations. For this reason, proper ventilation and a constant control of the concentration of EtO in the working areas is necessary. Because both the liquid and the gaseous phase diffuse very easily in natural and synthetic polymers, it is always necessary to take precautions to protect personnel and patients against EtO exposure. For that reason, instruments and devices sterilized with the EtO process must be aerated for 12–24h, which requires a relatively large inventory of medical devices. The residual level of EtO released into the environment must also be carefully monitored to assure that no environmental problems are created.

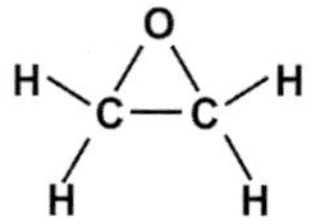

EtO is a poisonous gas that attacks the cellular proteins and nucleic acids of microorganisms. In detail, EtO inactivates all types of microorganisms, including spores of bacteria and viruses. Two carbon and one oxygen atoms (Fig. 5.3) are bonded together in an unstable ring; under certain conditions, the ring opens and initiates an alkylation reaction. Alkylation agents, such as EtO, react with many biologically important and vital cellular components for normal metabolism and reproduction. Some of these components include the constituents of nucleic acids and proteins.

EtO process cycle time is usually more than 14h, and the temperatures range from 25°C to 55°C. A lower temperature results in a less

**FIG. 5.3** Chemical structure of ethylene oxide.

**TABLE 5.4** Advantages and Disadvantages of Ethylene Oxide Sterilization

| Advantages | Disadvantages |
|---|---|
| Low temperature | Excessively long cycle |
| Short aeration time | Safety concerns—carcinogenic to humans |
| High efficiency: it destroys microorganisms including resistant spores | Toxicity issues: toxic residues on surgical instruments and tubing |
| Large sterilizing volume/chamber capacity | EtO is flammable |
| Noncorrosive to: plastic, metal, and rubber materials | Requires special room conditions, safety equipment, and separate ventilation system |

efficient process, which leads to a longer exposure time (Table 5.4).

There are at least three stages in a typical EtO sterilization cycle (Fig. 5.4):

– *Preconditioning*: This step prepares the chamber environment to meet the ideal conditions for temperature, pressure, and humidity. First, air is removed from the chamber to allow for gas penetration; then, steam is injected into the chamber, which humidifies the load as EtO is only effective in a humid environment. The chamber is heated by either steam or hot water, which is present in the jacket of the sterilizer.

– *Sterilization*: This step is the actual sterilization process. The EtO enters the chamber via evaporation with a certain amount of steam to keep the humidity level up as well as to make sure the EtO is reaching all parts of the load. When the required concentration in the chamber and load is achieved, the actual sterilization step starts. As EtO is absorbed by many kinds of plastic materials, it is important to keep the concentration at the right level. The appropriate concentration level of EtO in the chamber has to be ensuring, to achieve effective and safe sterilization.

– *Aeration (degassing)*: It is the most important and longest part of the EtO sterilization cycle. In fact, materials such as plastics and rubbers absorb gas and, if applied to patients, the toxic gas could damage human body tissue.

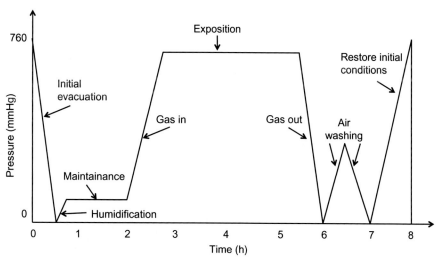

**FIG. 5.4** Scheme of ethylene oxide sterilization process.

For this reason, it is very important to have an excessive aeration stage to remove any remaining EtO gas and to allow absorbed gas to evaporate again from the sterilized items. This step is performed by circulating High Efficiency Particulate Air (HEPA)-filtered air over the load at a temperature between 30°C and 50°C. Sometimes the sterilized items can be placed in a special aeration cabinet for this stage. A commonly used time period for this aeration treatment is at least 48 h.

It is most commonly used to sterilize instruments with long lumens, such as endoscopes and all materials that have to be sterilized but cannot withstand higher temperature. Moreover, EtO can be used for the sterilization of solutions, glass bottles, sterilizer clothes, rubber, and wrapped or unwrapped materials. EtO is used to sterilize a wide range of medical products, including surgical sutures, intraocular lenses, neurosurgery devices, and heart valves.

Because of potential toxicity/carcinogenicity, residual EtO and its by-products (e.g., ethylene chlorohydrin), maximum allowable limits are no longer expressed as parts-per-million (ppm) in a medical product, but rather as a maximum allowable dose delivered to a patient, as reported in the standard ISO 10993-7 (Table 5.5).

## 5.1.3 Radiation

There are two general types of radiation used for sterilization: nonionizing radiation and ionizing radiation. Ionizing radiation is the use of short wavelength, high-intensity radiation to destroy microorganisms. This radiation can come in the form of gamma or X-rays that react with DNA resulting in a damaged cell. Nonionizing radiation uses a longer wavelength and lower energy. As a result, nonionizing radiation loses the ability to penetrate substances and can only be used for sterilizing surfaces. The most common form of nonionizing radiation is ultraviolet (UV) light, which is used in a variety of manners throughout the industry.

These forms of energy (e.g., UV and ionizing radiation, in particular $\gamma$-rays and X-rays, Fig. 5.5) are used for sterilization, especially for heat-sensitive materials.

**TABLE 5.5**  Examples of EtO Residue Limits on Medical Devices

| Device Category | Ethylene Oxide Limits | Ethylene Chlorohydrin Limits |
|---|---|---|
| Limited (<24 h) | 4 mg | 9 mg |
| Prolonged (>24 h to <30 days) | 60 mg/30 days | 60 mg/30 days |
| Permanent (>30 days) | 2.5 g/lifetime | 10 g/lifetime |
| Tolerable contact limit | 10 $\mu$g/cm$^2$ or negligible irritation | 5 mg/cm$^2$ or negligible irritation |
| Intraocular lens | 0.5 $\mu$g/lens/day<br>1.25 $\mu$g/lens | 4 × EtO limits suggested |
| Blood cell separator | 10 mg | 22 mg |
| Blood oxygenators | 60 mg | 45 mg |
| Cardiopulmonary bypass devices | 20 mg | 9 mg |
| Blood purification devices | 4.6 mg | 4.6 mg |
| Drapes contacting intact skin | 10 $\mu$g/cm$^2$ or negligible irritation | 5 mg/cm$^2$ or negligible irritation |

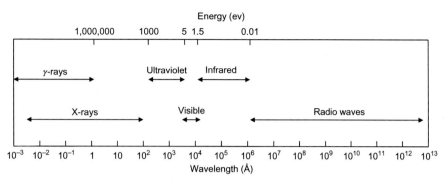

FIG. 5.5 Wavelength and energy of main electromagnetic radiation.

### 5.1.3.1 UV Radiation

The full spectrum of UV radiation can damage microbes, but only a small part is responsible for the so-called germicidal effect. Very strong "germicidal" effect can be achieved around 265 nm, because maximum UV absorption of DNA occurs at this wavelength. The main cause of cell death is the formation of pyrimidine dimers in nucleic acids. Bacteria are able to repair their nucleic acid after damage using different mechanisms; however, beyond a certain level of damage, the capacity of the enzyme system is not enough, and the accumulation of mutations causes death (Table 5.6).

TABLE 5.6 Advantages and Disadvantages of Ultraviolet Radiation Sterilization

| Advantages | Disadvantages |
| --- | --- |
| It does penetrate air, effectively reducing the number of airborne microorganism and killing them on surfaces | The radiation is not very penetrating |
| | It may not be effective against all bacterial spores |
| | UV light can damage human eyes, and prolonged exposure can cause burns and skin cancer in humans |

UV (germicidal) lamps are widely used in hospitals and laboratories (e.g., in biological safety cabinets) for decontamination of air and any exposed surfaces. The disadvantage of the use of UV radiation is that it does not penetrate through glass, dirty films, water, and other substances.

When microorganisms are subjected to UV light, cellular DNA absorbs the energy by purines and pyrimidine bases, and adjacent thymine molecules link together. Linked thymine molecules are unable to encode adenine on messenger RNA molecules during the process of protein synthesis. The damaged organism can no longer produce critical proteins or reproduce, and it quickly dies. UV light is especially effective in inactivating viruses. However, it kills far fewer bacteria than one might expect because of DNA repair mechanisms. In fact, once DNA is repaired, new molecules of RNA and protein can be synthesized to replace the damaged molecules.

### 5.1.3.2 Ionizing Radiation

Radiation sterilization doses are lethal. Terminal sterilization doses typically range from 8 to 35 kGy (typical units are kJ/kg, i.e., kilogray or kGy). An acute lethal dose to man is approximately 0.01 kGy, requiring an exposure time of only a fraction of a second in some processes. To ensure worker safety in these high radiation environments, significant shielding, robust interlocks, and the utmost care are required in radiation processing facilities.

Sterilization by radiation sterilization is simple. Fully packaged medical devices are exposed to a validated dose from a radiation source that emits electrons or photons that penetrate through the final packaging and inactivate the device's microbial load. One parameter, radiation dose, correlates directly with microbial kill, and this is easily measured to provide process control. The microbial kill mechanism involves radiation-induced scission of DNA chains to stop microbial reproduction. Although radiation destroys the ability of most microorganisms to reproduce, the resistance of viruses, such as HIV-1, remains a concern.

International radiation sterilization standards (ISO 11137-1) call out three radiation sterilization modalities: gamma, electron beam (e-beam), and X-rays. Gamma and e-beam dominate the radiation sterilization markets. X-rays and gamma rays have wavelengths shorter than the wavelength of UV light (Fig. 5.5). X-rays, which have a wavelength of 0.1–40 nm, and gamma rays, which have an even shorter wavelength, are forms of ionizing radiation, so named because they can dislodge electrons from atoms, creating ions. These forms of radiation also kill microorganisms and viruses, and ionizing radiation damages DNA and produces peroxides, which act as powerful oxidizing agents in cells. This radiation can also kill or cause mutations in human cells if it reaches them. Different penetration capacities of the radiations, particularly e-beam and gamma radiations, have an inversely proportional effect on the release rate of the energy and therefore on the exposure times. The resistance of various microorganisms to radiation changes with the species, as shown in Table 5.7.

### GAMMA RAYS

Cobalt-60 gamma sterilization accounts for approximately 80% of the radiation sterilization market and 40% of the overall terminal sterilization industrial market. Among the

**TABLE 5.7** Lethal Doses of Ionizing Radiation for Different Microorganisms

| Microorganism | Dose (kGy) |
| --- | --- |
| Vegetative bacteria<br>Animal virus (>75 nm)<br>*Bacillus anthracis* (spore) | 0.5–5 |
| Clostridium Sp. (spore)<br>Animal virus (20–75 nm)<br>Animal virus (>20 nm) | 5–20 |
| Bacterial virus<br>*Bacillus pumilus* (spore)<br>*Micrococcus radiodurans* | 20–60 |

high-energy ionizing radiation, $\gamma$-rays from radioactive nuclide $^{60}$Co are generally used for sterilization of disposable needles, syringes, bandages, medicines, and certain food (e.g., spices). The advantage of gamma radiation is its deep penetration through the packaging. Its disadvantage is the scattering in all directions, which requires special circumstances for application. $^{60}$Co is a radioactive element that undergoes nuclear decay producing useful gamma radiation. In particular, gamma rays are produced by $^{60}$Co, an isotope of $^{59}$Co, at two fixed wavelengths, characterized by two energy values, 0.33 and 1.17 MeV. An atom of $^{60}$Co (instable) contains 27 electrons, 27 protons, and 33 neutrons (compared to the 32 neutrons of $^{59}$Co). A neutron, composed by a proton and an electron, spontaneously emits a $\beta$-particle (electron) turning in a proton. Hence, the following reaction occurs:

$$Co_{27}^{60} \rightarrow Ni_{28}^{60} + \beta_{-1}^{0} + \gamma \, \text{rays}$$

Gamma rays are produced by transitions of nuclear energy during the transformation of $^{60}$Co to $^{60}$Ni. Because the half-life of $^{60}$Co is 5.3 years (equal to an emission reduction factor of about a 10% per year), new material must be regularly added to the source. Anyway, there is a certain availability of $^{60}$Co, being a waste product of nuclear fission.

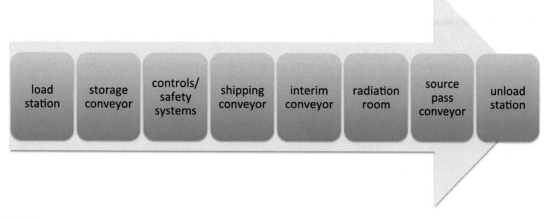

**FIG. 5.6** High volume production gamma irradiation facility (MDS Nordion, Canada).

It has to be pointed out that, although the gamma source itself is radioactive, the product sterilized by the exposure to the radiation does not become radioactive in the slightest measure.

High product volume $^{60}$Co processing plants are relatively simple. Small $^{60}$Co pellets are doubly encapsulated in rods, which are arranged into racks and stored in a pool of water, with water acting as shielding to keep the constantly emitted gamma radiation away from the processing room when product is not being processed. When the room is cleared, interlocks satisfied, and the facility is ready to process product, a mechanical elevator system raises the radiation source from the pool into the room. A conveyor system moves many totes of fully packaged product into the room and around the racks of $^{60}$Co, often passing by multiple times, as seen in Fig. 5.6.

$^{60}$Co gamma particles have no charge and penetrate uniformly through a very long distance of material. This is a significant processing benefit. Configurations of product within shipper boxes and processing load configurations are rarely a limitation due to high gamma penetration. Product is then brought back out of the processing room. Product is released for

distribution after dosimeters are read and documented to confirm that the product received the proper dose. For gamma radiation produced by $^{60}$Co, the energy is fixed at two values "by nature," and, once selected exposition time, relation between absorbed dose and depth of the device is linear (Fig. 5.7). The relation reported in Fig. 5.7 shows that gamma radiation, considering a material with a uniform density of $1\,g/1\,cm^3$, have a penetration power of about

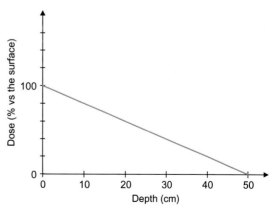

**FIG. 5.7** Gamma radiation dose versus depth of the device.

50 cm, 10 times greater than in the case of e-beam sterilization.

The beneficial sterilization effects of $^{60}$Co gamma rays (along with any deleterious material effects) come as a result of the energy of the photons being deposited into the microbes and materials. By definition, the energy deposited per unit weight is the radiation dose. Energy deposition from a gamma ray into a material occurs through a Compton Scattering interaction that generates high-energy electrons (0.5 MeV). These primary electrons travel through the material and deposit their energy in 60–100 eV bundles as they generate secondary electrons. It is these secondary electrons that cause all microbial kill. They do so by forming oxidizing free radicals. In addition to causing scission of DNA, the radicals can also cause scission or cross-linking of polymer chains.

### ELECTRON BEAM RADIATION

E-beam sterilization accounts for approximately 20% of the radiation sterilization market, 10% of the overall terminal sterilization industrial market. High-energy electrons for e-beam sterilization are generated by a variety of technologies that take 200 V electrons from the power grid and accelerate them to 0.2 million to 10 MeV (0.2–10 MeV, typically 5–10 MeV). At these energy levels, no part of the e-beam processing plant (or the product irradiated) is or becomes radioactive; e-beam accelerators turn off like a light bulb. Before turning on the accelerator, however, the processing room must be cleared and safety interlocks satisfied to avoid human exposure to the high doses of radiation and ozone that are generated. Accelerated electrons for device sterilization are typically magnetically focused into a 1–5 cm diameter beam and magnetically scanned at high frequency across the conveyor width, typically 30–122 cm.

The conveyor system transports products through the shielding and in front of the beam. Typically, one product, one shipper box, or one thin tote of product at a time passes by the

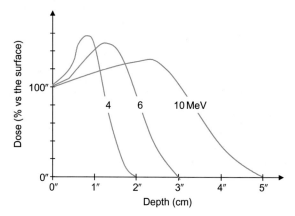

**FIG. 5.8**    Electron beam dose versus depth of the device.

beam's scan horn. Electrons from accelerators do not penetrate nearly as far as photons from gamma sources (Fig. 5.8).

A rule of thumb is that the maximum penetration (in centimeters) from a two-sided e-beam process is 0.8 times the beam energy (in MeV) divided by the density of product (in g/cm$^3$). Resulting maximum penetration distances are only a fraction of a meter, with higher beam energies resulting in higher penetration. Limited penetration can lead to inefficient processing and/or large distributions of delivered dose. In practice, this requires significant dosimeter work to develop load configurations that ensure an appropriate range of e-beam doses. Due to the limited penetration of accelerated electrons, products need to be carefully "dose-mapped," as their size and shape provide shielding, scattering interfaces, and incident angles that impact dose distribution. These differences provide a less homogeneous dose profile than that of $^{60}$Co.

Product is then brought back out of the processing room. Product is released for distribution after dosimeters are read and documented to confirm that the product received the proper dose. E-beam processing facilities can either be in a processing room as previously described for a gamma facility or be of a self-shielded design, as reported in Fig. 5.9.

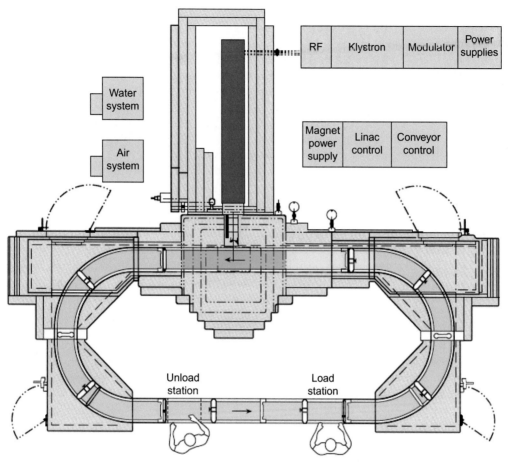

**FIG. 5.9** Medium volume self-shielded production electron beam irradiation facility (New Electron Beam Sterilization Systems: Surebeam On-Site System).

The mechanism of energy deposition from e-beams is that the high-energy electrons travel through the material and deposit their energy in 60–100 eV bundles as they generate secondary electrons. Like gamma processing, the resulting secondary electrons cause microbial kill, as well as all product material effects. The main advantage of e-beam in respect to gamma in terms of material compatibility is that the dose is delivered very quickly. Products being treated in gamma sterilization processes often have dwell times of several hours, whereas products typically dwell in an e-beam process for only minutes. The short

processing time limits oxidative degradation, because oxygen does not have time to diffuse into the product. The main disadvantage of e-beams versus $^{60}$Co is that the negatively charged electrons do not penetrate deeply into a material compared with neutral gamma rays, which penetrate through the product being sterilized.

## X-RAY RADIATION

X-ray sterilization accounts for a small fraction of the radiation sterilization market. It is a hybrid between gamma sterilization and e-beam sterilization. Radiation is generated from high-energy electrons from accelerators, typically 5 MeV electrons, impinging on a high atomic number ("Z") target, often tungsten, to generate X-rays. The X-ray photons behave nearly identically to photons from gamma sources in terms of energy deposition and high penetration capabilities. Like e-beam plants, no part of the X-ray processing plant is or becomes radioactive if X-rays are generated by electrons with energies no greater than 5 MeV.

## 5.1.4 Low Temperature Plasma

A low-temperature method of sterilization is cold plasma, which involves the exposure of the surface of the devices to a gas that has been ionized by an energy field.

Plasma is the fourth state of matter (i.e., solid, liquid, gas, and plasma) and is created when a gas is heated sufficiently or exposed to a strong electromagnetic field. When a gas becomes a plasma, it reaches an unstable state of matter in which the number of electrons are increased or decreased, thus producing ions, electrons, molecules, and radicals that are positively or negatively charged. In other words, plasma is an ionized gas that has special properties not seen in any other state of matter. Common examples of manmade plasmas include neon signs, fluorescent light bulbs, plasma displays used for televisions and computers, plasma lamps, and nuclear

fusion. Naturally occurring plasmas include fire, lightning, the sun, stars, auroras, tails of comets, and the Northern Lights.

Plasma sterilizes, killing germs, by a process called oxidation. The plasma produces a chemical reaction in which all microorganisms are deactivated (Fig. 5.10). The high heat turns the molecules of the hydrogen peroxide into free radicals, which are highly unstable. In their search for returning to a stable state, they latch onto the microorganisms in the load, thus effectively destroying the components of their cells, such as enzymes, nucleic acids, and DNA.

The process starts when liquid hydrogen peroxide is inserted into the sterilizer. The liquid is heated up in a vaporizer to turn it into gas. Once that has been accomplished, the hydrogen peroxide gas is heated to an even higher temperature, at which point it turns into plasma. Then the plasma is dispersed inside the sterilizer chamber to oxidize all microorganisms on the load. Most hydrogen peroxide plasma sterilization cycles run for less than an hour, with the average cycle running 35–45 min, depending on the size of the sterilizer, and size and contents of the load (Table 5.8).

Advanced Sterilization Products (Johnson & Johnson) have designed and commercialized Sterrad-100S sterilizers since the 1990s, and their use has already been introduced in the clinical field. The sterilization process is based on the use of a combination of hydrogen peroxide

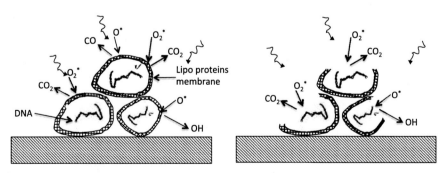

**FIG. 5.10** Killing microorganisms in a plasma sterilization process.

**TABLE 5.8** Advantages and Disadvantages of Cold Plasma Sterilization

| Advantages | Disadvantages |
| --- | --- |
| No chemical residues | Inability to sterilize: liquids, powders, and strong absorbers |
| Safety of handling | Requires specific synthetic packaging of the load |
| Safety for the environment | Sterilization chamber is relatively smaller than that of an EtO sterilizer |
| Short aeration time | |

($H_2O_2$) and low temperature gas plasma to sterilize medical devices and instruments quickly, efficiently, and without any toxic residues. $H_2O_2$, at low concentration levels, is a common disinfecting agent sold in pharmacies. At higher concentration levels, it is used as a sterilant in many industries. Hydrogen peroxide plasma has the serious advantage of safety, both for the environment (including the sterilizer operator) and the contents of the load. Moreover, $H_2O_2$ does not produce toxic fumes; there are no long aeration/degassing times in the cycle. At the end of the cycle, the plasma is "cracked" into the nontoxic by-products of water and oxygen, which safely evaporate into the air.

Although $H_2O_2$ may induce surface oxidation of some medical-grade elastomers, numerous research studies have reported no significant changes in the structural properties of the polymeric materials. Cold plasma sterilization process is not applicable for liquids and powders, cotton and paper, instruments that do not resist vacuum conditions, or instruments with an inner long and narrow lumen. It is, however, particularly suitable for the sterilization of heat- and moisture-sensitive devices and materials (e.g., polymeric materials), electrocautery instruments, Doppler, laser probes, defibrillator paddles, thermometers, ophthalmic lenses, laryngoscopes and their blades, fiber optic light cables, and endoscopes, such as rigid and flexible endoscopes.

## 5.1.5 Ozone

An innovative method of low-temperature sterilization is based on the use of ozone (i.e., $O_3$) as a sterilizing agent. Ozone, also known as trioxygen, is an inorganic molecule; it is a pale blue gas, with an unpleasant, distinctive smell similar to chlorine. It is well known as an antimicrobial agent capable of reacting with organic compounds (such as proteins, bacteria, fungi, etc.) by oxidizing the double bonds of the carbon atoms in their structure.

In the ozone sterilizer, ozone is produced inside the sterilizer from medical grade oxygen, which is commonly available in hospitals, by applying electrical energy combining $O_2$ with O to form $O_3$. Actually, ozone can be quite dangerous; in fact, it is toxic, corrosive, and flammable, but because the ozone is produced and broken down within the sterilizer, chances of exposure to it are quite minimal. The ozone sterilization process (Fig. 5.11; Table 5.9) uses two identical half-cycles. After the chamber is loaded with instruments, the door is closed, and the cycle begins. First, a vacuum is created within the chamber, followed by a humidification phase. Ozone is then injected into the chamber and the sterilization process begins. After the half-cycle is reached, the previous steps are

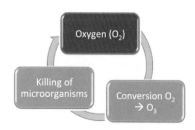

**FIG. 5.11** Scheme of the ozone sterilization process: starting from oxygen, it is converted by electrical discharge into ozone, able to kill the microorganisms. After that, by means of a catalyst, ozone can be transformed into oxygen.

**TABLE 5.9**　Advantages and Disadvantages of Ozone Sterilization

| Advantages | Disadvantages |
|---|---|
| Needs only medical grade oxygen, which is not a dangerous gas to handle or transport | Ozone itself is a toxic and flammable gas |
| Does not leave toxic fumes or residue that must be aerated | Cycle time is longer than hydrogen peroxide plasma sterilization |
| Even very tiny amounts of ozone could be detected from its pungent smell | |

repeated. A final ventilation phase is used to remove ozone from the chamber and the packaging within it. During the sterilization cycle, a catalytic agent is used to bring ozone back to a state of oxygen that can be safely expelled into the air. There are no toxic/hazardous residues or waste products associated with the process, and no toxic gas is vented into the environment.

Ozone sterilization is well suited for sterilizing heat- and moisture-sensitive medical devices, such as endoscopes, that cannot withstand the high heat and humidity of standard steam autoclaving. In particular, many devices containing the following materials can be sterilized with ozone: rigid polyvinyl chloride (PVC), nylon, polypropylene (PP), polytetrafluoroethylene (PTFE), silicone, polymethylmethacrylate (PMMA, e.g., Plexiglas), stainless steel, low-density polyethylene (LDPE), high-density polyethylene (HDPE), Pyrex glass, and anodized aluminum. Items that should not be used in this process include those containing: natural rubber, polyurethanes, textile fabrics, copper, brass, bronze, zinc, and nickel. Moreover, items that cannot withstand a vacuum, glass or plastic ampoules, and liquids are not recommended.

Approved by the FDA in 2003 as a new sterilization process for low-temperature sterilization, TSO3 (Québec, QC, Canada) developed a commercially available ozone sterilizer; in particular, some Canadian hospitals have been piloted to introduce this sterilizer into the clinical setting.

## 5.1.6 Other Sterilization Methods

### 5.1.6.1 Filter Sterilization

The most commonly used mechanical method of sterilization is filtration (Table 5.10). During filtration, liquids or gases are pressed through a filter, which, depending on its pore size, retains or adsorbs microbes, thereby the filtrate becomes sterile. Vacuum that is created in the receiving flask helps gravity pull the liquid through the filter. As fluid passes through the filter, organisms are trapped in the pores of the filtering material, as reported in Fig. 5.12.

Earlier Seitz-type asbestos or different glass filters were commonly used for the filtration of microorganisms. The pore diameter of filters should be chosen carefully, so that bacteria and other cellular components cannot penetrate (Table 5.11). The modern membrane filters are usually composed of high tensile-strength polymers (e.g., cellulose acetate, cellulose nitrate, or polysulfone); in particular, to remove bacteria, membrane filters with pore size of 0.22 μm are the best choice. Their operation is based partly on the adsorption of microbes and partly on a

**TABLE 5.10**　Advantages and Disadvantages of Filter Sterilization

| Advantages | Disadvantages |
|---|---|
| Except for those with the smallest pore sizes, membrane filters are relatively inexpensive, do not clog easily, and can filter large volumes of fluid reasonably rapidly | Many filters allow viruses and some mycoplasmas to pass through |
| Filters can be autoclaved or purchased already sterilized | Filters may absorb relatively large amounts of the filtrate and may introduce metallic ions into the filtrate |

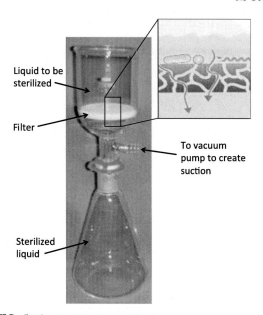

Liquid to be sterilized

Filter

To vacuum pump to create suction

Sterilized liquid

**FIG. 5.12** Sterilization by filtration.

mechanical sieve effect. The pure sieve-based filters can be beneficial because they do not change the composition of the filtered solution.

Membrane filters are used to sterilize heat-sensitive materials such as media, special nutrients that might be added to media, enzymes, vaccines, and pharmaceutical products such as drugs, sera, and vitamins. They are also used to sterilize liquids such as beverages, intravenous solutions, and bacteriological media. Some operating theaters and rooms occupied by burn patients receive filtered air to lower the number of airborne microbes.

Some filters can be attached to syringes so that materials can be forced through them relatively quickly. Filtration can also be used instead of pasteurization in the manufacture of beer. When using filters to sterilize materials, it is important to select a filter pore size that will prevent any infectious agent from passing into the product.

## 5.1.7 Procedures of Disinfection

Any process aimed at destroying or removing the infectious capability of pathogenic microbes that generally occur on inanimate objects is called disinfection. The chemicals used for disinfection can be classified according to their chemical structure and their mode of action.

Among the alcohols, ethanol and isopropanol are widely used as disinfectants. A 50%–70% aqueous solution has excellent antiseptic properties. The action mechanism of alcohols depends on the applied concentration. Due to the solubility of lipids in 50%–95% ethanol solutions, biological membranes are disintegrated. Alcohols pass through the cell membrane with altered permeability, denature the proteins inside the cell, and have a dehydration effect as well. Absolute alcohol (100% ethanol) provides the best dehydration effect but does not coagulate the intracellular proteins. A 70% dilution of alcohol is the most effective way to kill the vegetative forms of bacteria and fungi but is less effective against spores and lipid-enveloped viruses.

**TABLE 5.11** Examples of Pore Sizes of Membrane Filters and Particles That Pass Through Them

| Pore Size in (µm) | Particles That Pass Through Them |
|---|---|
| 10 | Erythrocytes, yeast cells, bacteria, viruses, molecules |
| 5 | Yeast cells, bacteria, viruses, molecules |
| 3 | Some yeast cells, bacteria, viruses, molecules |
| 1.2 | Most bacteria, viruses, molecules |
| 0.45 | A few bacteria, viruses, molecules |
| 0.22 | Viruses, molecules |
| 0.10 | Medium-sized to small viruses, molecules |
| 0.05 | Small viruses, molecules |
| 0.025 | Only the very smallest viruses, molecules |
| Ultrafilter | Small molecules |

A phenol called carbolic acid was first used as a disinfectant by Lister. Phenol denatures proteins and irreversibly inactivates the membrane-bound oxidases and dehydrogenases. Due to the unfavorable physical, chemical, and toxicological properties, phenol is no longer used. However, substituted (alkylated, halogenated) derivatives are often used in combination with surfactants or alcohols (e.g., cresol, hexachlorophene, chlorhexidine).

The halogens (F, Cl, I, Br) and their derivatives are very effective disinfectants and antiseptic agents; mainly their nonionic forms have antimicrobial activity. Chlorine gas is used almost exclusively for the disinfection of drinking water or other waters. In addition, different compounds (e.g., chloride of lime, chloramine-B, sodium dichloroisocyanurate) are among the most widely used disinfectant agents. Sodium hypochlorite ("household bleach" is a mixture of 8% NaClO and 1% NaOH) is one of the oldest high bleaching and deodorizing disinfectants. The basis of the effect of chlorine and its derivatives is that, during decomposition in aqueous solution, a strong oxidant, nascent (atomic state) oxygen ("O"), is released. Nascent oxygen is very reactive and suitable to destroy bacteria, fungi, and their spores as well as viruses.

Iodine is also a widely used disinfectant and antiseptic agent. There are two known preparations: tincture of iodine (alcoholic potassium iodide solution containing 5% iodine) and iodophors (aqueous solutions of iodine complexes with different natural detergents). It is applied in alcoholic solution to disinfect skin or in aquatic solution for washing prior surgery.

Aldehydes, such as formaldehyde and glutaraldehyde, are broad-spectrum disinfectants. They are used for decontamination of equipment and devices. Formalin is the 34%–38% aqueous solution of formaldehyde gas. Its effect is based on the alkylation of proteins.

Heavy metals such as mercury, arsenic, silver, gold, copper, zinc, and lead, and a variety of their compounds are highly efficient disinfectants, but they are too damaging to living tissues to be applied. They can be used as disinfectants at very low concentrations. Inside the cell, they bind to the sulfhydryl groups of proteins. Primarily, organic and inorganic salts of silver- and mercury-containing products are commercially available, which have bactericidal, fungicidal, and virucidal effect.

Detergents or surfactants are amphiphilic organic molecules, which have a hydrophilic "head" and a long hydrophobic "tail." Detergents can be nonionic, anionic, or cationic according to the charge of the carbon chain. Nonionic surfactants have no significant biocidal effect, and anionic detergents are only of limited use because of their poor efficiency. The latter group includes soaps, which are long-chain carboxylic acids (fatty acids) of sodium or potassium salts. They are not disinfectants on their own but are efficient cleaning agents due to their lipid-solubilizing effect. Cationic detergents, such as quaternary ammonium salts, are the best disinfectants.

### 5.1.7.1 Determination of the Microbiological Efficacy of Disinfectants

Microbiological laboratories, especially those involved in epidemiological, medical, or industrial processes (pharmaceutical companies, food industry), use disinfectant solutions for preventive, continuous, or terminal disinfection. In the case of an actual epidemiological event (epidemic, accumulation of infection), the effectiveness of these disinfectants is systematically inspected. The principle of efficacy testing is that the relevant disinfectant is incubated with a test bacterium for a defined time interval, and the treated bacteria are subsequently spread onto the surface of a suitable nutrient medium. Following the incubation period, based on the growth of the test microbe, conclusions can be drawn whether the disinfectant within the exposure time interval effectively killed the test microbe.

## 5.1.8 Control of The Sterilization Efficacy

To monitor the efficacy of sterilization equipment, several methods are available: instrumental monitoring, the use of chemical indicators, and biological monitoring with spore preparations.

By instrumental monitoring, the vapor pressure, temperature, and exposure time can be monitored inside the sterilizing equipment. In general, color changes of chemical indicators on packaging show that temperature and duration are sufficient for effective sterilization. The original Browne-type sterilization control glass tubes contain a red indicator solution, which turns yellow during inadequate heat treatment, and turns green in the case of sufficient sterilization. Another chemical indicator, the indicator tape, should be stuck onto the outer surfaces of the load (e.g., glassware or aluminum foil packaging) to be sterilized. The strips change color or a marking appears (e.g., "OK" or "STERILE"), which indicates that sterilization has taken place. The use of chemical indicators is recommended only in equipment previously qualified by biological tests.

The use of biological indicators is the most reliable method for the certification and periodic monitoring of sterilizing equipment. For this purpose, standardized bacterial spore products, the so-called "spore preparations" are required. Test organisms (e.g., spores of *Geobacillus stearothermophilus*) are usually more resistant to heat sterilization than most microorganisms. If the efficiency of sterilization is inadequate, test microbes remain viable (spore germination is maintained). For the microbiological control of steam sterilizers, standardized bioindicators made of the spores of the type strain of *G. stearothermophilus* ATCC 7953, containing $1.2 \times 10^6$ CFU equivalent endospores are used. This amount of spores can be destroyed by an efficient autoclaving cycle. The destruction or survival of microorganisms can be detected by culturing in broth. The spore preparations should be placed, equally distributed, with the load within the chamber of the autoclave at the characteristic technical points.

### 5.1.8.1 Sterility Assurance Level

The sterilization process can be evaluated, taking into account the ability to significantly reduce microbial contamination levels on a product/device/material, by means of high process control to meet regulatory requirements for sterility. The word sterile is defined as "free from viable microorganisms." Sterilization is defined as a "validated process used to render product free from viable microorganisms" (ISO/TS 11139, 2006). Sterilization processes are based on microbial inactivation, which is exponential in nature, in most cases, and follows first order kinetics. Therefore the sterility of a product/device/material has to be expressed in terms of probability. Although the probability can be reduced to a very low number, it can never be reduced to zero. Hence, the probability of a nonsterile unit associated with inactivation of a microbial population is quantified by the term Sterility Assurance Level (SAL).

Logarithmic microbial reduction to achieve various SALs is reported in Fig. 5.13. The SAL required for regulatory reasons is $10^{-3}$ or $10^{-6}$, the probability of 1 in 1000 units as nonsterile or 1 in 1,000,000 units as nonsterile, respectively. The SAL of $10^{-6}$ is a lesser number but provides a greater assurance of sterility. Hence, the SAL determination requires the counting of the number of viable microorganisms on the product (i.e., usually the number of replicates is between 10 and 30 samples), just before the sterilization process. By using conventional microbiological techniques, the number of microorganisms can be determined. After that step, packed product samples are exposed to a fraction of the sterilization process dose. Then, samples are tested for sterility, and the results are graphically reported to estimate the exposure time required to achieve the adequate SAL (i.e., $10^{-6}$). It is

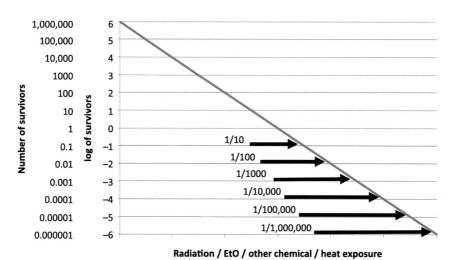

**FIG. 5.13**   Sterility assurance levels: logarithmic microbial reduction.

important to highlight that when there is not more than one microorganism per product sample, there is not 0.01 of an organism on each sample, but a probability of 1 in 100 that any given product unit is no-sterile.

## 5.2  DEGRADATION

### 5.2.1  Polymeric Materials (Plastics)

Plastics are susceptible to physical and chemical attack with consequent degradation phenomena, which are mainly due to modification of the chemical structure.

Degradation is not only due to the action of chemically aggressive substances, such as strong acids or solvents; in fact, water can cause hydrolysis and oxygen oxidation.

Heat, mechanical stresses and radiation can also cause degradation. During processing, the material is subjected to the first two phenomena and, for this reason, it is necessary to incorporate stabilizers and antioxidants during the processing to maintain the structural properties of the material. These additives also help delay subsequent degradation phenomena.

As already mentioned, the type of morphology influences susceptibility to degradation; crystalline plastics have better resistance to degradation than amorphous ones.

The following sections offer indications on the general causes of polymer degradation.

#### 5.2.1.1  Environmental Aging

It occurs due to the combined effect of moisture absorption and exposure to UV or cosmic radiation in general.

Moisture absorption (water) can have a plasticizing effect, with increased flexibility, but the elimination of water causes embrittlement.

UV and radiation in general cause breakdown of chemical bonds in the polymer chains. Loss of color and transparency may also take place.

#### 5.2.1.2  Oxidation

Many important commercial polymers are susceptible to oxidation when exposed to atmospheric oxygen, both during processing and during storage and use. Oxidation is a chain reaction (Fig. 5.14) that begins and propagates through the formation of free radicals that combine with molecular oxygen to form a peroxide radical

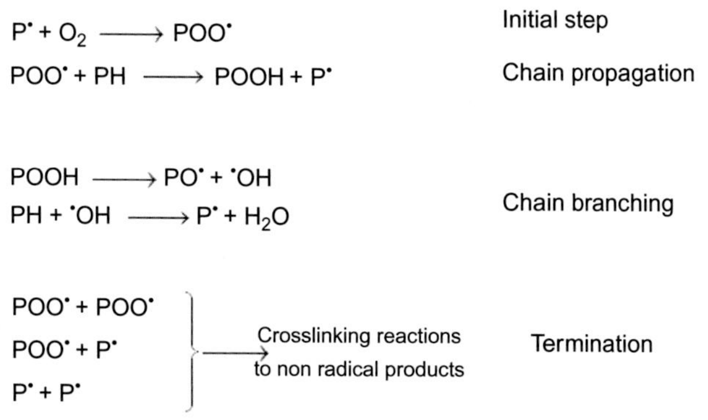

$$P^\bullet + O_2 \longrightarrow POO^\bullet \qquad \text{Initial step}$$

$$POO^\bullet + PH \longrightarrow POOH + P^\bullet \qquad \text{Chain propagation}$$

$$POOH \longrightarrow PO^\bullet + {}^\bullet OH$$
$$PH + {}^\bullet OH \longrightarrow P^\bullet + H_2O \qquad \text{Chain branching}$$

$$\left.\begin{array}{l} POO^\bullet + POO^\bullet \\ POO^\bullet + P^\bullet \\ P^\bullet + P^\bullet \end{array}\right\} \xrightarrow[\text{to non radical products}]{\text{Crosslinking reactions}} \quad \text{Termination}$$

**FIG. 5.14** Oxidation reaction steps: a free radical in polymer reacts with oxygen to produce a polymer peroxy radical POO$^\bullet$ (initial step). This reacts with a polymer molecule to generate polymer hydroperoxide POOH and a new polymer alkyl radical P$^\bullet$ (chain propagation). Polymer oxy radicals PO$^\bullet$ and hydroxy radicals HO$^\bullet$ are formed by photolysis (chain branching)— also at this stage chain scission reactions can occur. Cross-linking results from the reaction of different free radicals with each other (termination step). *From Rabek, J.F., 1990, Photostabilization of Polymers: Principles and Application. Elsevier Sci Publ Ltd, England.*

(POO$^\bullet$). The consequence on polymers is the modification of the molecular weight due to chain cleavages or cross-links, which cause deterioration of physical properties such as softening or embrittlement. The higher the degree of unsaturation (i.e., the presence of double bonds C=C), the more susceptible the polymer is to oxidation with consequent degradation.

Among saturated polymers, those that contain CH$_3$ groups (or better, tertiary carbon atoms) such as *PP* are more easily oxidized. In PP, oxidation causes cleavage of the main chain; in polymers such as *HDPE*, cross-linking predominates. In *PVC* there is a breakdown of bonds with the release of HCl (hydrochloric acid). *Polyesters, polyurethanes, polyamides,* and *polycarbonates* generally turn yellow by oxidation yet their physical properties do not change much. However, at high temperatures and for prolonged exposure to UV radiation, physical properties are deteriorated.

Certain rubbers, especially those that contain double bonds, such as natural rubber or styrene-butadiene copolymers, are very sensitive to oxidation. When the rubber is under stress, cracks appearing perpendicular to the direction of the stress can lead to breakage of the material. Light,

especially UV radiation, increases the action of oxygen, which causes deterioration of mechanical strength.

To prevent or avoid oxidation, antioxidant agents are added during processing of the polymers. Antioxidants are compounds that inhibit or delay oxidation and belong to various chemical classes.

### 5.2.1.3 Photoinduced Degradation

Photodegradation is the most important aspect of the climate aging of polymers, which generally contain groups capable of absorbing UV radiation from sunlight, as in the case of *cis*-1-4-polyisoprene, that is to say, of natural rubber (Fig. 5.15).

The formed macroradical can produce, following the reaction with the oxygen in the air, a cross-linking with consequent hardening, embrittlement, and loss of elasticity of the rubber.

In many cases, photodegradation, which is often combined with oxidation, causes a decrease in the molecular mass and loss of mechanical properties, but because the polymers are generally not very permeable to UV rays, photochemical degradation is generally limited to the surface.

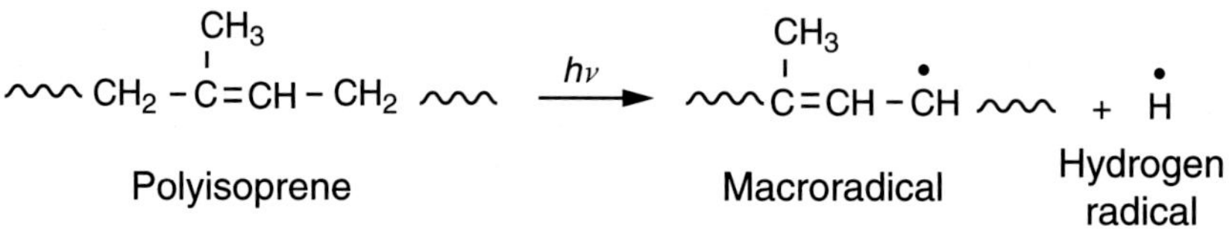

**FIG. 5.15**  Photodegradation of polyisoprene.

A large number of reactions that occur in thermal and photochemical degradation of polymers require the intervention of reactive mediators (i.e., free radicals). The stabilizers (e.g., antioxidants, UV stabilizers) allow both to prevent the formation of free radicals and to destroy them, thus greatly increasing the longevity of the polymers. The use of pigments (e.g., carbon black), which make the polymers opaque, considerably increases the durability of organic materials, for example, that of automobile tires.

### 5.2.1.4 Pyrolysis and High Temperature Degradation Mechanism

Thermal resistance of polymers is very low compared to that of other categories of materials. At very high temperatures, the polymers undergo pyrolysis.

Two mechanisms must be considered:

- *statistical degradation*, which derives from a homolytic rupture of chain links (Fig. 5.16). It generally causes a decrease in the molecular mass and makes the material brittle;
- *depolymerization*, starting from a radical formed by the homolytic rupture of a less stable bond and often located at one end of the chain (Fig. 5.17);

Pyrolysis of polymers at high temperature is generally a combination of the previous two mechanisms, and the proportion of monomer formed in volatile products ranges from zero to almost 100%, depending on the chemical structure. The formation of oligomers (e.g., dimers, trimers) and, in some cases, of nonvolatile graphitic structures is also observed. Pyrolysis can occur in both solid and melt state during molding.

The degradation of the polymer at high temperature is accelerated by the presence of oxygen, which in some cases causes its combustion.

### 5.2.1.5 Enzymatic and Bacterial Attack

In case of polymers in contact with the human body, biological fluids (in particular, blood, serum, and plasma), or food and natural extracts, degradations can occur due to enzymes (e.g., oxidation, hydrolysis).

The aggression of the polymeric material can also occur due to bacteria and microorganisms. The occurrence of this phenomenon constitutes a danger not only for the chemical structure of the material but also for the surrounding environment. Let us take in consideration, for example, food packaging or containers for collecting and storing biological fluids.

Adding specific substances of pharmacological or antibacterial activity, choosing polymeric structures resistant to these phenomena or, finally, using sterilization, can avoid the enzymatic and bacterial attack. In case of sterilization, as previously described, it must be ensured that the sterilization procedure does not damage the polymeric material.

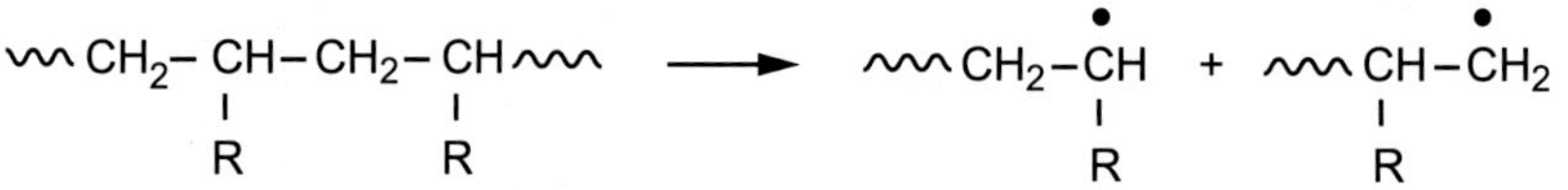

**FIG. 5.16**  Mechanism of statistical degradation.

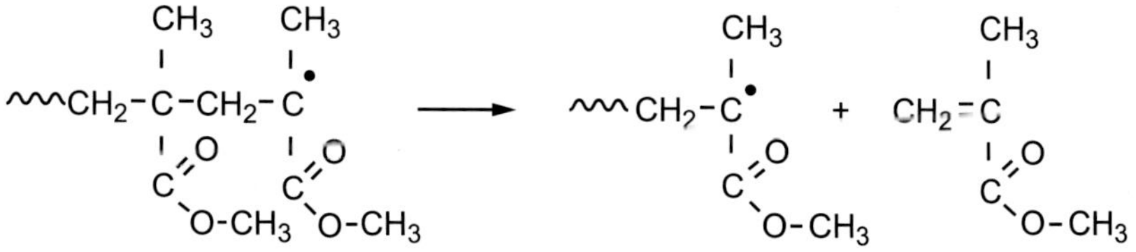

**FIG. 5.17** Mechanism of depolymerization.

### 5.2.1.6 *Chemical Attack*

This type of behavior depends on the chemical nature and the configuration of the monomeric units. The chemical attack is more often internal to the polymer and involves softening, swelling, and loss of mechanical strength. A general rule to predict chemical resistance is that the similar attracts whereas the nonsimilar rejects. Thus a polymer is more soluble in a solvent of similar chemical structure. Polymers having polar groups such as hydroxyl (OH) and carboxyl (COOH) generally swell or dissolve in polar liquids such as water or alcohols but are resistant to nonpolar solvents such as petrol, benzene, and carbon tetrachloride.

Polymers with nonpolar groups such as methyl ($CH_3$) and phenyl ($C_6H_5$) resist polar solvents but swell or dissolve in nonpolar ones. Furthermore, polymers with an aromatic character are soluble in aromatic solvents whereas those with an aliphatic character dissolve in the aliphatic ones. Polymers can be attacked by solvents in many ways: dissolution, swelling, permeability, environmental stress cracking (ESC[1]), or crazing. The dissolution process takes place in two stages: the diffusion of the solvent in the polymer causes swelling, followed by dissolution.

By increasing the molecular weight of the polymer, its solubility or tendency to swell decreases. In general, high molecular weight polymers give very viscous solutions. Molecular

[1] ESC is a degradation phenomenon that occurs in the environmental conditions of use and in the presence of mechanical stresses. It consists in the formation of deep cracks, which eventually lead to breakage in operation of the piece/device.

symmetry also influences solvent resistance. More crystalline polymers show greater chemical resistance than less crystalline ones with a similar structure; this is due to the greater packing of the molecular chains, which makes penetration of the solvent and other substances within the material more difficult. The degree of cross-linking greatly affects the solubility, so much so that even a low degree of cross-linking is sufficient to render the material insoluble. The influence of the ramifications is less, even if it tends to decrease the dissolution rate. The behavior of a strongly branched polymer is similar to that of a weakly cross-linked polymer, and it is difficult to distinguish between the two types in terms of solubility. The cross-linked polymers do not dissolve but only swell when they have affinity for a solvent.

Polymers generally have better resistance to acids and alkalis than metals; however, they may contain vulnerable polar groups that can make them susceptible to attack. As a consequence, alkalis, especially in high concentration and at high temperatures, can saponify the ester groups of the polymers containing them (e.g., cellulose acetate, polyesters, polyvinyl acetate). Nonoxidizing acids can hydrolyze (by breaking covalent bonds) these materials in a similar way. Polyamides (i.e., nylons) and polyurethanes, which have —NH—CO and NH—COO— bonds, respectively, are susceptible to attack by strong acids and alkalis. The chemical resistance in these cases depends on the length of the chains of methylene groups between these bonds. All polyolefins, PVC, ABS copolymers, fluorinated, polystyrene, and the like have excellent resistance to acids and bases.

### 5.2.1.7 Mechanical Degradation

Stretching, grinding, milling, and any type of polymer shearing process produces free radicals as a result of main chain fracture. Upon warming, these radicals attack the polymer matrix and lead to further scission reaction through radical rearrangement reactions. In the melt, it is difficult to separate the combined degradative effects of force, time, and temperature.

The mechanical action applied to a polymeric material can lead to the homolytic cleavage of the bonds of the macromolecules of which it is made, unlike what happens in low molecular weight substances. In fact, the small molecules can relax the stress exerted on them, by sliding until a new equilibrium position is reached. In the case of polymers, the flow of macromolecules is hindered by entanglements that can become rigid constraints to their movement, causing them to break ("mechanodegradation").

The mechanodegradation becomes more probable with increasing rate of application of the mechanical action and with increasing the molecular weight of the polymer. The homolytic splitting always concerns the bonds that make up the macromolecular chain with formation of terminal macroradicals and the probability of maximum cleavage toward the center of the macromolecules.

### 5.2.1.8 Effects of Sterilization

The limiting factor in selecting the sterilization method is often the resistance of the material to the sterilizing agent, whether it is heat, radiation, or reactive gas (see Section 5.1). Although metal or ceramic materials can tolerate autoclaving, most polymeric materials cannot sustain the high temperatures typical of sterilization in hot air and steam. Moreover, some polymeric materials can undergo to degradation with a reduction in mechanical properties as a result of radiation; other polymers, if sterilized with chemical agents, retain EtO or other reactive substances (e.g., formaldehyde) in unacceptable quantities, due to their cytotoxicity.

Chemicals such as EtO may react with some groups in the material structure or may be trapped inside it (e.g., porous device), with the risk of being released from the material in the body site. In addition, being necessarily aggressive substances, if they are released in significant quantities (i.e., related to their toxicity), they can cause toxic reactions in organic tissues.

Steam sterilization in an autoclave is the most used procedure for the sterilization of surgical instruments (e.g., metals). Heat and vapor can drastically change the properties of many polymeric materials, especially when they are treated at temperatures above their glass transition temperature. Moreover, many polymers, if exposed to steam, can degrade due to hydrolysis.

### 5.2.1.9 Effect of High-Energy Radiations

High-energy radiation leads to degradation and/or to cross-linking, depending on the chemical structure of the polymeric material. If cross-linking effect prevails, there is an increase in the molecular weight related to the applied dose, with the formation of a three-dimensional network; on the other hand, if the degradation effect occurs, an increase in the radiation dose causes random breaks in the bonds on the macromolecular chains and, consequently, a decrease in the molecular weight. Therefore mechanical, physical, and chemical properties of the polymer are largely influenced by irradiation sterilization and by the predominance of cross-linking or the degradation effect on the material structure.

Polymers, which tend to cross-link (e.g., polyethylene, polyamides, polyurethanes, and polyesters), generally show an increase in some mechanical properties (except impact strength and elongation at break), which grow with increasing the radiation dose at which they are subjected to.

Moreover, the irradiation of polymers may promote the development of gas, the formation of double bonds or free radicals with consequent discoloration, or changes in mechanical properties.

The effect of the sterilization strongly depends on the external environment and, in particular, on the presence of species able to react with the free radicals generated by sterilization in the material.

In the presence of oxygen or low molecular weight molecules, oxidations and reactions take place (as seen in Fig. 5.14), leading to the formation of macromolecules shorter than the original ones, with a molecular weight decrease; on the other hand, in inert atmosphere, radicals tend to bond with each other, hence cross-linking the structure. Therefore the extent of cross-linking is a function of the intensity of the radiation and of the environment in which it takes place. The effect can also be different in the amorphous and crystalline domains of polymeric structure. In fact, in the crystalline domains, the strong compaction and the low mobility of the molecules causes the radicals produced by the breaking of the covalent bonds to remain very close to each other, with a high probability that bonds will be reformed. In the amorphous domains, however, this event does not happen, and oxidation and cross-linking can take place with substantial variations of the molecular weight.

### 5.2.1.10 Degradation in the Physiological Environment

The human body and its fluids represent a very aggressive environment, in some cases comparable to the marine environment. Considering that the mechanical stresses to which tissues and components of the human body are subjected are considerable and repeated cyclically over time, it is not surprising if high structural and mechanical properties are requested to materials used to repair and replace parts of the human body.

It is necessary to distinguish between two main objectives, that is, if a high stability in time is required for the materials with which a device is manufactured, or if the objective is that the device be degraded and reabsorbed once it has performed its task in contact or within the human body. Therefore it is necessary to select "biostable" biomaterials in the first case and "bioabsorbable" biomaterials in the second one.

As already mentioned, degradation mechanisms can happen in the physiological environment: ESC, enzymatic attack, hydrolysis, and oxidation.

In this chapter, the main degradation mechanisms are presented and discussed, whereas some specific information will be provided in Section 6.3.

The susceptibility to degradation also depends on the treatments experienced before the implant, such as sterilization and processing, which, for example, can cause the formation of free radicals that, over time and in the presence of agents typical of the physiological environment, may evolve to produce massive degradation of the device and loss of the features essential to its functionality.

### 5.2.1.11 Environmental Stress Cracking

This phenomenon is defined as the generation of crazing and deep fissures (cracks) in response to particular aggressive environments and in the presence of residual stresses in the material (Fig. 5.18). In the human body, the aggressive environment is represented by the inflammatory and immune response to the foreign material implant (see Chapter 6), with the recruitment of phagocytic cells that not only surround the material but adhere to its surface, activating and releasing lithic enzymes, ions, and radicals, thus creating a strong acid environment (pH$=2$) at the points of adhesion to the surface. Especially susceptible materials of ESC are elastomers, specifically segmented polyurethanes (PTS), in which the soft segment component is proportionally high and containing ether-type bonds ($-O-$).

It is believed that the ESC is due to the synergistic effect of four factors:

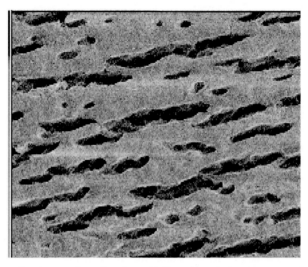

**FIG. 5.18**    SEM image of a poly-ether-urethane (hardness 80A) explanted after 2-years implantation at 400% strain in the animal model (rabbit) showing severe stress cracking. *From Ward, R., et al., 2006. J. Biomed. Mater. Res. 79A, 836–845.*

**(1)** surface oxidation with loss of soft ether segments, by free radicals, enzymes, and oxidizing substances produced by the activation of macrophages;

**(2)** presence of residual stresses in poly-ether-urethane (derived from processing and processing procedures) and localized stresses (generated by in vivo mechanical stress);

**(3)** presence of some proteins (e.g., 2-macroglobulin and ceruloplasmin) acting as catalysts of the degradative attack of the polyether segments;

**(4)** presence of ether bonds on the surface.

In some biomedical devices, oxidative degradation can be a contributing mechanism. Implanted polyether-urethane devices that contain metallic components have been subject to bulk oxidation, initiated by corrosion products catalyzed by the aqueous ionic environment of the implant site (Stokes and Cobian, 1982).

The ESC occurrence can be avoided by increasing the hardness of the elastomer and substituting the polyether soft segments with other types (e.g., polycarbonate). In addition, it is very advantageous to eliminate residual stresses with an annealing treatment (i.e., heat treatment under controlled conditions) and improve design and processing technologies.

### 5.2.1.12 Bulk and Surface Erosion in Bioabsorbable Polymers

Bioabsorbable polymers (or degradable in the human body) can undergo different degradation mechanisms.

Basically, these mechanisms can be divided into: *bulk erosion*, or mass degradation, where degradation begins within the material; *surface erosion*, or surface degradation, in which the degradation starts from the surface; and *bioresorption*, in which the material simply dissolves in the body fluids without degradation of molecular weight. Bulk and surface erosion are displayed in Fig. 5.19.

In general, polymers belonging to the group of poly-α-hydroxy acids, such as PLA and PGA, undergo mass degradation. In the case of PGA and PLA the biodegradation mechanism (due to hydrolysis of the ester bonds) leads to the release, respectively, of glycolic acid and lactic

Bulk degradation     Surface degradation

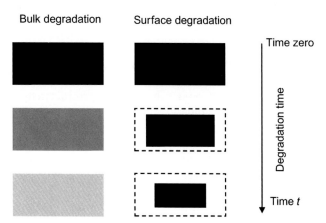

FIG. 5.19   Schematic representation of surface and bulk erosion. *Reproduced from Int. J. Pharm. 418 (2011) 28–41, with permission.*

acid, metabolites normally present in the human body, whose physiological path leads to elimination in the form of water and carbon dioxide.

However, the loss in weight accompanied by the release of acidic degradation products, in vivo, causes an intense inflammatory reaction (see Chapter 6). If the ability to eliminate these degradation products by adjacent tissues, due to limited vascularity or poor metabolic activity, is insufficient, the chemical composition of the products may create local disturbances, such as an increase in osmotic pressure or a decrease in pH, with fluid accumulation and increased inflammatory response.

In water, the molecular weight of the polymer begins to decrease after 1 day (PGA, PDLA) or after a few weeks (PLLA), but weight loss does not occur until the molecular chains are reduced to a size that allows them to be freely released outside the polymer matrix. This phenomenon, widely studied, accelerates the rate of degradation and reabsorption until it compromises the integrity of the polymer matrix.

### 5.2.1.13 Mineralization and Calcification

The failure of some implanted devices, especially in the cardiovascular system, is frequently caused by the formation of deposits of calcium phosphates or other calcium compounds. The process is known as *calcification* or *mineralization* and is the leading macroscopic cause of failure for most prosthetic heart valves and blood pumps, limiting the functional lifetime of the device by loss of elasticity.

Although the deposition of calcium mineral salts is a normal process in bone and tooth, the calcification of biomaterials must be avoided because it can interfere with the functionality of the implanted device. The calcification phenomenon of a biomaterial is therefore abnormal or pathological and, in the second case, it can be dystrophic or metastatic.

*Dystrophic calcification* consists of the deposition of calcium salts in damaged or suffering tissues of individuals with normal calcium metabolism. On the contrary, *metastatic calcification* consists of the deposition of calcium salts in previously intact tissues due to a disturbance of calcium metabolism (usually elevated blood levels of calcium).

The two processes can be synergistic; in the presence of abnormal calcium metabolism, the calcification associated with biomaterials or damaged tissues is increased.

Calcification associated with biomaterials can affect various prosthetic implants either in the circulatory system and in the connective tissue or in other sites (e.g., urinary system). For example, dystrophic calcification phenomena are due to the degeneration of bioprosthetic or homologous heart valves (of biological origin), as well as to mineralization of intrauterine contraceptive devices, prostheses of the urinary tract, and soft contact lenses.

Calcification can be associated with biomaterials of both biological and synthetic origin. The mature mineral phase of many calcification phenomena of biomaterials is a slightly crystalline calcium phosphate, known as *apatite* (with a structure similar to that of bone hydroxyapatite).

In general, the causes of mineralization of biomaterials include both factors related to the metabolism of the host organism and to the structure and chemical properties of the implanted device. Furthermore, the mineralization of the biomaterial is generally increased at sites of intense mechanical stress, such as the zones of cardiovascular devices subjected to flexural stresses.

Calcification can be enhanced in the presence of implant infections. Furthermore, calcification can occur at the surface of an implanted device (*extrinsic* calcification) where it is often associated with adherent tissues or cells, or within structural components of the implant (*intrinsic* calcification).

## 5.2.2 Metallic Materials

Corrosion may be defined as the deterioration of a metal as a result of chemical reactions between it and the surrounding environment. Both the type of metal and the environmental conditions determine the form and rate of deterioration.

All metals can corrode; some, like pure iron, corrode quickly. Stainless steel, however, which combines iron and other alloy elements, is slower to corrode and is therefore used more frequently. Moreover, small group of metals, called the noble metals, is much less reactive than others. They include copper, palladium, silver, platinum, and gold.

To ensure long and trouble-free operation in a medium, it is of utmost importance to have knowledge about corrosion and the effect it can have on the product in the operating environment.

### 5.2.2.1 *Corrosion Mechanism*

Corrosion may be broadly divided into two categories:

- wet corrosion: it takes place in presence of a liquid;
- dry corrosion: it occurs due to a corrosive gases.

In this chapter, only wet corrosion will be considered, as only this category is relevant in case of metallic biomaterials.

The body environment is very aggressive in terms of corrosion because it is not only aqueous but also contains chloride ions and proteins. A variety of chemical reactions occur when a metal is exposed to an aqueous environment (Fig. 5.20).

The electrolyte, which contains ions in solution, serves to complete the electric circuit

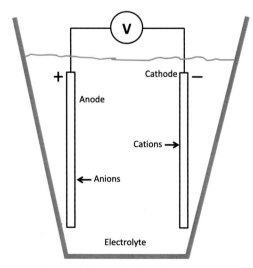

FIG. 5.20 Electrochemical cell.

(Fig. 5.20). In the human body, the required ions are plentiful in the body fluids. Anions are negative ions, which migrate toward the anode, and cations are positive ions, which migrate toward the cathode. In the corrosion process, an oxidation and a reduction process take place simultaneously and at equal rates. The general oxidation reaction of a metal (M) can be described in Eq. (5.1), leading to metal corrosion and release of metal ions into the surrounding environment:

$$M \rightarrow M^{+n} + ne \qquad (5.1)$$

The complementary reduction reaction can occur in different forms and takes place involving chemical species present in the surrounding environment:

– hydrogen evolution, as described in Eq. (5.2):

$$2H^+ + 2e \rightarrow H_2 \qquad (5.2)$$

– oxygen reduction in neutral or basic solutions, as described in Eq. (5.3):

$$O_2 + 2H_2O + 4e \rightarrow 4OH^- \qquad (5.3)$$

– oxygen reduction in acidic solutions, as described in Eq. (5.4):

$$O_2 + 4H^+ + 4e \rightarrow 2H_2O \qquad (5.4)$$

The tendency of metals to corrode is expressed most simply in the standard electrochemical series of Nernst potentials, shown in Table 5.12. The metal potentials are obtained in electrochemical measures in which one electrode is a standard hydrogen electrode formed by bubbling hydrogen through a layer of finely divided platinum black. The potential of this reference electrode is defined to be zero. Noble metals are those, which have a potential higher than that of a standard hydrogen electrode; a great majority of metals have lower potentials.

An example is represented by the corrosion of iron in presence of an aqueous medium (Fig. 5.21) that dissolves oxygen. The oxidation (5.5a) and the reduction (5.5b) processes of iron corrosion are represented in the following equations:

**TABLE 5.12**  Standard Potentials or Electromotive Force (EMF) Series for Metals

| Metal | Metal Ion | Electrode Potential (V) Versus Hydrogen Electrode at 25°C | |
|---|---|---|---|
| Gold | $Au^+$ | +1.68 | |
| Platinum | $Pt^{+2}$ | +1.2 | |
| Silver | $Ag^+$ | +0.80 | Noble or cathodic |
| Mercury | $Hg_2^{+2}$ | +0.79 | |
| Copper | $Cu^+$ | +0.16 | |
| *Hydrogen* | *$2H^+$* | *0.00* | |
| Lead | $Pb^{+2}$ | −0.13 | |
| Nickel | $Ni^{+2}$ | −0.23 | Active or anodic |
| Cobalt | $Co^{+2}$ | −0.28 | |
| Iron | $Fe^{+2}$ | −0.44 | |
| Chromium | $Cr^{+2}$ | −0.56 | |
| Zinc | $Zn^{+2}$ | −0.76 | |
| Titanium | $Ti^{+3}$ | −1.63 | |
| Aluminum | $Al^{+3}$ | −1.66 | |
| Magnesium | $Mg^{+2}$ | −2.36 | |
| Sodium | $Na^+$ | −2.71 | |
| Potassium | $K^+$ | −2.92 | |

Under ideal conditions, the metal potential can be measured and referenced against the hydrogel electrode. The potential for the hydrogen electrode is defined as zero. When two metals with different EMFs are connected, a possible current flow can occur. Immersing the two metals in an ionic solution leads to a completion of the electrical loop and corrosion ensues.

$$2Fe \rightarrow 2Fe^{+2} + 4e \qquad (5.5a)$$

$$2Fe^{2+} + 4OH^- \rightarrow 2Fe(OH)_2 \qquad (5.5b)$$

Since ferrous hydroxide $[Fe(OH)_2]$ is unstable (5.5b), this by-product is further oxidized to

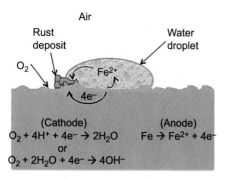

**FIG. 5.21** Example of corrosion of iron in presence of oxygen and water. Rust is the by-product of the corrosion phenomenon.

form ferric hydroxide [$Fe(OH)_3$], known as rust (5.6):

$$2Fe(OH)_2 + H_2O + \frac{1}{2}O_2 \rightarrow 2Fe(OH)_3 \qquad (5.6)$$

Each process of oxidation or reduction is characterized by a potential in the thermodynamic scale of the potentials (Table 5.12): the higher the potential, the less the metal is corrodible, from a thermodynamic point of view. If the potential of the oxidation process is lower than that of the complementary reduction process, the corrosion process can take place (i.e., the difference between the two potentials is called available motor work), otherwise the metal, in a specific environment, is in conditions of thermodynamic immunity. The potential of the oxygen reduction process (corresponding to the aggressiveness of the human body) is quite high; in fact, only some noble metals (i.e., gold) have a high oxidation potential and therefore are immune to corrosion in the human body. All other metals, including metals considered to be highly resistant to corrosion (e.g., stainless steels and titanium), have oxidation potentials lower than that of oxygen reduction in the human body and are therefore thermodynamically susceptible to corrosion.

In biomedical applications, metallic corrosion is an important issue; in fact, the in vivo environment is complex and the implants are exposed to a multitude of chemical moieties. The severity and the rate of the corrosion phenomenon depend on the chemical composition of the environment as well as other variables, for example, temperature, pH value, and stress level.

## PASSIVITY

Metallic passivity is an important and very complex phenomenon; its mechanism is out of the scope of this book; hence, it is not deeply explained in this section.

Usually, as the oxidizing power of a solution is increased, the rate of corrosion increases exponentially; this behavior is called active zone of the corrosion phenomenon (Fig. 5.22A).

However, in some metals, such as iron, titanium, chromium, and their alloys, an increase in the oxidizing power beyond a certain critical point results in a decrease in corrosion rates (Fig. 5.22B). This decrease in corrosion rate can be important, and the metal behaves as inert noble metal (very low corrosion rate). Hence, the active zone is followed by a passive region where increasing the oxidizing power of the solution has no effect on the corrosion rate. In fact, a hard and stable oxide layer is formed on the surface of the metal, acting as a passive layer, able to protect the metal from corrosion. However, further increase in the oxidizing power beyond the passive zone results in an increase in a higher corrosion rate; this zone is called the transpassive region.

### 5.2.2.2 Types of Corrosion

There are many different reasons for metal corrosion. Some can be avoided by adding alloys to a pure metal. Others can be prevented by a careful combination of metals or management of the metal's environment. Some of the most common types of corrosion are briefly described in Table 5.13, and some of them will be deeper described as they are related to metallic biomaterials.

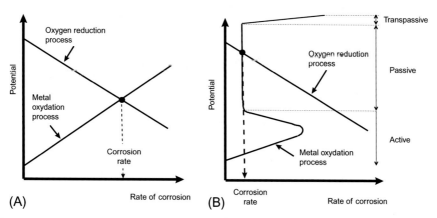

FIG. 5.22  Corrosion behavior of (A) an active metal and (B) of an active-passive metal as a function of electrode potential.

TABLE 5.13  Common Types of Corrosion in Metals

| Uniform corrosion | Uniform on all the surface of the metallic product |
|---|---|
| Localized corrosion | Pitting corrosion<br>Crevice corrosion<br>Filiform corrosion<br>Stress cracking corrosion<br>Intergranular corrosion<br>Fretting corrosion<br>Erosion corrosion |
| Galvanic corrosion | Two different metals are located together in a liquid electrolyte |
| Environmental cracking | Environmental conditions are stressful enough, some metal can begin to crack, fatigue, or become brittle and weakened |

## UNIFORM ATTACK

This very common form of corrosion attacks the entire surface of a metal structure, and it is the most prevalent of all forms of corrosion; its most common example is rust. Uniform corrosion occurs uniformly on the surface exposed to a solution that makes oxidative and reductive chemical reaction possible. Although general attack corrosion can cause a metal to fail, it is also a known and predictable issue. As a result, it is possible to plan for and manage uniform attack corrosion.

## CREVICE CORROSION

Crevice corrosion occurs in restricted area where there is fluid stagnation, for example in cracks, modular press-fit joints, or beneath screw heads (Fig. 5.23). Also in this form, corrosion is driven by reaction described in Eqs. (5.1), (5.3). With time, oxygen content in the crevice is depleted due to limited fluid flow, and the reduction reaction slows down in that site. The reduction reaction continues unfettered in areas

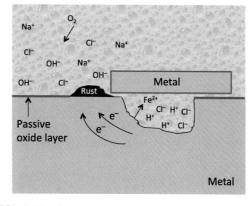

FIG. 5.23  Crevice corrosion mechanism: the crevice is formed where, due to capillary effects, the electrolyte accumulates, generating a mixture of electrolyte and corroded material that acts as a catalyst.

external to the crevice, and the number of metal ions going into solution does not decrease significantly to maintain balance between the two processes in the crevice. This leads to buildup of positive metal ions in the crevice, attracting negatively charged chloride ions. The presence of chloride ions in the crevice further accelerated the dissolution of the metal ions in the crevice, causing even more chloride ions to migrate and setting up an autocatalytic process. Crevice corrosion is very localized and causes severe localized damage whereas the surround areas of the metal remain relatively untouched.

## PITTING CORROSION

Pitting corrosion is similar to crevice corrosion in its mechanism and can be initiated by scratches, inclusions, or other damages. Usually, it starts on horizontal surfaces, and its growth follows gravity. Once initiated, the corrosion quickly forms surface pits and can grow at ever increasing rates, causing holes in the metal. The pits are usually numerous in number and small in size. Improved manufacturing processes have led to fewer inclusions in implants and a reduction in pitting corrosion.

## FRETTING CORROSION

In many materials, the surface reacts with the oxygen to form oxides. These oxides then form the protective layer, preventing the metal underneath from further chemical attack (i.e., passivation layer). Fretting corrosion occurs at contact areas between metal components subject to vibration or slip. The relative movement causes mechanical wear of the components and repeatedly removes the protective oxide layer, which forms every time that new metal surface is exposed and reacts chemically (Fig. 5.24). This cyclic process can cause rapid damage, even when the two materials are subjected to cyclic relative movements of very small amplitude (e.g., $10^{-8}$ cm). Fretting corrosion can be observed at the interface of screws with fracture fixation plates and intramedullary rods, and between press-fit components of modular hip prostheses.

## GALVANIC CORROSION

Galvanic corrosion can occur when two different metals are located together in a liquid electrolyte such as salt water. Metals have a characteristic electrical potential, as reported for different metals in Table 5.12. At the top of the list (Table 5.12) noble materials, also called cathodic materials, are reported, and the anodic metals are listed at the bottom. If the metals or alloys are in their passive state, they are listed further up in the series, closer to noble metals.

The further apart the two metals or alloys are in the galvanic series (Table 5.12), the greater the difference will be in their potential. The greater the potential difference, the higher rate of

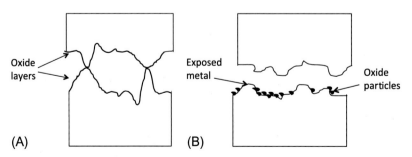

**FIG. 5.24** Fretting corrosion phenomenon: breakdown of oxide passivation layer due to the relative motion caused by repetitive movement or vibration (A) and oxide particle formation (B).

corrosion will be when these two metals are in contact or connected. On the other hand, materials that are very close to each other in the galvanic series pose little danger of galvanic corrosion.

Implants made of more than one metal, or implanted metal devices of different metals but in contact one each other, can undergo galvanic corrosion. This phenomenon can occur in implants such as modular hip joints, dental implants, fracture fixation plate and screw combinations, or an intramedullary nail in contact with a cerclage wire holding pieces of bone together. Moreover, an inclusion or an impurity introduced into a metal implant during fabrication can act as the second metal and galvanic corrosion occurs.

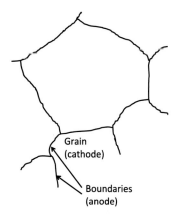

**FIG. 5.25** Intergranular corrosion: grain boundaries are anodic with respect to the grain interior.

### INTERGRANULAR CORROSION

Cast or forged metals and alloys structures are composed of grains; the size and structure of grains depend on the material and also on heat treatment. Moreover, the grain boundaries often contain material compositions different from the grains and may contain more impurities (e.g., the chromium content is depleted at the grain boundaries in stainless steel). These variations in composition, coupled with the high surface area of the grain boundaries, render them more reactive and hence subject to corrosion (Fig. 5.25). Such localized corrosion at or near the grain boundaries with limited corrosion in the grains is called intergranular corrosion and can lead to a decrease in mechanical properties.

### 5.2.2.3 Corrosion in Biomedical Implants

For biomedical applications, in particular for implants, only the metals able to form a passive oxide layer are used; in fact, they have an extremely low generalized corrosion rate ($<0.03\,\mu g/cm^2 \times day$). However, corrosion is already present and causes a limited passage of metal ions from the metallic implanted devices to the surrounding tissues. Therefore

metallic materials for biomedical applications have to be biocompatible, that is, the ions of the metals of the alloys, which are inevitably released, may be tolerated by the human body considering problems of local irritation, allergic reaction, carcinogenicity, and mutagenicity. The need to minimize the extent of generalized corrosion basically derives from the risk of the increase in content of various metals that this can cause in the human body.

Assuming a corrosion rate of $0.03\,\mu g/cm^2 \times day$, it is possible to easily calculate the quantity of metal ions released by a fixation plate or a prosthesis (Table 5.14). These values are rather low so that there is no risk of causing significant problems. In addition, the presence of localized corrosion, such as crevice corrosion (e.g., stainless steel fixation plate) or fretting corrosion (e.g., titanium alloy implants), can significantly increase the amount of ion release (Table 5.14).

Some of the released metal ions can be easily eliminated through physiological mechanisms (e.g., iron); other ions (e.g., chromium, nickel, cobalt) tend to concentrate in target organs (i.e., liver, kidneys, spleen). In particular, the increase in nickel content in metallic alloys can cause problems once implanted or put in contact

**TABLE 5.14** Metallic Ion Release in Human Body Caused by Orthopedic Implants

| Fixation Plate (Austenitic Stainless Steel, AISI 316L) | | Femoral Stem Prosthesis (Titanium Alloy, Ti6Al4V) | |
|---|---|---|---|
| Metal ion release, 1 year, in presence of passive oxide layer | 500 μm | Metal ion release, 1 year, in presence of passive oxide layer | 400 μm |
| Metal ion release, 1 year, in presence of crevice corrosion | 50 mg | Metal ion release, in presence of fretting corrosion | 50 mg |

with human body. In fact, nickel can cause allergic sensitization phenomena; it is estimated that about 30% people manifested skin allergy phenomena by contact with objects (e.g., jewelry, watch cases, glasses) that contain nickel. The use of fixation plates or prostheses containing nickel, or other elements with similar effects, can cause allergic reactions in sensitive patients or even cause sensitization in patients that were not allergic.

Among the possible corrosion mechanisms, crevice corrosion can occur in austenitic stainless steels or in cobalt alloys. In particular, stainless steels are particularly susceptible to this form of corrosion even few months after implantation, and can worsen over time (Fig. 5.26). The interfacial contacts do not need to be stressed, even if an increase in corrosion rate can occur in this case. Cobalt alloys are subject to crevice corrosion only when the interfacial contacts are subjected to stress. Titanium and titanium alloys are not subjected to crevice corrosion attacks.

The susceptibility to fretting corrosion is the main limitation of titanium and its alloys; in particular, fatigue fracture or brittle ones often start at the corroded area. Cast cobalt-chromium-molybdenum alloys can undergo to fretting corrosion below the screw heads in fixation plate.

Galvanic corrosion does not occur for all the metallic couplings. For example, with regard to commonly implanted materials, cobalt alloy/titanium alloy combination does not cause acceleration of corrosive phenomena because the corrosion potential is included in the passivity range of both the metals. Vice versa, titanium or cobalt alloys combined with stainless steel can cause galvanic attack because the potential assumed by the stainless steel is increased by this combination and can favor localized corrosion phenomena.

## 5.2.3 Ceramic Materials

The rate of degradation of ceramics within the body can differ considerably from that of metals, in that they can be either highly degradation resistant or highly soluble. As a general rule, we should expect to see a very significant resistance to degradation with ceramics and glasses. In fact, the interatomic bonds in a ceramic material, being ionic and covalent, are strong directional bonds, and large amounts of energy are required for their disruption. Thus ceramics such as $Al_2O_3$, $ZrO_2$, $TiO_2$, $SiO_2$, and TiN should be stable under normal conditions, and it is widely observed in clinical practice. If some variations in structure are made, there is evidence that problems may arise if incorrect structural forms are used. For example, zirconia is often used in the transformation toughened state; as with most oxide ceramics, the material is inherently brittle, but toughness can be improved by the use of doping elements that cause the formation and retention of a metastable state. This is achieved by the addition of small amounts of the rare earth oxide yttria to the zirconia.

Alternatively, there will be many ceramic structures that, although stable in the air, will dissolve in aqueous environments. Consideration of the ionic ceramic structure NaCl and its dissolution in water demonstrates this point. It is possible, therefore, on the basis of the chemical structure, to identify ceramics that will

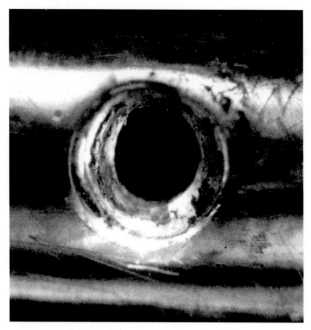

FIG. 5.26   Example of crevice corrosion in a fixation plate.

dissolve or degrade in the body, and the opportunity exists for the production of structural materials with controlled degradation. In that case, it is necessary to select anions and cations that are not cytotoxic and can be harmlessly incorporated into metabolic processes. Calcium phosphates and calcium carbonates biodegradable ceramics as well as bioactive glasses based on Ca, Si, Na, P, and O are examples of biodegradable ceramics. The degradation of such compounds will depend on chemical composition and microstructure.

## 5.3  WEAR PHENOMENA

Wear can be defined as a progressive loss of material in particulate form, as a result of mechanical interaction between two surfaces in contact, lubricated or not. In general, these areas are in relative motion (sliding or slipping) and with applied loads. Two materials placed together under load will only contact on a small area related to higher peaks or asperities on their surface. Electrorepulsive and atomic binding interaction may occur at each contact and, when the two surfaces slide relative to one another, these interactions are broken. This phenomenon causes the release of material in the form of particles (i.e., wear debris) that can be lost from the coupling, transferred to one of the counterface surface, or remain between the two sliding surfaces.

Various authors characterize wear mechanisms differently but, according to the *Modern Tribology Handbook* (2001), there are four main forms of wear: adhesive, abrasive, corrosive, and by fatigue (Fig. 5.27), as well as some side-line cases often classified as wear forms. Oxidation, erosion, and erosion by cavitation and impact are sometimes classified as types of wear, although some researchers found that, in fact, none of them are forms of wear. In this book, the main four mechanisms are considered

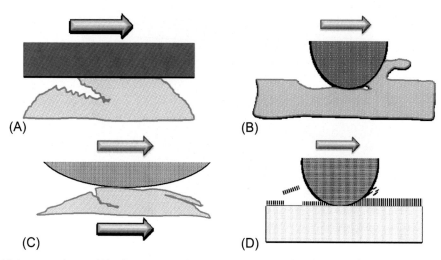

**FIG. 5.27**    Main types of wear: (A) adhesive wear, (B) abrasive wear, (C) fatigue wear, (D) corrosion wear.

as representative of the wear mechanisms occurring in in vivo implants.

Wear behavior of a material is not an intrinsic characteristic but a systemic response; each material can, in fact, wear out according to different mechanisms as a function of temperature, load conditions, surface condition, or environmental factors in which the material is working. Furthermore, several mechanisms can be active at the same time.

The dominant wear mode may change from one to another for reasons that include changes in surface material properties and dynamic surface responses caused by frictional heating, chemical film formation, and wear. Hence, in general, wear does not take place through a single wear mechanism, so understanding each wear mechanism in each mode of wear becomes important.

## 5.3.1 Types of Wear

### 5.3.1.1 Adhesive Wear

Adhesive wear and abrasive wear are wear modes generated under plastic contact. In case of plastic contact between similar materials, the contact interface has adhesive bonding strength.

When fracture is supposed to be essentially brought as the result of strong adhesion at the contact interface, the resultant wear is called adhesive wear, without particularizing about the fracture mode. Hence, adhesive wear occurs when the atomic forces occurring between the materials' surfaces under relative load are stronger than the inherent material properties of either surface. Continued motion of the surfaces causes breaking of the bond junctions. Each time a bond junction is broken, a wear particle is formed, usually from the weaker material (Fig. 5.28). Surface roughness also contributes to adhesion wear mechanism. The fragments removed from one surface may adhere to the other one, be expelled from the bearing coupling, or remain in the bearing joint causing abrasive phenomena.

In hip joint implants, adhesive wear usually occurs when small portions of the ultra high molecular weight polyethylene (UHMWPE) surface adhere to the opposing metal or ceramic bearing surface. The removal of UHMWPE results in pits and voids so small that they may not be evident on visual inspection of the joint surface. The adhesive wear performance of both acetabular hip and tibial knee components (i.e., UHMWPE insert) has been related to the plastic flow behavior of UHMWPE. In the acetabular cup, for example, the generation

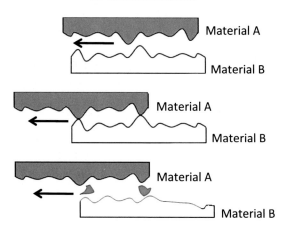

FIG. 5.28 Adhesive wear mechanism.

of submicron wear particles has been associated with local accumulation of plastic strain under multiaxial loading conditions until a critical or ultimate strain is reached.

### 5.3.1.2 Abrasive Wear

In the case of plastic contact between hard and sharp material and soft material, the harder material penetrates into the surface roughness of the softer one. When the fracture is supposed to be brought about in the manner of microcutting by the indented material, the resultant wear is called abrasive wear, recognizing the interlocking contact configuration necessary for cutting, again without particularizing about adhesive forces and fracture mode.

There are two common types of abrasive wear: two-body and third-body abrasion.

Two-body abrasion wear (Fig. 5.29A) is caused by surface roughness, for example, sandpaper. In particular, it refers to surfaces that slide across each other where one material (i.e., hard material) will dig in and remove some particles of the other material (i.e., soft material). The phenomenon can be limited taking into account the surface finishing of the harder material. Two-body abrasion is reduced by smoother surface roughness. For example, a metallic or

ceramic femoral head bearing against a softer material (e.g., UHMWPE acetabular cup) will slide against a harder drive shaft with little to no abrasion due to the surface finish. Using materials with similar hardness is generally not advised.

Third-body abrasion wear (Fig. 5.29B) is a form of abrasive wear occurring when hard particles become embedded in a soft surface; an example is represented by abrasive paste. Moreover, hard particles arriving from outside the bearing coupling (e.g., bone fragments, PMMA debris) or from one of the contact components dragged between the two surfaces in relative motion with consequent removal of other material (i.e., the softer one). The mechanism of migration of the wear particles is related to the type of abrasive particle, the pressure, and the characteristics of the coupling materials. Example of third-body wear includes metallic or bone particles or PMMA debris embedded in a UHMWPE bearing surface. The particles acts much like the asperity of a harder material in abrasive wear, removing material in its movement. Hard third-body particles such as bone cement can produce damage to both the UHMWPE surface and the metal or ceramic bearing counterface. Moreover, the extent of

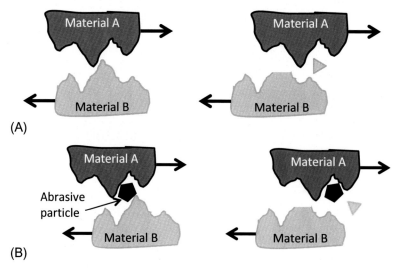

**FIG. 5.29**    Abrasive wear mechanisms: (A) two-body abrasive wear, (B) third-body abrasive wear.

abrasive wear of UHMWPE, metals, and ceramics has been shown to be a function of the surface roughness of the metal or ceramic counterface and the presence of absence of hard third-body particles.

### 5.3.1.3 *Fatigue Wear*

In case of contact in the running-in state, fatigue fracture is generated after repeated friction cycles. When surface failure is generated by fatigue, the resultant wear is called fatigue wear. Fatigue wear can be generated in both plastic and elastic contacts.

Fatigue wear occurs when surface and subsurface cyclic shear stresses or strains in the softer material of an articulation exceed the fatigue limit for that material. In the case of orthopedic joint implants, as UHMWPE is the weaker of the two materials in the bearing combination, fatigue wear damage to the UHMWPE component dominates. Under the repeated and cyclic loading conditions in joint implant, subsurface delamination and cracking can occur, eventually leading to the release of UHMPWE particles (Fig. 5.30). Fatigue damage can range from small areas of pitting, not apparent on visual inspection, to macroscopic pits several millimeters in diameter to large areas of delamination that can encompass an entire tibial insert.

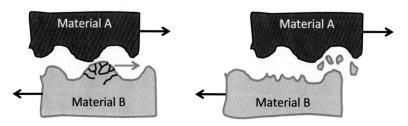

**FIG. 5.30**    Fatigue wear mechanisms.

### 5.3.1.4 Corrosion Wear

Corrosive wear is an indirect wear mechanism. It can be assumed to be a form of third-body wear, as the corrosion debris acts as an abrasive third body in the joint. Corrosive wear can also be considered an accelerating mechanism of corrosion itself, because the motion of the two counterparts can remove corrosive products and the protective passive layer sooner than interfaces with no relative motion. Liberation of corrosive products exposes a greater surface with less protection against corrosion, to further corrosion, hence accelerating the removal of even more material.

## 5.3.2 Measure of Wear Rate

During an initial "wearing in" step, the relative motion of the surfaces can cause a large number of asperities to break determining a high wear rate. After this first step, the contact area between the two counterparts increases, the wear rate decreases because of a sort of adaptation between the surfaces. Hence, over time, wear rate decreases and may become linearly dependent on the contact force and sliding distance, as reported in the steady-state wear equation (Eq. 5.7):

$$V = K \times F \times x \qquad (5.7)$$

where:

- $V$ is the volumetric wear (measured in mm$^3$/year),
- $K$ is a materials constant related to the material coupling,
- $F$ is the contact force (N),
- $x$ is the distance of the relative motion (mm).

To investigate the in vivo wear rate of orthopedic biomaterials, radiographic follow-up studies are carried out to measure and quantify the wear during implantation time. Radiographic wear measures are expressed as linear wear rates. In that case, the only possible measures consist in the evaluation of acetabular cup thickness reduction carried out on the basis of periodic control radiographs or in case of retrieval after explant surgery. The interpretation of the obtained data is important but some parameters have to be taken in consideration, for example, systemic nature of wear, materials of the coupling, or patient's clinical conditions.

In vitro studies can be also considered, measuring wear rates in terms of volumetric wear. Volumetric wear can be directly related to the number of wear particles released into periprosthetic fluids. These in vitro wear tests are certainly less problematic, using appropriate apparatus (e.g., pin-on-disc or pin-on-flat configurations) or hip simulators, but the obtained results strongly depend on the test conditions and on the ability to adequately mimic the physiological environment.

## 5.3.3 Wear in Biomedical Applications

The wear mechanisms in biomedical implants, particularly articular joints are found to be a function of the following variables: type of materials used, contact stresses, lubricants and clearance, surface hardness and roughness, type of articulation due to motion, number of cycles, solution particle count and distribution, and oxidative wear.

In general, the softer of the two bearing surfaces wears more rapidly. In most joint replacements, represented by metal-on-polymer coupling, exclusively polymer wears. The in vitro wear rates for the socket (in hip joint simulator tests) range from 0 to 3000 mm$^3$/year depending on the type of coupling and the environment (e.g., lubricant).

Titanium alloys and stainless steels result in increased wear rates compared to ceramics (e.g., alumina) and cobalt/chromium alloys. Stainless steels paired with UHMWPE produce higher wear rates than cobalt/chromium alloy on UHMWPE, and ceramic (e.g., alumina) on

UHMWPE produced the lowest rates of the materials. Taking this into consideration, the most common wear coupling for joint arthroplasty currently in use is a cobalt-base alloy head (most commonly a CoCrMo alloy) bearing on a UHMWPE cup or liner. The wear rates of this coupling are generally on the order of 0.1 mm/ year, with particulate generation of about $1 \times 10^6$ particles per step or per cycle. Clinically, implant wear rates have been found to increase when some of the following parameters occur:

**a.** physical activity,
**b.** weight of the patient,
**c.** size of the femoral head,
**d.** roughness of the metallic counterface,
**e.** oxidation of UHMWPE.

## References

International Standards Organization (ISO), ISO/TS 11139, 2006.

Stokes, K., Cobian, K., 1982. Polyether polyurethanes for implantable pacemaker leads. Biomaterials 3, 225–231.

## Further Reading

Bundy, K.J., 1994. Corrosion and other electrochemical aspects of biomaterials. Crit. Rev. Biomed. Eng. 22 (3/4), 139–251.

Callister, W.D., Rethwisch, D.G., 2007. Materials Science and Engineering: An Introduction. John Wiley & Sons, New York, NY, USA.

Fonseca, C., Barbosa, M.A., 2001. Corrosion behaviour of titanium in biofluids containing $H_2O_2$ studied by electrochemical impedance spectroscopy. Corros. Sci. 43, 547–559.

Fontana, M.G., Greene, N.D., 1986. Corrosion Engineering. McGraw Hill Higher Education, New York, NY, USA.

Hayashi, K., 1987. Biodegradation of implant materials. JSME Int. J. 30 (268), 1517–1525.

International Standards Organization (ISO) ISO 11137, 1994.

Jacobs, J.J., Roebuck, K.A., Archibeck, M., Hallab, N.J., Glant, T.T., 2001. Metal release and excretion from cementless titanium alloy total knee replacements. Clin. Orthop. 358, 173–180.

Karlsson, S., Albertsson, A.C., 2002. Techniques and mechanisms of polymer degradation. In: Scott, G. (Ed.), Degradable Polymers. Springer, Dordrecht.

Polymer Degradation and Stability Journal, Elsevier, ISSN: 0141-3910.

Pinchuk, L., 1994. A review of the biostability and carcinogenicity of polyurethanes in medicine and the new generation of 'biostable' polyurethanes. J. Biomater. Sci. Polym. Ed. 6, 225–267.

Rabek, J.F., 1990. Photostabilization of Polymers: Principles and Application. Elsevier Sci Publ Ltd, England.

Schoen, F.J., Harasaki, H., Kim, K.M., Anderson, H.C., Levy, R.J., 1988. Biomaterial-associated calcification: pathology, mechanism, and strategies for prevention. J. Biomed. Mater. Res. 22, 11–36.

ISO/TS 11139 (Ed.), 2006. Sterilization of Healthcare Products—Vocabulary. .

EN 556-1 (Ed.), 2001. Sterilization of Medical Devices—Requirements for Medical Devices to be Designated "STERILE"—Part 1: Requirements for Terminally Sterilized Medical Devices. .

Szycher, M., 1991. Biostability of polyurethane elastomers: a critical review. In: Blood Compatible Materials and Devices. Technomic publ, Lancaster, PA, pp. 33–85 (Chapter 4).

Williams, D.F., 2009. On the nature of biomaterials. Biomaterials 30, 5897–5909.

# 6

# Interactions Between Biomaterials and the Physiological Environment

## 6.1 PHYSIOLOGICAL STRUCTURES AND MECHANISMS

The body has levels of organization that build on each other. Cells make up tissues, tissues make up organs, and organs make up organ systems. Organs are structures made up of two or more tissues organized to carry out a particular function, and groups of organs with related functions make up the different organ systems. For instance, the heart and the blood vessels make up the cardiovascular system. They work together to circulate the blood, bringing oxygen and nutrients to cells throughout the body and carrying away carbon dioxide and metabolic wastes. Another example is the respiratory system that, through the nose, mouth, pharynx, larynx, trachea, and lungs, brings oxygen into the body and gets rid of carbon dioxide. Just as the organs in an organ system work together to accomplish their task, the different organ systems also cooperate to keep the body running. Again, the respiratory and the circulatory systems work closely together to deliver oxygen to cells and to get rid of the carbon dioxide the cells produce.

## 6.1.1 The Eukaryotic Animal Cell

The concept of a *cell* dates back to 1665 with microscopic observations of dead cork tissue by scientist R. Hooke. Without realizing their function or importance, Hooke coined the term *cell* based on the resemblance of the small subdivisions in the cork to the rooms that monks inhabited, called cells. About a decade later, A. van Leeuwenhoek became the first person to observe living cells under a microscope. In the century that followed, the theory that cells represented the basic unit of life would develop. A cell (from Latin *cella* meaning storeroom or chamber) is today defined as *the basic structural, functional, and biological unit of all known living organisms* (Fig. 6.1). The cell is sometimes called the *building block of life* because it is the smallest life unit capable of independent reproduction. Cells share many common features, yet they may look wildly different. Still, as different as these cells are, they all rely on the same basic strategies to keep the outside out, allow necessary substances in and permit others to leave, maintain their health, and replicate themselves.

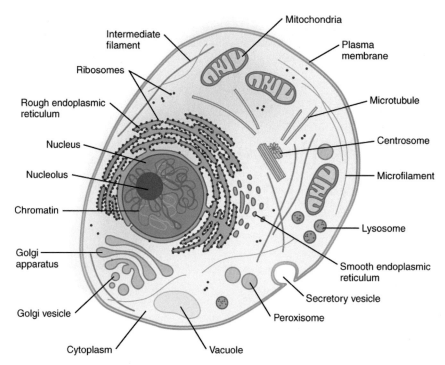

**FIG. 6.1**    Prototypical animal cell. Although this image is not indicative of any one particular cell, this is an example of a cell containing the primary organelles and internal structures. *From https://opentextbc.ca/anatomyandphysiology/.*

In fact, these traits are precisely what make a cell a cell. Broadly speaking, there are two kinds of cells, namely prokaryotic and eukaryotic cells. Prokaryotes (i.e., eubacteria and archaea) do not have a nucleus, that is, their DNA is not enclosed in any subcellular compartment surrounded by a membrane. Therefore only eukaryotic cells have a nucleus (Goodman, 2008).

All living cells in multicellular organisms are surrounded by the *plasma membrane* that, much like the house walls, serves as a physical boundary between the internal and the external environments. The *cell membrane* is sometimes also referred to as *plasma membrane*. Besides, every eukaryotic cell contains a cytoplasmic compartment, and at least one nucleus.

*Biological membranes* or *biomembranes* are thin and flexible envelopes separating cells from each other and cell compartments from the environment. Different membranes have different properties determined largely by the unique set of associated proteins, and all share a common architecture. The phospholipid bilayer is the basic structure of all biomembranes, which also contain proteins, glycoproteins, cholesterol and other steroids, and glycolipids (Fig. 6.2). *Phospholipids* are amphipathic molecules or amphiphiles (i.e., they have a hydrophilic (water-loving) and a hydrophobic (fat-loving) moiety), which spontaneously form 3–4 nm-thick bilayer structures in water because hydrocarbon fatty acyl tails in each leaflet are oriented toward one another to form a hydrophobic core and exclude water molecules. This structural arrangement is stabilized by tail-tail and head-head interactions. The asymmetry of biomembranes is reflected in the specific orientation of each type of integral and peripheral membrane protein with respect to the cytosolic

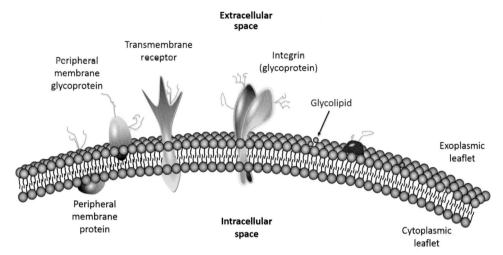

**FIG. 6.2** The cell membrane, also known as the plasma membrane or cytoplasmic membrane, is a biomembrane that separates the interior of all cells from the outside environment. It consists of a lipid bilayer with embedded proteins. The lipid bilayer is composed of phospholipids, with some glycolipids, and sterols (not shown in figure). Transmembrane glycoproteins, such as transmembrane receptors and integrins, span the entirety of the biological membrane to which they are permanently attached. Peripheral proteins are loosely held only at the surface of the plasma membrane.

and exoplasmic faces or leaflets. The presence of glycolipids exclusively in the exoplasmic leaflet contributes to the plasma membrane asymmetry as well.

Thermal motion allows phospholipids and glycolipids to rotate freely around their long axes and to diffuse laterally within the biomembrane leaflet, giving it the properties of a *fluid mosaic*. Because such movements are lateral or rotational, the fatty acyl chains are buried in the hydrophobic, low viscous interior of the membrane. Cells adjust their phospholipid composition to maintain bilayer fluidity. Depending on the cell type, it has been reported that from 30% to 90% of all integral proteins in the plasma membrane, such as phospholipids, float quite freely within the biomembrane plane. Instead, immobile proteins are permanently attached to the underlying cytoskeleton.

The various phospholipids differ in the charge carried by the polar head group at neutral pH; some phosphoglycerides (e.g., phosphatidylcholine (PC) and phosphatidylethanolamine (PE)) have no net electric charge; others (e.g.,

phosphatidylserine (PS)) are overall anionic (Fig. 6.3). In phosphoglycerides (i.e., PC, PE, and PS), which are the major phospholipid class, glycerol is esterified at two out of three hydroxyl groups with fatty acyl chains, whereas the third is esterified with a phosphate. The phosphate can be also esterified to a hydroxyl group on another hydrophilic compound, such as choline, to give PC, or to alcohols such as ethanolamine, serine, and the sugar derivative inositol in other phosphoglycerides. All such polar moieties interact strongly with water. Both of the fatty acyl chains in a phosphoglyceride may be saturated or unsaturated, or one chain may be saturated and the other one unsaturated. Sphingomyelin (Fig. 6.3) contains sphingosine instead of a glycerol backbone, an amino alcohol with a long unsaturated hydrocarbon chain. Cholesterol and its derivatives constitute another important class of membrane lipids, that is, the steroids. Although cholesterol is almost entirely hydrocarbon in composition, it is amphipathic because its hydroxyl group can interact with water.

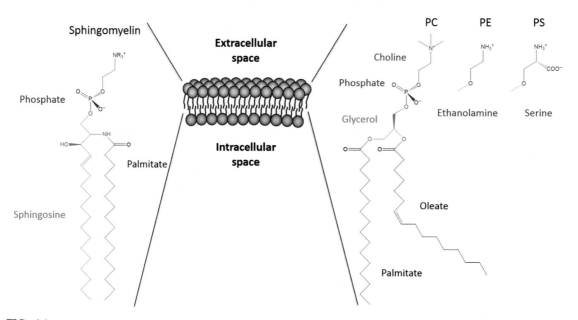

**FIG. 6.3** Diversity of membrane lipids. Structures of representatives from the key lipid types found in biomembranes. Phosphatidylcholine (PC), phosphatidylethanolamine (PE), and phosphatidylserine (PS) are phosphoglycerides (right) because glycerol (*green*) is esterified with two fatty acids (e.g., palmitate and oleate, in *black*), whereas the third hydroxyl group on glycerol is bound to phosphoric acid (*in blue*) through a phosphate ester bond. In addition, there is usually a complex amino alcohol (i.e., choline, ethanolamine, serine; in *red*) attached to the phosphate through a second phosphate ester bond. Instead, sphingomyelin is a sphingolipid that typically makes up 10%–20% of plasma membrane lipids. Sphingomyelin (left) displays a polar head (*in red*), a fatty acid (palmitate, in *black*), linked to a sphingosine molecule (*in green*).

Most kinds of phospholipids, as well as cholesterol, are generally present in both membrane leaflets, although they are often more abundant in one or the other. For instance, cholesterol is especially plentiful in the plasma membrane of mammalian cells but is absent from most prokaryotic cells, whereas sphingomyelin is generally found in plasma membranes.

In every cell, the plasma membrane has several essential functions. These include transporting nutrients into, and metabolic wastes out of the cell; preventing unwanted materials in the extracellular milieu from entering the cell; preventing loss of needed metabolites; and maintaining the proper ionic composition, pH ($\approx 7.2$), and osmotic pressure of the cytosol. To carry out such complex functions, the plasma membrane is equipped with specific transport proteins that allow the passage of certain small molecules but not others. Several of these proteins use the energy released by adenosine triphosphate (ATP) hydrolysis to actively pump ions and other molecules into or out of the cell against their concentration gradient. In fact, small charged molecules such as ATP and amino acids can diffuse freely within the cytosol but are restricted in their ability to leave or enter it through the plasma membrane. Besides, although specialized areas of the plasma membrane contain proteins and glycolipids that form specific contacts and junctions between cells to strengthen tissues and to allow the direct cell-to-cell exchange of metabolites, some plasma membrane proteins act as anchoring points for many of the cytoskeletal fibers that permeate the cytosol, imparting shape and strength to cells.

On occasion, it is necessary to distinguish between the *cytosol* and the *cytoplasm* because there is often much confusion between them. The *cytosol*, also known as intracellular fluid (ICF) or cytoplasmic matrix, is the jelly-like substance within the cell that provides the fluid medium necessary for biochemical reactions. Human cytosolic pH ranges between 7.0 and 7.4, but it is usually higher if a cell is growing. It is defined as all the material in the cytoplasm, often comprising more than 50% of a cell volume, and contains large amounts of macromolecules, which can alter how molecules behave, through macromolecular crowding. The cytosol does include the *cytoskeleton*, the ribosomes, and the centrosome, together with all the other macromolecules (e.g., protein complexes) and solutes outside the nucleus and the lumen of the various cytoplasmic organelles, and it thus excludes any part of the cytoplasm contained within organelles. Therefore the cytosol is basically an important part or element of the *cytoplasm*, which is

composed by the organelles and cytosol together (Goodman, 2008). The cell nucleus and its content are not considered part of the cytoplasm.

Much like the bony skeleton structurally supports the human body, the *cytoskeleton* helps the cells maintain their shape and internal organization. It also provides mechanical support that enables cells to carry out essential functions like division and movement. There is no single cytoskeletal constituent. Rather, several different components work together to form the cytoskeleton. The cytoskeleton of eukaryotic cells consist of three major classes of fibrous proteins that differ in size and composition (Fig. 6.4). *Microtubules* are the largest type of filament, with a diameter of ≈25 nm. They are composed of the tubulin protein. *Actin filaments*, also called *microfilaments*, are the smallest type, with a diameter of only ≈7 nm. They are made of actin. *Intermediate filaments*, as their name suggests, are mid-sized, with a diameter of ≈8–12 nm. Unlike actin filaments and microtubules, intermediate

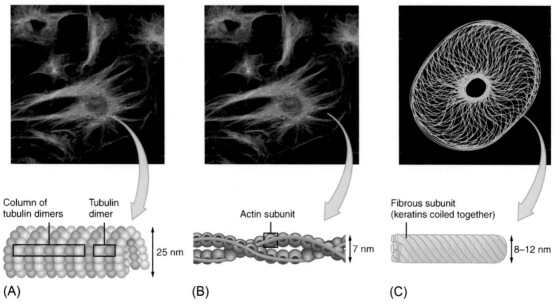

**FIG. 6.4** The cytoskeleton consists of (A) microtubules, (B) microfilaments (actin filaments), and (C) intermediate filaments. The cytoskeleton plays an important role in maintaining cell shape and structure, promoting cellular movement, and aiding cell division. *From https://opentextbc.ca/anatomyandphysiology/.*

filaments are constructed from a number of different subunit proteins. They are also structural components of the nuclear envelope.

An *organelle* (meaning *little organ*) is one of several different types of membrane-enclosed bodies in the cell, each performing a unique function critical to cell survival. Although each organelle performs a specific function in the cell, all of the organelles work together in an integrated fashion to meet the overall cell needs. Just as the various bodily organs work together in harmony to perform all human functions, the many different organelles work together to keep the cell healthy and performing all of its important functions. Any organelle is separated from the rest of the cellular space by a membrane, in much the same way that interior walls separate the rooms in a house. The membranes that surround the eukaryotic organelles are composed of lipid bilayers that are similar in composition, though not identical, to the plasma membrane. Yet, like the plasma membrane, they keep the inside *in* and the outside *out*. This partitioning allows different kinds of biochemical reactions to take place simultaneously but separately in different parts of the same cell.

A set of three major organelles form a system within the cell called the *endomembrane system*. These organelles work together to perform various cellular tasks, including the task of producing, packing, and exporting certain cellular products. The organelles of the endomembrane system include:

- the *endoplasmic reticulum* (ER), which is a system of channels continuous with the nuclear envelope (i.e., the nuclear membrane) covering the nucleus and composed of the same lipid bilayer. The ER provides passages throughout much of the cell that function in transporting, synthesizing, and storing materials. It can exist in two forms (Fig. 6.5): *rough ER* (RER) and *smooth ER* (SER), which perform some very different functions and can be found in very different amounts depending on the cell type. RER is so-called because its membrane is dotted with embedded granules called ribosomes

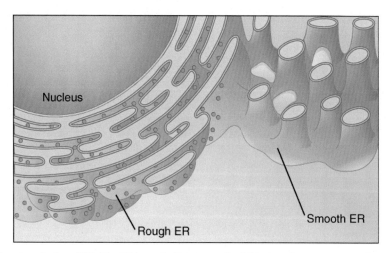

**FIG. 6.5** Endoplasmic reticulum (ER). The ER is a winding network of thin membranous sacs in close association with the cell nucleus. The smooth and rough endoplasmic reticula (SER and RER, respectively) are very different in appearance and function. RER is studded with ribosomes, which are sites of protein synthesis. SER synthesizes phospholipids, steroid hormones, regulates the concentration of cellular $Ca^{2+}$, metabolizes some carbohydrates, and breaks down certain toxins. *Modified from https://opentextbc.ca/anatomyandphysiology/.*

that give it a bumpy appearance. A ribosome is the site where the protein synthesis or translation happens. It is composed of two ribosomal RNA subunits that wrap around mRNA at the beginning of protein synthesis. The primary role of the RER is the synthesis and modification of proteins destined to the cell membrane or for export from the cell (i.e., secretion). For this protein synthesis, many ribosomes attach to the ER and give it the studded appearance of RER. Typically, a protein is synthesized within the ribosome and released inside the RER lumen. The neosynthesized protein is next glycosylated, that is, the covalent link of sugars to the protein, before it is transported within a vesicle to the following stage in the packing and shipping process. SER lacks ribosomes instead. One of its main functions is in the synthesis of phospholipids and steroid hormones. For this reason, cells that produce large quantities of such hormones, such as those of the female ovaries and male testes, are fully packed with SER;

- the *Golgi apparatus*, which is responsible for sorting, modifying, and shipping off the products that come from the RER, much like a post office. It looks like stacked, flattened, membranous discs. The Golgi apparatus has two distinct sides, each with a distinct role. One side of the Golgi receives products in vesicles. These products are sorted through the apparatus, then they are released from the opposite side after being repackaged into new vesicles. If the product is to be exported from the cell, the vesicle migrates to the cell surface and fuses with the cell membrane, and the cargo is secreted;
- *vesicles.*

Other cell organelles are:

- the *lysosome*, which functions as the digestive system of the cell. Lysosomes contain an array of $\approx 50$ different hydrolytic enzymes capable of breaking down any type of

biological polymer and micromolecule such as lipids, and can digest unneeded components of the cell itself such as damaged organelles. Autophagy (meaning *self-eating*) is the process of a cell digesting its own structures. Hydrolysis happens because some of the proteins packaged by the Golgi are digestive enzymes meant to remain within the cell to break down certain materials. The enzyme-containing vesicles released by the Golgi may form new lysosomes, or fuse with existing lysosomes. Generally speaking, lysosomes are at the intersection between the secretory pathway, through which lysosomal proteins are processed, and the endocytic pathway, through which extracellular molecules enclosed in vesicles are taken up by endocytosis at the plasma membrane. All of the lysosomal enzymes are acid hydrolases, which are active at the acidic pH ($\approx 5$) maintained within lysosomes, but not at the neutral pH ($\approx 7.2$) characteristic of the cytosol. The requirement of these lysosomal hydrolases for acidic pH provides double protection against uncontrolled digestion of the cytosol contents; even if the lysosomal membrane were to break down, the acid hydrolases released would be inactive at the cytosolic pH;
- the *peroxisome*, like the lysosome, is an organelle that contains mostly enzymes. Peroxisomes perform a couple of different functions, including lipid metabolism and chemical detoxification. In contrast to the digestive enzymes found in lysosomes, the enzymes within peroxisomes serve to transfer hydrogen atoms from various molecules to oxygen, producing hydrogen peroxide ($H_2O_2$). In this way, peroxisomes neutralize some poisons such as alcohols;
- the *mitochondrion* (plural = mitochondria) is a bean-shaped or oval-shaped organelle that is the *energy factory* or *energy transformer* of the cell responsible for making ATP, the primary, usable, energy-carrying molecule of cells.

These organelles are thought to have evolved from bacteria that developed a symbiotic relationship in which they lived within eukaryotic cells, according to the so-called endosymbiotic theory. A mitochondrion consists of an outer lipid bilayer membrane as well as an additional inner bilayered membrane, separated by the intermembrane space (Fig. 6.6). The inner membrane is highly folded into windings called *cristae*, which display a great deal of surface area and extend into the interior (or matrix). Every component plays a distinct functional role. It is along this inner membrane that a series of proteins and other molecules perform the biochemical reactions implied in cellular respiration, which is the energy conversion stored in nutrients, such as glucose, into ATP. Oxygen molecules are required during cellular respiration, which is why we must constantly breathe it in. Because a lot of energy, that is, ATP, is required to sustain muscle contraction, muscle cells are fully packed of mitochondria. Yet, the internal matrix contains the mitochondrial

genome, which is circular DNA molecules similar to those found in bacteria and present in multiple copies per organelle, as well as the enzymes responsible for the central reactions of oxidative metabolism (i.e., the Krebs cycle, named after H. Krebs).

Apart from membrane-enclosed organelles, every cell is characterized by the presence of a *nucleus* (plural = nuclei) that is the defining feature of eukaryotes. Although there is typically only one nucleus per cell, some cells in the body, such as muscle cells, are known as multinucleated because they do contain more than one. Conversely, other cells, such as mature mammalian red blood cells (RBCs), do not contain nuclei at all. The genomic DNA is surrounded by the nuclear envelope, which is a double membrane that constitutes the outermost portion of the nucleus. Both the inner and outer membranes of the nuclear envelope are made of a phospholipid bilayer, with a thin fluid space in between. Spanning these two bilayers are nuclear pores, which are tiny passageways

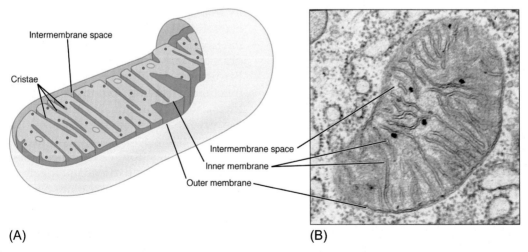

(A)                                         (B)

**FIG. 6.6**   The mitochondria are the energy-conversion factories of the cell. (A) A mitochondrion is composed of two separate lipid bilayer membranes. Along the inner membrane are various molecules that work together to produce adenosine triphosphate (ATP). (B) An electron micrograph of mitochondria (EM ×236,000). *From https://opentextbc.ca/anatomyandphysiology/.*

between the nucleus and the cytoplasm for the passage of proteins, RNA, and solutes. Proteins called nuclear pore complexes (NPCs) lining the pores regulate the passage of materials into and out of the nucleus. The nucleoplasm is the semisolid fluid within the nucleus where there is the chromatin. Some chromosomes have sections of DNA that encode ribosomal RNA. A darkly staining area within the nucleus called the *nucleolus* (plural=nucleoli) aggregates the ribosomal RNA with associated proteins to assemble the ribosomal subunits that are then transported out to the cytoplasm through the NPCs (Fig. 6.7). The nucleus is generally considered the control center of the cell because it stores all of the genetic instructions for manufacturing proteins.

## 6.1.2 Tissue Types

The term *tissue* is used to describe a group of cells found together in the body. Of note, the cells of a given tissue share a common embryonic origin. Microscopic observation revealed that the tissue cells share morphological features and are arranged in an orderly pattern to achieve the tissue-specific functions.

Tissues are not made up solely of cells. A substantial part of their volume is extracellular space, which is largely filled by an intricate network of macromolecules constituting the extracellular matrix (ECM).

Although there are many cell types in the human body, they are organized into four broad tissue categories, which are *epithelial*, *connective*, *muscle*, and *nervous* (Fig. 6.8). Each tissue type

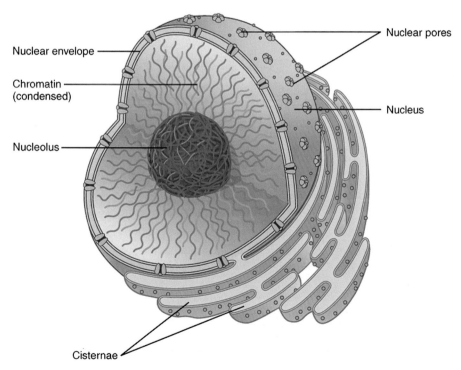

**FIG. 6.7**  The nucleus is the control center of the cell. It contains the cell genome that determines the entire structure and functions of that cell. In contrast to the plasma membrane, the nuclear envelope consists of two phospholipid bilayers, that are the outer membrane and the inner membrane. *From https://opentextbc.ca/anatomyandphysiology/.*

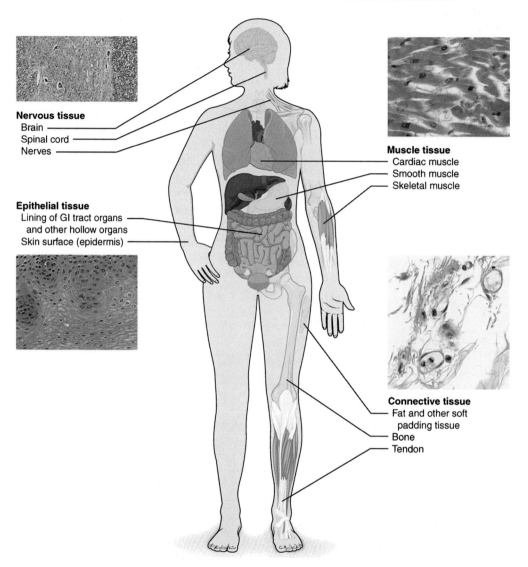

**FIG. 6.8**  The four types of tissues in the body are exemplified in nervous tissue, epithelial tissue, muscle tissue, and connective tissue. Clockwise from nervous tissue (LM ×872, LM ×282, LM ×460, LM ×800). *From https://opentextbc.ca/anatomyandphysiology/.*

is characterized by specific functions that contribute to the overall health and maintenance of the body.

- The *epithelial tissue*, also known as *epithelium*, refers to the sheets of cells that cover exterior body surfaces, line internal cavities and passageways, and form much of the glandular tissue as well. It also functions in absorption, transport, and secretion. All epithelia share some specific structural and functional features. In fact, this tissue is highly cellular, with little or no ECM present between cells. Adjoining cells form specialized intercellular connection called cell junctions. The epithelial cells exhibit polarity with differences in structure and function between the exposed or apical facing surface of the cell and the basal side close to the underlying body structures. The basal lamina, a mixture of glycoproteins and collagen, provides an attachment site for the epithelium and separates it from underlying connective tissue. Epithelial tissues provide the body first line of protection from physical, chemical, and biological wear and tear. The cells of an epithelium act as gatekeepers of the body controlling permeability and allowing selective transfer of materials across a physical barrier.

Epithelial tissues are classified according to the shape of the cells and number of the cell layers formed (Fig. 6.9). Cell shapes can be: *squamous*, that is, flattened and thin cells; *cuboidal*, that is, box-like cells (as wide as they

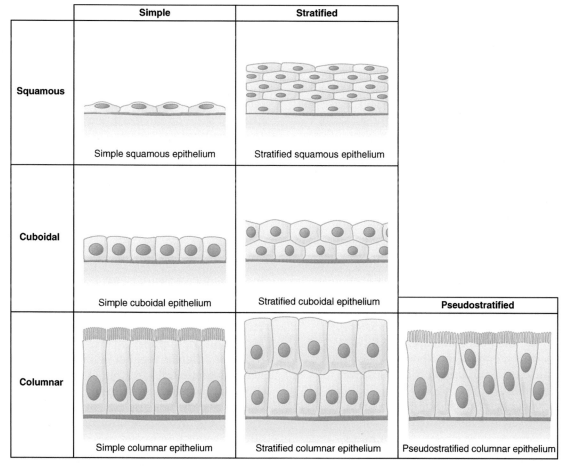

**FIG. 6.9** Epithelial tissue cells. Simple epithelial tissue is organized as a single layer of cells and stratified epithelial tissue is formed by several cell layers. *From https://opentextbc.ca/anatomyandphysiology/.*

are tall); or *columnar*, that is, rectangular, tall column-like cells (taller than wide). Likewise, the number of cell layers in the tissue can be one, where every cell rests on the basal lamina, which is a *simple epithelium*, or more than one, which is a *stratified epithelium*, and only the basal layer of cells rests on the basal lamina. *Pseudostratified* (pseudo- = *false*) describes tissue with a single layer of irregularly shaped cells that give the appearance of more than one layer. Transitional describes a form of specialized stratified epithelium with variable cell shape.

- The *muscle tissue* is characterized as being excitable, responding to stimulation. It is contractile, meaning that it can shorten and generate a pulling force. Indeed, when attached between two movable objects (i.e., bones), contractions of the muscles cause them to move. Some muscle movement is voluntary, which means it is under conscious control. For example, a person decides to walk or run. Other movements are involuntary, meaning they are not under conscious control, such as the pupil contraction in bright light. Muscle tissue occurs as three major types: skeletal (voluntary) muscle, smooth muscle, and cardiac muscle.

    *Skeletal muscle* is attached to bones, and its contraction makes possible locomotion, facial expressions, posture, and any other voluntary body movement. Forty percent of our body mass is made up of skeletal muscle. Skeletal muscles generate heat as a by-product of contraction and thus participate in thermal homeostasis. The muscle cell, or *myocyte*, develops from myoblasts derived from the mesoderm. Myocytes and their numbers remain relatively constant throughout life. Skeletal muscle tissue is arranged in bundles surrounded by connective tissue. Under the light microscope, muscle cells appear striated with many nuclei squeezed along the membranes. The striation is due to the regular alternation of the contractile proteins actin and myosin. The cells are multinucleated as a result of the fusion of the many myoblasts that fuse to form each long muscle fiber. Unlike skeletal muscle, *smooth muscle* tissue contraction is responsible for involuntary movements in the internal organs. It forms the contractile component of the digestive, urinary, and reproductive systems, as well as the airways and arteries. Each cell is spindle-shaped with a single nucleus and no visible striations. On the other hand, *cardiac muscle* forms the contractile heart walls. The cardiac muscle cells, known as *cardiomyocytes*, also appear striated under the microscope. Unlike skeletal muscle fibers, cardiomyocytes are single cells typically with a single centrally located nucleus. The principal characteristic of cardiomyocytes is that they contract on their own intrinsic rhythms without any external stimulation. Cardiomyocytes attach to one another with specialized cell junctions called intercalated discs. Intercalated discs have both anchoring junctions and gap junctions. Attached cells form long, branching cardiac muscle fibers that are, essentially, a mechanical and *electrochemical syncytium* allowing the cells to synchronize their action. The attachment junctions hold adjacent cells together across the dynamic pressure changes of the cardiac cycle.

- The *nervous tissue* is excitable, allowing the propagation of electrochemical signals in the form of nerve impulses that communicate between different regions of the body. The nervous tissue cells are specialized to transmit and receive impulses. Two main classes of cells make up nervous tissue: the *neurons*, that propagate information via electrochemical impulses called action potentials, which are biochemically linked to the release of chemical signals; the *neuroglia* or *glial cells*, that play an essential role in

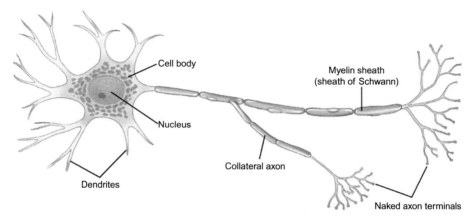

**FIG. 6.10** The neuron. The cell body of a neuron, also called the soma, contains the nucleus and mitochondria, together with other organelles. The dendrites transfer the nerve impulse to the soma. The axon carries the action potential away to another excitable cell through the synapse. *From Muscolino, J., 2010. Kinesiology: The Skeletal System and Muscle Function, second ed. Elsevier.*

supporting neurons and modulating their information propagation.

*Neurons* (Fig. 6.10) display a distinctive morphology, well suited to their role as conducting cells, with three main parts: the *cell body*, also called the *soma*, including most of the cytoplasm, the organelles, and the nucleus; the *dendrites* branch off the cell body and appear as thin extensions. They receive chemical signals from other neurons and transfer the nerve impulse to the soma; and the *axon*, as a long tail, extends from the neuron body and can be wrapped by a *myelin* sheath, a lipid derivative that acts as an insulator and speeds up the transmission of the action potential. The axon carries the action potential away to another excitable cell: when a neuron is sufficiently stimulated, it generates an action potential that propagates down the axon toward the synapse. The synapse is the gap between two nerve cells, or between a nerve cell and its target, for example, a muscle or a gland, across which the impulse is transmitted by chemicals known as neurotransmitters. If enough neurotransmitters are released at the synapse

to stimulate the next neuron or target, a response is generated.

*Neuroglia* or *glial cells* (glia comes from the Greek word for glue) mostly have a simple support role; the most important are: *astrocyte cells*, named for their distinctive star shape, are abundant in the central nervous system and have many functions, including regulation of ion concentration in the intercellular space, uptake and/or breakdown of some neurotransmitters, and formation of the blood-brain barrier, the membrane that separates the circulatory system from the brain; *microglia*, which protect the nervous system against infection but are not nervous tissue because they are related to macrophages; *oligodendrocytes*, responsible of myelin production in the central nervous system (brain and spinal cord), whereas the *Schwann cells* produce myelin in the peripheral nervous system; *ependymal cells*, also named *ependymocytes*, which line the spinal cord and the ventricular system of the brain. These cells are involved in the creation and secretion of cerebrospinal fluid (CSF) and beat their cilia to help circulate the CSF and make up the blood-CSF barrier. They are also thought to act as neural stem cells; and *satellite cells*, which are small cells that surround neurons in sensory, sympathetic, and parasympathetic ganglia. These cells help regulating the external chemical environment.

- The *connective tissue* (Fig. 6.11) that, as the name implies, binds the cells and other tissue types together and functions in the protection, support, and integration of all body parts. This tissue serves to hold in place, connect, and integrate organs and systems.

When considering the features that make a tissue a connective tissue, we should consider the following:

- Connective tissues are normally very vascularized. Some exceptions, such as tendons, ligaments, and cartilages, are less vascularized, but overall, connective tissues possess a great blood supply than the epithelial tissue previously discussed.
- They are made up of many types of specialized (resident or migrant) cells together with a large amount of nonliving material referred to as the ECM, composed of ground substance and fibers. Typically, this matter is synthesized and secreted by specific connective tissue cells. Of note, variations in the ECM composition as well as in cell

content and source determine the properties of the connective tissue itself.

A large part of connective tissue proper is composed of ECM produced primarily by fibroblasts. The extracellular material consists of a gel-like substance called the ground substance, and fibers. The properties of connective tissue (strength, distensibility, flexibility, etc.) are determined by the type of fibers present, the fiber orientation, and the relative amounts of fibers and the ground substance.

1. The *ground substance*, which is composed of proteoglycans (PGs, i.e., proteins carrying large, unbranched, and highly charged polysaccharide side chains) and glycoproteins (proteins carrying shorter, branched, and more neutral polysaccharide side chains). Because of their charge, PGs bind water, giving the ground substance its gel-like character. The components of the ground substance are highly water-soluble and are usually extracted by the solvents used for fixation, leaving clear areas in the stained tissues.

2. The *fibers*. There are three major types of fibers in connective tissue:

   a. the *collagen fibers* are composed of bundles of fibrils (i.e., linear arrays of type I collagen molecules and intermolecular

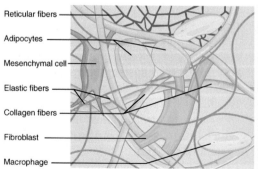
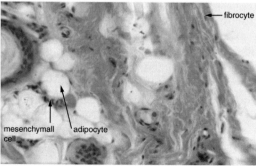

FIG. 6.11   Connective tissue. Fibroblasts produce this fibrous tissue. Connective tissue proper includes the fixed cells fibrocytes, adipocytes, and mesenchymal cells (LM ×400). *From https://opentextbc.ca/anatomyandphysiology/.*

cross-linking in between them) secreted into the extracellular space. They provide high tensile strength to the ECM;

**b.** the *reticular fibers* are short, fine collagenous fibers (type III collagen) that branch extensively to form a delicate network;

**c.** the *elastic fibers* are long, thin microfibril (consisting of numerous proteins such as microfibrillar-associated glycoproteins, fibrillin, and fibullin) and amorphous elastin that form branching network in the ECM. They allow connective tissue to stretch and recoil to its original shape.

Connective tissues can be classified into three broad categories according to the ECM features (i.e., ground substance and the types of fibers found within the matrix):

**1.** the *connective tissue proper*. It is characterized by a variety of cell types and protein fibers suspended in a viscous ground substance. It includes:

**a.** *the loose connective tissue*, which is highly cellularized and contains a loose arrangement of fibers (finer elastin fibers and thicker collagen fibers, leaving large spaces in between) and a moderately viscous fluid matrix. It is highly vascularized, and it includes areolar, which fills the spaces between muscle fibers, surrounds blood and lymph vessels, and supports organs in the abdominal cavity; adipose, which stores fat and cushions and insulates the body; and reticular, which is a supportive framework for soft organs such as the spleen and the liver;

**b.** the *dense connective tissue*, which is characterized by a dense woven network of collagenous and some elastic fibers in a viscous matrix that provide tensile strength, elasticity (i.e., greater resistance to stretching) and protection. There are two major categories of dense connective tissue: regular, in which fibers are parallel to each other in order to enhance tensile strength and resistance to stretching in the direction of the fiber orientations. Ligaments and tendons are made of dense regular connective tissue; and irregular, with randomly distributed fibers that give the tissue greater strength in all directions. Examples are the skin dermis (i.e., dense irregular connective tissue rich in collagen fibers) and the arterial wall (i.e., dense irregular elastic tissues).

**2.** The *supportive connective tissue*. It is characterized by a few distinct cell types and densely packed fibers in a matrix. It provides structure and strength to the body. The supportive connective tissue includes cartilage, a nonvascular form of connective tissue composed of chondrocytes embedded in a gelatinous matrix of chondroitin sulfate and various types of fibrillar collagen. It is usually found at the end of joints, the rib cage, the ear, the nose, in the throat, and between intervertebral discs; and bone, in which a rigid matrix, composed by fibrillar collagens and hydroxyapatite (HA or HAp) crystals, is found, described as calcified because of the deposited calcium salts (mineral bone matrix). It is a specialized connective tissue that is the main constituent of the skeleton. The principal cellular component of bone is comprised of osteoblasts, osteocytes, and osteoclast.

**3.** *Fluid connective tissue*. It is characterized by various specialized cells circulating in an aqueous fluid containing salts, nutrients, and dissolved proteins (i.e., a liquid ECM) with no fibers. It includes blood, with plasma and blood cells, and lymph, the interstitial fluid that is in the lymphatic system.

Overall, the connective tissue is heterogeneous, with many cell shapes and different tissue architectures. Structurally, all connective tissues contain cells embedded in an ECM stabilized by proteins. The chemical nature and physical layout of the ECM and proteins vary enormously among tissues, thus reflecting the variety of functions that connective tissue fulfills in the body. Indeed, the ECM is predominant in tissues with a mechanical function (ligaments, tendon, and bone), whereas cells are the predominant feature in tissues specialized for protection or metabolic maintenance.

### 6.1.2.1 Connective Tissue Cells

The cells of the connective tissue are classified in two categories: *resident cells*, which are normally present in relatively constant numbers; and *migrant cells*, that is, white blood cells (WBCs) including granular leukocytes (which are basophils, eosinophils, and neutrophils) as well as nongranular leukocytes (which are lymphocytes and monocytes), which migrate into and through the connective tissue from the vascular system and are present in unusually high numbers during inflammation. Hematopoiesis refers to the process by which various blood cells are formed (Fig. 6.12).

**i.** *Resident cells:*
  **a.** *Fibroblasts*, which are the most abundant cell type in the connective tissue proper. They are responsible for production of both the ground substance and fibers (i.e., the ECM).
  **b.** *Osteoblasts*, which originate from immature mesenchymal stem cells (preosteoblasts), make *osteoid*, a bone precursor mostly made of type I collagen that serve as a template for the subsequent mineral

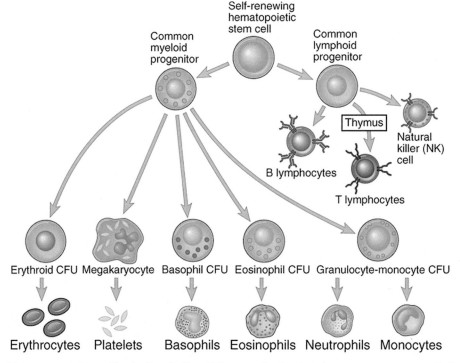

**FIG. 6.12** Hematopoiesis. Every blood cell, including all those involved in the immune response, arise in the bone marrow through various differentiation pathways from hematopoietic stem cells. *From Abbas, A., Lichtman, A.H., Pillai S., 2017. Cellular and Molecular Immunology, ninth ed. Elsevier.*

deposition in the form of HAp. Upon completion of bone matrix formation, some mature osteoblasts remain entrapped in bone as *osteocytes* (mature osteoblasts), some others flatten to cover quiescent bone surfaces as bone-lining cells (Hochberg et al., 2015). Osteocytes are the primary cells responsible for the adaptation of bone to mechanical forces through the control of a number of interacting signaling pathways that regulate bone formation and resorption (Prideaux et al., 2016).

c. *Fat cells.* They differentiate early in life. Their chief function is to store fats, usually triglycerides, as droplets that fill most of the cytoplasm.

d. *Mast cells.* They are bone marrow-derived, granulated cells functionally similar to blood basophils. Like the basophils, mast cells have large secretory granules that store pharmacologically active substances (e.g., heparin, histamine) and can rapidly secrete other biomolecules (i.e., chemotactic factors). Unlike basophils, mast cells usually remain in the tissues and do not circulate in the blood; although they can be found almost everywhere, they are most abundant in the skin, the gastrointestinal tract, and line up along small blood vessels. Mast cells have cell surface receptors for immunoglobulins E (IgE) (see Section 6.2.3) and release their chemical storehouse when specific cognate ligands (i.e., antigens) bind to them.

e. *Macrophages*, which are professional phagocytic cells, an essential component of the immune system (see Section 6.2.3), derived from blood monocytes. When stimulated, macrophages release cytokines, small proteins that act as chemical messengers. They represent a stable population in all connective tissues, except in the case of inflammation when their number increases because cytokines recruit other macrophages to the inflamed site and stimulate their activity. In some circumstances, $\leq 20$ macrophages fuse to form multinucleate *giant cells* (or *foreign body giant cells* (FBGCs)) that surround the foreign material too large to be phagocytosed by individual macrophages.

ii. *Migrant cells* (Fig. 6.12):

   a. *Neutrophils* (or *polymorphonucleocytes* (PMNs)), so-called because stainable by means of neutral dyes, are granular leukocytes having a nucleus with three to five lobes connected by slender threads of chromatin, and a cytoplasm containing fine inconspicuous granules. PMNs are the most abundant phagocytic immune cell type in the blood: they are rapidly recruited to the infection area or injured tissue where they directly produce toxic antimicrobials. The driving force of attraction that determines the direction neutrophils move towards is known as *chemotaxis*.

   b. *Eosinophils* (or *acidophils*) are a type of granulocytes with a nucleus displaying two lobes connected by a slender thread of chromatin, and a cytoplasm containing coarse, round, and eosin-stainable granules that are uniform in size. Similar to neutrophils, eosinophils are actively phagocytic and motile, responding to chemotactic signals released at the site of cell disruption. Eosinophils play a role in infections, allergic reactions, and asthma allergic diseases: following activation by an immune stimulus, eosinophils degranulate to release an array of cytotoxic proteins capable of inducing tissue damage and dysfunction.

   c. *Basophils* are the fewest granulocytes and constitute $< 1\%$ of the circulating leukocytes. They are characterized by a relatively pale-staining by basic dyes (e.g., hematoxylin), lobate nucleus, and cytoplasm containing coarse dark-stained granules variable in size. Having the capacity for chemotaxis, basophils are recruited into inflamed tissues. Basophils are not phagocytic; rather, when stimulated, they release the chemicals contained in their granules, such as heparin, histamine, proteolytic enzymes, and other inflammatory mediators that all have a role in immediate hypersensitivity, inflammatory reaction, mediate vasodilation, and infiltration of immune cells (i.e., they mediate eosinophils chemotaxis). In addition, basophils have high-affinity IgE receptors.

d. *Lymphocytes.* As part of the immune response to foreign substances in the body, these cells constitute $\approx 28\%-42\%$ of the WBCs. Lymphocytes regulate or participate in the acquired immunity to foreign cells and antigens. They are responsible for immunologic reactions to invading organisms, foreign cells such as those of a transplanted organ, and foreign proteins and other antigens not necessarily derived from living cells. The two classes of lymphocytes are not distinguished by the usual microscopic examination but rather by the type of immune response they elicit. The *B lymphocytes* (or *B cells*) are involved in what is called *humoral immunity* (see Section 6.2.3). Upon encountering a foreign substance (or antigen), the B cell differentiates into a *plasma cell*, which secretes immunoglobulins (i.e., antibodies) that recognize and bind invading bacteria, viruses, and toxins. The second class of lymphocytes, the *T lymphocytes* (or *T cells*), are involved in regulating the antibody-forming function of B lymphocytes (*T-helper* ($T_H$)) as well as in directly attacking foreign antigens (*T-suppressor*). T lymphocytes participate in what is called the *cell-mediated immune response* (see Section 6.2.3). The T cells destroy the body's own cells that have themselves been taken over by viruses or become cancerous (*T-killer*). T cells also participate in the rejection of transplanted tissues and in certain types of allergic reactions.

e. *Monocytes.* These are the largest cells of the blood, and they make up $\approx 7\%$ of the leukocyte population. Monocytes are actively motile and phagocytic, capable of ingesting infectious agents as well as RBCs and other large particles, but they cannot replace the function of the PMNs in the removal and destruction of bacteria. Monocytes usually enter areas of inflamed tissue later than the granulocytes. Monocytes move from the bone marrow and circulate in the bloodstream. After a period of hours, the monocytes enter the tissues and develop into macrophages. They are often found at the chronic infection sites.

### 6.1.2.2 *Extracellular Matrix*

Tissues are not made up solely of cells. Although cell boundaries are defined by the plasma membrane, cells in tissues are often surrounded by an insoluble array of secreted macromolecules (Cooper, 2000). The ECM is the noncellular component that fills the space between cells and binds cells together in tissues, providing not only essential physical scaffolding for the cellular constituents but also initiating the crucial biochemical and biomechanical cues required for tissue morphogenesis, differentiation, and homeostasis.

Although the ECM is fundamentally composed of water, proteins, and polysaccharides, each tissue has an ECM with a unique composition and topology, generated during tissue development through a dynamic and reciprocal biochemical and biophysical dialogue between the various cellular components (e.g., epithelial, fibroblast, adipocyte, endothelial cells (ECs)) and the evolving cellular and protein microenvironment. Indeed, besides being tissue-specific, the ECM is markedly heterogeneous. The ECM is a highly dynamic structure that is constantly being remodeled, either enzymatically or nonenzymatically, and its molecular components are subjected to a myriad of posttranslational modifications. Through such physical and biochemical features, the ECM generates the biochemical and mechanical cues specific to every organ, such as tensile and compressive strength and elasticity, and it also mediates protection by a buffering action that maintains extracellular homeostasis and water retention. The biochemical and biomechanical protective and organizational properties of the ECM can vary greatly from one tissue to another (e.g., lungs vs skin vs bone) and even within one tissue (e.g., renal cortex vs renal medulla), as well as from a physiological state to a pathological condition (normal vs cancerous) (Frantz et al., 2010).

The ECM is composed of a variety of macromolecules secreted locally and assembled into an organized complex meshwork in close association with the surface of the cells that have produced them. The main classes of macromolecules are:

i. the *glycosaminoglycans* (GAGs), or *mucopolysaccharides*, which are a heterogeneous group of anionic polysaccharides usually linked to proteins as PGs (with the exception of hyaluronic acid). They occupy a large volume of the extracellular interstitial space and form a hydrated gel. GAGs display a wide variety of functions that reflect their unique buffering, hydration, binding, and force-resistance properties. On the other hand, PGs are also found on the cell surface, where they behave as coreceptors and help cells respond to secreted signal proteins. These proteins enable cells to recognize and adhere to the ECM and one another. Such interactions are key to the organizational behavior of cell

populations and contribute to the formation of embryonic tissues and the function of normal tissue in the adult;

ii. the *fibrous ECM proteins*, which are embedded in gels formed from GAGs, confer both structural and adhesive functions to the matrix. The major ECM proteins are collagens, elastins, fibronectins, and laminins.

*Collagen* is the most abundant protein in animal tissues accounting for up to 30% of the total protein mass of a multicellular organism. Collagens, which constitute the main structural element of the ECM, provide tensile strength, regulate cell adhesion, support chemotaxis and migration, and direct tissue development (see Section 1.6) (Rozario and DeSimone, 2010). To date, 28 types of collagen have been identified in vertebrates (Gordon and Hahn, 2010) (see Section 1.6.3).

*Elastin* molecules form an extensive cross-linked network of fibers and sheets that can stretch and recoil, imparting elasticity to the matrix (see Section 1.6.3).

*Fibronectin* and *laminin* are examples of large, multidomain ECM glycoproteins. Fibronectin displays binding sites for both collagen and GAGs, helping to organize the ECM network. Besides, it has a distinct site specifically recognized by some cell surface receptors such as *integrins*, allowing cell adhesion to the ECM. Cell-ECM interactions are also mediated by transmembrane PGs on the surface of a variety of cells and that bind to the ECM components. In addition, integrins mediate focal adhesion, junctions that link the intracellular actin bundles to the ECM (Fig. 6.13).

The ECM organization can reciprocally influence the cytoskeleton organization and can mechanically influence cell spreading.

From a general point of view, the process of cell adhesion occurs in four phases (Fig. 6.14). First, the cell attaches to the ECM (which in some instances can also be a neighboring cell)

**FIG. 6.13**  Focal adhesion. Integrins connect the extracellular matrix (ECM) to the cytoskeleton, allowing the growing actin network to push the plasma membrane and the contractile cables to pull the cell body.

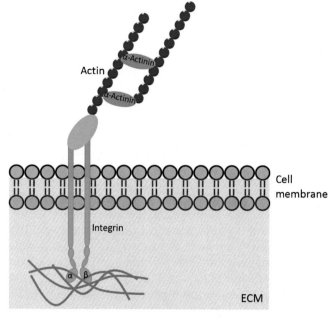

**FIG. 6.14**  A eukaryotic cell spreads over a substrate in distinct stages, with the earliest events characterized by passive adhesion and cell deformation, whereas the later stages of cell spreading do involve the mechanisms of cell crawling such as actin polymerization and myosin contraction. *Modified from McGrath, J.L., 2007. Cell spreading: the power to simplify. Curr. Biol. 17 (10), R357–R358.*

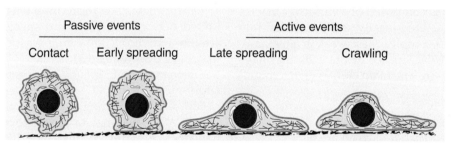

(Fig. 6.14, Contact). The very first point of contact is established by a receptor-ligand pair and is integrin-mediated. Next, the cell flattens, bringing the transmembrane integrin proteins in contact with the substrate that the cell aims to attach to (Fig. 6.14, Early and Late spreading). Over time, additional receptors and ligands come into contact and form connections, thus strengthening the adhesion. Finally, cells project *pseudopodia* and spread its actin skeleton to the cell edges in an effort to further its boundaries (Fig. 6.14, Crawling). This reorganization enables cells to make the maximum number of bonds with the substrate surface, and thus achieving stable adhesion.

Focal contacts not only tether cells together or to the ECM, but they also transduce signals in and out of the cell, influencing a variety of cellular behaviors, such as proliferation and migration. Cells use a multitude of resources to cruise a varying landscape of ECM. For instance, they degrade matrix locally by means of specific enzymes such as matrix metalloproteinases (MMPs), and use both integrin-mediated adhesion to the ECM and actomyosin-mediated contraction to move forward.

The matrix also influences cell behavior by binding to cell-surface receptors that activate intracellular signaling pathways. Indeed, the ECM directs essential morphological organization and physiological function by binding growth factors (GFs) and interacting with cell-surface receptors to elicit signal transduction and regulate gene expression. Such ECM-bound GFs differentially modulate cell growth and migration and, when released, comprise part of a tightly controlled feedback circuit that is essential for tissue homeostasis as well as cell differentiation.

For instance, *bone remodeling* is the continuous turnover of bone matrix and minerals by bone resorption (*osteoclast* activity) and formation (*osteoblast* activity) in the adult skeleton (see Section 6.3). The mechanical environment plays an essential role in the regulation of bone remodeling in intact bone and modeling during bone repair. Reduced loading during long-term immobilization or microgravity can result in significant bone loss. Conversely, mechanical loading enhances bone formation and directs the newly formed bone along the local loading direction. Besides, the mechanical environment regulates osteogenesis during bone repair as well. Bone cells respond directly or indirectly to the local strains engendered in their neighborhood by external loading activity. Bone lining cells and osteocytes act as sensors of local bone strains that are due to bending and compressive forces. The mechanical stimulus leads both to a direct activation of osteoblasts, which react with increased expression of matrix proteins, as well as to an indirect activation of osteoblasts due to increased

expression of GFs and release of other soluble biomolecules by osteocytes. *Integrins*, transmembrane heterodimers transducing signals from the ECM to intracellular components and vice versa, and *cadherins*, which interlink the cytoskeletons of neighboring cells, are the main receptors transducing physical stimuli. This happens because the deformation of the cell membrane of osteoblasts and shear stress induced by the fluid flowing in the canaliculi of osteocytes lead to alterations in the links between the integrins and the ECM, and those between cadherins of neighboring cells (Kamkin and Kiseleva, 2005).

Bone repair is extremely sensitive to the mechanical conditions at the site of the repair process. In contrast to rigid fixation, flexible fixation results in indirect healing characterized by periosteal callus formation and enchondral bone formation. Quite a number of mechanical factors affecting the fracture healing process have been identified. The most noteworthy features are the fracture geometry, the type of fracture, as well as magnitude, direction, and history of the interfragmentary movement. All these global factors determine the local strain distribution and thereby provide the mechanobiological signals for the repair processes and the cellular reactions.

## 6.2 DEFENSE AND REPAIR MECHANISMS

As soon as you are injured, your body tries to heal itself naturally and restore tissue homeostasis.

To gain insight of the cellular and molecular mechanisms underpinning tissue repair, first you need to learn about *blood vessels*. This is because molecules, such as signaling factors and antibodies, and the immune system cells are normally conveyed between cells and throughout the body in the blood through the vascular bed. Therefore, blood and vessels are central to defense and repair mechanisms.

An *artery* is a blood vessel that carries blood away from the heart, where it branches into ever-smaller vessels. Eventually, the smallest arteries (i.e., arterioles), further branch into tiny capillaries where nutrients and wastes are exchanged, and then combine with other vessels that exit capillaries to form venules. Thease are small veins generally 8–100 μm in diameter that carry blood to a vein, a larger blood vessel that returns blood to the heart. Different types of blood vessels vary slightly in their structures, but they share the same general features: arteries and arterioles have thicker walls than veins and venules because they are closer to the heart and receive blood surging at a far greater pressure. Any type of vessel has a lumen, which is a hollow passageway through which the blood flows. In this regard, arteries have smaller lumens than veins. This helps maintaining the pressure of blood moving through the system. Together, the thicker wall and smaller diameter give the arterial lumen a more rounded cross-section appearance than the vein lumen (Fig. 6.15). Arteries and veins display the same three distinct layers, called *tunics*, which, from the most interior layer to the outer are:

- the *tunica intima*, also called *tunica interna*, is the innermost and the thinnest layer composed of epithelial and connective tissue layers. Lining the tunica intima is the specialized simple *squamous epithelium* called *endothelium*, which is continuous throughout the entire vascular bed, including the lining of the chambers of the heart. ECs are glued by a polysaccharide intercellular matrix. Next to the endothelium is the basement membrane, or *basal lamina*, that effectively binds the endothelium to the connective tissue underneath it. The basement membrane provides strength while maintaining flexibility. Besides, it is permeable, allowing materials to pass through it. The endothelium

releases local biochemicals called *endothelins* that induce the smooth muscle within the middle layer of the vessel wall to constrict, increasing blood pressure;

- the *tunica media* is the middle layer. It is generally the thickest layer in arteries, and it is much thinner in veins. The tunica media consists of layers of vascular smooth muscle cells (vSMCs), most of which are arranged in circular sheets, supported by connective tissue primarily made up of elastic fibers. Towards the outer portion of the tunic, there are also layers of longitudinal vSMCs. Contraction (i.e., vasoconstriction) and relaxation (i.e., vasodilation) of the circular muscles decrease and increase the diameter of the vessel lumen, thus increasing and decreasing blood pressure, respectively;

- the *tunica adventitia*, also called *tunica externa*, is the outmost layer. It is the thickest layer in veins. The *tunica adventitia* is entirely made of connective tissue and, in the larger blood vessels, also contains nerves and capillaries (*vasa vasorum*, literally *vessels of the vessel*) to provide vassal cells with nourishment and drain waste.

A capillary is a microscopic channel that supplies blood to the tissues themselves through a process called perfusion. Exchange of gases and other substances occurs in the capillaries between the blood and the surrounding cells and their tissue fluid, called interstitial fluid. The diameter of a capillary lumen ranges from 5 to 10 μm; the smallest are just barely wide enough for an erythrocyte to squeeze through. Flow through capillaries is often described as microcirculation. The most common type of capillary, the continuous capillary, is found in almost all vascularized tissues. Continuous capillaries are characterized by a complete endothelial lining with tight junctions between ECs. Although tight junctions are usually impermeable and only allow for the

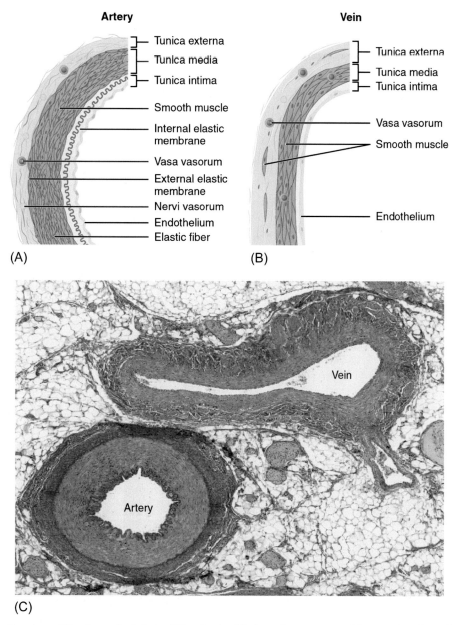

**FIG. 6.15** Structure of blood vessels. Arteries (A) and veins (B) share the same general features, but the wall of arteries is much thicker because of the higher pressure of the blood that flows through them. (C) A micrograph shows the relative differences in thickness (LM ×160). *From https://opentextbc.ca/anatomyandphysiology/.*

passage of water and ions, they are often incomplete in capillaries. leaving intercellular clefts that allow for exchange of water and other very small molecules between the blood plasma and the interstitial fluid. Substances that can pass between cells include metabolic products, such as glucose, water, and small hydrophobic molecules like gases and hormones, as well as various leukocytes. Capillaries consist of little more than a layer of endothelium and occasional connective tissue.

## 6.2.1 The Hemostatic System

*Hemostasis*, or *hemostasis*, with hemo- meaning blood, and -stasis meaning stopping, is the physiological process that stops bleeding at the injured site through the formation of a plug while maintaining blood fluid and normal blood flow elsewhere in the circulation.

In fact, the endothelium in intact blood vessels displays an anticoagulant surface that serves to maintain blood fluid, but when the vessel is damaged, components of the subendothelial matrix are exposed to the blood and hemostasis takes place (Gale, 2011). Noteworthy, this process is tightly regulated such that it is activated within seconds of the injury but must remain localized to the injured site.

The term hemostasis implies a balance between the extremes of hemostatic dysfunction: too little hemostasis results in *hemorrhage* and too much hemostasis results in *thrombosis*. Imbalance resulting in too little hemostasis results in some component of significant blood loss due to hypocoagulation at the injury site, while imbalance resulting in too much hemostasis results in hypercoagulation or thrombosis, obstruction of vascular blood flow, and distal organ hypoxia and injury. Conversely, when hemostasis is in balance, rapid clotting at the injury and appropriate healing occur.

The process of hemostasis is an interplay among three components:

a. the *vascular system*. It depends on the size and amount of vSMCs within their walls and integrity of the EC lining;
b. the *platelets*, also called *thrombocytes*. Resting platelets are small anuclear cell fragments of 1.5–3 μm in size that bud off from the cytoplasm of bone marrow megakaryocytes, specialized large polyploid blood cells (i.e., they contain more than two homologous sets of chromosomes) (Schulze and Shivdasani, 2005) (Fig. 6.10). There are roughly 1.5 to $4 \times 10^8$ platelets/mL of blood that circulate for about 10 days in the bloodstream;
c. the *coagulation cascade*.

Failure or deficiencies in any system and stage involved in hemostasis can lead to varying degrees of uncontrolled hemorrhaging or clotting.

During hemostasis three stages progress in rapid sequence (Fig. 6.16), followed by clot retraction and fibrinolysis:

1. *vascular spasm* is the first response of a blood vessel to injury. Vasoconstriction reduces the amount of blood flow through the area and allows less blood to be lost. This response is triggered by factors such as a direct injury to vSMCs, biochemicals released by ECs and platelets, and reflexes initiated by local pain receptors. Of note, vasoconstriction is most effective in small blood vessels;
2. *platelet plug formation* or *primary hemostasis* refers to the localized platelet aggregation and platelet plug formation to temporarily seal a small vascular breach. In essence, plug formation buys time while more sophisticated and durable body repairs are being made. Within 20s of an injury in which the epithelial wall is disrupted, primary hemostasis is initiated. The primary hemostasis is set through thromboregulation. In healthy blood vessels and under normal blood flow, resting platelets do not adhere to ECs or aggregate with each other. However,

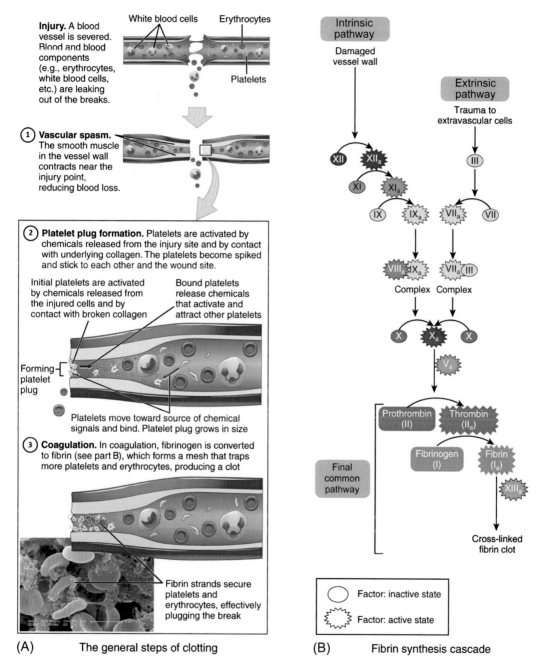

**(A)** The general steps of clotting

**(B)** Fibrin synthesis cascade

**FIG. 6.16** Hemostasis. (A) An injury to a blood vessel initiates the process of hemostasis. Blood clotting involves (1) vascular spasm, which constricts the flow of blood and (2) the formation of a platelet plug that temporarily seal small openings in the vessel. (3) Coagulation then enables the repair of the vessel wall once the leakage of blood has stopped. (B) The synthesis of fibrin in blood clots involves either an intrinsic or an extrinsic pathway, both of which lead to a common pathway. *Modified from https://opentextbc.ca/anatomyandphysiology/.*

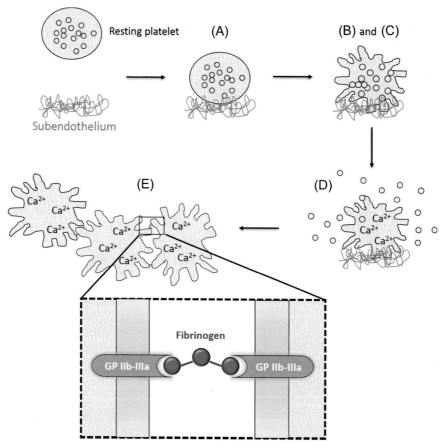

**FIG. 6.17**   In response to vessel wall injury, resting platelets rapidly undergo the process of (A) adhesion to the suben-dothelium, (B) activation, (C) shape change (spiny sphere), (D) secretion, and (E) aggregation through a series of exquisitely coordinated responses that culminate in the formation of a precisely localized hemostatic plug. The inset displays an integrin-fibrinogen-integrin bridge between two neighboring platelets that allows them to stick together (Kroll and Schafer, 1989).

in the event of injury, platelets are exposed to subendothelial matrix comprising thrombogenic collagen, and a platelet plug is readily formed. The formation of the platelet plug is a multifaceted process comprising five subphases (Fig. 6.17).

**a.** *Platelet adhesion* to the subendothelium (Fig. 6.17A). Platelets do express certain glycoprotein receptors used to adhere to the thrombogenic subendothelium. Collagen, which is characteristically found almost everywhere except in the blood

vessel lumen, fibronectin, other adhesive proteins, and von Willebrand factor (vWF) play distinct roles in platelet adhesion. Under high shear conditions, vWF forms a bridge between subendothelial collagen and the glycoprotein receptors (GP Ib-IX-V) on the platelet membrane. Exposed collagen also binds directly to platelet receptors (GP Ia-IIa and GP VI) (Yun et al., 2016).

**b.** *Platelet activation* (Fig. 6.17B). Platelet activation is triggered by some platelet-bound molecules such as the collagen signaling receptor exposed and involved in platelet activation, and local prothrombotic factors such as tissue factor (TF). When platelets are activated, they express more and more glycoprotein receptors (Yun et al., 2016). Of note, multiple pathways almost invariably lead to an increase in the intracellular calcium concentration ($[Ca^{2+}]_i$), which in turn results in a number of structural and functional changes of the platelet, as follows.

c. *Shape changes*. Activated platelets undergo organelle centralization and rapid cortical cytoskeleton rearrangement, resulting in a morphological transition from a discoid shape to a fully spherical form, which then convert into spiny sphere (Fig. 6.17C). The generated spikes dramatically increase the platelet surface area (Aslan et al., 2012) (Fig. 6.18). Platelet shape change is regarded as a prerequisite for platelet aggregation and cohesion with other platelets or erythrocytes, ultimately contributing with fibrinogen to the clot formation.

d. *Platelet degranulation* or *release reaction*. Activated platelets contain a number of distinguishable granules a plethora of

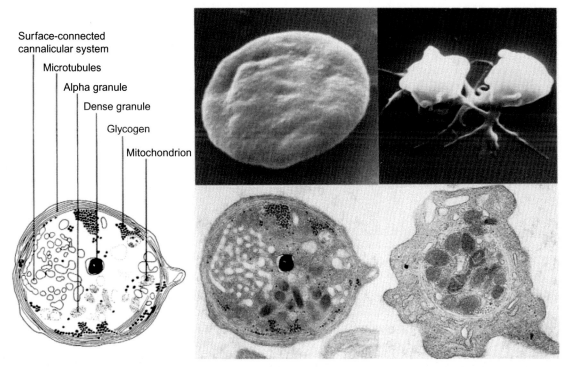

**FIG. 6.18** Electron micrographs of resting and activated platelets. Top right photographs are scanning electron micrographs demonstrating the disc shape of normal circulating platelets (left, ×20,000) and the more spherical form of activated platelets with many protrusions (right, ×10,000). Lower left photograph is a transmission electron micrograph of the cross-section of a resting platelet (×21,000) with a matched drawing labeling the normal subcellular structures. Lower right photograph (×30,000) of an activated platelet with centralized granules and some protrusions. *From George, J.N., 2000. Platelets. Lancet 355 (9214), 1531–1539.*

effector molecules that undergo release at sites of vascular injury (White and Estensen, 1972) (Fig. 6.17D). To date, more than 300 distinct molecules have been detected in platelet releasates. Some of them are produced by megakaryocytes and packaged into granules during platelet generation. Other cargoes (e.g., fibrinogen) are thought to be endocytosed by circulating platelets and stored in granules. Degranulation leads to the release of adenosine diphosphate (ADP) and serotonin. A plethora of proteins, such as fibrinogen, vWF, GFs, and protease inhibitors that supplement thrombin generation, and which act in an autocrine or paracrine fashion, are secreted as well (Polasek, 2006). Other kinds of granules insted contain small chemical messengers, such as serotonin and $Ca^{2+}$, which recruit more platelets to the injured site (Coppinger et al., 2004). As more chemicals are released, more platelets stick and release their content, creating a platelet plug and keeping the process in a positive feedback loop.

   e. *Platelet aggregation.* The long membrane projections brought about by shape-change reaction allow the platelets to interact with one another and form aggregates. Platelet-to-platelet adhesion requires the platelet membrane glycoprotein integrin receptor (GP IIb-IIIa complex), which, at least under low fluid shear stress conditions, is involved in $Ca^{2+}$-dependent interplatelet bridging through binding to plasma fibrinogen (Fig. 6.17E and inset). vWF substitutes for fibrinogen as a bridging molecule between GPIIb-IIIa during platelet aggregation induced by high shear conditions;

3. *coagulation* or *clot formation* or *secondary hemostasis* refers to the deposition of an insoluble fibrin polymer generated through the proteolytic coagulation cascade, to give rise to a stable, sticky fibrin-platelet plug at the injury site (Fig. 6.16B). Coagulation reinforces the platelet plug with fibrin threads that act as a *molecular glue*. It takes approximately 60 seconds until the first fibrin strands begin to intersperse among the wound. After several minutes, the platelet plug is completely formed through fibrin cross-linking.

   The coagulation cascade is a series of reactions in which a zymogen, that is, an inactive enzyme precursor or proenzyme, is activated, and in turn, it activates a downstream zymogen of the cascade. The 12 enzymes involved in the coagulation cascade are called *clotting factors* or *coagulation factors*. They are all primarily secreted by the liver and platelets, and are indicated by Roman numerals from I to XIII, because factor VI is now known to be identical to V, with a lowercase "a" appended to indicate an active form. The overall process is initiated along two basic pathways, whose division is mainly artificial and relies on historical reasons:

- the *extrinsic pathway* (Fig. 6.16B, right), also known as the *TF pathway*, which is triggered by a trauma to the surrounding tissues. It is more direct and quicker responding than the other one: the events in the extrinsic pathway are completed in a matter of seconds. Upon contact with blood plasma, the damaged extravascular cells, which are extrinsic to the bloodstream, release factor III. Sequentially, factor VII, activated by factor III to give factor $VII_a$, leads to activation of factor X, which is part of the downstream (common) pathway.

- the *intrinsic pathway* (Fig. 6.16B, left), also known as the *contact activation pathway*, which involves factors that are intrinsic to (i.e., present within) the bloodstream. It is more complex and longer, becoming completed in a few minutes after activation. The intrinsic pathway is initiated when factor XII comes into contact with foreign materials, such as negatively charged molecules or surfaces. This is an issue of considerable practical importance for modern surgery, as some materials are required to make substitutes (prostheses) for heart valves and blood vessels. In addition, open-heart surgery requires pumping of blood through equipment that does not have to activate the blood-clotting process. Similarly, blood filtration of waste products during kidney dialysis must not lead to the generation of fibrin clots. In all these cases, the formation of clots can lead to serious or even fatal complications. Factor XII sets off a series of reactions that, in turn, activates factor XI, then factor IX. Finally, factor $VIII_a$ from the platelets and ECs combines with factor $IX_a$ to form an enzyme complex that activates factor X, which belongs to the downstream (common) pathway.

Both pathways converge to a common downstream pathway, referred to as the *common pathway* that leads to fibrin formation and the assembly of fibrin molecules into a stable, multimeric, cross-linked net. The cross point of the two upstream pathways is the factor X. Once activated, the factor $X_a$ interacts with the factor $V_a$ and some cofactors, to form a complex that converts the zymogen prothrombin (factor II) into active thrombin (factor $II_a$). This, in turn, catalyzes the cleavage of fibrinogen (factor I) at the wound site to give fibrin (factor $I_a$). The latter is the major protein component of the blood clot. Excessive generation of fibrin due to activation of the coagulation cascade leads to *thrombosis*, which is the blockage of a vessel by an agglutination of RBCs, platelets, polymerized fibrin, and other components. Ineffective generation or premature lysis of fibrin increases the likelihood of a *hemorrhage*. Therefore, central to coagulation is *fibrin* and its precursor, the *fibrinogen* molecule. Fibrinogen is composed of $2\alpha$, $2\beta$, and $2\gamma$-chains linked by 29 disulfide bonds, which are arranged in a dimer of bilateral symmetry (Undas and Ariens, 2011) (Fig. 6.19). The central E region contains the *N*-termini of all six chains. Each of the two D regions is composed of the $\beta$- and $\gamma$-chain *C*-termini arranged in a rod-like shape. The central nodule is linked with the lateral portions via two triple-helical coiled-coil connectors. The thrombin-catalyzed cleavage of four peptide bonds in fibrinogen induces the release of fibrinopeptide A (FPA) and B (FPB), and leads to formation of monomeric fibrin.

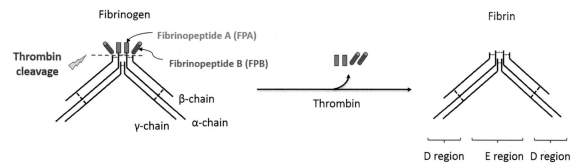

**FIG. 6.19**   Fibrin formation. Thrombin converts fibrinogen to fibrin monomers by cleaving fibrinopeptides A and B from the $\alpha$ and $\beta$ chains, respectively. Fibrin monomers display two lateral D regions, each comprising an $\alpha$- and a $\gamma$-chain, and a central E domain, which is constituted by $\alpha$-, $\beta$-, and $\gamma$-chains.

Noncovalent self-assembly of monomeric fibrin in a half-staggered manner forms two-stranded fibrin oligomers that elongate up to the length of a protofibril comprising 20–25 monomeric units (Fig. 6.20). When sufficiently long, protofibrils aggregate laterally and get packed into a fiber with a regular 22.5 nm periodic cross-striation due to the half-staggered molecular structure and regular protofibril arrangement (Litvinov and Weisel, 2016). Somewhere during the formation of the protofibrils, or their lateral assembly into thicker fibers, branches form and a 3D protein network results.

Loosely bound fibrin threads entangle platelets, building up a spongy mass (i.e., the *blood clot*) that is gradually stabilized and hardened by factor XIII$_a$. This is because factor XIII$_a$, the thrombin-activated form of the zymogen factor XIII, is a transglutaminase enzyme that introduces covalent cross-links between Lys and Gln side chains of α- and γ-chains of fibrin-fibrin strands to form stable homopolymers (Fig. 6.21). Accordingly, deficiencies in factor XIII do not interfere with

clot formation, but resulting clots are unusually fragile and therefore less effective in maintaining hemostasis.

The fibrin clots are very open and porous networks, which is important for their function in hemostasis, fibrinolysis, and wound healing. The mechanical properties of fibrin are unique, in that it is a viscoelastic polymer, which means that it has both reversible elastic characteristics and irreversible plastic or viscous properties. It undergoes strain stiffening, or increasing stiffness at high strains, which helps to prevent damage under harsh conditions, as in arterial shear. In addition, fibrin clots possess extreme extensibility and compressibility, such that they can deform greatly without rupture (Litvinov and Weisel, 2016).

4. *Clot retraction and fibrinolysis.* Clot retraction, also called clot shrinking, is important for promoting clot stability and maintaining blood vessel patency. This process refers to the mechanisms whereby activated platelets transduce contractile forces onto the fibrin network of the clot, which over time increases

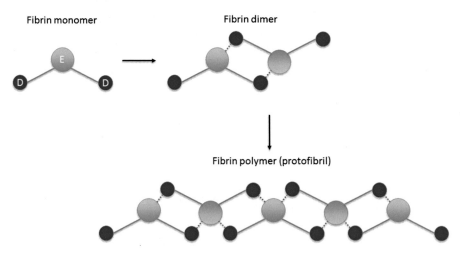

**FIG. 6.20** Fibrin polymerization. Formation of fibrin dimers through noncovalent bonds *(red dotted lines)* between D and E domains of adjacent fibrin molecules. A number of dimers bind to each other to form two-stranded fibrin oligomers and protofibrils. Protofibrils eventually aggregate into fibers. Every E domain of fibrin can interact with up to four D regions of distinct fibrin molecules, forming a long, fibrous latticework (not shown).

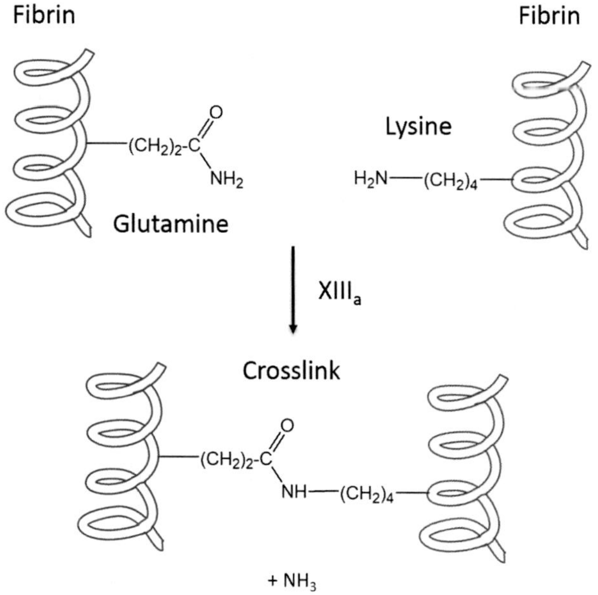

**FIG. 6.21** Factor $XIII_a$ catalyzes transglutamination between adjacent fibrin monomers. Factor XIII has two forms: a plasmatic form that flows freely in the blood plasma and a cellular form carried within platelet granules.

clot density and decreases size (Samson et al., 2017). Fibrinolysis is the process that occurs following clot retraction. It allows clot fibrin to be degraded, whereas macrophages remove platelets to prevent possible thromboembolism (Bagoly et al., 2017). The main enzyme involved in fibrinolysis is plasmin, which hydrolyzes the fibrin mesh at various places to give fibrin degradation products (FDPs or fibrin split products) such as D dimers. D dimer testing is of clinical use for the diagnosis of some cardiovascular diseases.

Plasmin is produced in an inactive form, called plasminogen, in the liver. Although plasminogen cannot cleave fibrin, it still has an affinity for it and becomes entrapped into the clot when it is formed. A couple of enzymes (tissue plasminogen activator (t-PA) and urokinase) convert plasminogen to the active plasmin, thus allowing fibrinolysis to occur. t-PA is released into the blood very slowly by the damaged endothelium of the blood vessels, such that, when the bleeding has stopped (i.e., after several days), it converts plasminogen to plasmin that hydrolyzes the fibrin mesh, and the clot is slowly broken down. A big picture of activators and inhibitors of fibrinolysis is given in Fig. 6.22.

## 6.2.2 The Inflammatory Reaction

The *inflammatory response* or *inflammation* is a first line defense mechanism that evolves through a series of local cellular and vascular responses, triggered by a variety of stimuli, to contain and repair the damage to the injured tissue once it has occurred. It is worth noting that inflammation does not have to be necessarily initiated by an infection but can also be caused by tissue injuries.

It is something everyone has experienced. For instance, when a bee stings you, a foreign body gets in your eye, or you nick your skin, inflammation occurs to protect you.

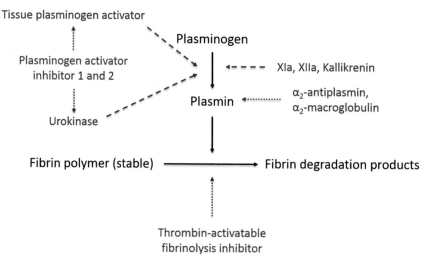

FIG. 6.22    Simplified view of fibrinolysis. *Dashed blue arrows* denote fibrinolysis stimulation, and *dotted red arrows* inhibition.

Causes of inflammation can be:

1. *exogenous stimuli*, meaning that they are external to the body. They can be further classified as:
   a. mechanical/physical stimuli, such as traumatic injury, or thermal or freezing injury;
   b. chemical stimuli, such as caustic agents, venoms, and poisons;
   c. biological stimuli, such as bacteria and viruses;
2. *endogenous stimuli*, meaning that they come from inside the body. They are categorized as:

   a. *metabolic disorders*, which happen when abnormal chemical reactions in the body alter the normal metabolic process. Examples are iron metabolism disorders, disorders of calcium metabolism, lipid metabolism disorders, diabetes, and mitochondrial diseases;
   b. *immune alterations*, such as those caused by human immunodeficiency virus (HIV) and hepatitis C virus (HCV) infections.

The inflammatory response can be either acute or chronic. *Acute inflammation* is a short-term inflammatory response to a body insult. This response starts to occur within a few seconds or minutes after tissue is injured and typically lasts only a few days. Acute inflammation allows immune cells such as PMNs to migrate into affected tissue through the capillary wall and respond to the offending agent. If the agent causing inflammation cannot be eliminated, the stimulus is not removed, or if there is some interference with the healing process, an acute inflammatory response may progress to the chronic stage. Repeated episodes of acute inflammation can also give rise to *chronic inflammation*, this means long-term inflammation that can last for weeks, months, and even years. Chronic inflammation occurs in two stages: the first stage is acute inflammation, as previously explained; simultaneous healing and destruction of cells characterize the second stage. During chronic inflammation, the inflammation itself may become the problem rather than the solution to infection, injury, or disease. If cell damage is severe and extensive enough, chronic inflammation can be self-propagating even if the irritant is removed. Chronically inflamed tissues continue to generate signals that attract leukocytes from the bloodstream. Of note, the physical extent, duration, and effects of chronic inflammation vary with the cause of the injury and the ability of the body to alleviate the damage. The hallmark of chronic inflammation is the infiltration of the tissue site by macrophages, lymphocytes, and plasma cells (mature antibody-producing B lymphocytes). Symptoms of chronic inflammation present in different ways. These can include fatigue, mouth sores, fever, rash, loss of appetite, muscle stiffness, aches, and pains, among others.

Although acute inflammation is usually beneficial, it often causes unpleasant sensations. Discomfort is usually temporary and disappears with the end of the inflammatory response. The five cardinal signs or symptoms of acute inflammation are:

- *localized redness* (from Latin *rubor*) is caused by the dilation of small blood vessels in the injured area;
- *localized swelling* (from Latin *tumor*), also called edema, is caused primarily by the accumulation of fluid outside the blood vessels;
- *localized warmth* (from Latin *calor*) results from increased blood flow through the area and is experienced only in peripheral parts of the body. The increased heat makes the local environment unfavorable for microorganisms eventually present;
- *localized pain* (from Latin *dolor*) results in part from the distortion of tissues caused by edema. The mechanical pressure exerted on neurons alerts you to the injury;
- *loss of function* (from Latin *function lesa*) results from pain that inhibits mobility or from severe swelling that prevents movement in the area.

Of note, these five acute inflammation signs only apply to inflammations of the skin. If inflammation occurs deep inside the body, such as in an internal organ, only some of the signs may be noticeable.

The primary physical effect of the inflammatory response is for blood circulation to increase around the affected area: gaps appear in the vessel walls, allowing the immune cells to extravasate from blood vessels and colonize the injured site. As a result of the increased blood flow, the immune presence is therefore strengthened. All of the different cell types that constitute the immune system congregate at the inflammation site, along with a large supply of immune defense proteins that, in turn, fuel the immune response. The release of damaged cellular contents into the site of injury is enough to stimulate such response, even in the absence of breaks in physical barriers that would allow pathogens to enter. The inflammatory reaction brings in phagocytic cells to the damaged area to clear cellular debris and set the stage for wound repair.

The most important events that lead to the acute inflammatory response are:

- *tissue injury, vasodilation, and increased vascular permeability*. When tissue is first injured, the small blood vessels in the damaged area constrict shortly, a process called *vasoconstriction*.

  Following this transient event, which is believed of little practical importance to inflammation, the content of the injured cells stimulate the release of mast cell granules with their potent inflammatory mediators such as histamine, leukotrienes, and prostaglandins. Histamine induces vasodilation of local blood vessels, and an increase in blood flow. Vasodilation may last from 15 min to several hours. Histamine also increases the permeability of local capillaries, which normally allow only water and salts to pass through easily, causing plasma to leak out and form interstitial fluid. A protein-rich fluid, called exudate, is now able to exit and move into tissues, causing the

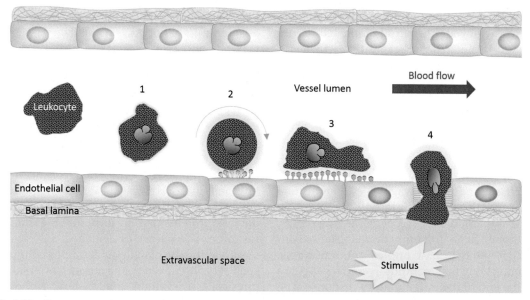

**FIG. 6.23** Recruitment of phagocytes. Leukocyte margination (1) is mediated by vasodilation and slowing of blood flow. Rolling adhesion (2) results from initial interaction of leukocyte and endothelial adhesion molecules. There is an increase in expression of these molecules on cells as the cells are activated by inflammatory cytokines. Thigh adhesion (3) occurs following the interaction with other types of adhesion molecules, such as intergrins. This leads to transmigration of leukocytes (4) through the vessel wall.

swelling associated with inflammation. Additionally, injured cells, phagocytes, and basophils are sources of inflammatory mediators that attract PMNs from the bloodstream, increase vascular permeability, and cause vasodilation by inducing vSMC relaxation;

- *recruitment of phagocytes.* Leukocyte extravasation occurs mainly in post-capillary venules, where hemodynamic shear forces are minimized. This process can be summarized in the four steps outlined here in below (Fig. 6.23):

  1. *margination.* As fluid and other substances leak out of the blood vessels, blood flow becomes more sluggish and leukocytes fall out of the axial stream in the center of the vessel to flow nearer the vessel wall. This happens because macrophages resident in the affected tissue release soluble biochemicals, called *cytokines*,

which in turn cause the ECs of blood vessels near the affected site to express cell adhesion molecules, including selectins.

*Cytokines* are small secreted proteins released by cells that have a specific effect on the interactions and communication between cells. Cytokines may act directly on the cells that secrete them (i.e., autocrine action), on nearby cells (i.e., paracrine action), or in some instances on distant cells (i.e., endocrine action). It is common for different cell types to secrete the same cytokine or for a single cytokine to act on several different cell types (i.e., they exert a pleiotropic effect). Besides, cytokines are redundant in their activity, meaning that different cytokines may stimulate similar functions. They are often produced in a cascade, as one cytokine stimulates target cells to make additional cytokines. Cytokine is a general name; other names include *lymphokine* (i.e., a cytokine made by lymphocytes), *monokine* (i.e., a cytokine made by monocytes), *interleukin* (i.e., a cytokine made by one leukocyte and acting on other leukocytes), and *chemokine* (i.e., a cytokine with chemotactic activities) (Zhang and An, 2007). Circulating PMNs are localized towards the injured site due to the presence of chemokines;

2. *rolling adhesion.* Like velcro, carbohydrate ligands displayed on the surface of circulating leukocytes bind with marginal affinity to selectin molecules on the inner vessel wall. This causes the leukocytes to slow down and begin rolling along the wall surface. During this rolling motion, transitory bonds between selectins on the endothelium and their cognate ligands on leukocytes are continously formed and broken;

3. *tight adhesion.* At the same time, chemokines released by macrophages activate the rolling leukocytes and cause surface molecules, called integrins, to become activated. In this way, integrins bind tightly to cognate receptors displayed by ECs and cause the immobilization of leukocytes, despite the sheer forces of the ongoing blood flow;

4. *transmigration.* The leukocyte cytoskeleton is reorganized in such a way that the leukocytes are spread out over the ECs. In this form, leukocytes extend pseudopodia, pass through gaps existing between ECs, and penetrate the basement membrane. The entire process of blood vessel escape is known as *diapedesis.* Once in the interstitial fluid, leukocytes migrate along a chemotactic gradient towards the site of injury or infection.

*Chemoattractants* can be exogenous, in other words, made by something other than the host body, or endogenous, meaning that they are made by the body own cells. Most exogenous chemotactic factors are microbial products (e.g., endotoxin), whereas chemokines are endogenous chemoattractants. They bind to receptors on the cell surface and activate intracellular messenger systems that cause intracytoplasmic calcium ($Ca^{2+}$) to increase. The calcium then interacts with the cytoskeleton resulting in active cell movement.

The recruitment of PMNs is followed, in the first 24–28 h, by the arrival of monocytes, which eventually mature into macrophages. More and more macrophages are next recruited at the damaged site to clean up the debris left over. The macrophages become more prevalent at the injury site only after days or weeks and are a cellular hallmark of chronic inflammation;

• *healing and repair.* Once acute inflammation has begun, a number of outcomes may follow. These include healing and repair, and chronic inflammation. The outcome depends on the type of tissue involved and the amount of tissue destruction that has occurred, which are in turn related to the cause of the injury.

During the healing process, damaged cells capable of proliferation regenerate. Different cell types vary in their ability to regenerate. Some of them (e.g., epithelial cells) regenerate easily, some others (liver cells, also called hepatocytes) do not normally proliferate but can be stimulated to do so after damage has occurred, whereas still other cells (i.e., neurons) are incapable of regeneration. Of note, the simpler the tissue structure, the easier the process to reconstruct it. In fact, flat surfaces (e.g., the skin) are easily rebuilt, whereas the complex architecture of a gland is not.

The failure to replicate the original framework of an organ can sometimes lead to disease. For example, in liver cirrhosis, the regeneration of the damaged tissue results in shaping abnormal structures that can lead to hemorrhaging and death. Repair, which occurs when tissue damage is substantial or the normal tissue architecture cannot be regenerated successfully, results in the formation of a fibrous scar. Through the repair process, ECs give rise to new blood vessels, and fibroblasts deposit a loose framework of connective tissue. This delicate vascularized connective tissue is called *granulation tissue*. Its name is derived from the small red granular areas that are seen in the healing tissue, such as the skin beneath a scab, when it is incised and visually examined. This is because numerous new capillaries endow the new ECM with its granular appearance. As repair progresses, new blood vessels re-establish blood circulation in the healing area, and fibroblasts produce some collagen that confers mechanical strength to the growing tissue. A scar consisting almost completely of densely packed collagen is eventually formed. The volume of the scar tissue is usually less than that of the tissue it replaces, which can cause an organ to contract and become distorted. The most dramatic cases of scarring occur in response to severe burns or trauma.

Natural acute wound healing is a highly dynamic process that involves complex interactions of ECM molecules, soluble mediators, various resident cells, and infiltrating leukocyte subtypes. It proceeds through some largely overlapping phases that involve an inflammatory response and associated cellular migration, proliferation, ECM deposition, and tissue remodeling. Interruption or deregulation of one or more phases of the wound-healing process leads to a nonhealing, chronic wound. During the early phase of wound repair that takes a few days, the ECM of a healing wound undergoes rapid changes. In fact, the fibrin

clot and the fibrin-rich provisional matrix is replaced by a collagenous scar, and a granulation tissue begins to form. This ECM transition is highly orchestrated and tightly regulated both spatially and temporally.

Restoring blood flow in injured tissue, also called wound angiogenesis, is a basic requirement for successful repair response, as vessels support cells with nutrients and oxygen at the wound site. *Angiogenesis* is defined as the sprouting of capillaries from existing blood vessels during tissue repair (Eming et al., 2007). During wound healing, capillary sprouts invade the fibrin/fibronectin-rich wound clot and within a few days organize into a microvascular network throughout the granulation tissue. As collagen accumulates in the granulation tissue to produce the scar, the density of blood vessels decreases. A dynamic interaction occurs among ECs, angiogenic cytokines, macrophages, and the ECM proteins. In the form of developing capillary sprouts, ECs digest and penetrate the underlying vascular basement membrane, invade the ECM stroma, and form tube-like structures that continue to extend, branch, and create networks, pushed by EC proliferation from the rear and pulled by chemoattractants from the front. Besides, arrival of peripheral blood monocytes and their subsequent activation to macrophages ensures continual synthesis and release of GFs (Tonnesen et al., 2000).

## 6.2.3 The Immune System

The vertebrate *immune system* is a complex multilayered host system for defending against external and internal threats to the integrity of the body. It is a collection of barriers, cells, and soluble proteins that interact and communicate with each other in extraordinarily complex ways to neutralize or destroy microorganisms that would otherwise cause a disease or death. The immune system comprises two overlapping

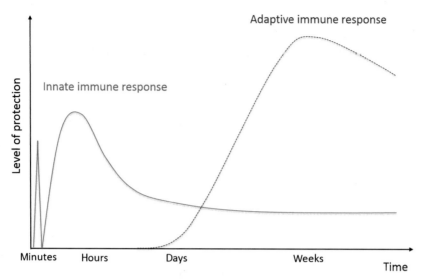

FIG. 6.24    Innate and adaptive immunity. The mechanisms of the innate immunity *(blue line)* provide the initial defense against infections. The early sharp spike is due to anatomical barriers. The adaptive immune response *(red dashed line)* develops later. The kinetics of the innate and adaptive immune responses are approximations and shall vary between infections.

mechanisms of action, or phases, with specific timing of appearance and effects (Fig. 6.24):

- the *innate immune system* is the first line of defense that comprises a variety barriers, defense mechanisms, and general immune responses, which is present from birth onwards to prevent and eradicate pathogenic invasion into the body. The innate immune system is relatively rapid but nonspecific and thus not always effective;
- the *adaptive immune system* comprises many cell types and soluble proteins, and is primarily controlled by a subset of leukocytes known as lymphocytes. The adaptive immunity is slower to develop during an infection but is highly specific and effective in recognizing a wide variety of pathogens. By convention, the terms immune response and immune system generally refer to the adaptive immunity.

Because one of the most difficult intellectual exercises in immunology is to try to understand the global organization and control of the immune system, we provide you with a simplified overview of the immune system, with special focus on the cell and protein components, as well as on recognition and activation mechanisms.

### 6.2.3.1 The Innate Immune System

The *innate immune system*, also known as the *nonspecific immune system* or *in-born immunity*, is an important subsystem of the overall immunity that comprises physical barriers, defense mechanisms, and general immune responses. It is always ready in healthy individuals (hence the term innate) to halt the entry of invading pathogens and rapidly fight and eliminate those that do succeed in entering body tissues. In some cases, it is unfortunately ineffective in eliminating pathogens completely from the host. The innate immune response is general, or nonspecific, meaning that anything identified as foreign or nonself is a target for the innate immune response.

Of note, *nonself* refers to anything not made by our body and is recognized as potentially harmful. Conversely, the term *self* refers to proteins and other molecules, which are a part of or made by the body itself. This means that something that is self should not be targeted and destroyed by the immune system. Such nonreactivity of the immune system to self-entities is called tolerance.

The primary barrier to the entry of microorganisms into the body is the (1) *intact skin*, which is a continuous layer of dead, keratinized epithelium (see Section 6.1.2). The (2) *mucus membranes* and *glands* that, together with the skin, constitute the anatomical barriers of the body (the spike in Fig. 6.24) represent another barrier. Defense mechanisms are the production of (3) *saliva* in the mouth, which is rich in lysozyme and destroys bacteria by digesting the cell wall; the *gastric acids* and *bile*, which are fatal to many microorganisms; the *mucus layer* of the gastrointestinal, respiratory, and reproductive tracts, eyes, ears, nose; *tears* and *sweat* that trap microbial cells and debris, and facilitate their removal. The barrier defenses and defense mechanisms are not a response to infections, but they are continuously working to protect against a broad range of pathogens. Besides, the general immune responses comprise (4) *antimicrobial proteins*, (5) the *fever*, and (6) the *inflammation*. Among antimicrobial proteins, it is worth mentioning the *complement system* (see Section 6.2.4), which is a biochemical cascade that helps, or complements, the ability of antibodies to clear pathogens or mark them for destruction by other cells, cytokines, and interferons, which have specific antiviral activity. The elevation in core body temperature, also known as fever, pyrexia, or hyperthermia, is caused by the action of thermoregulatory pyrogens on the host hypothalamus after infective insult. Besides, inflammation is one of the first responses of the innate immune system to harmful stimuli, such as pathogens, damaged cells, or irritants. It is stimulated by chemical factors released by injured cells and serves to promote healing of any damaged tissue (following the clearance of pathogens). It is important to note that inflammation does not have to be initiated by an infection but is also caused by tissue injuries. Another component of the innate immune system are (7) *natural killer* (NK) cells and *phagocytic cells* (or phagocytes, meaning eating cells). NKs are cytotoxic lymphocytes that play a major role in the host-rejection of both neoplastic and virally infected cells. As the name suggests, they do not require prior recognition to recognize different cellular and pathogen-associated ligands and to perform their effector functions (Mandal and Viswanathan, 2015), therefore allowing for a much faster immune reaction. Small granules in their cytoplasm contain special proteins, such as *perforin*, and proteases, known as granzymes. Upon release in close proximity to a cell slated for killing, perforin forms pores in the cell membrane of the target through which the granzymes and associated molecules can enter and induce cell death. Professional phagocytes are macrophages and PMNs, cells able to engulf particles, debris, worn-out cells, or invading pathogens in a phagosome, which then fuses with a lysosome. Within the phagolysosome, enzymes and toxic peroxides degrade their content. Macrophages (see Section 6.1.2.1) are the most versatile phagocytes because they participate in the innate immune responses and have also evolved to cooperate with T and B lymphocytes as part of the adaptive immune response. Macrophages are amoeboid, irregularly shaped cells that move through tissues and squeeze through capillary walls using protrusions called pseudopodia. When a pathogen breaches the body barriers, macrophages are the first defense line. A monocyte is a circulating precursor cell that differentiates into either a macrophage or dendritic cell, which can be rapidly attracted to the areas of infection by signal molecules of inflammation. Whereas macrophages act like sentries, always on guard against infection,

PMNs can be thought of as military reinforcements called into a battle to hasten the destruction of the enemy.

### 6.2.3.2 The Adaptive Immune System

The *adaptive immune system* is formed by lymphocytes and their products, such as antibodies. The adaptive immunity thus consists of defenses against infections that are activated immediately upon a pathogen attack, in other words, it requires expansion and differentiation of lymphocytes in response to microorganisms before it can provide an effective defense. There are two types of adaptive immunity, called (1) *humoral immunity* and (2) *cell-mediated immunity*, which are mediated by different cells and factors, and provide defense against extracellular and intracellular microorganisms, respectively.

An important feature of the adaptive immune response is the ability to distinguish between self-antigens, those that are normally present in the body, and foreign antigens, those that may arise from a potential pathogen.

The terms *antigen* (Ag) and *immunogen* are often used interchangeably, but they describe two types of interactions between a given molecule and the immune system. Any substance specifically recognized by lymphocytes or antibodies (Ab), which are the end products of the adaptive immune response, is called antigen. Antigens are typically proteins and glycoproteins that pathogens display. Each antigen may consist of many different antigenic determinants, also called *epitopes*, every one being defined as one of the small antigen regions specifically recognized and bound by a cognate receptor molecule. Epitopes usually consist of six or fewer amino acid residues in a protein, or one or two sugar moieties in a carbohydrate antigen. The cells and antibodies of the adaptive immunity each have the theoretical capacity to recognize $10^9$–$10^{11}$ distinct antigenic determinants. Conversely, an immunogen refers to a molecule capable of eliciting the adaptive immune response. In light of that, an immunogen is necessarily an antigen, but an antigen may not necessarily be an immunogen. As lymphocytes undergo maturation, there are specific mechanisms in place that prevent them from recognizing self-antigens, avoiding a damaging immune response against the body itself. The difficulty of such discrimination task between self and nonself is given by the fact that the immune system can sometimes make some mistake, leading to autoimmune diseases.

The cells taking part to the adaptive immunity serve different roles in host defense.

The primary cells that control the adaptive immune response are the *T and B lymphocytes*, *macrophages*, and *dendritic cells*. T and B lymphocytes are critically important, because they control a multitude of immune responses directly and indirectly. T cells are distinguishable from the other lymphocytes because of the presence of the T-cell receptors (TCRs) they display on the surface. Instead, B lymphocytes function in the humoral immunity response by secreting antibodies (see herein below). Besides, professional *antigen-presenting cells* (APCs), such as *macrophages* and *dendritic cells*, detect the presence of microorganisms in body tissues and react against them (Hume, 2008). APC is any cell capable of displaying a *nonself* antigen on its surface. When an APC engulfs a microorganism or a part thereof (e.g., a peptide or carbohydrate), it partially digests it and displays some fragments on its cell surface, bound to major histocompatibility complex (MHC) molecules, like a diamond (i.e., the antigen) sitting on a ring (i.e., the MHC). In fact, it is the combination of the MHC molecule and the fragment of the original antigen that is actually physically recognized by a TCR on a $T_H$ cell. The MHC is a set of cell surface proteins essential for the adaptive immune system to recognize foreign molecules. The main function of MHC molecules is to display *nonself* antigens on the cell surface for recognition by the appropriate T cells. This process is known as *antigen presentation*. There are two major types of MHC proteins very similar in function, called class I (MHC-I) and class II (MHC-II). Of note, MHC-I molecules span the membrane of almost every cell in an organism, whereas MHC-II is confined to professional APCs only. The binding of the TCR with MHC II-antigen fragment complexes on APCs and the presence of costimulatory proteins activate naïve *T helper* ($T_H$) lymphocytes and push them to differentiate into either a $T_H1$ or a $T_H2$ cell subtypes. Once a $T_H1$ or a $T_H2$ effector cell develops, it inhibits the differentiation of the other types of $T_H$ cells through the release of specific sets of cytokines. In this way, the innate response dictates which kind of effector cell a $T_H$ cell will develop into and thereby determine the adaptive immune response elicited, as reported in the following sections. Each $T_H$ cell that has been activated had a specific receptor *hard-wired* into its DNA, and all of its progeny will have identical DNA and TCRs displayed, forming a clone of the original $T_H$ cell.

In the case of the HIV, the $T_H$ cells are directly attacked by the virus and killed off, so that there are no more $T_H$ cells available to elicit the immune response after HIV infection. Because of this, the person who has contracted HIV does not die of the virus itself but can die of a common cold, because the immune system is no more able to counteract it.

1. *Humoral immunity*. It is mediated by soluble effectors called antibodies produced by B cells. Secreted antibodies enter the circulation and mucosal fluids, and neutralize and eliminate microorganisms and microbial toxins present outside the host cells, such as the blood, and in the lumens of mucosal organs such as the gastrointestinal and respiratory tracts. One of the most important functions of antibodies is to stop microorganisms present on the mucosal surface and in the blood from gaining access to and colonizing tissues. In this way,

antibodies prevent infections from ever being established

Antibodies (also called immunoglobulins, or Igs) are Y-shaped glycoproteins that bind to specific antigens (Fig. 6.25). Each antibody molecule is constituted by two identical heavy chains ($_H$) and two identical light chains ($_L$), which are composed of constant (C) and variable domains (V), connected together through disulfide bonds. Noncovalent interactions between the $V_L$ and $V_H$ domains and between the $C_L$ and $C_H1$ domains may also contribute to the association of $_H$ and $_L$ chains. The Y-shape of an antibody can be divided into three sections: two identical antigen-binding regions (F(ab)) and a constant region (Fc). Each F(ab) region contains the variable domain, called *paratope*, that binds to cognate epitope on an antigen. The molecular forces involved in the Fab-epitope interactions are invariably weak (i.e., electrostatic forces, hydrogen bonds, hydrophobic interactions, and van der Waals forces).

The Fc fragment provides a binding site for endogenous Fc receptors (FcRs) on the surface of B lymphocytes, macrophages, dendritic cells, NKs, PMNs, eosinophils, basophils, platelets, and mast cells that contribute to the protective functions of the immune system, thus allowing the antigen-antibody complex to mediate different roles depending on which FcR it binds. In mammals, antibodies are divided into five isotypes based on the number of Y units and the type of Fc: IgA, IgD, IgE, IgG, and IgM (Fig. 6.26).

Depending on the kind of antigen displayed by the MHC-II (i.e., whether it comes from extracellular pathogens) and the specific cytokine profile secreted by APCs evoked by the offending agent, a naïve $T_H$ cell undergoes differentiation into a $T_H2$ cell, secreting in turn a specific set of cytokines.

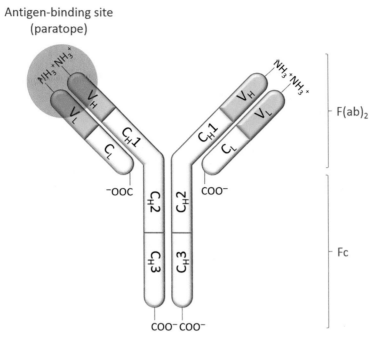

**FIG. 6.25** Structure of a general antibody. Fc is the crystalline fragment, whereas F(ab) is the antigen-binding fragment. The variable light chain ($V_L$) and the adjoining variable heavy ($V_H$) chain are held together by disulfide bridges to form a unique antigen-binding site or paratope *(highlighted in red)*, which binds to a specific epitope on a given antigen. C stands for the constant domains of the light chain ($C_L$) or of the heavy chain ($C_H$).

A $T_H2$ lymphocyte stimulates B cells to make more classes of antibodies. Briefly, a microorganism is recognized by the surface immunoglobulins on a B cell through a B cell antigen receptor (BCR), and it is next internalized and degraded into antigens. These are returned to the B cell surface bound to MHC-II molecules that are recognized by $T_H$ cells. Of note, B cells are another class of professional APCs displaying MHC-II. Once activated, the $T_H2$ cell can in turn activate a B cell that specifically displays the same foreign antigen-MHC-II complex on its surface. The selected B cell will proliferate and differentiate into a *memory B cells,* and *plasma cells* secreting a specific type of antibody during the late part of the adaptive immune response. Bound antibodies sometimes serve as tags, called *opsonins,*

enhancing phagocytosis (see Section 6.2.4). Finally, once the microorganism is removed, negative feedback mechanisms become predominant, turning off the humoral immune response. However, a residual population of long-lived lymphocytes specific for the offending antigen remains behind.

It is worth noting that the adaptive immunity mounts larger and more effective responses to repeated exposures to the same antigen. This feature of the adaptive immune response is called *immunological memory*: it implies that the immune system remembers a previous exposure to an antigen. The first exposure to a pathogen is called *primary adaptive response*, which is initiated by naïve lymphocytes that are encountering a given antigen for the first time

| The Five immunoglobulin (Ig) classes | | | | | |
|---|---|---|---|---|---|
| | IgM pentamer | IgG monomer | Secretory IgA dimer | IgE monomer | IgD monomer |
| | 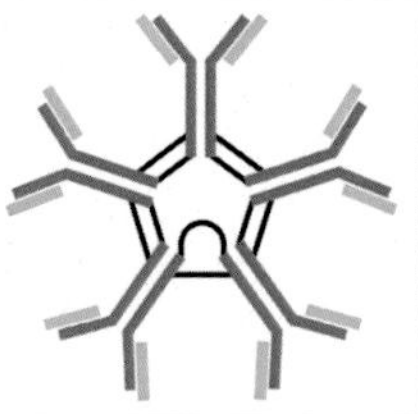 | 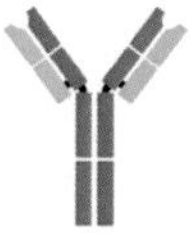 | 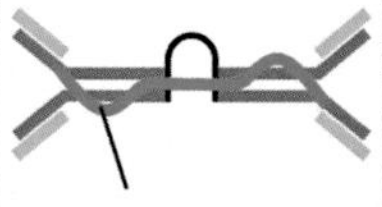 Secretory component | 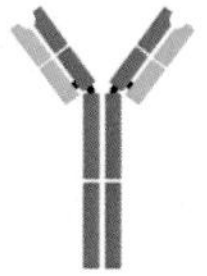 | 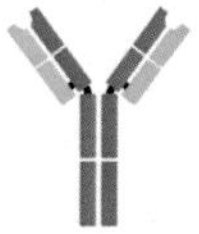 |
| Heavy chains | $\mu$ | $\gamma$ | $\alpha$ | $\varepsilon$ | $\delta$ |
| Number of antigen binding sites | 10 | 2 | 4 | 2 | 2 |
| Molecular weight (Daltons) | 900,000 | 150,000 | 385,000 | 200,000 | 180,000 |
| Percentage of total antibody in serum | 6% | 80% | 13% | 0.002% | 1% |
| Crosses placenta | No | Yes | No | No | No |
| Fixes complement | Yes | Yes | No | No | No |
| Fc binds to | | Phagocytes | | Mast cells and basophils | |
| Function | Main antibody of primary responses, best at fixing complement; the monomer form of IgM serves as the B cell receptor | Main blood antibody of secondary responses, neutralizes toxins, opsonization | Secreted into mucus, tears, saliva, colostrum | Antibody of allergy and antiparasitic activity | B cell receptor |

FIG. 6.26   The five isotypes of antibodies. The isotypes differ in their biological properties, functional locations, and ability to deal with different antigens. *From https://opentextbc.ca/anatomyandphysiology/.*

and are therefore immunologically inexperienced. Because it takes time for an initial adaptive immune response to a pathogen to become effective, the symptoms of a first infection, called primary disease, are always relatively severe. Upon re-exposure to the same antigen (i.e., a pathogen), a *secondary adaptive response* is generated, which is stronger and faster that the primary one. The secondary response is the result of the activation of memory T and B lymphocytes, which are long-lived cells induced by the primary immune response. The secondary adaptive response does recognize the antigen and often eliminates a pathogen before it can cause significant damage or any symptoms and protects us from getting sick repeatedly from the same pathogen. By this mechanism, the exposure to pathogens early in life spares later the person from these diseases. Memory is also the reason why vaccines confer long-lasting protection against infections.

2. *Cell*-mediated *immunity* refers to protective mechanisms that are not primarily characterized by antibodies. $T_H$ cells are arguably the most important cells in adaptive immunity, as they are required for all adaptive immune responses. Specifically, $T_H1$ cells secrete interferon gamma (IFN-γ) and help activating *cytotoxic T cells* ($T_C$ or CTL), known as *T-killer cells*, that kill specifically damaged and dysfunctional cells, including tumor cells, and virus-infected cells. All viruses (and some bacteria) multiply in the cytoplasm of infected cells; indeed, viruses are highly sophisticated parasites that have no biosynthetic apparatus and, as a consequence, can replicate only intracellularly. Unfortunately, these pathogens are not accessible to antibodies once inside cells, and therefore escape or bypass antibody defenses. The only way to eliminate these pathogens is to kill the host cells. Thus the role of $T_C$ cells is to identify cells that are synthesizing nonself proteins through the binding of TCRs with antigens associated with MHC-I molecules that are present on the surface of any nucleated cell, professional APCs included.

In order to generate long-lasting memory $T_C$ cells and to allow the activation of $T_C$ cells, an APC interacts physically with and activates both a $T_H$ cell and a $T_C$ cell. During this process, the $T_H$ cell enhances the costimulatory activity of the APC through the secretion of specific cytokines that act on the $T_C$ cell. Once activated, the $T_C$ cell undergoes clonal expansion to give rise to a subpopulation of memory $T_C$ cells and activated $T_C$ cells that, when exposed to infected/dysfunctional cells, kill them. When an activated $T_C$ binds the target through a TCR, the granules content is discharged by exocytosis only in the direction of the target cell, ultimately resulting in its death. This is because the $T_C$ cells have cytoplasmic granules containing the proteins perforin and granzymes, very similar to those found in NK cells.

### 6.2.4 The Complement System

The *complement system*, also called the *complement cascade*, is a mechanism that complements the other components of the immune response. Typically, the complement system acts as a part of the innate immune system, but it can work together with the adaptive immune system if necessary. The complement cascade consists of a large number of distinct small proteins found in the blood, mostly produced by hepatocytes and that circulate as inactive precursors (i.e., zymogens), and membrane regulators and receptors.

The native components have a simple number designation, for example, C1 and C2, but unfortunately, the components were numbered in the order of discovery rather than their exact position in the sequence of reactions, which is actually C1, C4, C2, C3, C5, C6, C7, C8, and C9. The products of the cleavage reactions are designated by adding lower-case letters, the larger fragment being designated "b" and the smaller "a"; for instance, C4 is cleaved to C4b, the large fragment of C4 that binds covalently to the surface of a pathogen, and C4a, the smaller fragment with proinflammatory properties.

There are three distinct pathways, also known as the early events of the complement activation, through which the complement system can be activated (Fig. 6.27):

a. *the classical pathway*, which typically requires antigen-antibody complexes for activation (specific immune response). It gives rise to a large number of activated complement proteins that bind covalently to pathogens,

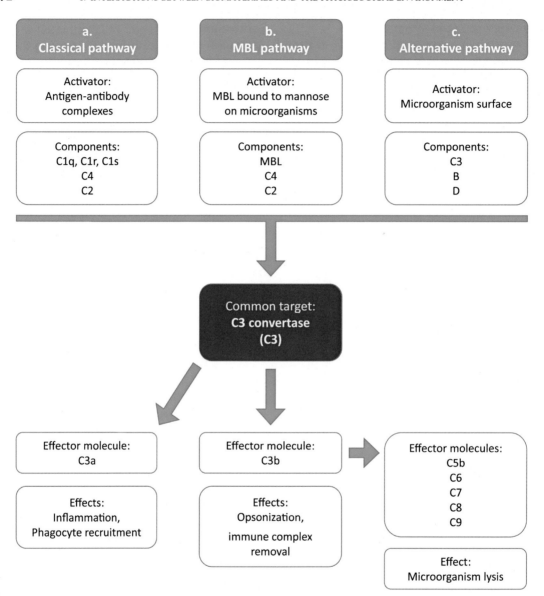

**FIG. 6.27** Simplified diagram of the complement activation pathways and function. All three pathways of complement activation involve a series of cleavage reactions that culminate in the formation of an enzymatic activity called a C3 convertase *(in the red box),* which cleaves complement component C3 into C3b and C3a. The production of the C3 convertase is the point at which the three pathways converge, and the main effector functions of complement are generated.

opsonizing them. *Opsonization* refers to the coating of targets with complement ligands to promote their elimination through immune adherence and phagocytosis by cells bearing complement receptors. The classical pathway is initiated by the binding of the first protein in the complement cascade, called C1q, directly to the pathogen surface. It can also be activated during an adaptive immune response by the binding of C1q to antibody-antigen complexes (i.e., *immunocomplexes*), and is thus a key connection between the effector mechanisms of innate and adaptive immunity;

b. the *mannan-binding lectin* (MBL) *pathway*, which is initiated by the binding of the serum protein MBL to the mannose-containing carbohydrates present on many bacterial cell wall. Microorganisms inducing MBL pathway are bacteria, such as *Salmonella*, *Listeria*, and *Neisseria*, some fungi and some viruses including HIV-1;

c. the *alternative pathway*. It is initiated by cell surface constituents that are foreign to the host, such as lipopolysaccharides (LPS) displayed by some bacteria. Unlike the classical pathway, the alternative pathway does not require the immunocomplex for the initiation of complement cascade.

Although these pathways depend on different molecules for their initiation, they all converge to give the same set of effector molecules. In fact, they generate the protease C3 convertase, which leads to the cleavage of the central component C3 into C3a and C3b, and causes a cascade of further cleavage and activation events. C3b binds covalently to microorganisms and opsonizes them, enabling their internalization by phagocytes. Instead, C3a mediates local inflammation.

The end products of the complement activation or complement fixation cascade trigger the following immune functions (Janeway et al., 2001):

- *opsonization and phagocytosis*. C3b, once bound to immune complex or coated on the surface of pathogen, activates phagocytic cells that engulf the cargo through the binding with specific receptors;
- *membrane attack and cell lysis*. The *membrane attack complex* (MAC) formed by different complement components (i.e., from C5b to C9) creates pores that disrupt the membrane of target cells, leading to their lysis and death;

- *chemotaxis*. Complement fragments attract phagocytes (i.e., PMNs and macrophages) to the area where the antigen is present. These cells have surface receptors for some complement components (e.g., C3a) and move towards the inflammation site;

- *antibody production*. B cells have receptor for C3b. They sense its presence and start secreting more antibodies. Thus C3b is also an antibody-producing amplifier that makes it an effective defense mechanism to destroy invading microorganism;

- *enhancement of inflammation*. The fragment C3a induces acute inflammation by activating mast cells and inducing their degranulation, with the release of vasoactive mediators such as histamine;

- *immune clearance*. Complement proteins promote the solubilization of immune complexes and their clearance by phagocytes in the spleen and liver. They thus exert anti-inflammatory functions as well.

# 6.3 INTERACTIONS BIOMATERIAL/HUMAN BODY (BIOCOMPATIBILITY)

When a biomaterial is implanted into the human body, the nonliving and living components come into contact, and *host-to-implant* and *implant-to host* responses take place (Fig. 6.28).

The issues arising from implant-to-host interactions are basically divided into:

- local effects, when they are confined to the implant site. Some examples are local cytotoxicity, local inflammation, infections, and carcinogenesis;
- distal and/or systemic effects, which happen when some biomaterial by-products

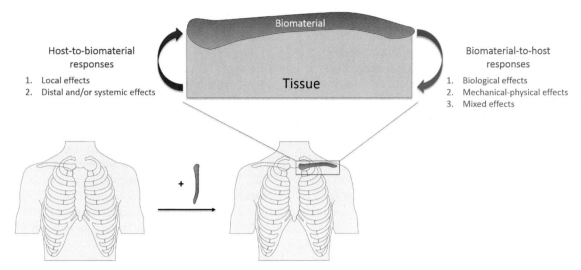

**FIG. 6.28**  Every time a biomaterial is implanted into the body, *host-to-implant* and *implant-to-host* responses take place.

(e.g., leaching substances) or host-to-material interaction products (e.g., blood clots) impact target organs such as liver or kidneys, or induce general inflammation, sensitization, immune activation, hemotoxicity, or general toxicity.

Conversely, host-to-implant responses are categorized as:

- biological effects, such as the adsorption of biomolecules at the biomaterial surface, enzymatic degradation, and pathological calcification;
- mechanical-physical effects, such as wear, fatigue, corrosion, degradation, or surface cracking;
- mixed effects, such as environmental stress cracking (ESC).

In general, there are three terms in which a biomaterial may be described in or classified into representing the tissues responses:

- *bioinert (bio)materials.* The term bioinert refers to any material that, once placed into the human body, has minimal interaction with

surrounding tissues; examples of these are stainless steel, titanium, and ultra-high molecular weight polyethylene (UHMWPE). Generally a fibrous capsule might form around bioinert implants, and hence biofunctionality relies on tissue integration through the implant;
- *bioactive (bio)materials.* By definition, bioactive is a material that elicits a specific biological response. This results in the formation of a bond between the tissue and the material itself. Nowadays, the term bioactive materials broadly refer to biomaterials that induce and conduct the response to the biological system upon interaction (Zhao et al., 2011). They also:
  - i. stimulate cell proliferation and differentiation;
  - ii. stimulate tissue regeneration;
  - iii. release bioactive molecules to restore and repair the impaired functionality of organs.

  At present, bioactive materials are used in almost every medical field, including regenerative medicine, orthopedics, gene therapy, drug delivery, plastic surgery, and diagnostics such as biosensors. Historically, medical applications of bioactive materials have more or less been focused on bone tissue repair and implant replacement. In this specific context, bioactive refers to a material, which upon being placed within the human body typically interacts with the surrounding bone. This occurs through a time-dependent kinetic modification of the surface, triggered by the implantation in the living bone. An ion-exchange reaction between the bioactive implant and the bone results in the formation of a biologically active carbonate apatite (CHA) layer on the implant, which is very similar to the mineral bone phase. The property of establishing chemical bonding with bone tissue is known as *osseointegration (or osteointegration)*. Prime examples of these materials are synthetic HAp ($Ca_{10}(PO_4)_6(OH)_2$) and bioactive glasses (i.e., bioglasses, surface reactive glass-ceramic materials). Bioactive materials are classified into two types:

i. class A, *osteoproductive* materials. Osteoproduction has been defined by Wilson as *the process whereby a bioactive surface is colonized by osteogenic stem cells free in the defect environment as a result of surgical intervention*. Class A bioactivity occurs when a material elicits both intracellular and extracellular responses at the interface. Class A bioactive glasses can bond with both bone and soft tissues (Cao and Hench, 1996);

ii. class B, *osteoconductive* materials. The osteoconductive implant simply provides a biocompatible interface along which bone migrates. Osteoconductive bioactivity occurs when a material elicits only an extracellular response at its interface, and stable bonding with soft tissues does not occur (Cao and Hench, 1996);

• *bioresorbable (bio)materials*. Bioresorbable refers to a material that, upon implantation in the human body, starts to dissolve (i.e, is resorbed) and is slowly replaced by advancing tissue (such as bone). Common examples of bioresorbable materials are tricalcium phosphate ($Ca_3(PO_4)_2$), calcium oxide, and polylactic-polyglycolic acid copolymers.

## 6.3.1 Events Following Implantation

The implantation of a biomaterial is an inherently invasive event that results in injury to tissues or organs. Such an injury, together with the persistence of the biomaterial and the consequent perturbation of homeostatic mechanisms, lead to the cellular cascade of impaired wound healing. Altough biomaterials are generally not immunogenic and are generally not rejected like a transplanted organ, they typically elicit the *foreign body reaction* (FBR), also known as the host response to implanted biomaterials. This is a special form of inflammatory response characterized by the infiltration of inflammatory cells into the area where the material was implanted, followed by the repair or regeneration of the injured tissue. However, if the foreign material cannot be phagocytosed and removed, the inflammatory response persists until it becomes encapsulated in a dense layer of fibrotic connective tissue that shields and isolates it from the surrounding tissues (Eberli, 2011).

Such response to the implanted biomaterial is dependent on multiple factors including the extent of the surgical injury, blood-material interactions, provisional matrix formation, the extent or degree of cell death, and that of the inflammatory response (Anderson, 2001).

It is obvious that the specific response to an implanted biomaterial depends very much on some features, first and foremost on its surface. One of the major current goals of biomaterial research is therefore to understand, predict, and intentionally influence the reaction of our body to a biomaterial (Klopfleisch and Jung, 2017). This section gives a summary on the

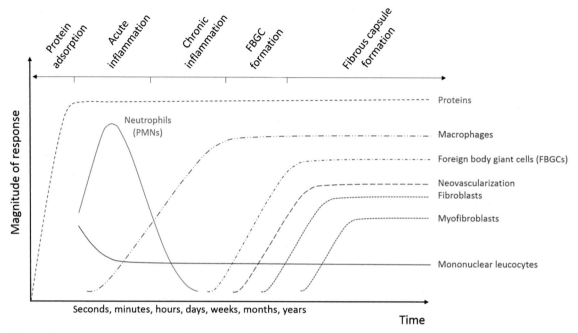

**FIG. 6.29** The temporal variation of the foreign body reaction (FBR) to an implanted biomaterial through the protein adsorption, the acute and chronic inflammatory responses, foreign body giant cell (FBGC) formation, and fibrous capsule formation. The magnitude of response and time variables are dependent upon the extent of injury created in the implantation and the size, topography, and chemical and physical properties of the biomaterial (Narayan, 2016).

current state of knowledge on the general mechanisms of FBR against biomaterials and puts a focus on how topography as well as chemical and physical features of biomaterial surface may influence the extent of the reaction, which occurs along a continuum from which some discrete intervals are described as typical phases (Fig. 6.29).

The concept of the FBR against biomaterials is divided into five phases:

1. *protein adsorption*. From a wound healing prospective, the blood protein deposition on a biomaterial surface is described as *provisional matrix* formation. Such provisional matrix is a 2–5-nm-thick protein layer that may be viewed as a naturally derived system in which bioactive agents are released to control the subsequent phases of

wound healing and FBR, furnishing structural, biochemical, and cellular components. Of note, the presence of bioactive molecules such as cytokines within the provisional matrix provides for a rich milieu of substances capable of modulating macrophage activity, along with the proliferation and activation of other cell populations in the inflammatory and wound healing responses (Anderson et al., 2008). Besides, the intrinsic system is catalyzed by biomaterial contact-induced autoactivation of factor XII, followed by a downstream cascade of reactions resulting in the release of thrombin. Yet, it is well known that the complement system is activated upon contact with biomaterials predominantly through the classical and alternative pathway (Franz et al., 2011) (Fig. 6.30);

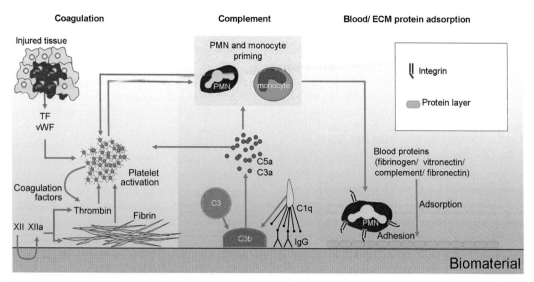

**FIG. 6.30** Adsorption of blood proteins and activation of the coagulation cascade, complement, and platelets result in the priming and activation of neutrophils (PMNs), monocytes, and resident macrophages. *Modified from Franz, S., Rammelt, S., Scharnweber, D., Simon, J.C., 2011. Immune responses to implants—a review of the implications for the design of immunomodulatory biomaterials. Biomaterials 32 (28), 6692–6709.*

2. *acute inflammation.* Neutrophils (PMNs), monocytes, and mast cells characterize the acute inflammation (Fig. 6.30). Upon arrival and adhesion of PMNs at the implantation site, they undergo degranulation. Histamine release and fibrinogen adsorption mediates the acute inflammatory response to implanted biomaterials.

This is a short-lived reaction that usually resolves in less than 1 week and whose severity primarily depends on the extent of the injury during implantation. Conversely, there is considerably less research on the influence of the biomaterial surface on this phase than in protein adsorption and fibrosis;

3. *chronic* inflammation. The chronic inflammation is confined to the implant site and is less uniform histologically than the acute phase response. Macrophages are the driving force in perpetuating immune responses.

In biomaterials science, chronic inflammation describes a rather short period of time postimplantation that lasts no longer than a couple of weeks. The persistence of inflammation beyond 3 weeks is instead a common sign of infection. This definition is in contrast to the common pathologic nomenclature, which uses chronic inflammation to describe everything that is postacute. The most prominent cells in this phase are macrophages due to the great number of biologically active products they produce, and because they presumably attempt to phagocytize the material, altough the degradation is difficult. The chronic inflammation comprises the extravasation and infiltration of circulating mononuclear cells, such as monocytes/macrophages and lymphocytes, in response to platelet-, PMN-, and mast cell-derived chemoattractants at the implantation site (Fig. 6.30), and does not include the formation of FBGCs and the fibrous capsule.

The biomaterial surface chemistry and the topography play a pivotal role in macrophage infiltration, adhesion, activation. For instance, hydrophobic and cationic surfaces seem to promote greater macrophage adhesion with respect to hydrophilic, anionic surfaces. In addition, nickel, magnesium, and corroding metals in general as well as hydroxyl and amino groups stimulate a more intense inflammation characterized by higher numbers of infiltrating macrophages and lymphocytes. The 3D surface topography does significantly influence the chronic FBR inflammation as well. In fact, it was shown that biomaterials displaying 30–40-mm-wide pores exhibited the highest number of infiltrating macrophages (Klopfleisch and Jung, 2017).

Until phase 3, the FBR has many overlaps with the common wound healing process;

4. *FBGC formation.* This phase is considered the hallmark of the FBR, and is a turning point from chronic inflammation. FBGC formation is the process of macrophage fusion, which leads to large, up to several hundred μm-large, multinucleated giant cells with several nuclei. Although the exact mechanisms that lead to macrophage fusion have not been fully elucidated, multiple macrophages adhering to the inedible biomaterial surface come into contact with each other though specific receptors (e.g., mannose receptors, integrins) and fuse to form FBGCs. This happens because, when macrophages encounter a foreign body too large to be phagocytosed, such as an implant or large wear debris, they experience *frustrated phagocytosis* and fuse to form larger FBGCs composed of many individual macrophages. Giant cells secrete degradative agents (e.g., superoxides, free radicals), causing a localized damage to the implanted material.

The quantity and quality of the adsorbed proteins (e.g., vitronectin, fibronectin) in the provisional matrix formed on the biomaterial, the chemical properties (e.g. hydrophilicity,

cationicity), and topographic features (e.g. smoothness, flatness) may prompt or inhibit FBGC formation;

5. *encapsulation* or *fibrous capsule formation* (and possibly *contraction*). Successful integration of the implanted biomaterial into the surrounding tissue with full regeneration is often the desired outcome. However, the establishment of the chronic inflammation and the FBGC formation may finally lead to the formation of a fibrotic, collagenous capsule that surrounds the biomaterial with its interfacial FBR and sealed it off from the local tissue.

Some profibrotic and proangiogenic factors secreted by macrophages and other cells activate and attract fibroblasts and ECs to the biomaterial surface. These, in turn, deposit ECM proteins to form the *granulation tissue* (Anderson, 2001) that next matures into a less cellular and more collagenous, peripheral fibrous capsule, which may lead to mechanical impairment or failure of the biomaterial to interact with the surrounding tissue. In addition, during capsule formation, some of the fibroblasts differentiate into myofibroblasts under the influence of a specific cytokine (i.e., TGF-β), which can contract the fibrous capsule and thus lead to deformation, mechanical stress, and sometimes aesthetic problems. *Capsular contracture* is a major issue to deal with when implants are used in both breast augmentation and breast reconstruction surgery. In fact, when capsular contracture does develop, there are currently only a limited number of surgical options including capsulotomy (i.e., the surgical procedure consisting in scoring the capsule with small incisions), capsulectomy (i.e., the surgical procedure to remove the capsule and the implant which have formed around it) with or without reimplantation, or reconstruction with autologous tissue (Malahias et al., 2016).

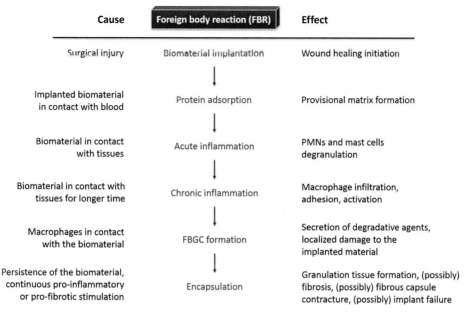

| Cause | Foreign body reaction (FBR) | Effect |
|-------|------------------------------|--------|
| Surgical injury | Biomaterial implantation | Wound healing initiation |
| Implanted biomaterial in contact with blood | Protein adsorption | Provisional matrix formation |
| Biomaterial in contact with tissues | Acute inflammation | PMNs and mast cells degranulation |
| Biomaterial in contact with tissues for longer time | Chronic inflammation | Macrophage infiltration, adhesion, activation |
| Macrophages in contact with the biomaterial | FBGC formation | Secretion of degradative agents, localized damage to the implanted material |
| Persistence of the biomaterial, continuous pro-inflammatory or pro-fibrotic stimulation | Encapsulation | Granulation tissue formation, (possibly) fibrosis, (possibly) fibrous capsule contracture, (possibly) implant failure |

**FIG. 6.31** Temporal sequence of events involved in foreign body response (FBR) over different scale lengths. The left column displays the cause(s) whereas the right column reports typical effect(s) of every FBR phase.

In successful wound healing, the formation of the granulation tissue culminates in the *resolution phase*, which is characterized by the programmed cell death (i.e., the apoptosis) and/or the senescence (i.e., the cell stasis) of myofibroblasts and fibroblasts, the regression of the neovasculature, and the decrease in collagen content by fibrinolytic macrophages. During FBR, this resolution phase is unfortunately delayed, if not missing, probably because of the persistence of the biomaterial, and the continuing proinflammatory or profibrotic stimulation of cells nearby (Fig. 6.31).

Studies aimed at shed light on the influence of the biomaterial chemistry on capsule formation are rather rare. Polyurethane (PU) including silicone and polyethylene oxide (PEO) moieties showed decreased encapsulation when compared to pure PU. Besides, amino (-NH$_2$) and hydroxyl (-OH) groups on hydrophilic surfaces have been reported to induce the thickest capsules as compared to other chemistries. Besides, the surface topography, and especially the surface porosity, does play a role in the capsule formation. In fact, increased porosity has been associated with better healing and decreased fibrosis and encapsulation (Klopfleisch and Jung, 2017).

The most schematic possible view of the overall FBR process is given in Fig. 6.31.

## 6.3.2 Surface Phenomena After Biomaterial Implantation

Implanted materials very quickly acquire a layer of proteins from the host, well before the arrival of inflammatory cells (Hu et al., 2001). This happens because commonly used biomaterials have high affinity to a wide variety of proteins that accumulate at the interface between the material and the body fluids, a property that can be both a practical asset and a problem (Hlady and Buijs, 1996).

Shortly after implantation, proteins begin to adhere to the solid surface through a process known as *protein adsorption*. As the biomaterial interfaces with a biological environment, it becomes thus covered with one or more layers of host proteins, typically albumin, fibrinogen, IgG, fibronectin, and vWF (Thevenot et al., 2008). Adsorption is the process of association of solutes (e.g., proteins, polysaccharides, salts, etc.) to a material surface at a solid-liquid interface. Instead, absorption is when the solution is taken up by the material itself. The kind and the quantity of proteins adsorbed on a biomaterial surface determine the type and the surface density of the bioactive sites that may be available for cell interaction, and the orientation, confirmation, and packing density of the adsorbed proteins determine whether the available bioactive sites are presented in a manner such that they can be recognized by the cell receptors as they interrogate the protein layer(s). Accordingly, when living cells approach the biomaterial surface, they do not actually get in contact with the material surface itself, but rather they adhere and interact with the adsorbed protein layer(s), which in turn affects the cell fate (Bowlin and Wnek, 2005). In fact, by the time cells arrive, the material surface has already been coated in a monolayer of proteins; hence, the host cells do not *see* the material but *see* instead the dynamic protein layer(s) (Fig. 6.32).

Although a comprehensive, rigorous protein adsorption theory is still pending, some parts of the adsorption process are becoming more and more accurately modeled. Protein adsorption on a solid biomaterial surface is energetically favorable. The kinetic of protein adsorption to a solid material surface typically consists of a very rapid initial phase, followed by a slower phase upon approaching to the steady-state phase. Initially, the proteins adsorb as quickly as they arrive at the largely empty surface. Later on, it is more and more difficult for the arriving proteins to find and fit into an empty spot on the surface. This is why the dynamics slows down.

The adsorption from a complex protein mixture, which happens when body fluids come into contact with biomaterials, is selective and leads to the protein enrichment of the surface. Of note, enrichment means that the fraction of the total mass of the adsorbed protein layer(s)

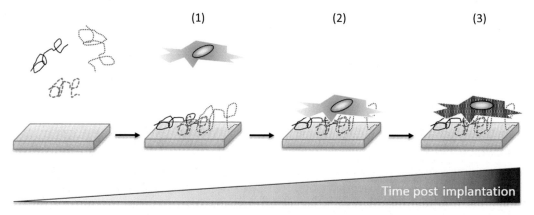

FIG. 6.32   The evolution of a solid biomaterial surface upon implantation. Cell adhesion to the surface of a biomaterial is a major factor mediating its biocompatibility and integration. The following events succeed each other at the biomaterial-biological fluids interface: (1) *protein adsorption* on the solid material surface happens first ($t < 1\,s$). This may lead to the production of a multilayer ($1\,s < time < 10\,s$) and eventually unfold ($10\,s < time < 30\,s$); (2) *cell adhesion* to protein layer ($30\,s < time < minutes$); (3) protein-mediated cell-biomaterial interactions may lead to *cell activation* (soon after).

corresponding to a given protein is often much higher than the fraction of the same protein in the liquid bulk phase mixture in which was dispersed. Of note, proteins that adsorb to solid surfaces can undergo *conformational changes*, because of the relatively low structural stability of proteins and the tendency to unfold to allow further noncovalent bonding with the surface. This means that *soft* proteins (i.e., those with low thermodynamic stability) do adsorb more readily and more tenaciously to surfaces than *hard* proteins and it implies that less stable proteins are more adsorptive. Yet, the conformational changes happening upon protein adsorption induce the exposure of hidden domains and sequences, which are typically hydrophobic, that may serve as receptor sites for inflammatory cells, which then trigger the FBR and may initiate other adverse responses, such as inflammation and coagulation. For instance, when the zymogen XII is exposed to a neutral or anionic artificial surface, it undergoes autoactivation to give rise to factor XIIa, which is the first component of the classic intrinsic pathway of coagulation (Tankersley and Finlayson, 1984). It has been shown that such autoactivation has a great practical impact on the host coagulation during contact between the blood and artificial materials such as the glass.

Protein adsorption is initially strongly influenced by protein diffusion, but protein affinity for the surface becomes soon critically important. Accordingly, the more concentrated and smaller proteins adsorb to the surface first and are then displaced by larger, more strongly interacting proteins that arrive at the surface at a later time. This leads to the *compositional changes* of the protein layer(s) at the biomaterial surface. The state of the final adsorbed protein layer is determined by the manner in which the proteins are able to reorganize themselves on the surface, which depends on both protein-protein and protein-surface interactions in the presence of the surrounding aqueous environment. This means that the relative enrichment of a given protein when adsorbed from a complex mixture, such as plasma, is unique to each biomaterial surface.

Protein mixtures may also give rise to *multilayers*. This phenomenon is not surprising if one considers that, once the first monolayer is established, subsequent proteins have access and can adsorb to a completely different interface.

A schematic view of the final results of protein adsorption on a solid biomaterial surface is portrayed in Fig. 6.33.

The importance of the adsorbed proteins in biomaterials science is well-illustrated by the recent and ongoing efforts in developing protein-repellent materials, also referred to as *nonfouling* materials. Because fibrinogen adsorption is required for platelet adhesion, an obvious approach to improve the biomaterial hemocompatibility is to reduce this phenomenon through the design of protein-repellent biomaterials, such as various types of PEO coatings that bury the biomaterial itself.

## 6.3.3 Response to Wear Debris

Advancements in medicine and medical interventions have increased human life expectancy. This demands longer lifetime body support from the major body-bearing joints. Consequently, the incidence of total joint arthroplasty (TJA) is increasing, with more than four million primary total hip and knee arthroplasties (THA, TKA) projected annually by 2030 in the US alone, and revision rates are expected to double even earlier (Kurtz, 2007).

Providing a smooth, durable, and sliding surface that can afford sufficient and stable movement without pain over time, prostheses utilized in TJA serve to mimic the function of the natural joint and therefore face similar mechanical and chemical challenges. Because central to the longevity of orthopedic implants, wear debris and the subsequent tissue reaction to such particles are discussed in this section.

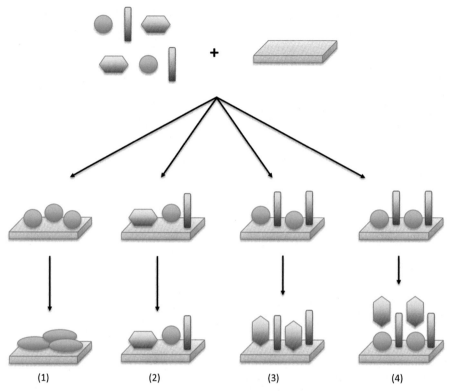

**FIG. 6.33** A schematic front view of protein absorption on a solid biomaterial surface and evolution over time. Differently *colored circles, rectangles, and hexagons* represent structurally diverse proteins. (1) *Conformational changes* may happen upon protein adsorption. (2) Depending on the relative composition of the protein mixture and affinity for the surface, a *protein monolayer* may form. (3) Over time, lower-affinity proteins can be replaced by higher-affinity proteins, leading to *compositional changes*. (4) Multilayer protein adsorption can be also observed for some protein mixture-surface combinations.

*Periprosthetic osteolysis*, which is the active resorption of bone matrix with bone erosion and loss around an orthopedic implant, is the primary concern associated with a local, low inflammatory response that evolves to implant aseptic loosening, and ultimately leads to implant failure (Nine et al., 2014). The general particulate debris features on which local inflammation has been shown to depend are:

- the *particle load*, which is the concentration of phagocytosable particles, in terms of size and volume, per tissue volume. Though it is obvious, if a given debris mass is comprised of small diameter particles ($\varnothing_{part}$), there will

be far greater numbers than if the same debris mass was composed of larger particles. Of note, greater particle load increases inflammation. Besides, also the particle size per se does matter;
- the *aspect ratio*, which relies on the particle shape. In this regard, fibers are known to be more proinflammatory than round-shaped debris. However, it remains unknown which length to width ratio is the most proinflammatory;
- the *chemical reactivity*. The more chemically reactive, the more proinflammatory debris are.

Currently, hip prostheses tend to comprise a titanium (Ti) or cobalt-chromium (CoCr) alloy femoral stem, either cemented with polymethylmethacrylate (PMMA) (see Section 4.2.2.3) or press fitted, which is connected to a CoCr alloy or ceramic head. The head articulates with a UHMWPE (see Section 4.2.1.1) or a ceramic or CoCr cup liner, which is cemented, screwed, or press fitted into the acetabulum (Sansone et al., 2013). UHMWPE has been utilized for long in TJA, as it is durable and has low antigenicity and well-established clinical success. Of note, UHMWPE usually wears approximately 120 μm a year (range, 90–300 μm) as a result of cyclic loading (i.e., human walking). UHMWPE wear-induced osteolysis is a well-documented degenerative process by which prosthetic debris induces a nonspecific and self-propagating immune response, with macrophages and osteoclasts being the main effectors of bone catabolism. This results in prosthetic loosening with eventual failure or fracture (Kandahari et al., 2016).

To shed light on the overall adverse biological reaction to prosthetic debris, which has not been fully deciphered yet, a basic understanding of physiologic bone metabolism is required. Normal bone maintenance relies on the balance of bone formation and breakdown, which involves the coordinated function of osteoblasts and osteoclasts, respectively. Therefore, either a decrease in osteoblastic bone formation or an increase in osteoclast bone resorption (or both) result in *osteolysis* (Fig. 6.34).

The size of UHMWPE wear debris is known to determine the nature of the cellular response to them (Affatato et al., 2001). Although highly variable, most of the UHMWPE wear particulates range from 0.1 to 1.0 μm in size (mean size of 0.5 μm) when recovered from the hip, and are much larger when recovered from the knee. Of note, debris in the range of 0.1–1.0 μm are thought to be the most biologically active, with reported critical size range comprised between 0.2 and 0.8 μm. UHMWPE particulate debris with size <10 μm are ingested by macrophages and dendritic cells, but they are unfortunately unable to degrade them. Of note, a particle should have a size < 10 μm (ranging from 150 nm to 10 μm) to undergo phagocytosis (Bitar, 2015). If ingested, UHMWPE wear debris are next released secondary to endosomal instability and are likely to be rephagocytosed by other macrophages and dendritic cells. The actual or attempted ingestion of particles, referred to as phagocytosis and frustrated phagocytosis, respectively, occur through nonspecific ingestion mechanisms and result in the activation of macrophages and dendritic cells. They release proinflammatory cytokines that, in

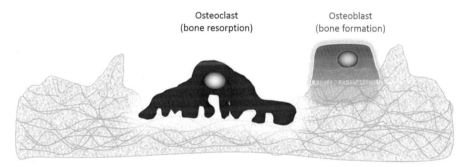

Osteoclast
(bone resorption)

Osteoblast
(bone formation)

**FIG. 6.34** Bone remodeling occurs throughout the human lifetime and is the process by which minerals and tissues are constantly added/removed from bone. The formation of new bone is called ossification, and the removal of bone is called resorption. Bone remodeling is important in regular growth and development of skeletal structure, but it is also necessary following injuries such as fractures. Mechanical demand on bones can also increase or decrease the rate of bone remodeling on specific bones. Osteoblasts do form bone matrix, whereas osteoclasts function in resorption, that is, the breakdown of bone matrix.

turn, inhibit the activity of the osteoblasts and activate osteoclasts (i.e., periprosthetic osteolysis) in a perpetual inflammatory local cascade (Kandahari et al., 2016). Instead, when macrophages are challenged by UHMWPE particles > 20 μm, FBGCs do form, and UHMWPE debris are trapped near the cell surface and fixed in the periprosthetic tissue, a fibrous and granulomatous area composed of a complex cell amalgam and particulate debris. The lymphatic transport is thought to be a major route for dissemination of wear debris. Wear particles may migrate via perivascular lymph channels as free or macrophage-phagocytosed particles. Within abdominal lymph nodes, the majority of disseminated particles are submicron in size, but UHMWPE debris as large as 30 μm have also been identified. These particles may further disseminate to the liver or spleen, but the effect of the systemic dissemination of UHMWPE wear debris is still controversial without established risk of toxicity and carcinogenicity to date (Bitar, 2015).

Because the biomaterials from which a prosthesis is formed have implications on this process, research is focusing on developing prostheses with decreased wear rates, and therefore that induce less osteolysis and aseptic loosening (Kandahari et al., 2016). In this regard, cross-linked UHMWPE has recently come into use, proven in vitro to decrease wear by up to 92%, and with promising short-term in vivo results (Jacobs et al., 2007; Thomas et al., 2011) (see Section 4.2.1.1).

## 6.3.4 Release of Toxic Products From the Biomaterial

The presence of biologically active substances within or upon the biomaterial, such as degradation products and toxic by-products that are prone to solubilization (i.e., *leachables*) in biological fluids and/or dispersion into the surrounding tissues, are the most frequent cause for a biomaterial incompatibility and device failure.

The examples of release of some toxic products from biomaterials are manifold, but one stands out, as described in the following. Although severe adverse reactions to implants are relatively rare, certain types of hip prosthesis, such as those that have direct contact between two metal joint components (MoM: metal-on-metal) have come under the spotlight amid evidence of adverse effects related to metal ion release. As this design was particularly attractive for its longevity, they have been widely used over the last few years, especially in young patients (Sansone et al., 2013). The most widely used metal implants are cobalt-chromium-molybdenum (CoCrMo) and cobalt-nickel-chromium-molybdenum (CoNiCrMo), then Ti alloys, and recently new zirconium and tantalum alloys (Drummond et al., 2015).

The adverse effects of such MoM implants rely on the presence of degradation products, which are primarily generated by corrosion and wear. The main degradation products are metal oxides (e.g., $CoO$), hydroxides (e.g., $Cr(OH)_3$), and metal phosphates (e.g., $CrPO_4$), and wear debris, whose size and shape changes with the severity of wear and the passage of time, as corrosion is a continuous process.

Of note, metal particles are markedly smaller than UHMWPE debris (<50 nm in size vs >0.1 μm) and more numerous (up to 13,500 times more in MoM hip arthroplasties) (Cobb and Schmalzreid, 2006).

Owing to their small size, nanoparticles are first ingested by macrophages in the periprosthetic tissue, leading to local inflammation, but may also disseminate systemically via lymphatics to lymph nodes, bone marrow, liver, and spleen, leading to systemic sequelae (Urban et al., 2000). In target organs, metal nanoparticles can penetrate cells mainly by diffusion or endocytosis. Once intracellularly located, the particles are exposed to the oxidative attack intended to destroy them but result in the generation of metal ions and free radicals that induce oxidative damage to the nucleic acids, proteins, and lipids (Sansone et al., 2013).

Likewise, metal ions may cause local toxic effects due to cell and tissue damage through intracellular mechanisms described herein. Histologically, inflammatory pseudotumor, which is the clinical term given to an aseptic solid or cystic mass in the periprosthetic soft tissue, may form in addition to localized osteolysis (Drummond et al., 2015). Metal ions are transported at distance from the orthopedic implant throughout the body (Doorn et al., 1998). Most in vitro and in vivo studies related to the systemic effects induced Co and Cr. Co toxicity typically affect many organs, and can cause various types of neurological (e.g., vertigo, deafness, blindness, convulsions), cardiological, and hematological symptoms. Instead, Cr seems to be less cytotoxic than Co.

## 6.3.5 Bacterial Adhesion to Biomaterials and Strategies to Evade It

It is estimated that at least 20 million people in the US have a biomaterial device implant. Implant device failure or implant-associated infections can have unfavorable consequences for the implant device function and the host. Although the risk of implant-associated infections is small (1%–7%), such infections are associated with considerable morbidity and require expensive health care. It is worth noting that the medical and surgical cost of treating certain device failures or implant-associated infections can average up to $50,000 per patient (Higgins et al., 2009).

In nature, bacteria may exist in the form of free-floating (i.e., planktonic) or sessile (i.e., fixed in one place) microorganisms that may grow upon living and inanimate surfaces such as tooth enamel, heart valves, lungs, and indwelling medical devices, respectively. The common feature of the nonmotile cell state is that bacteria may develop a *biofilm*. The biofilm is a *sessile* (i.e., fixed in one place) community of microbial cells associated with a typically abiotic surface (but it can be a biotic surface as well) at the interface between the biomaterial and the host environment, encased in an autoproduced matrix called *extracellular polymeric substance* (EPS, or *slime*) (Donlan, 2001). Sessile bacteria embedded in such slimy matrix are 10–1000 times more resistant to antibiotics than planktonic cells. The EPS protects them from assaults such as those by immune cells. Besides, it has been shown that phagocytic cells exhibit a reduced bactericidal capability following adhesion to a biomaterial surface. In part, this is related to a respiratory burst that occurs upon adhesion and leaves the adherent cell exhausted (Anderson et al., 2008). In light of this, biofilms frequently lead to prosthetic device failure.

Although it is common knowledge that biofilm formation occurs through multiple pathways, and that the spatial biofilm structure is species-dependent and environmental cues-dependent, the general process of biofilm formation can be broadly summarized in reversible and irreversible stages outlined here (Fig. 6.35):

1. *reversible adhesion*. This step involves the introduction of bacteria to a surface, a process that is at least in part stochastic, driven by Brownian motion and gravitational forces, and influenced by surrounding hydrodynamic forces. Biomaterial surface properties, along with bacterial cell-surface composition, affect velocity and direction towards or away from the contact surface. Upon intercepting the surface, adherence is mediated by adhesive cell appendages. However, the decision to stick is not absolute at this time; initial attachment is dynamic and reversible, during which bacteria can detach and rejoin the planktonic population if perturbed;
2. *irreversible adhesion*. Irreversible attachment is attained by bacteria that can withstand shear

## 5 stages of biofilm development

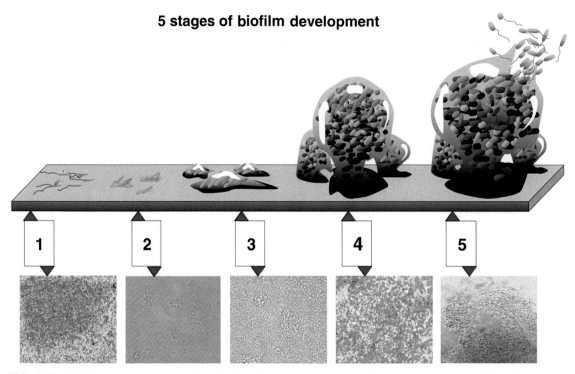

**FIG. 6.35**    Biofilm maturation is a complex developmental process involving five stages: stage 1, reversible adhesion; stage 2, irreversible adhesion; stage 3, maturation I; stage 4, maturation II; stage 5, dispersion. Each stage of development in the diagram is paired with a photomicrograph of a developing *Pseudomonas aeruginosa biofilm*. All photomicrographs are shown to same scale. *From Monroe, D., 2007. Looking for chinks in the armor of bacterial biofilms. PLoS Biol. 5 (11), e307.*

forces and maintain a steadfast grip on the surface. The switch from a solitary planktonic to a sticky bacterial phenotype involves a change in the bacteria so that they initiate to produce specific adhesive proteins called *adhesins*;

3. *maturation I* or *microcolony formation*. Surface contact triggers responses that lead to gene expression changes, up-regulating factors favoring sessility, such as those implicated in the formation of the EPS (Kostakioti et al., 2013). The slime serves as a 3D scaffold: it exerts structural functions and plays a role in a number of processes including cell attachment, cell-to-cell interactions that allow forming small bacteria

aggregates named *microcolonies*, and antimicrobial tolerance. Of note, microcolonies vary greatly in dimensions and grow rate depending on the biofilm-forming bacterial species;

4. *maturation II*. Within the mature biofilm, there is a bustling community that actively exchanges and shares products that play a role in maintaining biofilm architecture and providing a favorable environment for the resident bacteria. *Quorum sensing* (QS) is the regulation of gene expression in response to fluctuations in cell-population density through the release of biochemical mediators called *autoinducers*. Although the nature of the such chemical signals, the signal relay

mechanisms, and the target genes controlled by bacterial QS systems differ; however, the ability to communicate with one another allows bacteria to coordinate gene expression, and therefore the behavior of the entire community. Through QS, single-cell organisms assume a temporary multicellular lifestyle (Miller and Bassler, 2001);

5. *dispersion*. Dispersal of cells from the biofilm colony is an essential stage of the biofilm life cycle. Biofilm dispersal can be the result of several cues, such as alterations in nutrient availability, oxygen fluctuations and increase of toxic products, or other stress-inducing conditions. Besides passive dispersal, brought about by shear stresses, bacteria have evolved ways to perceive environmental changes and gauge whether it is still beneficial to reside within the biofilm or whether it is time to resume a planktonic lifestyle. This means that, as biofilms mature, dispersal may thus become an option, and EPS-degrading enzymes secreted by the sessile population contribute to bacterial detachment from the EPS (Kostakioti et al., 2013).

Because up to 50% of patients are catheterized during their stay in a hospital, the risk of infections induced by urinary catheters turns out to be a critical clinical issue. Microbial biofilm may develop at the catheterization site or elsewhere in the body, due to the migration of microorganisms through or around the catheter. The bacteria involved in urinary catheter infections are typically *Klebsiella pneumoniae* and *Escherichia coli*. Instead, *Streptococcus aureus* and some *Staphylococci* are typically responsible for infections in orthopedic implant patients (Widmer, 2001). Orthopedic prosthesis-related infections represent a serious concern in patients undergoing orthopedic device implantation. Although according to the type of infection (acute or chronic) and the type of pathogens involved, multiple antibiotic treatments are often necessary and effective in orthopedic implant patients, the implant removal is the only solution in some circumstances, such as severe sepsis.

Because biofilms, once formed, are hard to eradicate, academics and makers of medical devices have tried to stop them from forming in the first place, in other words, engineering the device surfaces to disrupt the initial adhesion (Monroe, 2007). As a rule of thumb, highly hydrated, hydrophilic, and, their opposite, highly hydrophobic, uncharged surfaces are generally regarded as poorly adhesive substrates (Tanzi and Faré, 2017). In addition, antibiotic-imbued surfaces have shown to delay biofilm growth, although this technique has the important downside of encouraging resistant bacteria. For this reason, scientists are going beyond simple modifications to incorporate biologically active agents into the surfaces of medical devices.

## 6.3.6 Calcification

In healthy subjects, $\approx 99\%$ of the body calcium is stored in bones and teeth where it contributes to their mechanical and structural properties, while the rest is in the blood, muscles, and other body tissues. *Calcification* is the process whereby calcium salts are deposited in an organic matrix.

Among calcifications, we can distinguish between the physiological process that happens in bones, teeth, and otoliths, and pathological one, happening elsewhere in the body, such as in:

- small and large arteries. Most individuals aged > 60 years have progressively enlarging deposits of calcium mineral in major arteries. Vascular calcification results in vessel stiffening, which impairs cardiovascular hemodynamics and is associated with substantial morbidity and mortality (Demer and Tintut, 2008). The deposition of calcium into the neointima and/or the tunica media of the blood vessels can occur during pathological conditions such as atherosclerosis;

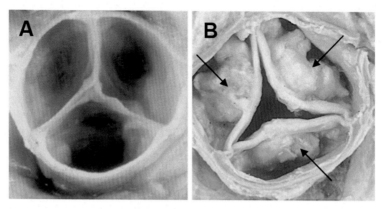

FIG. 6.36   Tricuspid or right atrioventricular valve, so-called because it displays three cusps and is placed between the right atrium and ventricle. (A) Gross photograph of a normal, healthy heart valve. (B) Calcified stenotic tricuspid valve characterized by marked leaflet thickening by nodular calcium deposits *(noted by arrows)*. Calcium deposits are more commonly found in areas of turbulent blood flow, such as the valve leaflets (doors) and the annulus (the ring on which the valve sits, not displayed in (B)). *Modified from Willis, M.S., Homeister, J.W., Stone, J., 2014. Cellular and Molecular Pathobiology of Cardiovascular Disease, first ed. Academic Press.*

- heart valves (Fig. 6.36);
- the brain, through a process also referred to as cranial calcification;
- joints and tendons, such as knee joints and rotator cuff tendons;
- soft tissues like breasts, muscles, and fat. Breast calcifications occur when calcium builds up within the breast soft tissue. Breast injuries, cell secretions, infections, and inflammation can all cause breast calcifications, which can be classified into two main types: macrocalcifications and microcalcifications, which are small and calcium buildups, respectively;
- kidney, bladder, and gallbladder. Nephrolithiasis is a very common condition in industrialized countries; 80% of kidney stones are made of calcium oxalates. Prostatic calculi are another common finding, although few investigations exist on their chemical composition.

The pathologic calcification, sometimes also referred to as *heterotopic calcification* or *calcinosis*, is the precipitation or the abnormal tissue deposition of calcium, primarily phosphates, which give rise to unorganized small white calcium lumps (Ratner et al., 2012).

Pathologic calcifications may be classified according to a mineral balance in:

1. *dystrophic calcification* refers to the calcification that forms in degenerating, diseased, and dead tissues, despite normal serum calcium and phosphate levels, and without systemic mineral imbalance. Therefore, the abnormal deposition of calcium salts occurs in any type of dead or dying tissue, with normal serum levels of calcium and normal calcium metabolism. The dystrophic calcification is affected by multiple factors, which include serum phosphorus and alkaline phosphatase levels, tissue pH, vitamin D concentration, hormonal balance, blood supply, and injury extent. Of note, most calcifications are dystrophic For instance, the calcification of the aortic valve that causes aortic stenosis in the elderly is dystrophic;
2. *metastatic calcification* that is the deposition of calcium salts in normal tissues, and reflects some derangement in calcium metabolism

and an elevation of calcium levels in the blood (*hypercalcemia*) and in other tissues.

In this context, calcification of soft tissues may be an unspecific local response or is only a symptom of a complex underlying disease.

Pathologic calcification is further classified depending on where it happens onto/into the implant material in:

1. *intrinsic calcification*, which occurs deep within the implant material;
2. *extrinsic calcification*, which is found at the material surface, and is generally associated with cells, proteins, thrombi, or vegetations that have attached to the implant surface.

Calcified degeneration of glutaraldehyde-pretreated porcine and bovine pericardial heart valves is a well-characterized, clinically relevant dysfunction due to intrinsic and extrinsic calcification and that typically cause bioprosthetic stenosis and failure.

# References

Affatato, S., Fernandes, B., Tucci, A., Esposito, L., Toni, A., 2001. Isolation and morphological characterisation of UHMWPE wear debris generated in vitro. Biomaterials 22 (17), 2325–2331.

Anderson, J.M., 2001. Biological responses to materials. Annu. Rev. Mater. Res. 31 (1), 81–110.

Anderson, J.M., Rodriguez, A., Chang, D.T., 2008. Foreign body reaction to biomaterials. Semin. Immunol. 20 (2), 86–100.

Aslan, J.E., Itakura, A., Gertz, J.M., McCarty, O.J., 2012. Platelet shape change and spreading. Methods Mol. Biol. 788, 91–100.

Bagoly, Z., Ariens, R.A.S., Rijken, D.C., Pieters, M., Wolberg, A.S., 2017. Clot structure and fibrinolysis in thrombosis and hemostasis. Biomed. Res. Int. 2017, 4645137.

Bitar, D., 2015. Biological response to prosthetic debris. World J. Orthod., 6(2).

Bowlin, G.L., Wnek, G.E., 2005. Encyclopedia of Biomaterials and Biomedical Engineering. CRC Press, Boca Raton, US.

Cao, W., Hench, L.L., 1996. Bioactive materials. Ceram. Int. 22 (6), 493–507.

Cobb, A.G., Schmalzreid, T.P., 2006. The clinical significance of metal ion release from cobalt-chromium metal-on-metal hip joint arthroplasty. Proc. Inst. Mech. Eng. H J. Eng. Med. 220 (2), 385–398.

Cooper, G.M., 2000. The Cell: A Molecular Approach, second ed. Oxford University Press, Sunderland, MA.

Coppinger, J.A., Cagney, G., Toomey, S., Kislinger, T., Belton, O., McRedmond, J.P., Cahill, D.J., Emili, A., Fitzgerald, D.J., Maguire, P.B., 2004. Characterization of the proteins released from activated platelets leads to localization of novel platelet proteins in human atherosclerotic lesions. Blood 103 (6), 2096–2104.

Demer, L.L., Tintut, Y., 2008. Vascular calcification: pathobiology of a multifaceted disease. Circulation 117 (22), 2938–2948.

Donlan, R.M., 2001. Biofilm formation: a clinically relevant microbiological process. Clin. Infect. Dis. 33 (8), 1387–1392.

Doorn, P.F., Campbell, P.A., Worrall, J., Benya, P.D., McKellop, H.A., Amstutz, H.C., 1998. Metal wear particle characterization from metal on metal total hip replacements: transmission electron microscopy study of periprosthetic tissues and isolated particles. J. Biomed. Mater. Res. 42 (1), 103–111.

Drummond, J., Tran, P., Fary, C., 2015. Metal-on-metal hip arthroplasty: a review of adverse reactions and patient management. J. Funct. Biomater. 6 (3), 486–499.

Eberli, D., 2011. Regenerative Medicine and Tissue Engineering. IntechOpen, London, UK.

Eming, S.A., Brachvogel, B., Odorisio, T., Koch, M., 2007. Regulation of angiogenesis: wound healing as a model. Prog. Histochem. Cytochem. 42 (3), 115–170.

Frantz, C., Stewart, K.M., Weaver, V.M., 2010. The extracellular matrix at a glance. J. Cell Sci. 123 (24), 4195–4200.

Franz, S., Rammelt, S., Scharnweber, D., Simon, J.C., 2011. Immune responses to implants—a review of the implications for the design of immunomodulatory biomaterials. Biomaterials 32 (28), 6692–6709.

Gale, A.J., 2011. Current understanding of hemostasis. Toxicol. Pathol. 39 (1), 273–280.

Goodman, S., 2008. Medical Cell Biology, third ed. Academic Press, Cambridge, US.

Gordon, M.K., Hahn, R.A., 2010. Collagens. Cell Tissue Res. 339 (1), 247–257.

Higgins, D.M., Basaraba, R.J., Hohnbaum, A.C., Lee, E.J., Grainger, D.W., Gonzalez-Juarrero, M., 2009. Localized immunosuppressive environment in the foreign body response to implanted biomaterials. Am. J. Pathol. 175 (1), 161–170.

Hlady, V., Buijs, J., 1996. Protein adsorption on solid surfaces. Curr. Opin. Biotechnol. 7 (1), 72–77.

Hochberg, M.C., Silman, A.J., Smolen, J.S., Weinblatt, M.E., Weisman, M.H., 2015. Rheumatology, sixth ed. Elsevier, New York City, US.

Hu, W.J., Eaton, J.W., Ugarova, T.P., Tang, L., 2001. Molecular basis of biomaterial-mediated foreign body reactions. Blood 98 (4), 1231–1238.

Hume, D.A., 2008. Macrophages as APC and the dendritic cell myth. J. Immunol. 181 (9), 5829–5835.

Jacobs, C.A., Christensen, C.P., Greenwald, A.S., McKellop, H., 2007. Clinical performance of highly cross-linked polyethylenes in total hip arthroplasty. J. Bone Joint Surg. Am. 89 (12), 2779–2786.

Janeway, C.A.J., Travers, P., Walport, M., Shlomchik, M.J., 2001. Immunobiology: The Immune System in Health and Disease, fifth ed. Garland Science, Oxford, UK.

Kamkin, A., Kiseleva, I., 2005. Mechanosensitivity in Cells and Tissues. Academia, Moscow.

Kandahari, A.M., Yang, X., Laroche, K.A., Dighe, A.S., Pan, D., Cui, Q., 2016. A review of UHMWPE wear-induced osteolysis: the role for early detection of the immune response. Bone Res. 4, 16014.

Klopfleisch, R., Jung, F., 2017. The pathology of the foreign body reaction against biomaterials. J. Biomed. Mater. Res. A 105 (3), 927–940.

Kostakioti, M., Hadjifrangiskou, M., Hultgren, S.J., 2013. Bacterial biofilms: development, dispersal, and therapeutic strategies in the dawn of the postantibiotic era. Cold Spring Harb. Perspect. Med. 3 (4), a010306.

Kroll, M.H., Schafer, A.I., 1989. Biochemical mechanisms of platelet activation. Blood 74 (4), 1181–1195.

Kurtz, S., 2007. Projections of Primary and Revision Hip and Knee Arthroplasty in the United States from 2005 to 2030. J. Bone Joint Surg. Am. 89(4).

Litvinov, R.I., Weisel, J.W., 2016. What is the biological and clinical relevance of fibrin? Semin. Thromb. Hemost. 42 (4), 333–343.

Malahias, M., Jordan, D.J., Hughes, L.C., Hindocha, S., Juma, A., 2016. A literature review and summary of capsular contracture: an ongoing challenge to breast surgeons and their patients. Int. J. Surg. Open 3, 1–7.

Mandal, A., Viswanathan, C., 2015. Natural killer cells: in health and disease. Hematol. Oncol. Stem Cell Ther. 8 (2), 47–55.

Miller, M.B., Bassler, B.L., 2001. Quorum sensing in bacteria. Annu. Rev. Microbiol. 55 (1), 165–199.

Monroe, D., 2007. Looking for chinks in the armor of bacterial biofilms. PLoS Biol. 5 (11), e307.

Narayan, R.J., 2016. Monitoring and Evaluation of Biomaterials and Their Performance In Vivo. Woodhead Publishing, Cambridge, UK.

Nine, M., Choudhury, D., Hee, A., Mootanah, R., Osman, N., 2014. Wear debris characterization and corresponding biological response: artificial hip and knee joints. Materials 7 (2), 980–1016.

Polasek, J., 2006. Three modes of platelet degranulation? Pathophysiol. Haemost. Thromb. 35 (5), 408–409.

Prideaux, M., Findlay, D.M., Atkins, G.J., 2016. Osteocytes: the master cells in bone remodelling. Curr. Opin. Pharmacol. 28, 24–30.

Ratner, B.D., Hoffman, A.S., Schoen, F.J., Lemons, J.E., 2012. Biomaterials Science: An Introduction to Materials in Medicine, third ed. Academic Press, Oxford, UK.

Rozario, T., DeSimone, D.W., 2010. The extracellular matrix in development and morphogenesis: a dynamic view. Dev. Biol. 341 (1), 126–140.

Samson, A.L., Alwis, I., Maclean, J.A.A., Priyananda, P., Hawkett, B., Schoenwaelder, S.M., Jackson, S.P., 2017. Endogenous fibrinolysis facilitates clot retraction in vivo. Blood 130 (23), 2453–2462.

Sansone, V., Pagani, D., Melato, M., 2013. The effects on bone cells of metal ions released from orthopaedic implants. A review. Clin. Cases Miner. Bone Metab. 10 (1), 34–40.

Schulze, H., Shivdasani, R.A., 2005. Mechanisms of thrombopoiesis. J. Thromb. Haemost. 3 (8), 1717–1724.

Tankersley, D.L., Finlayson, J.S., 1984. Kinetics of activation and autoactivation of human factor XII. Biochemistry 23 (2), 273–279.

Tanzi, M.C., Faré, S., 2017. Characterization of Polymeric Biomaterials. Woodhead Publishing, Cambridge, UK.

Thevenot, P., Hu, W., Tang, L., 2008. Surface chemistry influences implant biocompatibility. Curr. Top. Med. Chem. 8 (4), 270–280.

Thomas, G.E., Simpson, D.J., Mehmood, S., Taylor, A., McLardy-Smith, P., Gill, H.S., Murray, D.W., Glyn-Jones, S., 2011. The seven-year wear of highly cross-linked polyethylene in total hip arthroplasty: a double-blind, randomized controlled trial using radiostereometric analysis. J. Bone Joint Surg. Am. 93 (8), 716–722.

Tonnesen, M.G., Feng, X., Clark, R.A., 2000. Angiogenesis in wound healing. J. Investig. Dermatol. Symp. Proc. 5 (1), 40–46.

Undas, A., Ariens, R.A., 2011. Fibrin clot structure and function: a role in the pathophysiology of arterial and venous thromboembolic diseases. Arterioscler. Thromb. Vasc. Biol. 31 (12), e88–e99.

Urban, R.M., Jacobs, J.J., Tomlinson, M.J., Gavrilovic, J., Black, J., Peoc'h, M., 2000. Dissemination of wear particles to the liver, spleen, and abdominal lymph nodes of patients with hip or knee replacement. J. Bone Joint Surg. Am. 82 (4), 457–476.

White, J.G., Estensen, R.D., 1972. Degranulation of discoid platelets. Am. J. Pathol. 68 (2), 289–302.

Widmer, A.F., 2001. New developments in diagnosis and treatment of infection in orthopedic implants. Clin. Infect. Dis. 33 (Suppl. 2), S94–106.

Yun, S.H., Sim, E.H., Goh, R.Y., Park, J.I., Han, J.Y., 2016. Platelet activation: the mechanisms and potential biomarkers. Biomed. Res. Int. 2016, 9060143.

Zhang, J.M., An, J., 2007. Cytokines, inflammation, and pain. Int. Anesthesiol. Clin. 45 (2), 27–37.

Zhao, X., Courtney, J.M., Qian, H., 2011. Bioactive Materials in Medicine, first ed. Woodhead Publishing, Cambridge, UK.

## Further Reading

Abbas, A., Lichtman, A.H., Pillai, S., 2017. Cellular and Molecular Immunology, ninth ed. Elsevier, New York City, US.

George, J.N., 2000. Platelets. Lancet 355 (9214), 1531–1539.

McGrath, J.L., 2007. Cell spreading: the power to simplify. Curr. Biol. 17 (10), R357–R358.

Muscolino, J., 2010. Kinesiology: The Skeletal System and Muscle Function, second ed. Elsevier, New York City, US.

Willis, M.S., Homeister, J.W., Stone, J., 2014. Cellular and Molecular Pathobiology of Cardiovascular Disease, first ed. Academic Press, Oxford, UK.

CHAPTER

# 7

# Techniques of Analysis

## 7.1 INTRODUCTION

The analysis techniques include techniques for the characterization of biomaterials and investigation techniques for diagnostic purposes.

**(a)** *Characterization techniques*: those that allow us to describe and understand the properties of materials used for manufacturing biomedical devices, and how they have changed after in vivo implantation;

**(b)** *Analysis techniques for diagnostic purposes*: those that allow us to describe and understand the properties of organs, biological tissues, or substances in the human body for preventive and diagnostic purposes.

The physical principles on which both types of analysis techniques are based are often similar, and, in some cases, even the same, but the output signals are used in a different way, depending on the purpose of the analysis performed.

In the following paragraphs, we will learn some of the main analysis and investigation techniques currently used in the biomedical field.

## 7.2 BIOMATERIAL CHARACTERIZATION

As mentioned previously, the techniques used to characterize biomaterials and/or biomedical devices allow us to study their properties, helping us to:

- identify a material;
- check behavioral uniformity in operation and then evaluate the safety of the material/device;
- make predictions about in vivo performance, and
- carry out quality controls.

Biomaterial properties can be distinguished as either *mass properties* (also called *bulk* properties) or *surface properties*. Both types of properties are very important for any biomedical application, but especially for fabricating devices intended for contact with the biological system. In fact, these properties can control the dynamics at the interface with biological tissues after in vivo implantation.

As shown in Fig. 7.1, once a biomaterial is identified as optimal for a specific application, its mass properties must be verified as adequate for the application's required mechanical characteristics,

*Foundations of Biomaterials Engineering*
https://doi.org/10.1016/B978-0-08-101034-1.00007-4

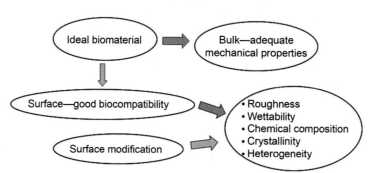

**FIG. 7.1** Schematic view of the role of bulk and surface properties of a biomaterial. It should be noted that surface modification can improve the properties at the interface with the biological system, but it is not always necessary.

while its surface properties must guarantee good biocompatibility. If necessary, it is possible to perform specific surface modifications to improve the interface between the material and biological tissue (for example, improving mineralization in the case of bone tissue, or lowering thrombogenicity if the biomaterial contacts blood).

Mass properties depend on the following structural characteristics:

– intermolecular forces;
– atomic structure (typical of the class of material);
– microstructure;
– interatomic bonds;
– average grain size (for metal and ceramic materials), and
– phase changes (e.g., in the case of semicrystalline or amorphous polymeric materials when the material is at a temperature lower or higher than $T_g$).

Surface properties are very important because they control the biomaterial-biological tissue interface (Fig. 7.2), determining the specific adsorption of proteins present in the biological fluids, and therefore, a different interaction (adhesion, activation, repulsion, etc.) between the cells present at the implantation site (see Section 6.3). The main surface properties are:

– topography/roughness;
– wetting/hydrophilicity;

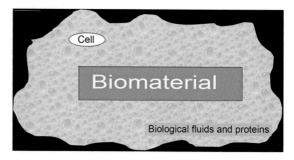

**FIG. 7.2** A biomaterial in the biological environment should present adequate surface properties. Therefore, these must be carefully investigated to predict the right in vivo outcome.

– surface mobility;
– crystallinity, and
– chemical composition.

The main characterization techniques available can be reviewed as shown in Table 7.1:

– thermal analyses;
– spectroscopic analyses;
– chromatography techniques;
– techniques for investigating mechanical properties (see Chapter 2);
– microscopy techniques, and
– other techniques (Table 7.1)

In the following, we will briefly illustrate some of these techniques. However, for the study of the mechanical properties of materials, see Chapter 2.

**TABLE 7.1** Main Characterization Techniques Divided Into Two Categories: Those That Allow Investigation of Mass Properties and Those That allow Investigation of Surface Properties.

| Type of Properties | Information Given |
| --- | --- |
| *Bulk properties* | |
| Differential scanning calorimetry (DSC) | Thermal properties |
| X-ray diffractometry (XRD) | Crystallinity<br>Elemental chemical analysis |
| Infrared (IR) spectroscopy | Chemical structure<br>Concentration of a functional group |
| UV-vis spectroscopy | Chemical structure |
| Nuclear magnetic resonance (NMR) spectroscopy | Chemical structure<br>Concentration of molecules |
| Size exclusion chromatography (SEC) | Molecular weights values and distribution |
| Uniaxial mechanical tests | Stress/strain behavior |
| Cyclic mechanical tests | Dynamic and fatigue behavior |
| *Surface properties* | |
| Optical microscopy (OM) | Surface morphology/topography<br>Grain size |
| Scanning electron microscopy (SEM) | Surface morphology/topography<br>Chemical composition |
| Atomic force microscopy (AFM) | Topography<br>(Surface) mechanical properties |
| Contact angle | Wettability/hydrophilicity |
| Profilometry | Roughness and roughness parameters |

## 7.2.1 Thermal Analyses

Thermal analysis refers to a variety of techniques in which the physical property of a sample is uninterruptedly measured as a function of temperature, while the sample is subjected to a predefined temperature program. Techniques for the study of thermal properties are used in every field of research and technology. The basic information that these techniques provide, such as crystallinity, specific heat and thermal expansion, is of fundamental importance for the research and development of new products and for quality controls.

The most frequently used of these techniques are:

- thermogravimetric analysis (TGA);
- differential thermal analysis (DTA);
- differential scanning calorimetry (DSC);
- thermomechanical analysis (TMA), and
- dynamic TMA (DTMA)

### 7.2.1.1 Thermogravimetric Analysis, TGA

In TGA, the mass of a sample is measured over time as the temperature changes (or as a function of time at a constant temperature), providing information about physical phenomena such as phase transitions, absorption, and desorption, as well as chemical phenomena including chemisorptions, thermal decomposition, and solid-gas reactions (e.g., oxidation or reduction).

The thermogravimetric analyzer itself is an instrument consisting of a sample pan supported by a precision balance, residing in a furnace, and heated or cooled during the experiment. The mass of the sample is monitored during the experiment, and a sample purge gas controls the sample environment.

TGA can interface with a mass spectrometer to identify and measure volatile substances generated during mass changes of the test sample. TGA determines temperature and weight change in decomposition reactions, facilitating quantitative composition analysis, and it may be used to determine water content or detect residual solvents in a material. TGA allows identification of organic materials by measuring the temperature of bond scission in inert atmospheres or of oxidation in air or oxygen.

Inorganic materials, metals, polymers and plastics, ceramics, glasses, and composite materials can be analyzed. Samples can be analyzed in the form of powder or small pieces to maintain an interior sample temperature close to the measured gas temperature.

An example of thermogravimetric analysis is provided in Fig. 7.3, in which the weight percent of each component of a polymer composite can be assessed (the components had been previously identified by use of complementary techniques such as microscopy, XPS, and FTIR).

### 7.2.1.2 Differential Thermal Analysis, DTA

DTA involves heating or cooling a test sample and an inert reference under identical conditions, while recording any temperature difference between the sample and reference. This differential temperature is then plotted against time, or against temperature, permitting the calculation of the heat flow difference between reference and sample, which are kept in almost identical environments (DTA furnace). Therefore, changes in the sample which lead to the absorption or evolution of heat can be detected relative to the inert reference.

The first major applications for this method were in the study of phase diagrams and transition temperatures, and in qualitative analyses of metals, oxides, salts, ceramics, glasses, minerals, and soils.

A DTA curve can be used as a fingerprint for identification purposes, for example, in the study of clays in which the structural similarity of different forms renders diffraction experiments difficult to interpret. For many problems, it is advantageous to use both DTA and TGA, because DTA events can be classified as involving mass change or not.

Fig. 7.4 shows both TGA and DTA curves concerning the analysis of an inorganic sample (calcium oxalate monohydrate).

### 7.2.1.3 Differential Scanning Calorimetry, DSC

This method of analysis is one of the most useful techniques for obtaining quantitative calorimetric measurement by studying enthalpy and entropy changes that come with the transformation of different materials.

Calorimetry holds a special place among other methods. In addition to the method's simplicity and universality, energy characteristics (heat capacity $C_P$ and its integral over temperature, enthalpy H) measured via calorimetry have a clear physical meaning, even though interpretation may sometimes be difficult.

The advantage of DSC over other calorimetric techniques lies in its broad dynamic range with regard to heating and cooling rates, including isothermal and temperature-modulated operation. Today, 12 orders of magnitude in scanning rate can be covered by combining different types of DSCs.

DSC examines how a material heat capacity ($C_p$) is changed by temperature. A sample of known mass is heated or cooled, and the

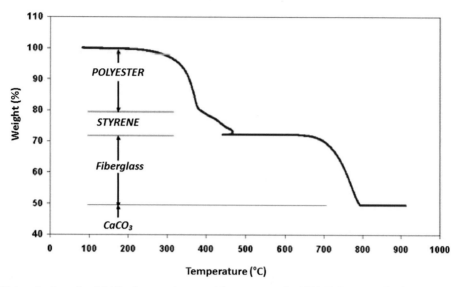

**FIG. 7.3** Determination of weight % of composite material components by TGA. Polyester and polystyrene (71% and 29%, respectively, of the polymeric matrix), fiberglass and $CaCO_3$ (22.9% and 49.3%, respectively, of the whole composite) were easily identified by their different temperatures of combustion or evaporation; the material remaining after TGA was confirmed by XPS to consist only of $CaCO_3$. *From www.andersonmaterials.com/tga.html.*

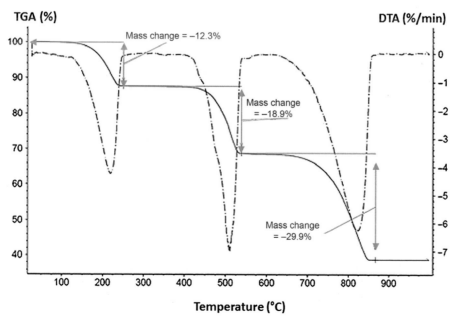

**FIG. 7.4** Thermal analysis (TGA and DTA patterns) of calcium oxalate monohydrate, $CaC_2O_4 \cdot H_2O$. *From www.ebatco.com/laboratory-services/chemical/thermogravimetric-analysis-tga/.*

changes in its heat capacity are tracked as changes in the heat flow. This allows the detection of transitions such as melts, glass transitions, phase changes, and curing. Because of this flexibility, and because most materials exhibit some sort of transition, DSC is used in many industries, including pharmaceuticals, polymers, food, paper, printing, manufacturing, agriculture, semiconductors, and electronics.

From a practical point of view, the difference in heat flow to or from a sample and to or from a reference sample is recorded as a function of temperature (Fig. 7.5). The obtained graphic representation is a *thermogram*, in which the horizontal axis shows the temperature (expressed in °C), and the ordinate shows the flow of heat absorbed by the material under examination.

Through DSC analyses, it is possible to determine the presence and concentration of a crystalline phase in a solid, as well as the melting point of the crystals. This is a technique useful for the characterization of metal alloys and composite materials; in particular, it allows us to obtain information about the phase diagram of alloys.

### 7.2.1.4 *Thermomechanical Analysis, TMA, and Dynamic TMA, DTMA*

Thermomechanical analysis (TMA) measures sample displacement (growth, shrinkage, movement, etc.) as a function of temperature, time, and applied force. Material deformation changes are measured under controlled conditions of force, atmosphere, time, and temperature. Force can be applied in expansion, penetration, compression, flexure, or tensile modes of deformation using specially designed probes. In TMA, a constant static force is applied to a sample, and its modifications are observed as temperature or time changes. Hence, TMA reports dimensional changes.

TMA measures intrinsic material properties (e.g., expansion coefficient, glass transition, Young's modulus), plus processing/product performance parameters (e.g., softening points). As an example, Fig. 7.6 shows expansion and penetration probe measurements of the $T_g$ and the softening point ($T_s$) of a synthetic rubber using a temperature ramp at constant applied force.

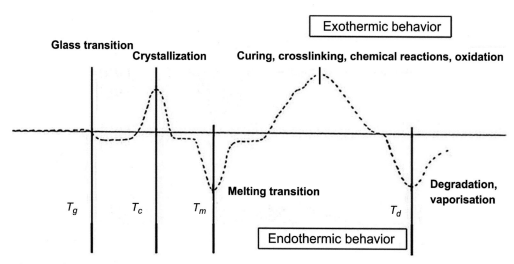

FIG. 7.5　Thermogram showing the main events that can be detected in polymeric materials by a DSC analysis. *From Tanzi, M.C., 2017. Characterization of thermal properties and crystallinity of polymer biomaterials, Fig. 6.3, In: Characterization of Polymeric Biomaterials, first ed., reproduced with permission of Elsevier.*

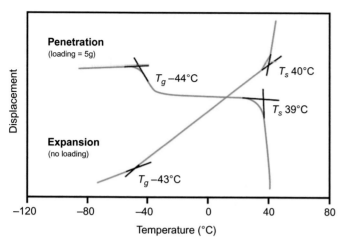

**FIG. 7.6** Expansion and penetration probe measurements of the $T_g$ and the softening point ($T_s$) by TMA analysis of a synthetic rubber using a temperature ramp at constant applied force. The large changes of the Coefficient of Thermal Expansion (CTE) in the expansion pattern indicate the transition temperatures. In penetration, the transitions are detected by the sharp deflection of the probe into the sample. *From www.tainstruments.com/wp-content/uploads/BROCH-TMA-2014-EN.pdf, p.84.*

In Dynamic TMA (DTMA), also named DMA (Dynamic Mechanical Analysis), a known sinusoidal stress and linear temperature ramp are applied to the sample, and the resulting sinusoidal strain is measured. DMTA is a technique where a small deformation (or stress) is applied to a sample in a cyclic manner, and the force exerted on the probe alternates automatically by the given frequency. This method allows study of the viscoelastic behavior of materials.

As an example, Fig. 7.7 shows a dynamic test in which a semicrystalline polyethylene terephthalate (PET) film is subjected to a fixed sinusoidal force in tension during a linear temperature ramp. The resulting strain and phase data are used to calculate the material's viscoelastic properties (e.g., $E'$, $E''$, and tan $\delta$). The plotted data shows dramatic modulus changes as the film is heated through its glass transition temperature.

Both TMA and DMTA methods detect thermal transitions, but DMTA is much more sensitive, allowing the detection of secondary thermal transition.

A more detailed description of the methods described above goes beyond the scope of this book. For a more in-depth understanding, please refer to specific texts and publications (a bibliography is provided at the end of the chapter).

## 7.2.2 Spectroscopic Analyses

Spectroscopy studies the interaction of light with matter and finds application both in physics and in various instrumental analytical techniques. *Spectroscopic physics* uses emitted, absorbed, or diffracted light to understand and study the mechanisms of a chemical system. *Analytical spectroscopy* uses the same identical processes to determine the content and concentrations of atomic and molecular species present in a chemical system.

Spectroscopic analyses are based on the interaction of different types of electromagnetic radiation with matter, and allow us to obtain a theoretical representation of a material or an object that must be interpreted based on models or assumptions.

This abstract representation can then be processed by using appropriate software to obtain an image (we will see later how this can be used as a diagnostic method of investigation) or a *spectrum*, which is a graph of the signal intensity issued by the analysis as a function of the frequency.

In particular, if we expose a substance to the action of electromagnetic radiation, the energy contained in it can be absorbed by the molecules

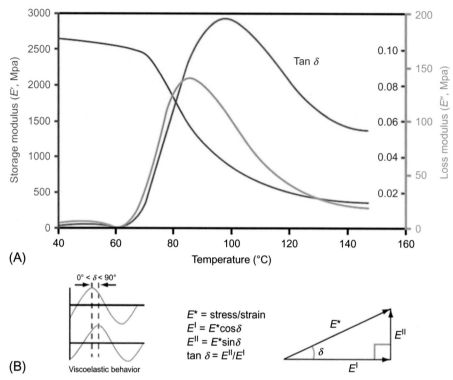

**(A)**

**(B)**

**FIG. 7.7**   (A) DMTA test in tension of a semicrystalline PET film subjected to a fixed sinusoidal force during a linear temperature ramp showing the material viscoelastic properties. (B) Sinusoidal strain, and sine wave phase difference ($\delta$) resulting from the sinusoidal force and linear temperature ramp applied to the sample are used to calculate the material viscoelastic properties ($E'$, $E''$, and tan $\delta$ parameters). *From the TA Instruments technical brochure "Thermal Analysis," www.tainstruments.com/wp-content/uploads/BROCH-TMA-2014-EN.pdf, pp. 81, 87.*

or atoms and then transformed into other forms of energy.

Fig. 7.8 illustrates the different types of radiation in the electromagnetic spectrum.

The effects of electromagnetic radiation on a material are listed in Table 7.2 and shown in Fig. 7.9.

### 7.2.2.1 *UV-Vis Spectroscopy*

The UV-vis spectrophotometric analysis exploits the radiation absorption of molecules in the ultraviolet and visible fields. The working principle of UV-vis spectroscopy is the following: the bonding electrons of a compound

(whether organic or inorganic) can perform transitions between the various electronic levels (which are five: $\sigma$, $\pi$, nonbonding n, antibonding $\pi^*$ and antibonding $\sigma^*$, represented in Fig. 7.10) at the expense of the energy supplied to them by the electromagnetic radiation. The possible transitions are different and their detectability in the UV-vis field depends on the difference in energy between the two levels involved. In general, there are observable transitions: those between $n$-$\pi^*$ and $\pi$-$\pi^*$ electronic levels.

The set of radiation absorptions at varying wavelengths is called the *absorption spectrum*. The intensity of the absorbed monochromatic

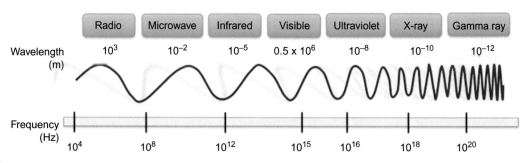

**FIG. 7.8** Different types of radiation in the electromagnetic spectrum.

**TABLE 7.2** Effects of Magnetic Radiation on the Atomic or Molecular Structure of the Material Under Examination

| Electromagnetic Radiation | Wavelength (l) | Frequency | Effect on Matter |
|---|---|---|---|
| Gamma rays | ≤1 pm | >300 EHz | Nucleus |
| X-rays | 1 pm–10 nm | 300 EHz–30 PHz | Internal electrons |
| UV | 10–390 nm | 30–768 THz | Valence electrons |
| Visible | 390–760 nm | 768–395 THz | |
| Infrared (IR) | 760 nm–1 mm | 395 THz–299 GHz | Rotation/vibration of chemical bonds |
| Microwaves, radar | 1–10 cm | 299–3 GHz | Molecular rotation and torsion |
| ESR (EPR) | ≅30 mm | 9–10 GHz | Electron spin (study of radicals formed in solid materials) |
| Radio waves, radio frequency (RF) | ≥10 cm | ≤3 GHz | |
| (*) NMR and MRI = measurement of resonant absorption of RF radiation by atomic nuclei in a strong magnetic field | 0.6–10 m | | Spin of atomic nuclei |

*ESR*, electron spin resonance (*EPR*, electron paramagnetic resonance); *MRI*, magnetic resonance imaging; *NMR*, nuclear magnetic resonance.

radiation is proportional to the concentration of the absorbent molecule and to the optical path according to the *Lambert-Beer law*:

$$A(\lambda) = \varepsilon c d$$

where A is the measured absorbance, $\varepsilon$ is the molar extinction coefficient specific for each substance (a wavelength-dependent absorptivity coefficient), **d** is the optical path length (cm),

and **c** is the analyte concentration (moles/l) (Fig. 7.11).

Lambert-Beer law is of fundamental importance for quantitative analysis, because it shows a linear dependence of the absorbance on the concentration of the analyte. Because many compounds absorb electromagnetic radiation in the regions of the visible (vis) and ultraviolet (UV), if the absorbance of a

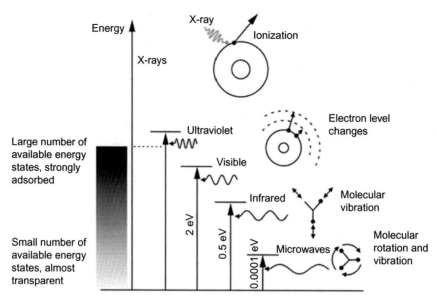

**FIG. 7.9**   The interaction of radiation with matter. *Modified from http://hyperphysics.phy-astr.gsu.edu/hbase/mod3.html#c1.*

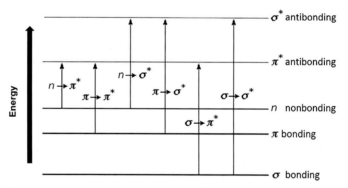

**FIG. 7.10**   Illustration of the various types of electronic excitation that may occur in organic molecules. Of the six transitions outlined, only the two lowest energy ones ($n$-$\pi^*$ and $\pi$-$\pi^*$) are achieved by energies available in the 200–800 nm spectrum (that of UV-vis wavelengths of interest).

reference compound in solution is known, it is possible to trace back the concentration of the sample under examination.[1]

---

[1] The light incident on the sample can instead be reflected, or in some cases diffused. In all these cases, Beer law is not respected. To avoid these problems, a clear and very diluted solution should be available.

Data are frequently reported in percent Transmission ($I/I_0 \times 100$) or in Absorbance [$A = \log_{10} (I_0/I)$], where $I$ is the radiation Intensity.

If the transitions occurred only between electronic levels, the UV spectrum would be composed of a series of lines, more or less high, each of which would represent a transition.

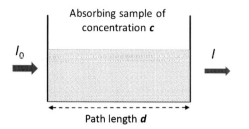

**FIG. 7.11** Schematic of the working principle for UV-vis spectroscopy: an unknown concentration of an analyte can be determined by measuring the amount of radiation that a sample absorbs and applying the Lambert-Beer law. If the absorptivity coefficient is not known, the unknown concentration can be determined by use of a calibration curve of absorbance versus concentration derived from standards.

However, because the levels of *electronic* energy are, in turn, composed of sublevels of *vibrational* energy, which, in turn, are subdivided even more finely into *rotational* sublayers, the UV-vis spectrum results in rather broad bands due to the multiple electrovibrorotational transitions, which are, in fact, possible.

Of these bands, for analytical purposes, the wavelength at which maximum absorption ($\lambda$ max) is reached, and the absorbance value recorded at this point are considered. Registering the absorption of a given sample requires a source of radiation (*lamp*), a device capable of selecting the most appropriate wavelengths (*monochromator*), and, downstream of the sector where the sample and the reference are placed

(*cuvettes*), another device able to measure the intensity of the outgoing radiation (i.e., a *detector*) (Fig. 7.12).

Finally, a suitable indicator provides the measured absorption values. The quantitative determination is carried out by comparison, building a calibration curve by measuring the absorption of the test substance at different known concentrations. UV-vis spectroscopy reveals the presence of aromatic systems and conjugated systems and is used for the quantitative determination of a molecule in solution. In the study of biomaterials, UV-vis spectroscopy is used to evaluate the presence of some functional groups considered important, for example, because the presence or absence of these functional groups allows evaluation of the quantity of a released substance.

As practical example, Fig. 7.13 shows the UV-vis spectrum of hexavalent chromium (Cr6 +) Cr(VI), used to quantitatively analyze at a wavelength of 540 nm the presence of this toxic element in a Fe—Ni alloy sample (Huaa et al., 2009). Because the threshold limit of Cr(VI) is 1000 ppm in a matrix, it is important to severely restrict the presence of such hazardous substances for companies producing stainless and alloy steels, pigments, and tanning agents, and the galvanizing, electroplating, and corrosion-resistant industries, etc., where chromium is used in large amounts.

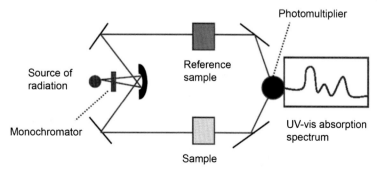

**FIG. 7.12** Drawing of the instrument for UV-vis analysis. The UV-vis spectrum is a graph of A in terms of $\lambda$, according to Beer law.

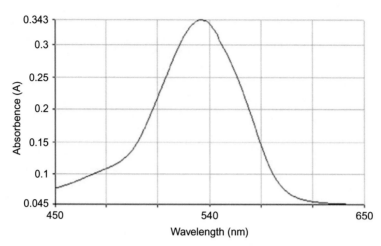

**FIG. 7.13**   The UV-vis spectra of Cr6+ in a Ni–Fe alloy matrix. *Reproduced with permission, Fig. 3 in Huaa, L., Chanb, Y.C., Wuc, Y.P., Wub, B.Y., 2009. The determination of hexavalent chromium (Cr6+) in electronic and electrical components and products to comply with RoHS regulations. J. Hazard. Mater. 163, 1360–1368.*

### 7.2.2.2 *Infrared (IR) spectroscopy*

The infrared radiation can be divided into three regions (near IR or NIR, mid IR or MIR, and far IR or FIR), but the region of greatest interest for studying materials' chemical structure is MIR, because it represents the bands related to the molecules' vibrational motion. The MIR region is between 2.5 and 50 μm wavelength (corresponding to vibrational frequencies, or wavenumbers,[2] between 4000 and 200 cm$^{-1}$).

Infrared spectroscopy is a nondestructive analysis and can be defined as qualitative when identifying the structure of a sample under examination, and quantitative when, depending on the absorption intensity, it is possible to derive the concentration of a component. From the chemical analysis point of view, it is, therefore, a useful technique for the recognition of

molecules or characteristic groups in molecules and for their dosage.

By directing infrared radiation at a molecule under certain conditions, the natural periodic variations (oscillations) in the interatomic distances and in the binding angles can be amplified. The phenomenon results in absorption of IR radiation.

For the molecule to absorb IR radiation, it is fundamental that the vibration movement also involves a change in the dipolar moment. In this way, the vibrating molecule produces an oscillating electric field, making it possible to exchange energy with electromagnetic waves. A bi-atomic molecule has a unique mode of vibration, in which the two atoms oscillate along a direction or degree of freedom (x), while the center of mass remains stationary. A molecule composed of **n** atoms has more modes of vibration.

**MODES OF VIBRATION**

*Stretching vibration*, due to rhythmical stretching along the bonding axis (with consequent increase and decrease of the interatomic distance);

---

[2] The frequency scale at the bottom of the chart is given in units of reciprocal centimeters (cm$^{-1}$) rather than Hz, because the numbers are more manageable. The reciprocal centimeter is the number of wave cycles in 1 cm.

*Bending vibration*, due to variation of the bonding angle (may be due to a variation of the angle in the bonds with a common atom, or to the movement of a group of atoms with respect to the rest of the molecule without moving the atoms in the group, with respect to one another).

In this case we have:

- vibrations on the plane → *scissoring* and *rocking*
- vibrations out of the plane → *wagging* and *twisting*

When these vibrations determine a change in the dipole moment of the molecule, then an active IR vibration occurs. When this variation occurs, the vibrating molecule produces an oscillating electric field, making it possible to exchange energy with electromagnetic waves:

$$\mu = q \cdot \vec{d}$$

$\mu$ = dipolar moment
$q$ = electric charge
$d$ = vector distance

We recall that the intensity of a band depends on the value of the dipolar moment of the bond to which it refers and, therefore, on the relative electronegativity of the atoms involved in the bond:

$$C - O > C - Cl > C - N > C - C - OH$$
$$> C - C - H$$

**EXAMPLES**

The triatomic $H_2O$ molecule can have 3 modes of vibration. These are schematized in Fig. 7.14, indicating with a dashed arrow the displacement of an atom in the course of vibration (stretching and bending).

The triatomic $CO_2$ molecule can have 4 modes of vibration; it is noteworthy that there are two possible ways of bending (Fig. 7.15).

In an organic molecule of a more complex structure than triatomic molecules, some normal modes can be assimilated to movements of individual functional groups, which have characteristic vibration frequencies of the group structure and not of the molecule, such as the stretching of the C=O carboxyl bond, which has a characteristic frequency in the range of $1640 \div 1780 \, cm^{-1}$.

Group frequencies are the basis of qualitative IR analysis of organic molecules. Instead, whatever cannot be considered localized is attributed to complex vibration movements of the molecule as a whole; the corresponding spectral transitions occur in the region of frequencies below $1500 \, cm^{-1}$. These modes are characteristic of the structure of the molecule and are also useful for identification of particular substances with known IR spectra in the low-frequency region; for this reason, the region is called (to use a police term) the *fingerprint region*.

Fig. 7.16 provides an example of the vibration modes related to different chemical bonds in the IR absorption range.

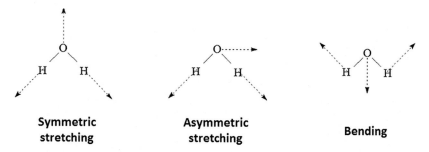

| Symmetric stretching | Asymmetric stretching | Bending |

**FIG. 7.14** Vibrational modes of a water molecule.

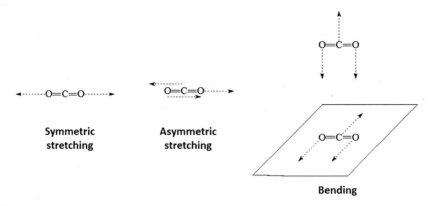

**FIG. 7.15**   Vibrational modes of the $CO_2$ molecule.

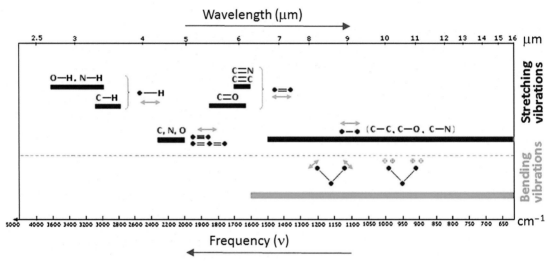

**FIG. 7.16**   Main vibration modes of different chemical bonds in the IR absorption range. *From www2.chemistry.msu.edu/faculty/reusch/virttxtjml/spectrpy/infrared/infrared.htm.*

The infrared spectrum appears as a sequence of absorption bands, recorded as a function of wavelength or wavenumber (Fig. 7.17).

When the analytical problem related to a substance consists in its identification within a group of known substances, it is possible to make a direct comparison of the spectra to the benefit of efficiency and simplicity. This is on the condition that the spectra to be compared are limited in number; otherwise, the operator's task becomes long and difficult, so much so as to compromise the result.

The recognition of substances in IR can be done in different ways, but all are based on the comparison of the spectrum under examination with spectra of known substances or, in any

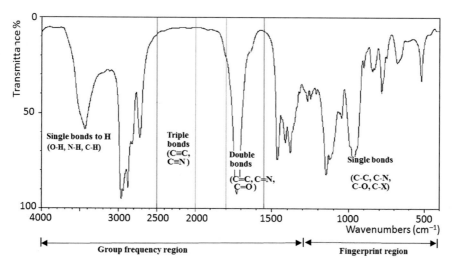

**FIG. 7.17** Infrared spectrum of an organic polymer. *Modified from https://ashleygolbuschemistry.weebly.com/ spectroscopy.html.*

case, with data obtained from them and collected in special tables or electronic databases.

### 7.2.2.3 IR Spectrophotometer

The analytical instrument used is a spectrophotometer. Infrared spectrophotometers record the relative amount of energy as a function of the wavelength/frequency of the infrared radiation when it passes through a sample. Therefore, chemical structures of different samples will reflect differences in the IR absorption spectrum allowing for identification of a sample.

An IR spectrophotometer allows to perform qualitative (and quantitative) investigations in the organic field. Like the UV spectrophotometer, the IR presents the components shown in Fig. 7.18.

There are two types of spectrophotometers: dispersive (Fig. 7.19) and Fourier-transform spectrophotometers (Fig. 7.20). Today, spectrophotometers of the second type are mostly used thanks to the speed of spectra execution.

Unlike a dispersive spectrometer, an FTIR spectrometer (or spectrophotometer) is used to simultaneously obtain spectral data of a sample. It does this by using an interferometer to collect the interferogram, also known as the raw data/

signal format, which can then be translated into the infrared spectrum of the sample by means of a Fourier transform algorithm.[3] As a result, there are many advantages including greater signal-to-noise ratio, high resolution, higher throughput, and a short wavelength limit. FTIR spectrometers can be used in a variety of industries, including environmental, pharmaceutical, and petrochemical.

### 7.2.2.4 Attenuated Total Reflection

An alternative to the transmission mode is a technique based on attenuated total reflection (ATR).

In this technique, the IR beam is guided in an IR transparent crystal by total reflection; due to quantum mechanical properties of the IR light, the electromagnetic field may extend beyond

---

[3] The Fourier transform, carried out by the instrument software once the beam has reached the detector, shows on the screen a traditional infrared spectrum, transforming the light intensity signal as a function of time (mirror displacement) in intensity signal as a function of the wave number.

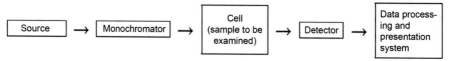

**FIG. 7.18**   Diagram showing the various components of an IR spectrophotometer.

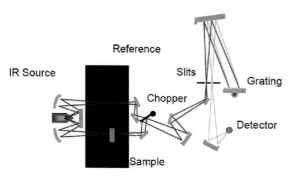

**FIG. 7.19**   Dispersive IR spectrophotometer: Radiation from the source is dispersed by the grating into single wavelength components. The sample beam rationed to the reference beam produces a plot of the sample spectrum. Time required for recording a spectrum is approximately 10 min. *Courtesy of Dr. Giuseppe Casassa, Thermo Fisher Scientific Spa, Italy.*

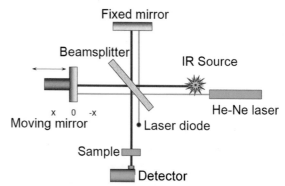

**FIG. 7.20**   Fourier-transform spectrophotometer: An incandescent light source emits a light beam in the IR field; this beam reaches a semireflecting mirror (beam splitter) which divides it in 50% to a fixed mirror and a 50% to the moving mirror. The rays reflected by these mirrors are sent once more to the semireflecting mirror, which joins the two rays and sends them to the detector. Even if the two rays have joined the detector together, they have made a different optical path. Depending on the difference in the optical path of the two beams, constructive or destructive interferences are created that create a signal to the detector proportional to the optical path difference of the two rays, and therefore, from the position of the moving mirror at that moment. When the combined beam is transmitted through the sample, it is detected as an interferogram and contains all infrared information on the sample. The IR spectrum is obtained from the interferogram by the mathematical process of Fourier transformation. *Courtesy of Dr. Giuseppe Casassa, Thermo Fisher Scientific Spa, Italy.*

the crystal surface for about 1 μ as a so-called evanescent wave.

An ATR accessory operates by measuring the changes that occur in a totally internally reflected infrared beam when the beam comes into contact with a sample; an infrared beam is directed onto an optically dense crystal (e.g., zinc selenide, ZnSe, or germanium, Ge) with a high refractive index at a certain angle. This internal reflectance creates an evanescent wave that extends beyond the surface of the crystal into the sample held in contact with the crystal (Fig. 7.21).

The evanescent wave extends only a few microns (0.5–5 μ) beyond the crystal surface and into the sample; consequently, there must be good contact between the sample and the crystal surface.

It is important that the refractive index of the crystal is significantly greater than that of the sample; otherwise, internal reflectance will not occur.

### 7.2.2.5 Nuclear Magnetic Resonance spectroscopy

Nuclear magnetic resonance (NMR) spectroscopy (or magnetic resonance spectroscopy, MRS), is a powerful analytical technique that provides detailed information on the molecular structure of the compounds under investigation.

NMR spectroscopy measures the absorption of electromagnetic radiation in the radio

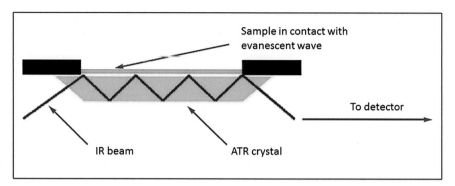

**FIG. 7.21** Working principle for ATR analysis.

frequency in molecules immersed in a strong magnetic field. This absorption takes place by the nuclei of particular atoms that contain odd atomic or mass number or both, such as protons ($^1$H), isotope $^{13}$C of carbon, fluorine ($^{19}$F), and the isotopes $^{14}$N and $^{15}$N of the nitrogen. Therefore, with NMR, the atomic nuclei (and not the electrons) are directly examined.

$^1$H and $^{13}$C NMR are the most commonly used techniques for elucidating the number of protons and carbon atoms in the compound. In the case of $^1$H NMR, it determines the type and number of hydrogen (H) atoms, while $^{13}$C NMR determines the type and number of carbon (C) atoms in the molecule.

All information on the chemical surrounding area is deduced by observing the behavior of the atomic nuclei. Only nuclei that have a *spin* nuclear magnetic moment are observable at the NMR.

In many atoms (as in $^{12}$C), the spins are all paired, one in opposition to the other, and then cancel each other, and the atomic nucleus has a resulting spin equal to zero. In some atoms, however (as in $^1$H and $^{13}$C), the nucleus has a resulting spin that is not zero.

When a nucleus with a spin is immersed in a magnetic field $B_0$, the nucleus, like the needle of a compass, is subjected to a couple of forces that make it rotate to align with the external magnetic field. For a nucleus with spin = 1/2, there are two permitted orientations: parallel to the

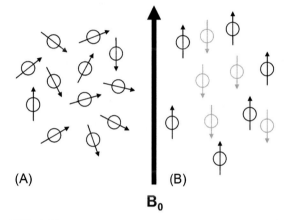

**FIG. 7.22** (A) Magnetic dipoles randomly oriented in all directions under normal conditions; (B) possible orientations in a magnetic field $B_0$ of nuclei with spin = 1/2: parallel to the field (low energy) and against the field (high energy).

field (low energy) and against the field (high energy) (Fig. 7.22).

If the oriented nuclei are now irradiated with electromagnetic radiation in the radio frequency (region 4–900 MHz), the lower energy states (aligned to the field) absorb a quantum of energy and rotate their spin to assume the high energy state (opposite to the field). When this spin transition occurs, it is said that the nuclei *resonate* with the applied radiation.

With an NMR instrument, the transition between the two states is stimulated with the help of a radiofrequency transmitter, and the absorbed

**FIG. 7.23** Representative NMR Instrument. *Reproduced with permission of Elsevier—Fig. 7.2 in Pradhan, S., Rajamani, S., Agrawal, G., Dash, M., Samal, S.K., 2017. NMR, FT-IR and Raman characterization of biomaterials. In: Tanzi, M.C., Farè, S. (Eds.), Characterization of Polymeric Biomaterials, first ed. Woodhead Publ., Elsevier Ltd. (Chapter 7).*

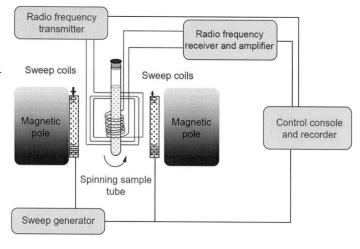

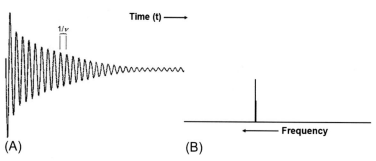

**FIG. 7.24** NMR analysis of CH₃I molecule: (A) signal that fades over time for the hydrogen atom; (B) NMR spectrum.

energy is further recorded by a radiofrequency receiver in the form of a resonance signal. Fig. 7.23 provides a schematic image of a NMR Instrument.

The NMR signal in the currently most used instruments is generated with the *impulse method*. With this technique, all the nuclei of a species are excited at the same time by a radiofrequency pulse, which contains the whole necessary frequency range. The signal collected is an oscillating signal with frequency $\nu$ that fades over time (Fig. 7.24A) and is called FID (Free Induction Decay).

The $^1$H spectrum can be obtained very quickly with a low amount of sample concentration; however, in $^{13}$C NMR, the minimum scan time

is longer, with the concentrated sample needed to acquire a good informative spectrum, due to only 1.1% natural abundance of 13C isotope.

### 7.2.2.6 $^1$H NMR

In a simple molecule such as CH₃I, the hydrogens are equivalent, and they will all have the same resonant frequency. In the graph, this frequency is easily identifiable by measuring the wavelength (the distance between two successive ridges) and calculating the reciprocal, according to the formula: $\nu = 1/\lambda$ (where $\nu$ is the frequency and $\lambda$ is the wavelength). The graph shown in Fig. 7.24B is thus obtained as a function of the frequencies, called the *NMR spectrum*, which shows the frequency absorbed by the hydrogen atoms in the $CH_3I$ molecule.

If the sample contains nuclei with different resonance frequencies, these are all excited at the same time by the radiofrequency impulse, and then the collected signal will be a complex curve, or interferogram, given by the combination of several FIDs, one for each frequency absorbed by the nuclei. To be able to track the single frequencies that, when combined, generate the complex path, it is necessary to apply the mathematical procedure called Fourier Transform which allows moving from the graph as a function of time (the FID) to the graph as a function of frequencies (the NMR spectrum) (Fig. 7.25).

## 7.2.3 Chromatographic Techniques

Chromatography, born as a separation technique and later developed as an analytical technique, is based on the fact that the various

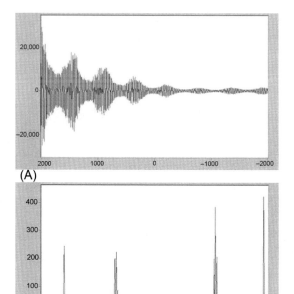

**FIG. 7.25** NMR spectrum for a complex sample: (A) interferogram; (B) NMR spectrum, resulting from a Fourier Transform analysis.

components of a mixture tend to distribute differently between two phases, depending on their affinity with each of them. If one of the phases is somehow immobilized (*stationary phase*), while the other one (*mobile phase*) is continuously kept flowing, it is possible to perform the extraction process in a continuous manner. A chemical species deposited on the stationary phase and introduced into the flowing mobile phase will be distributed dynamically between the two phases in proportion to the different affinity it has for them.

To schematically show this phenomenon, Fig. 7.26 represents a column filled with a suitable stationary phase, at the top of which, a small volume of a mixture of three different chemical species is deposited, dissolved in a suitable solvent. As the solvent (mobile phase) is introduced at the top of the column and allowed to drain through the stationary phase, there is a tendency for the various components of the mixture to run in a different measure, depending on the higher or lower affinity toward the two competing phases. Continuing the elution, the three species separate completely and can be collected one after the other out of the column.

In today's chromatographic techniques, the stationary phase can be a solid or a liquid, deposited on a surface or introduced into a column. As for the mobile phase, it can be a low viscosity gas or liquid which is made to run through a column, or (and this applies only to liquids) to rise or fall by capillarity along a stationary phase layer.

Depending on the physical state and the nature of the two phases, very different interactions occur with the sample: on this basis, a classification of the chromatographic methods can be made, also taking into account the method in which the stationary phase is positioned to conduct the separation (Table 7.3).

Among these chromatographic techniques, High Performance Liquid Chromatography (HPLC) and the technique of Gel Permeation Chromatography (GPC) will be briefly explained.

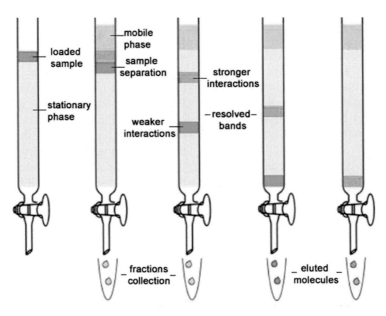

**FIG. 7.26** Chromatographic separation on a column of a mixture of three chemical species, indicated with three different colors. *Reproduced from https://prepgenie.com.au/gamsat/column-chromatography/.*

**TABLE 7.3**   Some of the Main Types of Chromatographic Techniques Currently Used

| Principle of Analysis | Physical State Mobile Phase | Type of Chromatography |
|---|---|---|
| *Absorption chromatography*: competition between an absorbent solid and a mobile phase | Gas | GC/GSC |
| | Liquid | LC/HPLC; TLC/PC |
| *Partition chromatography*: competition between a stationary liquid phase and a mobile phase | Gas | GC/GLC; SFC |
| | Liquid | LC/HPLC |
| *Ion exchange chromatography*: competition between a stationary phase in ion-exchange resin and a liquid mobile phase | Gas | GC/GSC |
| | Liquid | IEC/IC/HPIC |
| *Permeation chromatography*: competition between a polymeric matrix and a liquid mobile phase | Liquid | GPC |

### 7.2.3.1 High Performance Liquid Chromatography

High performance liquid chromatography (HPLC) is the most powerful of all the chromatographic techniques, often achieving separations and analyses that would be difficult or impossible with other types of chromatography.

HPLC is basically a highly improved form of *column chromatography*: the solvent is forced through the column(s) under high pressures (up to 400 atm), instead of being allowed to percolate through gravity, which makes the process much faster. This technique allows us to use a much smaller particle size for the column packing material, which gives a much greater surface area for interactions between the stationary phase and the molecules passing through it. This allows a much better separation of the components of the mixture under analysis.

The other major improvement over column chromatography concerns the detection methods which can be used. These methods are highly automated and extremely sensitive.

In general, chemical compounds can be separated by HPLC according to three primary characteristics:

- Polarity
- Electrical Charge
- Molecular Size

Concerning *polarity*, there are two variants in use in HPLC, depending on the relative polarity of the solvent and the stationary phase:

**(a)** *Normal phase HPLC*. The column is filled with tiny silica particles, and the solvent is non-polar (e.g., hexane). A typical column has an internal diameter of 4.6 mm (and may be less than that), and a length of 150–250 mm. Polar compounds in the mixture being passed through the column will stick longer to the polar silica than nonpolar compounds. Nonpolar compounds will, therefore, pass more quickly through the column.

**(b)** *Reversed-phase HPLC*. The column size is the same, but the silica is modified to make it nonpolar by attaching long hydrocarbon chains (typically with 8 ÷ 18 carbon atoms) to its surface. A polar solvent is used (e.g., a mixture of water/methanol). In this case, there will be a strong attraction between the polar solvent and polar molecules in the mixture being passed through the column. There will not be as much attraction between the hydrocarbon chains attached to the silica (the stationary phase) and the polar molecules in the solution. Polar molecules in the mixture will, therefore, spend most of their time moving with the solvent.

Nonpolar compounds in the mixture will tend to form attractions with the hydrocarbon groups because of van der Waals dispersion forces. They will also be less soluble in the solvent because of the need to break hydrogen bonds as they squeeze in between the water or methanol molecules, for example. They, therefore, spend less time in solution in the solvent, and this will slow them down on their way through the column. Therefore, in this case, the polar molecules will travel through the column more quickly.

Reversed-phase HPLC is the most commonly used form of HPLC.

Separations based on *electric charge* are called Ion-Exchange Chromatography (IEC). In this case, unlike separation based on polarity, opposites are attracted to each other. Stationary phases for ion-exchange separations are characterized by the nature and strength of the acidic or basic functions on their surfaces and the types of ions that they attract and retain. *Cation exchange* is used to retain and separate positively charged ions on a negative surface. Conversely, *anion exchange* is used to retain and separate negatively charged ions on a positive surface.

The components of a basic HPLC system are shown in the simple diagram of Fig. 7.27.

A *reservoir* holds the solvent (i.e., the *mobile phase*). A *high-pressure pump* (i.e., solvent delivery system) is used to generate and meter a specified flow rate of mobile phase, typically milliliters per minute. An *injector* (i.e., sample manager or autosampler) injects the sample into the continuously flowing mobile phase stream that carries the sample into the HPLC column. The column contains the chromatographic packing material needed to effect the separation. This packing material is called the *stationary phase* because it is held in place by the column hardware. High-pressure tubing and fittings are used to interconnect pump, injector, column,

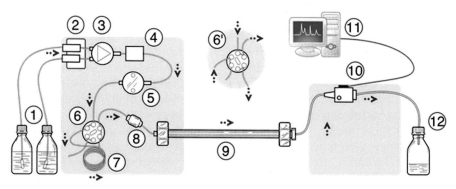

**FIG. 7.27**  Diagram of an HPLC system . (1) Solvent reservoirs, (2) Solvent degasser, (3) Gradient valve, (4) Mixing vessel for delivery of the mobile phase, (5) High-pressure pump, (6) Switching valve in "inject position," (6′) Switching valve in "load position," (7) Sample injection loop, (8) Precolumn (guard column), (9) Analytical column, (10) Detector (i.e., IR, UV), (11) Data acquisition, (12) Waste or fraction collector. *From https://en.wikipedia.org/wiki/High-performance_liquid_chromatography.*

and detector components to form the conduit for the mobile phase, sample, and separated compound bands.

A detector is needed to see the separated compound bands as they elute from the HPLC column. The mobile phase exits the *detector* and can be sent to waste, or collected, as desired. Because sample compound characteristics can be very different, several types of detectors have been developed. A UV-absorbance detector is used in the case of substances that can absorb ultraviolet light; in case of molecules that fluoresce, a fluorescence detector is used. The most powerful approach is the use of multiple detectors in series. For example, a UV detector may be used in combination with a mass spectrometer (MS) providing, from a single injection, more comprehensive information about a compound (or *analyte*). The practice of coupling a mass spectrometer to an HPLC system is called LC/MS.

The computer data station is the component that records the electrical signal needed to generate the chromatogram on its display and to identify and quantitate the concentration of the sample constituents.

When the mobile phase contains a separated compound band, HPLC provides the ability to collect this fraction of the eluate containing that purified compound for further study. This is called *preparative chromatography.*

The analysis in HPLC allows evaluating the possible release of additives or low molecular weight substances from a biomaterial or a device. This is very useful both in tests of (accelerated) aging and in situations simulating the final use. The analysis of possible released substances is very useful in the case of indirect cytotoxicity tests (see Section 7.3) when cytotoxic effects are observed from the eluates and can give indications on the degradation or aging of the material under examination.

### 7.2.3.2 Gel-Permeation Chromatography

In the 1950s, Porath and Flodin discovered that biomolecules could be separated on the basis of their size, rather than on their charge or polarity, by passing, or filtering, them through a controlled-porosity, hydrophilic, dextran polymer. This process was called *gel filtration.*

Later, an analogous scheme was used to separate synthetic oligomers and polymers using organic polymer packings with specific pore-size ranges. This process was called *gel-permeation chromatography* (GPC). Similar separations performed with controlled-porosity silica packings

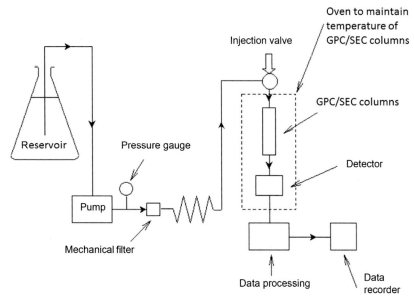

**FIG. 7.28** Scheme of the instrument for GPC analysis showing the various components. The pump moves the eluent through the system at a constant flow rate. The injection valve introduces the sample into the eluent stream at a known concentration. The column(s) are maintained at the selected temperature by the oven and perform the separation, resolving the sample on the basis of size in solution. Standard kits are used to calibrate the columns, to show the relationship between retention time and molecular weight. The detector detects resolved components of sample after elution from the column(s).

were called *size-exclusion chromatography* (SEC). Introduced in 1963, the first commercial HPLC instruments were designed for GPC applications.

GPC/SEC is a very useful technique for studying the molecular weights of thermoplastic polymeric materials. The principle behind this technique is the transport of particles, followed by their separation in accordance with their size, and consequently, with their molar mass. The same solvent is allowed to serve the role of the two phases (stationary and mobile) in a column with a microporous gel. In fact, part of the solvent is present in the stationary phase, and the other part constitutes the mobile phase in which the polymer is previously dissolved.

A scheme of the instrument for GPC analysis with the elements that compose it is provided in Fig. 7.28.

GPC is the method that uses an organic solution for the mobile phase; when, instead, an aqueous solution is used for the mobile phase, the technique is also known as gel filtration chromatography (GFC). The material most frequently used as the stationary phase is polysiloxane, and more recently, cross-linked polystyrene (PS). The use of PS as stationary phase has reduced the height of chromatography columns due to the reduced size of PS particles, which determines the porosity of the column. An important and satisfied condition of this method is that molecules do not interact with the gel material, that is, the stationary phase. Thanks to the basic principle, the method can also be used as a purification technique or as a method of studying the interaction between macromolecules.

According to the elution times, GPC can be used to obtain information about molecular weight parameters of polymers, namely, the

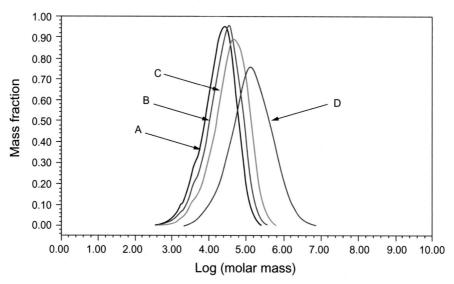

**FIG. 7.29**  Molar mass distribution obtained by GPC/SEC of PP samples after 48 days of natural weathering: (A) with no antioxidant; (B) with 100 mg/kg antioxidant; (C) with 300 mg/kg antioxidant. (D) PP sample with no antioxidant, without any exposure. *From Ojeda, T., Freitas, A., Birck, K., Dalmolin, E., Jacques, R., Bento, F., Camargo, F., 2011. Polym. Degrad. Stab. 96, 703–707, Fig. 1.*

number of the average molecular weight (Mn), weight average molecular weight (Mw), viscosity average molecular weight (Mv), average molecular weight (Mz), and higher average molecular weight (Mz + 1).

By plotting the concentration of each component as a function of time, a chromatogram is obtained, and from the analysis of the chromatogram, it is possible to obtain the aforementioned parameters. As an example, changes in molecular weights distribution of polypropylene (PP) after aging for 48 days in different conditions are shown in Fig. 7.29. Quantitative data can be obtained after calibration of the columns with polymer samples of known and very narrow distribution of molecular weighs, usually standard kits of polystyrene or polymethylmethacrylate.

Unlike other techniques, which require a very long time to determine molecular weight averages, using GPC, the determination of those values can be done in just 2 h, depending on the elution rate.

### 7.2.4 X-Ray Techniques for Crystallinity Analysis

In 1913, W.H. Bragg and his son, W.L. Bragg developed a relationship to explain why the cleavage faces of crystals appear to reflect X-ray beams at certain angles of incidence (theta, $\theta$):

$$n\lambda = 2d \sin \theta \ (\text{Bragg's law})$$

The variable $d$ is the distance between atomic layers in a crystal, and the variable lambda $\lambda$ is the wavelength of the incident X-ray beam; $n$ is an integer, referred to as the order of diffraction, and is often a unity value. This observation is an example of X-ray wave interference, commonly known as X-ray diffraction (XRD), which gave direct evidence for the periodic atomic structure of crystals postulated for several centuries.

In XRD analysis, a monochromatic X-ray beam is incident on the crystalline material, and the intensity of the elastically scattered

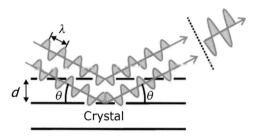

**FIG. 7.30** Schematic of X-ray diffraction from part of a single crystal. *Reproduced with permission of Elsevier, Fig. 6.5 in Tanzi, M.C., 2017. Characterization of thermal properties and crystallinity of polymer biomaterials. In: Characterization of Polymeric Biomaterials, first ed. Woodhead Publ., Elsevier Ltd. (Chapter 6).*

beam due to the periodic arrangement of atoms in the sample is measured as a function of the diffraction angle "$2\theta$"(Fig. 7.30).

X-rays scattered from successive planes will interact constructively when they eventually reach the X-ray detector, thus, registering the passage of an intense beam called the *diffracted beam*. Intensity and angular position of the diffracted beams (*diffraction peaks*) produced by a polycrystalline sample are unique for each crystalline material. In a case in which the test material is not known, it can be identified for

comparison with a database. Practically, the basic features for a typical XRD experiment are, sequentially, production of X-rays, diffraction, detection, and interpretation.

The XRD method provides a variety of information: phase identification, crystallite size, lattice strain, crystallographic orientation, and nanostructured materials. In addition, by measuring crystallite size in several different crystallographic directions, the crystallite shape can be determined. Remarkably, broadening of the diffraction peak occurs due to deviations from ideal crystallinity, such as decrease in crystallite size for nanomaterials, as shown in Fig. 7.31.

The X-ray diffractogram of a semicrystalline polymer sample can be considered as constituted by the superimposition of the effects due to both the crystalline fraction and the amorphous fraction, with intensities proportional to relative mass of the two components. The amorphous part produces a continuous distribution of intensity, devoid of peaks; the diffraction peaks produced by the crystalline part of the polymer overlap. As an example, Fig. 7.32 shows the diffraction spectrum of isotactic polypropylene. Because in a polymer the crystallites are generally very imperfect, the diffraction peaks appear much

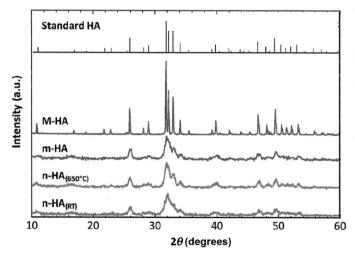

**FIG. 7.31** XRD diffractograms of microsized (M-HA and m-HA) and nanosized hydroxyapatite (n-HA) powders, obtained as described in Munarin et al. (2015). "Standard HA" is a sample of pure hydroxyapatite ($Ca_5(PO_4)_3OH$—JCPDS no: 090432), used for comparison. *Reproduced with permission of Elsevier—Fig. 6.8 in Tanzi, M.C., 2017. Characterization of thermal properties and crystallinity of polymer biomaterials. In: Characterization of Polymeric Biomaterials, first ed. Woodhead Publ., Elsevier Ltd. (Chapter 6).*

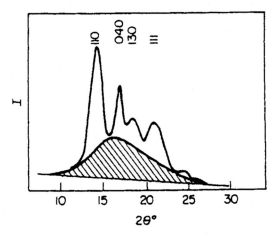

**FIG. 7.32** Diffractogram of isotactic polypropylene. The dashed area represents the contribution of the amorphous part. *Reproduced with permission of Elsevier—Fig. 6.16 in Tanzi, M.C., 2017. Characterization of thermal properties and crystallinity of polymer biomaterials. In: Characterization of Polymeric Biomaterials, first ed. Woodhead Publ., Elsevier Ltd. (Chapter 6).*

wider (spread) than those normally observed in the crystalline substances.

Conventional XRD is called Wide-Angle X-ray scattering (WAXS) and examines smaller structures (<1 nm) such as atoms and interatomic distances, which scatter toward large angles and provide information, as already mentioned, on phase state, crystal symmetry, and molecular structure.

Instead, small-angle X-ray scattering (SAXS) evaluates the scattering pattern of nanosized particles and domains (size range: 1–100 nm) at small angles to get information about their particle structure, that is, size, shape, and internal structure.

### 7.2.5 Microscopy Techniques

Microscopy techniques are the election methods to characterize morphology of thin (2D) samples or surfaces, comprising internal or fracture sections of 3D samples.

The microscopy techniques used for biomaterials include light microscopy (optical and stereo), Scanning Electron Microscopy (SEM), Transmission Electron Microscopy (TEM) and scanning probe microscopies, including Atomic Force Microscopy (AFM).

When discussing microscopy as a useful tool for investigating morphology of a sample, it is important to be conscious of the resolution limit. Fig. 7.33 provides the limits of resolution for optical and electron microscopy, considering the observation of a cell and its internal elements, down to the atoms. For the unaided human eye, the limit of resolution corresponds to 200 μm.

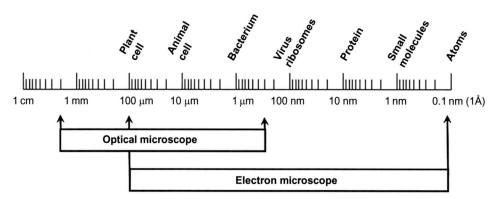

**FIG. 7.33** Scale of resolution of light and electron microscopy.

### 7.2.5.1 Optical (Light) Microscopy

A light microscope provides complementary information and is not a substitute for examination using the naked eye. This type of analysis uses electromagnetic radiation in the visible field to obtain enlarged images of objects and devices, the structure of materials, and biological structures.

The microscope must accomplish three tasks: create a magnified image of the sample under analysis, separate the details in the image, and make the details visible to the human eye or camera. Not only multiple-lens (compound microscopes) instruments with objectives and eyepieces can satisfy these tasks, but also very simple single lens instruments, which are often handheld (such as a magnifying glass).

Compound microscopes include a two-stage magnifying device built around separate lens systems, the *objective* and the eyepiece (commonly named *ocular*), which, in the simplest design, are mounted at opposite ends of a tube, known as the *body tube*. The objective forms a magnified real image (the intermediate image) of the sample under examination. The intermediate image is further magnified by the eyepiece. The magnification of the light microscope refers to the ratio of the dimension in the viewed image to the corresponding dimension in the original sample and is determined by multiplying the individual magnifications of the objective and ocular. Besides magnification, resolution, and contrast, the depth of field (vertical resolution, the ability to produce a sharp image from an irregular surface), is another important parameter.

In light microscopy, the sample under examination can be imaged by passing illuminating light through it (*transmission* microscopy) or by reflecting the light incident on the sample (*reflected* microscopy) to the detector. For transmitted light microscopy, the sample must be thin enough to allow the light to pass through it. Reflected light techniques are the method of choice for imaging specimens that remain

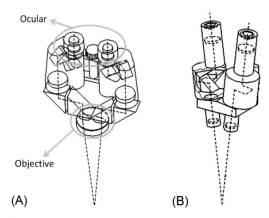

**FIG. 7.34** Operating principle of (A) a light microscope operating in reflected light, and (B) a stereo microscope.

opaque even when ground to a thickness of 30 μm (for example, metals). The working principle of a light microscope operating in reflected light is displayed in Fig. 7.34A.

Biological samples and living cells can be observed under transmitted light, but the substrate has to be transparent. Cells and tissues can be stained for a better recognition and visualization; dyes such as Coomassie blue, Neutral red, Methylene blue, and Oil red O, are organic compounds with selective affinity for specific cell phenotypes or their subcellular components. Staining introduces a contrast in the biological sample to be examined, which would be otherwise not visible. Thin sections (0.5 ÷ 100 μm) of biological tissues are usually stained for histological analyses to recognize cellular phenotypes or cell components, such as nucleus or actin-myosin organization in the cell cytoskeleton.

Some examples of images acquired by optical microscopy are provided in Fig. 7.35.

### 7.2.5.2 Stereo Microscopy

This type of microscope allows maintaining the tridimensionality of the object under observation.

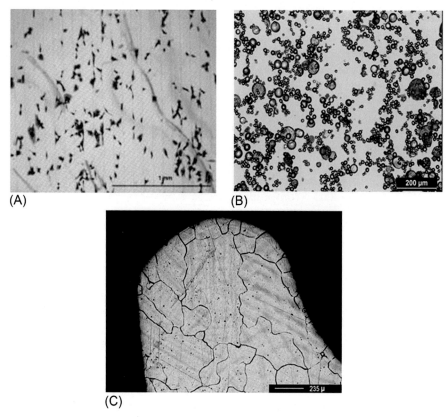

**FIG. 7.35**  OM images of (A) living cells stained with Coomassie blue; (B) gelatin microspheres; (C) grain structure of a component of a cobalt alloy knee prosthesis.

This is possible due to the stereoscopic effect that is obtained by transmitting two twin images, but inclined to a few degrees; the images take two separate and differently aligned optical paths. The human brain combines the two images while maintaining a high perception of depth (stereoscopic effect). However, the maximum magnification allowable is $50\times$ and, as a consequence, structures of few microns' size, such as cells, cannot be suitably observed.

Inexpensive stereomicroscopes are available commercially and can be effectively used to observe device components, 3D structures and scaffolds, and objects with complex geometries and solid surfaces, particularly opaque ones. As examples, Fig. 7.36 displays representative stereo microscope images.

### 7.2.5.3 *Fluorescence Microscopy*

This technique is one of the most widely used biological methods for observing living cells in culture or "fixed." In contrast to other types of optical microscopy that are based on macroscopic specimen features, fluorescence microscopy is capable of imaging the distribution of a single molecular species based exclusively on the properties of fluorescence emission. Therefore, using fluorescence microscopy, the precise location of intracellular components labeled with specific substances (called *fluorophores*[4]) can be monitored,

---

[4] Fluorophores are special molecules able to absorb light at a known wavelength (i.e., of a specific color, e.g., blue), reaching a high energy state that is subsequently re-emitted by the fluorophore as a higher wavelength light (e.g., green).

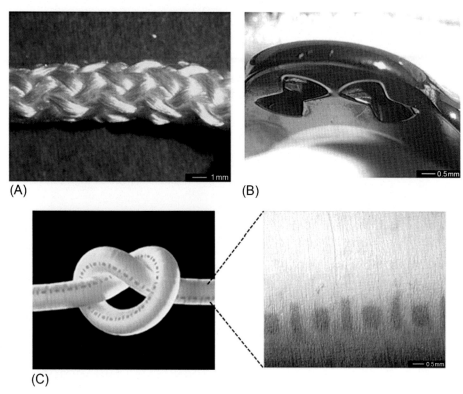

(A)

(B)

(C)

**FIG. 7.36** Stereo microscopy images of (A) textile structure in a natural polymer (silk fibroin) as a possible ligament substitute; (B) component of a mechanical heart valve prosthesis; (C) ePTFE vascular prosthesis and, on the right, a magnification of its surface morphology.

as well as their associated diffusion coefficients, transport characteristics, and interactions with other biomolecules.

Some fluorophores may be incorporated and concentrated into specific subcellular components which consequently become fluorescent. Therefore, the cellular structure can be dynamically observed in culture or examined if "fixed" at different culture times.

The modern fluorescence microscope combines the power of high performance optical components with computerized control of the instrument and digital image acquisition. Light is provided by a xenon or mercury lamp that emits light in the UV/visible range so as to excite the fluorophore at a certain wavelength and consequently image

the distribution of a specifically labeled molecular species.

The improvement of fluorescent microscopes and the introduction of more powerful focused light sources, such as lasers, have led to more technically advanced opportunities such as the *confocal laser scanning microscopes* (CLSM) and *total internal reflection fluorescence microscopes* (TIRF).

CLSM is an invaluable tool for producing high-resolution, 3D images of subsurfaces in specimens such as microbes. Its advantage is the ability to produce sharp images of thick samples at various depths by taking images layer by layer and reconstructing them with a computer rather than viewing whole images through an eyepiece.

Representative images are shown in Fig. 7.37.

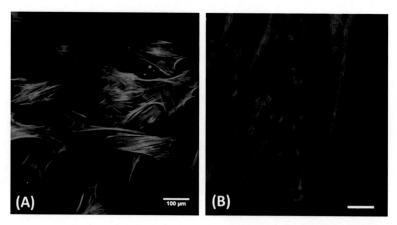

**FIG. 7.37**   (A) Confocal microscopy image of human bone marrow mesenchymal stem cells (hBM-MSCs) cultured for 24 h on a substrate (F-actin (Phalloidin, green) and Hoechst (nuclei, blue) staining) [Unpublished results]; (B) Immunofluorescence micrographs of murine C2C12 cells cultured on a PLLA–TMC substrate (Staining: blue, nuclei; red, myosin). *(B) From Altomare, L., Gadegaard, N., Visai, L., Tanzi, M.C., Farè, S., 2010. Biodegradable microgrooved polymeric surfaces obtained by photolithography for skeletal musclecell orientation and myotube development. Acta Biomater. 6, 1948–1957.*

### 7.2.5.4 *Electron Microscopy*

Electron Microscopy utilizes electron beams to illuminate materials. Compared to optical microscopy, higher magnifications of about 1–2 million times can be obtained, but, more importantly, higher resolution down to the subnanometric scale can be achieved.

There are two types of electron microscope: transmission electron microscopes (TEM) and scanning electron microscopes (SEM).

#### TRANSMISSION ELECTRON MICROSCOPY

TEM is one of the most powerful techniques to provide direct important information on size, shape, structure, and morphology of biological and nonbiological materials. In TEM, a beam of electrons generated from an electron gun is transmitted through an ultrathin specimen, interacting with the specimen as it passes through. An image is formed from the interaction of the electrons transmitted through the specimen, then the image is magnified and focused onto an imaging device, such as a fluorescent screen, on a layer of photographic film, or to be detected by a sensor such as a charge-coupled device camera. The whole trajectory from source to screen is under vacuum, and the specimen has to be very thin (thickness below 100–200 nm) to allow the electrons to penetrate.

#### SCANNING ELECTRON MICROSCOPY

In SEM, an electron beam of relatively low energy from few hundred eV to 40 keV, but with a very fine spot size, is scanned across the sample surface instead of being transmitted through the sample as in TEM. The image is formed pixel by pixel as the beam is scanned across the sample.

Resolution in a SEM depends not only on the beam spot size but also on interaction of the beam with the sample (the spreading of the signal in the interaction volume) and is much less than that of a TEM.

Scanning Electron Microscopy is used to examine morphology (physical features) of size ranging from many microns to a few nanometers. The advantages over an optical microscope are a greater depth of field and higher resolution. Sample preparation is relatively simple, but the sample must be electrically conductive. For polymeric materials and biological samples, anhydrous condition and thin conductive coating (using sputtering of gold or graphite) are required to make them conductive.

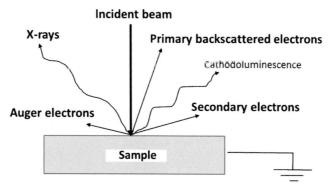

**FIG. 7.38** When an incident electron beam hits the sample under test, photons and electrons are emitted as signals. Primary backscatter electrons give information on the atomic number of the chemical species present and on the topography; cathodoluminescence provides electrical information; secondary electrons give topography information and X-rays on the composition in the first superficial layers. Auger electrons provide information on the surface composition.

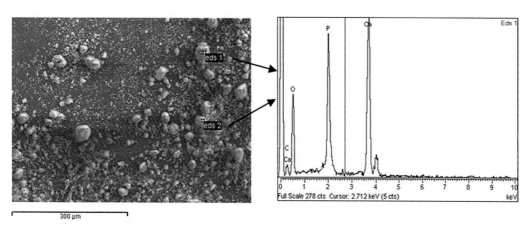

**FIG. 7.39** Left side: SEM image of a composite material (polymeric matrix loaded with calcium phosphates); right side: EDS microanalysis showing the presence of Ca and P signals.

As a consequence of the interaction between the electrons of the beam and the sample, the emission of electrons and photons from the sample occurs, as shown in Fig. 7.38. The electrons and photons from the sample are used for imaging and characterization. Secondary electrons, backscattered electrons, and characteristic X-rays are the signals mainly used in SEM observation and analysis and are widely used to characterize biomaterials. Chemical information is achieved by evaluating the emission of X-rays with an Energy Dispersive Spectrometer (EDS, or EDX) introduced in the SEM microscope.

SEM analysis is widely used to characterize biomaterials and scaffolds morphology, as well as cell adhesion and growth onto the scaffold surface. Examples are provided in Figs. 7.39 and 7.40, considering a composite material and a scaffold (a polyurethane foam) with cells attached and grown onto the foam pores.

### ENVIRONMENTAL SEM

The typical SEM instrument works under vacuum with dry and electrically conductive specimens. Nonconductive samples are usually sputter-coated with a thin layer of electrically

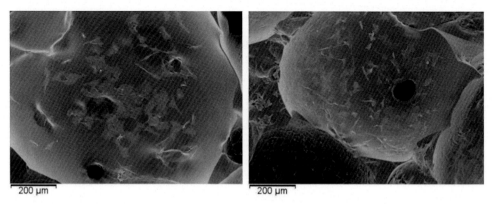

**FIG. 7.40**  SEM images of cells (from human osteosarcoma MG63 line) grown and spread onto a porous polyurethane scaffold after 1 day (left) and 7 days (right) post-seeding. *From* Zanetta, M., Quirici, N., Demarosi, F., Tanzi, M.C., Rimondini, L., Farè, S., 2009. Ability of polyurethane foams to support cell proliferation and the differentiation of MSCs into osteoblasts. Acta Biomater. 5(4), 1126–1136.

conductive materials such as gold or carbon. Hydrated and nonconductive samples (such as hydrogels or biological materials) need a series of pretreatments, such as structure fixing and drying, before being coated with the conductive layer. For most of these samples, the pretreatments can be difficult, and their "natural" state may be altered.

Variable Pressure-Environmental Scanning Electron Microscopy (VP-ESEM) uses the same working principle as conventional SEM but is specialized for imaging hydrated and nonconductive samples in their natural states. This instrument is equipped with specialized electron detectors and differential-pumping systems that allow for transferring the electron beam from the high vacuum in the gun area to the low vacuum attainable in the specimen chamber, thus, imaging specimens in their natural state.

### 7.2.5.5 Scanning Probe Microscopy

Scanning probe microscopy (SPM) includes several microscopy techniques for imaging surfaces with a physical probe that scans the specimen and measures surface features on a fine scale, down to the level of molecules and groups of atoms.

In all the SPM techniques, a sharp tip of atomic dimensions (typically of 3–50nm size with a tip spanning one to a few atoms) is scanned at few Angstroms above the surface of the sample, and the interaction between the tip and the sample is measured (Fig. 7.41). The tip is mounted on a cantilever, and the scan may cover a distance of about 100μm in the $x$ and $y$ directions and about 4μm in the $z$ direction.

As shown in Fig. 7.42, there are different Scanning Probe Microscopy techniques used for the characterization of 2D materials or surfaces.

### SCANNING TUNNELING MICROSCOPY

In Scanning tunneling microscopy (STM), a conducting tip is mounted on a piezoelectric transducer and scanned across the surface of the sample; a bias (voltage difference) allows electrons to tunnel through the gap in between, and the resulting "tunneling current" (typically nanoamperes) is a function of the tip position, applied voltage, and local density of states on the sample surface. A topographical map of the sample surface with atomic resolution is generated by keeping the height of the tip constant (constant height mode). Alternatively, the deflection of the tip is measured by keeping the current constant (constant current mode) to form a graded height map of the sample surface.

However, the samples for STM analysis should be conductive, and this is not the case with organic

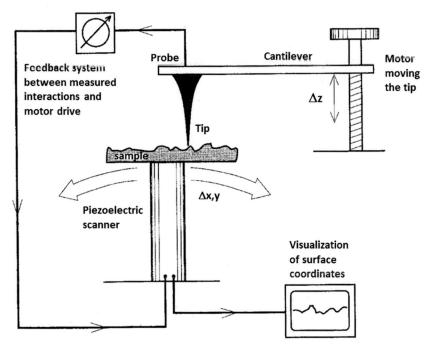

**FIG. 7.41**  Basic notion for Scanning Probe Microscopy techniques.

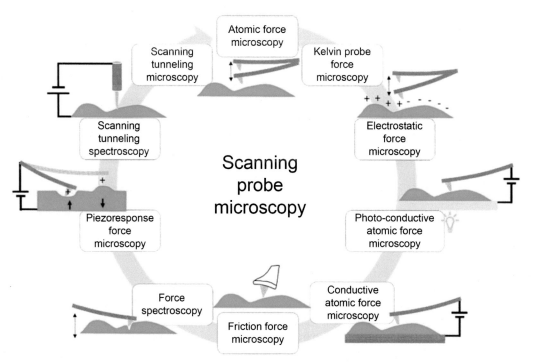

**FIG. 7.42**  Scheme showing the different Scanning Probe Microscopy techniques available for the characterization of two-dimensional (2D) materials. *From Musumeci, C., 2017. Advanced scanning probe microscopy of graphene and other 2D materials. Crystals 7(7), 216 (Fig. 1).*

or inorganic nonmetallic materials; even in metals such as titanium, the surface usually presents a native oxide layer which is too thick for electron tunneling. Therefore, the application of STM to biomaterials is presently limited, but STM is the preferred method for investigating the atomic structures of novel conductive materials such as carbon nanotubes and graphene.

### ATOMIC FORCE MICROSCOPY

Atomic force microscopy (AFM) was developed to overcome the limitations of STM and is now the most widely used member of the SPM family for imaging nonconducting surfaces and for operation in wet conditions for imaging biological samples.

Atomic Force Microscopy images surfaces using the force exerted between the AFM probe and the sample as the feedback parameter. To obtain an AFM topographic image, the sample is scanned by a tip mounted on a cantilever spring. While scanning, a feedback loop maintains the force between the tip and the sample constant by adjusting the scanner's height pixel by pixel, so that the image is obtained by plotting the height position versus its position on the sample. There are different modes of operation, which differ according to the nature of tip motion and tip-sample interaction. Interactions can be attractive or repulsive, ultimately setting the distance between the tip and the sample. In a static mode, (i.e., *contact mode*), the tip is raster scanned over the sample's surface by maintaining its deflection constant. In dynamic modes, such as *tapping*, *non-contact*, and *peak force tapping*, the tip oscillates, and feedback is given by the amplitude, frequency, or maximum force at the contact point.

## 7.2.6 Surface Analysis Techniques

The surface region of a solid can be thought of as a new phase of matter, as its chemical composition often differs from that of the bulk. Atomic and structures arrangements at (or near) the surface also differ from those in the solid bulk.

The precise behavior of atoms at the surface is crucial in many practical applications, including corrosion, adhesion, biocompatibility, and in-vivo performance of materials in clinical situations.

The success of processes and properties such as adhesion, durability, passivation, coating, and crystallinity depends critically on surface characteristics. Likewise, minimizing material or device failure (i.e., product reliability) can depend on the reduction of corrosion or wear, or the absence of phenomena such as embrittlement.

To adequately characterize a surface, it is often necessary to measure several and different aspects:

- type of atomic species present on the surface;
- arrangement or structure of surface atoms;
- electronic structure (principally of the valence electrons) of the surface atoms;
- motion of the atoms (atomic vibrations, surface diffusion, diffusion to and from the bulk, and evaporation);
- nature and distribution of defects on the surface;
- surface topography;
- nature and area of exposed surfaces;
- spatial distribution of foreign atoms on the surface, and
- depth distribution of different atomic species in the vicinity of interfaces.

The more parameters we can measure, the more we are able to fully describe a surface. A complete characterization requires the use of many techniques to collect all the necessary information, and presently, we can only partially recognize and specify the most important parameters necessary to understand, predict, and address the biological response. In addition, an extensive characterization of a surface can be hardly attempted on account of the cost and complexity involved.

All methods used for surface analysis can potentially alter the surfaces themselves. Because of the possibility of creating artifacts, we need to collect as much information with different methods of investigation. Compared to other classes of materials, organic materials and polymers can be damaged more easily by surface analysis methods.

The various surface analysis methods include:

1. Methods that evaluate a particular property of the surface (wettability, roughness, uniformity, thickness, optical constants, etc.) for example, contact angle measurement, laser profilometry, ellipsometry (ELS), and Optical Waveguide Lightmode Spectroscopy (OWLS)
2. Methods that evaluate the surface chemical composition such as the spectroscopic techniques, for example, Electron Spectroscopy for Chemical Analysis (ESCA), Auger Electron Spectroscopy (AES), Secondary Ion Mass Spectroscopy (SIMS), X-ray (e.g., EDS probe applied to SEM), Fourier Transform-Infrared Spectroscopy (FT-IR)
3. Methods that allow mass measurements to be combined with spectroscopic techniques, for example, Laser microprobe Mass Analysis (LMMA), Time of Flight (TOF)—mass analysis, Quartz Crystal Microbalance (QCM), and OWLS
4. Microscopy techniques which evaluate surface topography and morphology, for example, OM, SM, SEM, STM, AFM

Therefore, the information obtainable from surface analysis covers many aspects, including:

- the chemical composition,
- the number and type of functional groups on the surface (e.g., OH, C=O, COOH, SiOH, $NH_2$, etc.),
- stability (and lifespan) of the modified surfaces,
- any contamination present before or introduced during surface treatments with plasma, UV/ozone, etc.,
- the degree of wettability and adhesion in comparison with other materials,

- adsorption/desorption phenomena and rearrangement of biomolecules from the biological environment,
- the interactions between surface and cells.

We have already seen some of these techniques, in particular, infrared spectroscopy and techniques of microscopy. In the last part of Section 7.2 (covering biomaterials' characterization), we will examine the basic principles of the most important methods for surface characterization, such as wettability and spectroscopy for chemical analysis.

### 7.2.6.1 *Contact Angle for Wettability*

Contact angle is one of the simplest surface analysis techniques, and it allows evaluation of the wettability of a surface for different types of liquids. The lower the contact angle, the higher the degree of wettability of the surface under examination (Fig. 7.43).

The advantage of this technique is its simplicity, but its main disadvantage is the low precision (high degree of error) of the obtained measurement.

The contact angle is defined as the angle $\theta$ between the tangent to the drop profile at the point of contact between the three phases present (solid, gaseous (i.e., the air), and liquid) and the contact line between the liquid phase and the solid substrate (Fig. 7.44).

The theory underlying the formation of the contact angle can be summarized in the achievement of a thermodynamic equilibrium following the interaction of at least three phases.

The equilibrium condition to be satisfied, also known as the *Young-Dupré equation* (Eq. 7.1), can be expressed with:

$$\cos\theta = \frac{\gamma SV - \gamma SL}{\gamma LV} \qquad (7.1)$$

This means that, when a drop of liquid is placed on a solid surface, its perimeter moves, widening or contracting, until the angle $\theta$ assumes

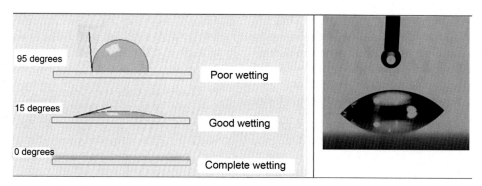

**FIG. 7.43**  Contact angle and wettability of a surface.

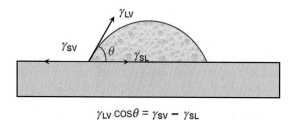

$$\gamma_{LV} \cos\theta = \gamma_{SV} - \gamma_{SL}$$

**FIG. 7.44** Balance of surface interaction energies. $\gamma_{SV}$ = interface energy between the solid phase $(S)$ and the gaseous phase $(V)$; $\gamma_{SL}$ = interface energy between the solid phase $(S)$ and the liquid phase $(L)$; $\gamma_{LV}$ = interface energy between the liquid phase $(L)$ and the gas phase $(V)$.

the value given by Eq. (7.1). In particular, if we agree to correlate the wettability with the measurement of the $\theta$ angle, considering water as the liquid, the four possible limit cases can be easily identified:

$\theta > 90$ degrees, poor wetting (i.e., hydrophobic surface);
$\theta < 90$ degrees, good wetting (i.e., hydrophilic surface);
$\theta = 0$ degrees, completely spread, maximum wetting, and
$\theta = 180$ degrees, completely round, no wetting.

The Young equation has a simple and effective formulation; however, it presents a problem: it is only possible to measure the contact

angle and the surface tension of the liquid. This means that there remain two unknowns (the solid-liquid and the solid-air tension) that cannot be determined with a single equation.

The *Zisman* method allows measurement of the *critical surface tension*, an approximation of the surface energy; the contact angles formed by drops of liquids having different surface tensions with the surface under examination are plotted as a function of the surface tension of the liquids, and the value corresponding to the 0° angle is then extrapolated to obtain the *critical surface tension* value.

### 7.2.6.2 Profilometry

Roughness is a property of the surface of an object, consisting of geometric microimperfections normally present on the surface or even resulting from mechanical processing; these imperfections generally appear in the form of grooves or scratches of variable shape, depth, and direction.

The roughness measurement procedure consists of recording the profile of the surface obtained along a determined measurement line (or scanning line); this profile is then analyzed by defining a numerical parameter that constitutes the measure of roughness. A fundamental part of the process of calculating the various roughness parameters is the filtering operation that allows us to obtain a measurement of the surface quality only, purified by the effects that the geometric errors of the piece have on the measured profile.

The measurement of the roughness $R_a$, expressed in microns, is the arithmetic mean of the distances of the points of the profile from the mean line (midline[5]) of reference.

However, the $R_a$ value is not sufficient to completely define the morphological characteristics of the surface, as profiles with different trends from the same arithmetic mean deviation will have the same value of $R_a$. For this reason, other parameters have been introduced, such as $R_z$ and $R_{max}$.

$R_z$ is the distance between two straight lines parallel to the mean line drawn at a distance equal to the average of the 5 highest peaks and the average of the 5 lowest valleys in the range of the base length. $R_z$ is then the average of the maximum roughness of 5 peaks and 5 valleys.

$R_{max}$ is the distance between two lines parallel to the midline, the first tangent to the highest peak and the second tangent to the lower valley. However, $R_{max}$ is not a very significant parameter because it may be strongly influenced by an accidental irregularity of the surface.

The roughness of a surface can be measured using *profilometers*. A profilometer is the instrument for the measurement and evaluation of micro-irregularities of a surface and is able to perform measurements with a precision that can reach a thousandth of a micron.

Measurement methods may vary and can be direct or indirect. A method of *direct* measurement is one that utilizes a *probe profilometer*, positioned at the end of a stylus. A method of *indirect* measuring is one that uses a sensor interpreting the variations in height measured along the axis

of acquisition as changes in position of the reflected beam on the surface to be probed. An *optical profilometer* is a method of noncontact measurement.

There are different techniques that can be used today, such as laser triangulation, confocal microscopy (for very small objects), low coherence interferometry, and digital holography.

Sophisticated techniques that provide information on topography, such as STM and SFM (Scanning Probe Microscopies) are useful as well.

### 7.2.6.3 Spectroscopic Techniques for Surface Analysis

As we already know, spectroscopic analyses are based on the interaction of different types of electromagnetic radiation with matter.

In the case of surface analysis, the principle is to hit the surface to be analyzed with a *probe beam* and to evaluate the output response, in reflection, rather than absorption (*analysis beam*), as shown in Fig. 7.45.

The sample under investigation is irradiated by a beam of energetic (primary) particles (i.e., photons, electrons or ions). When the primary particles penetrate the material, their interactions will give rise to various excitations in the sample, which then responds to the perturbation by the emission of secondary particles (usually electrons or ions).

The emitted secondary particles reflect various properties of the sample material; thus, by measuring a certain physical quantity of the

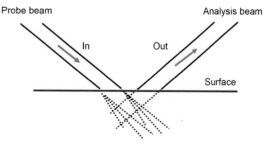

FIG. 7.45 Basic principle for all surface spectroscopy techniques.

[5] Midline = line for which the sum of distances is minimum to the square of the contour points from the same line.

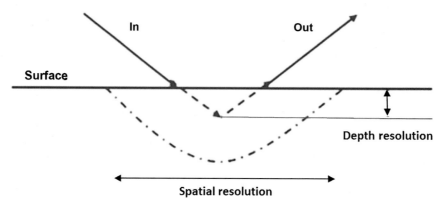

**FIG. 7.46**　Depth resolution and spatial resolution in surface analysis.

secondary particles, information about the sample can be obtained. The nature of the primary particles used determines the different types of secondary particles which will be emitted, yielding different kinds of information. All surface-sensitive spectroscopies have the common principle of employing primary and/or secondary particles which have short penetration (or escape) depths in solids. This guarantees that the information obtained originates mainly from the relevant surface region.

In general, each technical specification is distinguished by two typical variables: *depth resolution* and *spatial resolution*, as shown in Fig. 7.46.

**A.** *Electron spectroscopies*—in which the radius of analysis is represented by electron beams
　　Two important techniques of this type are:
- X-ray Photoelectron Spectroscopy (XPS) or ESCA (Electron Spectroscopy for Chemical Analysis)
- Auger Electron Spectroscopy (AES/SAM)

**B.** *Ion Spectroscopies*—in which the range of analysis is represented by ion beams and include
- Ion Scattering Spectroscopy (ISS)
- Secondary Ion Mass Spectrometry (SIMS)

Fig. 7.47 shows where surface analysis techniques are placed in terms of depth resolution and spatial resolution.

### 7.2.6.4 *Electron Spectroscopies*

### X-RAY PHOTOELECTRON SPECTROSCOPY (XPS) OR ESCA (ELECTRON SPECTROSCOPY FOR CHEMICAL ANALYSIS)

X-ray photoelectron spectroscopy (XPS) is one of the most widely used methods in the field of surface analysis spectroscopy techniques and is based on the analysis of the energy of photoelectrons emitted by a solid material hit with electromagnetic radiation in the X-ray field (incidents, typically Al Kα or MgKα).

The phenomenon that underlies this analysis technique is the *photoelectric effect* (Fig. 7.48); a material subjected to the bombardment with an electromagnetic radiation of sufficiently high energy $h_\nu$ emits electrons with kinetic energy $E_k$, which depends on the energy level from the which was extracted.[6]

This effect is based on the following relation proposed by Einstein:

$$E_k = h\nu - E_B$$

where:

$h\nu$ is the energy of the emitted photon,
$\nu$ is the frequency of the incident radiation,

---

[6] Photoemission was first detected by Hertz in 1887 and explained by Einstein in 1905. Einstein showed that light behaves like particles.

$E_B$ is the binding energy of the electron in the atom at a particular level, and
$E_k$ is the kinetic energy of the electron.

The XPS technique is particularly useful in determining the chemical species present on the surface of a sample and the nature of the bonds in which they are involved. $E_B$ is the most important physical quantity determined with the XPS technique because its determination provides information not only on the elements present, but also on their state of valence and on the surrounding electronic environment.

The information obtained is closely linked to the first atomic layers, which makes XPS one of the main surface analysis techniques with an average depth of exploration of electrons of about 100 Å. The excited electrons are, in fact, only close to the surface and manage to escape from the sample without undergoing interactions with consequent energy losses, thus preserving the information they carry.

XPS is a nondestructive technique; therefore, the chemical-physical properties are not altered during the measurement itself and can be applied to all solid materials, including insulating materials such as polymeric materials and glasses.

### AES (AUGER ELECTRON SPECTROSCOPY)

Auger electron spectroscopy is a surface analytical technique that utilizes the *Auger effect* to measure the elemental composition of surfaces.

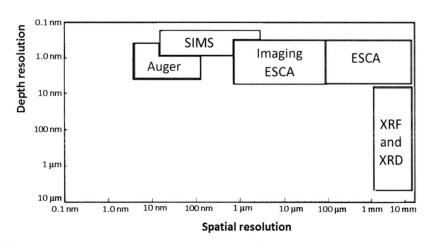

**FIG. 7.47** Spatial resolution values toward depth resolution for surface spectroscopies (XRF=X-ray fluorescence and XRD=X-ray diffraction).

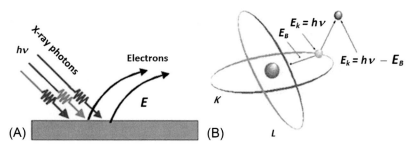

**FIG. 7.48** (A) Principle of ESCA and (B) the photoelectric effect.

A beam of energetic electrons, 3–25 keV, is used to eject a core level electron from surface atoms. To release energy, those atoms may emit Auger electrons from their induced excited state. The energy of the Auger electron, specific to the atom from which it originated, is measured, and the quantity of Auger electrons is proportional to the concentration of the atoms on the surface.

Auger electron spectroscopy can measure two dimensional maps of elements on a surface and elemental depth profiles when accompanied by ion sputtering.

AES is routinely used to analyze the chemical composition of surfaces, and Auger *depth profiling* is a valuable tool for characterizing thin film systems and processes. However, the high energy involved does not allow the analysis of temperature-sensitive materials such as synthetic polymers and biological environments.

### 7.2.6.5 Ion Spectroscopies

#### 7.2.6.5.1 SECONDARY ION MASS SPECTROSCOPY

SIMS is a mass spectrometry technique consisting of bombarding the sample with an ion beam (called *primary ions*) and analyzing the ions produced by the bombardment (*secondary ions*).

The SIMS technique guarantees a very high sensitivity in detecting substances present at trace levels (ppm-ppb) and an excellent depth resolution ($3 \div 10\,nm$).

Fig. 7.49 schematically shows the process of SIMS analysis.

By use of a mass spectrometer, the secondary ions removed from the surface and emitted are measured. Measurement requires high-vacuum or ultra-high-vacuum ($P \sim 10^{-5} \times 10^{-8}\,Pa$). Typical sampling depth ranges from a few atomic layers up to relatively thick samples.

In *Static SIMS* (SSIMS), the total dose of primary ions hitting the sample is $<1 \times 10^{13}\,ions/cm^2$. The method is nondestructive and is used to obtain chemical information by analyzing whole molecular fragments. It can be applied to conductive and nonconductive samples and to organic samples.

In *Dynamic SIMS* (DSIMS), the total dose of primary ions hitting the sample is $>1 \times 10^{13}\,ions/cm^2$. Secondary ions are generated by eroding the sample surface. The method is destructive, is used to recognize the elements present, and is applicable to inorganic samples.

*SIMS Imaging* is used for spatially-resolved elemental analysis.

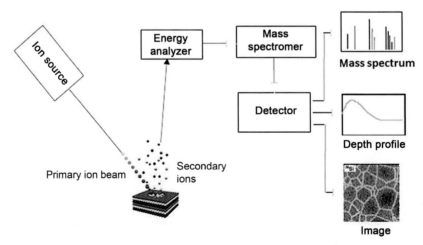

**FIG. 7.49**  Schematic procedure for a SIMS analysis.

The mass analyzer may be a quadrupole mass analyzer (with unit mass resolution), but magnetic sector mass analyzers or time-of-flight (TOF) analyzers are also often used providing substantially higher sensitivity and mass resolution, and a much greater mass range (although at a higher cost). In general, TOF is preferred for static SIMS, while the other two methods are preferred for dynamic SIMS.

## 7.3 DIAGNOSTIC TECHNIQUES

The terms *imaging*, *medical imaging*, or *imaging diagnostics* refer to the process by which an operator can observe and investigate a district of an organism that is not visible from the outside. In fact, most medical issues occur inside the body, so making a diagnosis can be a challenge. Medical imaging has made diagnosis far easier over the last century. Generally speaking, imaging is the technique of producing visual representations of areas inside the human body to diagnose medical problems and monitor treatment.

Imaging uses the energy provided by a specific source and measures the interaction between this energy and the target (e.g., organ, tissue) whose image has to be obtained. These techniques can be classified into two classes (Fig. 7.50):

1. techniques that use nuclear radiation (i.e., ionizing radiation);
2. techniques that do not use ionizing radiation.

Each technique is used in different circumstances. For example, radiography is often used when the clinician needs images of bone structures to look for breaks. MRI scanners are often used to take images of the brain or other internal tissues, particularly when high-resolution images are needed. Nuclear medicine is used when the doctor needs to look inside the digestive or circulatory systems, to look for blockages, for example.

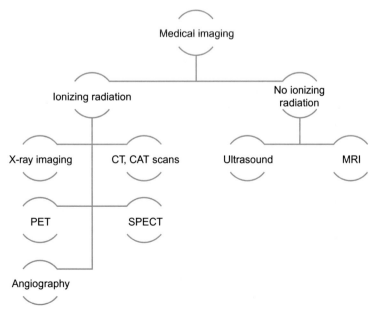

**FIG. 7.50** Classification of imaging techniques. CT, computed tomography; CAT, computed axial tomography; PET, positron emission tomography; SPECT, single-photon emission computed tomography; MRI, magnetic resonance imaging.

Ultrasound is used to look at fetuses in the womb and to take images of internal organs when high resolution is not necessary.

All forms of diagnostic imaging may be useful tools for pinpointing disease and providing optimum medical care when medically indicated. In many cases, the medical benefits of diagnostic investigations may outweigh their inherent risks. Nevertheless, radiation exposure from medical imaging may increase cancer risk, even at low levels, and may be cumulative over an individual's lifetime. Even if diagnostic radiation is an effective tool that can save lives, the higher the dose of radiation delivered at any one time, the greater the risk for long-term damage. In particular, the risks associated with the use of ionizing radiation in medical imaging include cancer, burns, and other injuries. Because of those possible drawbacks, many experts believe that ultrasound imaging and MRI should be preferred where they would provide similar diagnostic information at a quality level similar to radiation-based diagnostic exams.

In this section, mainly X-ray-based techniques and MRI will be more deeply described, whereas main aspects of other important medical imaging techniques are briefly reported in Table 7.4, as they are far from the final aim of this book.

## 7.3.1 X-Ray Investigation

### 7.3.1.1 X-Rays

X-rays were discovered in 1895 by the German physicist Wilhelm Conrad Röntgen, who earned the Nobel Prize in physics in 1901. From then until today, much research has been done to improve and widely use X-rays, as reported in Table 7.5, where the history of X-rays is briefly reported. As observable in Table 7.5, X-rays can be used for materials analysis, in particular in crystallography investigation, as well as in clinical diagnostic. In this section, the use of X-rays in clinical imaging applications will be described.

X-rays are a high-energy form of electromagnetic radiation, similar to visible light (Fig. 7.8). In particular, because the energy of electromagnetic waves is directly related to their frequency, X-rays are much more energetic and penetrating than light waves. Hence, X-rays can pass through most objects, including the human body. If X-rays can travel through the body, they can also pass through an X-ray detector on the other side of the patient, and an image will form, showing the shadows formed by the objects inside the body.

There are materials that allow X-rays to pass straight through them and materials that stop X-rays in their tracks. In fact, when X-rays enter a material, they have to fight their way through a huge crowd of atoms if they are going to emerge from the other side. What really get in their way are the electrons whizzing round those atoms. The more electrons there are, the greater the chance they will absorb the X-rays, and the less likely the X-rays are to emerge from the material. X-rays tend to pass through materials made from lighter atoms with relatively few electrons (such as skin, which is made of carbon-based molecules), but they are stopped in their tracks by heavier atoms with lots of electrons. Lead, a heavy metal with 82 electrons spinning round each of its atoms, is particularly good at stopping X-rays (that is why X-ray technicians in hospitals wear lead aprons and stand behind lead screens). The fact that some materials let X-rays travel through them better than others turns out to be very useful indeed.

To make X-rays, it is necessary to simply fire a beam of really high-energy electrons (accelerated using a high-voltage electricity supply) at a piece of metal (typically tungsten). What is reflected back, in this case, is neither light nor electrons, but a beam of X-rays. Generally speaking, the higher the voltage you use, the faster the electrons go, the more energetically they crash into the tungsten, and the higher the energy (and frequency) of the X-rays they produce.

**TABLE 7.4**  Main Medical Imaging Techniques and Characteristics

| | |
|---|---|
| X-rays | – Invisible beams of ionizing radiation |
| | – 2-dimensional images of various parts of the body, such as bones, lungs, and organ systems |
| | – They are not painful, and lead shields may decrease radiation exposure to areas that are not imaged |
| CT scans | – They use X-rays rotating around the body to create image "slices" that are transformed into 3-dimensional images |
| | – They provide superior data when compared to a single X-ray, but does so using more radiation |
| | – Patients undergoing a CT scan are required to lie still on a table while the table moves through the CT scanner |
| | – They are painless, but sedation may be required |
| | – CT contrast agents are iodinated solutions that are administered intravenously to enhance visualization of organs and blood vessels |
| PET | – It uses injectable radionuclide (radioactive) agents that emit gamma ray signals measured by PET scanners |
| | – It is often used to detect and evaluate cancerous tumors, neurological conditions, and cardiovascular disease |
| Nuclear SPECT | – It uses small amounts of radioactive drugs administered intravenously and emitted gamma rays detected by a camera, which then converts the rays to electronic signals to create an image |
| | – The amount of radiation exposure depends on the type of study being performed |
| | – SPECT studies are not painful, but require patients to lie still while images are being obtained |
| | – Sedation may be required for some patients |
| Angiography | – It is an invasive procedure that produces X-ray images of the inside of blood vessels to evaluate blockages or other damage |
| | – During an angiogram, a radio-opaque dye (which absorbs the X-rays) is injected into arteries using flexible catheter tubes and guide wires |
| Ultrasound | – It uses high-frequency sound waves to view inside the body |
| | – Because ultrasound images are captured in real time, they can also detect movement of the body's internal organs as well as blood flowing through the blood vessels |
| | – No ionizing radiation exposure is associated with ultrasound imaging |
| MRI | – It uses powerful magnetic fields and radio-frequency transmitters to produce 3-dimensional images throughout the body, including the brain, muscles, heart, and cancerous tumors |
| | – MRI procedures do not expose patients to ionizing radiation |
| | – MRI scans are painless, but some patients complain of claustrophobia and noise |
| | – MRI contrast agents used to enhance images may have associated risks |
| | – Pacemakers, implantable cardioverter-defibrillators, insulin pumps, cochlear implants, and certain other medical or biostimulation implants are generally considered contraindications for MRI imaging |

### 7.3.1.2 *Use in Medicine*

The oldest use of X-rays is in medical imaging; this application is known as medical radiography. A radiograph is the image produced using X-rays. Medical radiography is one of the most widely performed diagnostic procedures in medicine.

X-rays are an important medical tool, used mainly in diagnosis, but also in therapeutic treatment. Because X-rays are highly energetic, they can damage living tissue when they pass through it. On one hand, this means X-rays have to be used cautiously and quite selectively, and X-ray technicians have to take precautions to prevent absorbing too much of the radiation during their work. But on the other hand, X-rays can be used to sterilize medical equipment (i.e., they are able to destroy bacteria, viruses, germs)

**TABLE 7.5**   History of X-Rays

| | |
|---|---|
| 1895 | Wilhelm Röntgen discovers X-rays experimenting with cathode rays in a glass tube. Röntgen does not know what these rays are, so he calls them "X-rays." This discovery earns him the very first Nobel Prize in physics in 1901 |
| 1896 | Thomas Edison develops an X-ray viewer called a fluoroscope |
| 1906 | Charles Barkla shows that X-rays can be polarized in a way similar to beams of light. This provides important evidence that X-rays are light waves of different wavelength and frequency |
| 1912 | Max von Laue discovers he can measure the wavelength of X-rays, confirming the wavelength of X-rays and the regular atomic nature of crystals |
| 1913–14 | William Henry Bragg and his son Lawrence Bragg show how X-rays of known wavelength can be used to measure the atomic spacing of crystals, developing the field of X-ray crystallography. They earn the 1915 Nobel Prize in physics |
| 1913 | William David Coolidge develops the X-ray-making machine. He patents his invention in 1916. Most X-ray machines still work broadly this way today |
| 1922 | Arthur H. Compton studies the reflection of X-rays, discovering the phenomenon called the Compton effect (or Compton scattering): the scattered X-rays have less energy than the particles in the original beam, providing evidence for the particle-nature of electromagnetic radiation |
| 1953 | Francis Crick and James D. Watson work out the structure of DNA with help from X-ray diffraction images produced by Rosalind Franklin |
| 1972 | Godfrey Hounsfield invents the CT scanner, which makes 3D images of the human body using thin X-ray beams |
| 1980s | Powerful X-ray lasers are proposed that would produce X-rays through a process of stimulated emission |
| 1999 | The Space Shuttle launches the Chandra X-ray observatory, the most sensitive X-ray telescope to date |
| 2000s | CT X-ray scanners are used to improve baggage-screening security in airports |
| 2009 | Scientists at SLAC National Accelerator Laboratory (Menlo Park, CA, United States) produce a powerful X-ray laser described as "the world's brightest X-ray source" |

and can kill tumors in the treatment of cancer (i.e., radiation therapy and radiotherapy).

Hard materials such as bone and teeth are very good at absorbing X-rays, whereas soft tissues like skin and muscle allow the rays to pass straight through (Fig. 7.51). That makes X-ray images extremely useful for all kinds of medical diagnosis; they allow us to see broken bones, tumors, and lung conditions such as tuberculosis and emphysema. Dentists also use X-rays extensively to understand what is the problem in your mouth, inside your teeth and gums. Soft tissues allow the X-rays to pass through them, and when the X-ray hits the photographic plate, it turns it black. Dense tissues like bone absorb X-rays, so these are shown as white on the photographic plate.

There is a limit to what a physician can understand from a two-dimensional image of the human body (i.e., a three-dimensional structure), but 3D-scanning technology helps to overcome that. In fact, computer tomography (CT) or computerized axial tomography (CAT) scanners draw what are effectively 3D X-ray pictures on screens by firing thin beams of X-rays through a patient's body and use computer technology to turn lots of 2D pictures into a single 3D image (see Section 7.3.1.7).

In general, for imaging purpose, there is an X-rays generator that produces the radiation by

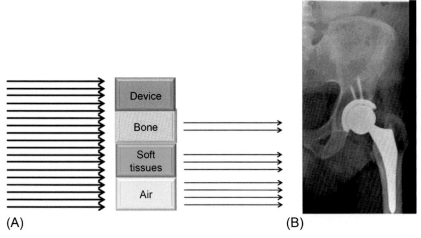

**FIG. 7.51** (A) Device and tissues differently absorb X-ray radiation, so that the resulting image is composed of gray scale related to the tissue composition; (B) Example of X-ray radiography in which it is possible to evidence the gray tones corresponding to different tissue. *Modified from Oliveira, C.A., Candelária, I.S., Oliveira, P.B., Figueiredo, A., Caseiro-Alves, F., 2015. Metallosis: a diagnosis not only in patients with metal-on-metal prostheses. Eur. J. Radiol. 2, 3–6.*

means of an X-ray tube as described above. In addition, there is a device that detects X-rays. Typically, this is a digital X-ray detector (e.g., a flat panel detector), but it could also be a photographic film. This device essentially serves to convert the X-rays into visible light to produce images that the human eye can see and analyze. Between the X-ray generator and the detector, the medical team will position the object or human body part that has to be investigated. The X-ray emitting device sends the X-rays toward the object, which casts varying "shadows" upon the X-ray detector behind it. The relative variance of the shadows depends on the density of the materials within the object or body part.

*Possible risks of X-ray diagnosis.* When used appropriately, the diagnostic benefits of X-ray scans significantly outweigh the risks. X-ray scans can diagnose possibly life-threatening conditions such as blocked blood vessels, bone cancer, and infections. However, X-rays produce ionizing radiation (i.e., a form of radiation that has the potential to harm living tissue). This is a risk that increases with the number of exposures added up over the life of the individual.

However, the risk of developing cancer from radiation exposure is generally small.

An X-ray analysis in a pregnant woman poses no known risks to the baby if the area of the body being imaged is not the abdomen or the pelvis. When imaging of the abdomen and pelvis is needed, doctors prefer to obtain information using examination methods that do not require radiation, such as MRI or ultrasound. However, if neither of those can provide the expected results, or there is an emergency or other time constraint, an X-ray investigation may be an acceptable alternative imaging option. Children are more sensitive to ionizing radiation and have a longer life expectancy, thus, they have a higher relative risk for developing cancer than adults. Sometimes, the X-ray machine can be set for imaging on children.

### 7.3.1.3 X-Ray Radiography

Radiograph is the type of X-ray image that first comes to mind for many people. Radiographs can help clinicians detect bone fractures, certain tumors and other abnormal masses, pneumonia, some types of injuries,

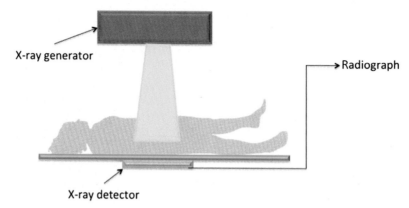

**FIG. 7.52**    Scheme of the acquisition of a radiograph.

calcifications, foreign objects, and dental problems. To obtain a radiograph, the patient is positioned so that the part of the body being imaged is located between an X-ray source and an X-ray detector (Fig. 7.52). When the machine is turned on, X-rays travel through the body, and different parts of the body absorb and scatter X-rays in dissimilar ways due to differences in tissue density, as well as differences in atomic number of the atoms that make up the tissues. Hence, X-rays can be absorbed by human tissues, deflected through impact with the body's atoms, or may pass undamaged. This property allows the technician to differentiate between various parts of the body in an image, because X-ray imaging distinguishes between the different physical parts of the body by converting the physical contrast between organs and tissues to image contrast (Fig. 7.51).

In particular, dense, calcium-rich bone absorbs X-rays to a higher degree than soft tissues so that more X-rays can pass through them en route to the detector, making X-rays very useful for capturing images of bone; bone is subsequently referred to as a radio-dense substance, in contrast to the softer tissues in the body that are considered more radiolucent. Moreover, X-ray imaging is also very good for capturing images of certain pathologies that affect other types of body tissue as well, for

example, breast and lung cancers. What makes X-rays so good at capturing images from our bodies is that the human body is composed tissues which are different in density and composition (Fig. 7.51). In projection radiography, there is much room for adjusting the energy level of the X-rays depending on the relative densities of the tissues being imaged, as well as how deeply the waves must travel through a patient's body in order to achieve imaging. For example, images of bone (e.g., to examine a fracture or for diagnostic measures related to bone conditions like osteoarthritis or certain cancers) require high-energy X-rays because of bone's high density. But images of soft tissues such as lungs, the heart, and breasts require relatively less energy from the X-rays in order to penetrate properly and achieve excellent images. X-rays are not very useful when trying to look at muscle or the brain. Below, some radiography applications for body imaging are reported.

*Bone X-ray.* Basic radiography is often the fastest and the easiest tool to diagnose problems in the skeletal system such as fractured bones, dislocated joints, and arthritis, and even to assist in the diagnosis of bone cancer. X-rays can also be used as a guide in orthopedic surgeries and to check for proper alignment of bones post-treatment. Foreign objects positioned in soft tissue around the bones will also show up in an X-ray, making X-ray imaging useful during the performance of medical procedures as well.

*Chest X-ray.* Chest X-rays are the most common type of X-ray imaging. They are usually the first analysis performed to evaluate patients presenting with shortness of breath, persistent cough, chest injury, and even fever. Chest X-rays can detect a variety of conditions such as pneumonia, heart failure, collection of air or fluid around the lungs, and even lung cancer. They are also used to check for proper placement of lines and tubes in the chest area. Even though chest X-rays are very useful tools, they are not without limitation. For example, some conditions such as blood clots in the lungs or small cancers may not show up on a chest X-ray.

*Bone mineral density measurement.* Dual Energy - X-Ray Absorptiometry (DXA) is often used to assess bone mineral density. Its major application is the quantification of the extent of osteoporosis in older patients by a DXA scan of the lumbar spine, proximal hip, and, less commonly, the forearm. The bone density data (i.e., areal bone mineral density) obtained from the scan is compared to a reference value to estimate the extent of osteoporosis in the patient.

### 7.3.1.4 Mammography

A radiograph of the breast is used for cancer detection and diagnosis. The two types of imaging currently used for mammography are:

- screen-film mammography in which X-rays are beamed through the breast to a cassette containing a screen and film that must be developed. The image is commonly referred to as a mammogram;
- full-field digital mammography in which X-rays are beamed through the breast to an image receptor (Fig. 7.53A). A scanner converts the information to a digital picture, which is sent to a digital monitor and/or a printer.

Tumors tend to appear as regular or irregular-shaped masses that are somewhat

brighter than the background on the radiograph (i.e., whiter on a black background or blacker on a white background). Mammograms can also detect tiny bits of calcium, called microcalcifications, which show up as very bright specks on a mammogram. While usually benign, microcalcifications may occasionally indicate the presence of a specific type of cancer. The benefits of mammography in detecting breast cancer at an early stage (i.e., under age 40) may outweigh the risks of radiation exposure. For example, a mammogram may reveal that a suspicious mass is benign, and, therefore, does not need to be treated. Additionally, if a tumor is malignant, the early detection of a breast lump by mammogram (Fig. 7.53B) can help the surgeon determine whether it can be removed before it spreads and requires more aggressive treatment such as chemotherapy.

*Risks.* Mammography uses X-rays to produce an image of the breast, and the patient is exposed to a small dose of ionizing radiation. The Mammography Quality Standards Act (MQSA) established baseline standards for radiation dose, personnel, equipment, and image quality. For most women, the benefits of regular mammograms outweigh the risks posed by this amount of radiation. The risk associated with this dose appears to be greater among younger women (<40 years).

### 7.3.1.5 Angiography

Angiography (Fig. 7.54A) is an imaging technique where contrast agent is injected into the bloodstream via a guided catheter from the femoral artery. In such a way, it is possible to enable X-ray imaging of the human circulatory system or investigating excretory pathways. The resulting angiogram can be either film (i.e., capturing movement, as in fluoroscopy) or still images. The contrast agent, such as iodine, has a density higher than human blood, enabling high quality images of human blood vessels. Although the term historically implies the use of projected radiography, CT angiography has emerged as another form of

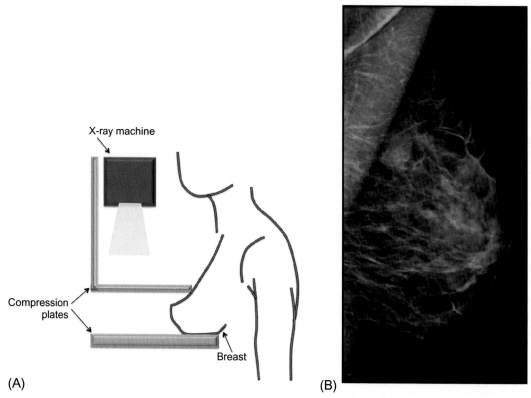

**FIG. 7.53** (A) Scheme of the setup for acquisition of breast X-ray imaging; (B) mammography can detect cancer lesions in breast tissue (identified by the *arrow*). *From Marino, M.A., Riedl, C.C., Bernathova, M., Bernhart, C., Baltzer, P.A.T., Helbich, T.H., Pinker, K., 2018. Imaging phenotypes in women at high risk for breast cancer on mammography, ultrasound, and magnetic resonance imaging using the fifth edition of the breast imaging reporting and data system. Eur. J. Radiol. 106, 150–159.*

angiography (see Section 7.3.1.7). Angiography has many applications and can be used to help diagnose or investigate a number of problems affecting the blood vessels (Fig. 7.54B). Among these, we can include the detection of a clot (or thrombus) in a vein, or a pulmonary embolism, evaluation of coronary artery disease, location of an aneurysm, or guiding the placement of a stent during a procedure. In particular, angiography has acquired an important role in preoperatory and intraoperatory radiology. Angiography is generally a very safe procedure, although minor side effects are common, and there is a small risk of serious complications. The test will only be performed if the

benefits of having the procedure are believed to outweigh any potential risk.

### 7.3.1.6 Fluoroscopy

Fluoroscopy uses X-rays and a fluorescent screen to obtain real-time images of movement within the body, or to view diagnostic processes, such as following the path of an injected or swallowed contrast agent. Thomas Edison coined the term fluoroscopy while experimenting with X-rays (Table 7.5). The same principles of radiography were combined with a fluorescent screen technology, and the result traditionally was a live visualization with movement. Now, the medical community uses the same digital detector screen

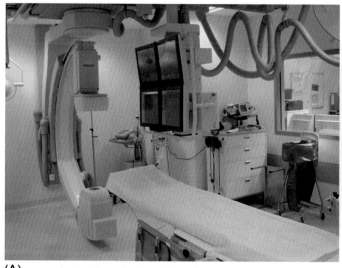

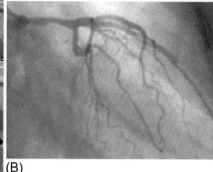

(B)

(A)

**FIG. 7.54** (A) Scheme of the setup for angiography (https://en.wikipedia.org/wiki/Angiography); (B) coronary angiography demonstrating an eccentric, focal, and severe narrowing of left main artery. *From Redondo, B., Gimeno, J.R., Pinar, E., Valdés, M., 2009. Unusual presentation of acute coronary syndrome. Bilateral coronary dissection after car accident. Am. J. Emerg. Med. 27(8), 1024.e3-5.*

technology for fluoroscopy as it uses to achieve standard projected radiography, significantly reducing the amount of radiation absorbed by the patient in the process.

For example, fluoroscopy is used to view the movement of the beating heart, and, with the aid of radiographic contrast agents, to view blood flow to the heart muscle as well as through blood vessels and organs. This technology is also used with a radiographic contrast agent to guide an internally threaded catheter during cardiac angioplasty, which is a minimally invasive procedure for opening clogged arteries that supply blood to the heart.

### 7.3.1.7 Computed Tomography, CT

Radiography allows for a 2D-imaging of a specific human body tissue or organ, and, as already exposed, the radiograph is similar to the shadow of the part that is hit and crossed by the X-ray beam (Fig. 7.55).

For that reason, the development in the use of X-rays in medical imaging led to the invention of

Computed Tomography (CT). In fact, CT scans produce 2D images of a "slice" or section of the body, but the data can also be used to construct 3D images. A CT scan can be compared to looking at one slice of bread within a whole loaf. Sometimes called a CAT (computerized axial tomography) scan, the Computed Tomography scan is a powerful diagnostic tool that involves many X-ray images of a body part captured simultaneously from various angles. The word tomography is derived from *tomos* = slice and *graphein* = to write; hence, a possible definition of CT is imaging of an object by analyzing its slices, called tomographic images, that contain more detailed information than conventional X-rays. Once a number of successive slices are collected by the machine's computer, they can be digitally "stacked" together to form a three-dimensional image of the patient site that allows for easier identification and location of basic structures as well as possible tumors or abnormalities. CT combines traditional X-ray technology with computer processing to

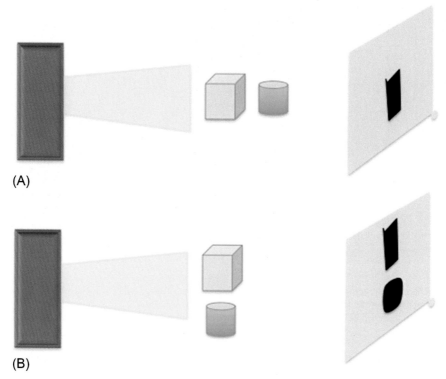

(A)

(B)

**FIG. 7.55**  Limit of X-ray radiography that does not allow the investigation of all tissues/organs on the same plane (A), whereas there is no problem if the tissues/organs under investigation are on two different planes (B).

generate a series of cross-sectional images of the body that can later be combined to form a 3D X-ray image. CT images are more detailed than plain radiographs and give doctors the ability to view structures within the body from many different angles.

Table 7.6 briefly reports the main steps of the development of CT. Although the X-ray's potential applications in medical imaging diagnosis were clear from the beginning, the first X-ray Computed Tomography system was implemented in 1972 by Godfrey Newbold Hounsfield (Nobel prize winner in 1979 for physiology and medicine), who constructed the prototype of the first medical CT scanner and is considered the father of Computed Tomography. CT was introduced into clinical practice in 1971 with a scan of a cystic frontal lobe tumor on a patient at Atkinson Morley Hospital in Wimbledon (United Kingdom). After this, CT was immediately welcomed by the medical community and has often been referred to as the most important invention in radiological diagnosis since the discovery of X-rays. In fact, before CT, entire body areas were inaccessible to radiography (e.g., brain, mediastinum, retroperitoneum) and diagnostic procedures showing better detail in these areas were potentially harmful or poorly tolerated by the patient (e.g., pneumoencephalography, diagnostic pneumomediastinum, diagnostic laparotomy). Different improvements to the CT instruments were performed over the years; however, CT refers to a computerized X-ray imaging

**TABLE 7.6** History of Computed Tomography, CT

| | |
|---|---|
| 1924 | J. Radon: Mathematical theory of tomographic image reconstructions |
| 1930 | A. Vallebona: Conventional tomography |
| 1963 | A. McLeod Cormack: Theoretical basis of CT |
| 1971 | G. Newbold Hounsfield: First commercial CT |
| 1974 | First third generation CT |
| 1979 | Cormack and Hounsfield: Nobel prize |
| 1989 | Single-row CT |
| 1994 | Double-row spiral CT |
| 2001 | 16-row spiral CT |
| 2007 | 320-row spiral CT |

procedure in which a narrow beam of X-rays hits the patient and quickly rotates around the body, producing signals that are further processed by the machine's computer to generate cross-sectional images (i.e., slices) of the part of the body under investigation. The slices (i.e., tomographic images) contain more detailed information than conventional X-ray diagnostic images, because they can be digitally stacked together, forming a 3D image allowing for easier identification and location of basic structures as well as possible tumors or abnormalities.

The classification of medical CT scanners according to their scanner geometry is divided into five generations that are briefly described in the next section.

In first-generation CT systems (Fig. 7.56), a single X-ray source has a linear shape (i.e., pencil-like) directing across the object and a single detector. Both the source and the detector translate simultaneously along a scan plane. This process is repeated for a given number of angular rotations. The advantages of this design are simplicity, good view-to-view detector matching, flexibility in the choice of scan parameters (e.g., contrast and resolution), and the ability to accommodate a wide range of

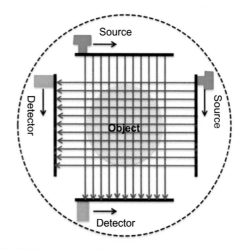

FIG. 7.56 Scheme of first-generation CT.

different object sizes. The disadvantage is longer scanning times.

Second-generation CT systems use the same type of movement (i.e., translation/rotation geometry) as the first generation, but a fan-shaped X-ray beam replaces the linear pencil-shaped beam (Fig. 7.57), and a row of multiple detectors substitutes for the single detector. With that configuration, a series of views can be acquired during each translation, which leads

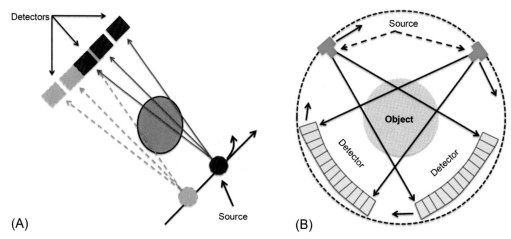

FIG. 7.57    (A) Scheme of second generation CT; (B) second generation scanning.

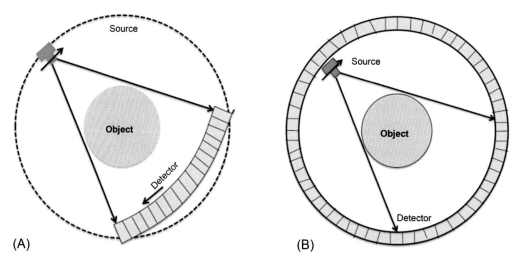

FIG. 7.58    Scheme of (A) third-generation CT scanning; (B) fourth-generation CT scanning.

to correspondingly shorter scanning times. So, body parts of a wide range of sizes can be easily scanned with the second-generation scanners.

Third (Fig. 7.58A) and fourth (Fig. 7.58B) generation CT improved both source and detector. In particular, the third-generation CT systems (Fig. 7.58A) use a full rotational X-ray scan and a complex detector array that allows a complete view to be collected during each sampling interval. Typically, third-generation systems are faster than second-generation systems. Their detector arrays incorporate a larger number of sensors. The fourth-generation systems (Fig. 7.58B) also use only an X-ray rotating scan motion, whereas the detectors are stationary. The system consists of a fixed ring with multiple detectors and a single X-ray source (fan beam), which rotates around the scanned object. The number of views is equal to the number of detectors. These scanners are more susceptible to scattered radiation.

Several other CT scanner geometries that have been developed and marketed do not precisely fit the above categories. The fifth-generation CT systems are different from the previous systems in that there is no mechanical motion involved. The scanner uses a circular array of X-ray sources that are electronically switched on and off. The sources project onto a curved fluorescent screen, so that when an X-ray source is switched on, a large volume of the body part is imaged simultaneously, providing projection data for a cone beam of rays diverging from the source (Fig. 7.59A). Here, a series of two-dimensional projections of a three-dimensional object is collected.

In previous generation detectors, the gantry had to be stopped after every slice, so that data acquisition could not be a continuous process. In order to have perpetual access to an energy source, the X-ray tube and the detectors were connected to an electrical source via wires, and

had to be stationary. This problem was solved in the 1990s when slip ring technology was introduced to the field of medical imaging. A slip ring allows electricity to be passed to rotating components, eliminating the requirement for stationary components. Using a slip ring allowed the gantry to rotate continuously through all of the patient slices, therefore, creating shorter scan times (i.e., as short as 30 s to scan the entire abdomen). This led to the development of the sixth-generation CT scanner, that is, the helical or spiral CT. This generation essentially combined the principles of the third and fourth generations with slip ring technology to create a system that could rotate continually around the patient without being limited by electrical wires. The main drawback of helical CT scanners lies in the way in which the data is collected. Because the data is acquired in a helical formation, no full slices of data are available, as the scanner is not producing planar sections. This problem can be compensated through the reconstruction process.

The seventh-generation systems, the most recent generation, are composed of a multiple detector array with a cone beam X-ray

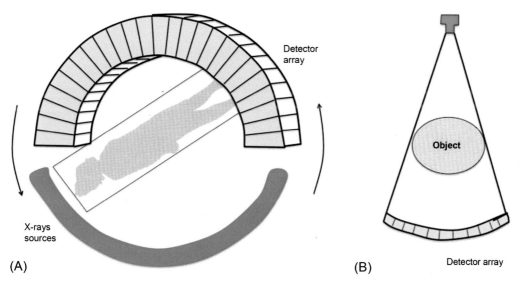

**FIG. 7.59** Scheme of (A) fifth-generation CT scanning; (B) seventh-generation CT scanning.

(Fig. 7.59B). Recall that when the CT scanner progressed from pencil beam geometry to fan beam geometry, the X-ray beam was used more efficiently. Not only did more of the X-ray beam interact with the detectors, but these wide-angled X-ray beams also allowed images to be acquired more rapidly. Unlike the pencil beam and fan beam, the cone beam does not pass through a narrow collimator. Therefore, the intensity of the initial X-ray beam is not as strongly reduced, and the X-ray can interact more efficiently and effectively with the detector array. In order to use a cone-beam X-ray geometry, the linear detector array found in previous generations of CT scanners had to be modified to make a flat panel detector or a multiple detector array. Essentially, the combination of the cone-shaped X-ray beam and the paneled detector allows for a very large number of slices to be acquired in a very short period of time. The seventh-generation CT scanner can acquire an outstanding amount of information in a very short time span, requiring a much higher level of sophistication in the reconstruction process.

Summarizing, a CT scan system uses a motorized X-ray source that rotates around the circular opening of a donut-shaped structure called a gantry. During the CT scan, the patient lies on a bed that slowly moves through the gantry while the X-ray tube rotates around the patient, shooting narrow beams of X-rays through the body. The body scan can be performed with two approaches (Fig. 7.60):

- sequential: sequence of complete gantry rotation followed by table movement with the patient;
- spiral: continuous gantry rotation and table movement.

In particular, the spiral CT is a relatively new technology that improves the accuracy and speed of CT scans. The beam takes a spiral path during the scanning, so it gathers continuous data with no gaps between images.

CT scanners use digital X-ray detectors located directly opposite the X-ray source so that, when the X-rays leave the patient, they are picked up by the detectors and transmitted to a computer. Each time the X-ray source

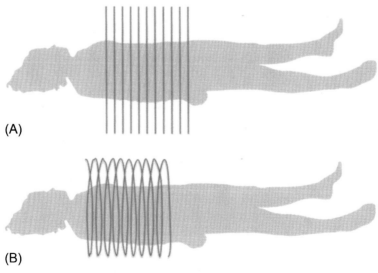

**FIG. 7.60** Scanning performed in (A) sequential mode; (B) spiral mode.

completes one full rotation, the CT computer uses sophisticated mathematical techniques to construct a 2D image slice of the patient. The thickness of the tissue represented in each image slice can vary depending on the CT machine used, but usually ranges from 1 to 10 mm. When a full slice is completed, the image is stored, and the motorized bed is moved forward incrementally into the gantry. The X-ray scanning process is then repeated to produce another image slice. This process continues until the desired number of slices is collected. Image slices can either be displayed individually or stacked together by the computer to generate a 3D image of the patient that shows the skeleton, organs, and tissues, as well as any abnormalities the physician is trying to identify. This method has many advantages including the ability to rotate the 3D image in space or to view slices in succession, making it easier to find the exact place where a problem may be located. This is the difference between a CT scanner and an X-ray machine, which sends just one radiation beam. In fact, the CT scanner produces a more detailed final picture than an X-ray image.

*Image acquisition.* Computed Tomography provides higher resolution images than traditional 2D radiography. In fact, the CT scanner's X-ray detector can "see" hundreds of different levels of density, investigating tissues within a solid organ. The varying densities of body tissues enable X-rays to provide the medical team with good 2D images, but sometimes tissues are so close in density that the traditional projected radiography fails to convey the differences. In these cases, CT scans can provide images with higher contrast resolution than X-ray radiography. In fact, as reported in Fig. 7.61A and B, different tissues are differentiated in gray tones so that among soft tissues, it is possible to discern between different composition (e.g., fat versus lung versus muscle).

*Investigations within the human body.* CT scans can be used to identify disease or injuries within various regions of the body, helping doctors diagnose a variety of medical problems. It is useful for obtaining images of:

- soft tissues;
- the pelvis;

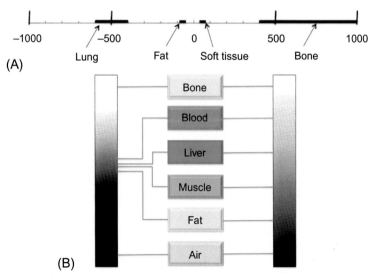

**FIG. 7.61** (A) Hounsfield scale: bone = +400 to +1000; soft tissue = +40 to +80; water = 0; fat = −60 to −100; lung = −400 to −600; air = −1000; (B) comparison between detection of different tissues in X-ray radiography and CT.

- blood vessels;
- lungs;
- the brain;
- the abdomen, and
- bones.

A CT scan of the head, for example, proves useful in detecting tumors, tissue death, bleeding in the brain, and damage to the skull. Moreover, CT has become a useful screening tool for detecting possible tumors or lesions within the abdomen. A CT scan of the heart may be ordered when various types of heart disease (Fig. 7.62) or abnormalities are suspected. CT can also be used to image the head in order to locate injuries, tumors, clots leading to stroke, hemorrhage, and other conditions. It can image the lungs in

order to reveal the presence of tumors, pulmonary embolisms (i.e., blood clots), excess fluid, and other conditions such as emphysema or pneumonia. A CT scan is particularly useful when imaging complex bone fractures, severely eroded joints, or bone tumors because it usually produces more details than would be possible with a conventional X- ray.

Moreover, as previously described, dense structures within the body (e.g., bone) are easily imaged, whereas soft tissues vary in their ability to stop X-rays and, thus, may be faint or difficult to see. For this reason, intravenous contrast agents have been developed to be highly visible in an X ray or CT scan and are safe to use in patients. Contrast agents contain substances that are better at stopping X-rays and, thus, are more

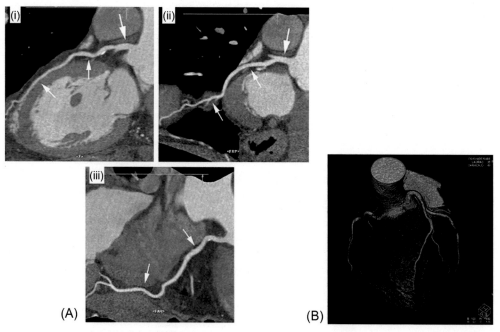

**FIG. 7.62**    (A) CT reconstructions of the coronary arteries: (i) left main (*large arrow*) and left anterior descending coronary artery (*small arrows*), calcified and slight amounts of noncalcified plaques are present in the proximal left anterior descending coronary artery; (ii) left main (*large arrow*) and left circumflex coronary artery (*small arrows*); (iii) right coronary artery (*arrows*), calcified plaque and small amounts of noncalcified plaque are present in the proximal vessel segment; (B) 3D reconstruction of the heart and coronary arteries. *Modified from Achenbach, S., Marwan, M., Schepis, T., Pflederer, T., Bruder, H., Allmendinger, T., Petersilka, M., Anders, K., Lell, M., Kuettner, A., Ropers, D., Daniel, W.G., Flohr, T., 2009. High-pitch spiral acquisition: a new scan mode for coronary CT angiography. J. Cardiovasc. Comput. Tomogr. 3(2), 117–121.*

visible on an X-ray image. For example, to examine the circulatory system, a contrast agent based on iodine is injected into the bloodstream to help illuminate blood vessels. This type of test is used to look for possible obstructions in blood vessels, including those in the heart. If a 3D image of the abdomen or digestive system, including the esophagus and stomach, it is required an oral contrast agent, such as barium-based compounds, is used for imaging. The barium appears white on the scan as it travels through the digestive system.

*Possible risks.* CT scans use X-rays that produce ionizing radiation. Ionizing radiation has the potential to cause biological effects in living tissue. This is a risk that increases with the number of exposures added up over the life of an individual. However, the risk of developing cancer from radiation exposure is generally small; in fact, the chance of developing cancer as the result of a CT scan is thought to be <1 in 2000. A CT scan in a pregnant woman poses no known risks to the baby if the area of the body being imaged is not the abdomen or pelvis. Citing the American College of Radiography, the American Pregnancy Association (APA) points out that "no single diagnostic X-ray has a radiation dose significant enough to cause adverse effects in a developing embryo or fetus." However, the APA notes that CT scans are not recommended for pregnant women, "unless the benefits clearly outweigh the risk." In general, if imaging of the abdomen and pelvis is needed, doctors prefer to use exams that do not use radiation, such as MRI or ultrasound. However, if neither of those can provide the answers needed, or there is an emergency or other time constraint, CT may be an acceptable alternative imaging option. In some patients, contrast agents may cause allergic reactions, or in rare cases, temporary kidney failure. Intravenous contrast agents should not be administered to patients with abnormal kidney function because they may induce a further reduction of kidney function, which may sometimes become

permanent. Children are more sensitive to ionizing radiation and have a longer life expectancy, and, thus, a higher relative risk for developing cancer than adults. As metal interferes with the workings of the CT scanner, the patient will need to remove all jewelry and metal fastenings.

### 7.3.1.8 X-Rays for Treating Disease

Radiation therapy in widely used for cancer treatment. X-rays and other types of high-energy radiation can be used to destroy cancerous tumors and cells by damaging their DNA. The radiation dose used for treating cancer is much higher than the radiation dose used for diagnostic imaging. Therapeutic radiation can come from a machine outside of the body or from a radioactive material that is placed in the body, inside or near tumor cells, or injected into the blood stream.

## 7.3.2 Magnetic Resonance Imaging

Nuclear magnetic resonance imaging (MRI) is based on the principles of nuclear magnetic resonance (NMR), a spectroscopic technique used to obtain microscopic chemical and physical information about molecules.

NMR is a relatively new field of research (Table 7.7), having only first been described by Isidor Rabi in the 1930s. Following his work measuring NMR in molecular beams, Felix Bloch and Edwards Mills Purcell were able to expand on this work and measured NMR in liquids and solids in 1946. The main observation these scientists made was that certain magnetic nuclei were able to absorb radiofrequency energy when placed in a magnetic field and forced to precess at a specific rate. Different molecules precess at different rates depending on the amount of energy absorbed. Scientists were then able to determine chemical and structural properties of these molecules. The differences in the chemical properties and their influence on atomic precession would eventually lead to the invention of MRI. However, it was not until the 1970s that this physical principle could be

**TABLE 7.7** History of MRI

| | |
|---|---|
| 1924 | Pauli suggests that nuclear particles may have angular momentum (spin) |
| 1937 | Rabi measures magnetic moment of nucleus. Coins "magnetic resonance" |
| 1946 | Purcell shows that matter absorbs energy at a resonant frequency |
| 1946 | Bloch demonstrates that nuclear precession can be measured in detector coils |
| 1959 | Singer measures blood flow using NMR (in mice) |
| 1972 | Damadian patents idea for large NMR scanner to detect malignant tissue |
| 1973 | Lauterbur publishes method for generating images using NMR gradients |
| 1973 | Mansfield independently publishes gradient approach to MR |
| 1975 | Ernst develops 2D Fourier transform for MR |
| 1980 | NMR renamed MRI |
| 1985 | Insurance reimbursements for MRI exams begin |
| 1990 | Ogawa and colleagues create functional images using endogenous, blood-oxygenation contrast |

utilized to create images of biological tissue. The primary reason for this is that, during an NMR experiment, the signal came from the entire chemical sample, and any spatial differences were simply averaged. In 1973, Paul Lauterbur expanded upon Herman Carr's original idea to use gradients in the main magnetic field to produce spatial localization, and the first NMR image was then produced. In 1977, the first MRI exam was performed on a human being; it took 5h to produce one image. MRI was first put into practice between the late 1970s and the early 1980s. Because then, MRI has progressed immensely, taking over the field of medical imaging. Additionally, this imaging technique has had such a profound impact on the medical field that, in 2003, Paul Lauterbur and Peter Mansfield received the Nobel Prize in physiology or medicine for their role in the development of MRI.

MRI is a noninvasive imaging technology that produces three-dimensional, detailed anatomical images used by physicians to diagnose medical conditions. It is often used for disease detection, diagnosis, and treatment monitoring as explained in the following. It is based on sophisticated technology that excites and detects the change in the direction of the rotational axis of protons found in the water that makes up living tissues.

MRI does not use ionizing radiation as X-rays do, but uses a large magnet and radio wave pulses to look at organs and structures inside your body. Health care professionals use MRI scans to diagnose a variety of conditions, from torn ligaments to tumors, within the chest, abdomen, and pelvis, and to examine the brain and spinal cord. Moreover, MRI can be used to obtain both functional and anatomical information.

MRI allows us to obtain sectional images of a patient in a noninvasive and bloodless way, without the need to administer any substance, following the interaction of radio-frequency electromagnetic waves with the human body. MRI is based on the absorption and emission of energy in the radiofrequency (RF) range of the electromagnetic spectrum. It produces images based on spatial variations in the phase and frequency of the RF energy being absorbed and emitted by the imaged object. A number of biologically relevant elements, such as hydrogen, oxygen-16, oxygen-17, fluorine-19, sodium-23, and phosphorus-31 are potential candidates for producing MR images. In fact, some nuclei of elements in the human body behave differently (i.e., magnetic dipoles or spin) following the application of an intense external magnetic field. The human body is primarily fat and water, both of which have many hydrogen atoms, making the human body approximately 63% hydrogen atoms. Hydrogen nuclei have an NMR signal, so, for these reasons, clinical MRI primarily images the NMR signal from the hydrogen nuclei given its abundance in the human body. In particular, MRI uses the body's natural magnetic properties to produce detailed images from any part of the body. For imaging purposes, the hydrogen nucleus (i.e., single proton) is used because of its abundance in water and fat. Hence, for every unit volume of tissue (voxel, Fig. 7.63), there is a number of cells, and these cells contain water molecules (Fig. 7.63); each water molecule contains one oxygen and two hydrogen atoms (Fig. 7.63). Each hydrogen atom contains one proton in its nucleus with a +/− ½ spin (Fig. 7.63).

A hydrogen atom consists of a proton nucleus with positive charge and a single electron with negative charge equal in magnitude to that of a proton. Because the hydrogen nuclei do not contain a neutron, it is often called a proton. Protons carry properties of magnetic moment and electrical charge. The nucleus of the hydrogen atom carries a small magnetic dipole moment which behaves like a tiny magnet. These protons are found not only in water (2H), but also in organic macromolecules such as glucose (12H), and lipids, making them an ideal target for MR imaging. Different tissues thus produce different images based on the amount of their hydrogen atoms producing a signal.

### 7.3.2.1 How MRI Works

The hydrogen proton can be likened to the planet Earth, spinning on its axis, with a north-south pole. In this respect, it behaves like a small bar magnet. Under normal circumstances, these hydrogen proton "bar magnets" spin in the body with their axes randomly aligned (Fig. 7.64A). When the body is placed in a strong magnetic field (i.e., magnetic field strengths usually between 0.2 and 3T), such as an MRI scanner, the protons' axes (i.e., proton spin) all line up (Fig. 7.64B). This uniform alignment creates a magnetic vector oriented along the axis of the MRI scanner. Hence, inside the bore of the scanner, the magnetic field runs down the center of the tube in which the patient is placed, so the hydrogen protons will line up in either the direction of the feet or the head. The majority will cancel each other out, but the net number of protons is sufficient to produce an image.

When additional energy in the form of a radio wave is added to the magnetic field, the magnetic vector is deflected. The radio wave frequency that causes the hydrogen nuclei to absorb the

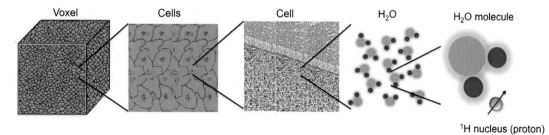

**FIG. 7.63** Hydrogen nucleus considered for MRI analysis.

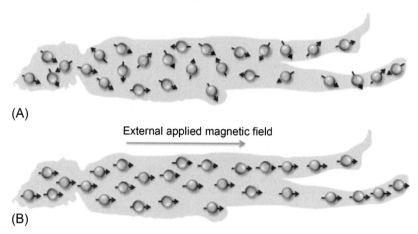

FIG. 7.64    (A) Without an external magnetic field, protons assume a random orientation of magnetic moments; (B) a strong magnetic field forces protons in the body to align with that field.

energy from the magnetic field and to flip their spins is dependent on the element sought (hydrogen in this case) and the strength of the magnetic field. In particular, radiofrequency pulses realign hydrogen atoms that naturally exist within the body while the patient is in the scanner without causing any chemical changes in the tissues. The strength of the magnetic field can be altered electronically from head to toe using a series of gradient electric coils, and by altering the local magnetic field by these small increments, different slices of the body will resonate as different frequencies are applied.

When the radiofrequency source is switched off, the magnetic vector of hydrogen atoms returns to its resting state (i.e., their usual alignment), and this causes the emission of a signal (also a radio wave) with different amounts of energy varying according to the type of body tissue from which they come. This process is called precession (i.e., a change in the direction of the axis of a rotating object). It is this radio signal that can be measured by receivers in the scanner, and which is used to create the MR images. The intensity of the received signal is then plotted on a gray scale, and cross-sectional images are built up.

Multiple transmitted radiofrequency pulses can be used in sequence to emphasize particular tissues or abnormalities. A different emphasis occurs because different body tissues relax (i.e., return to their normal spins) at different rates when the transmitted radiofrequency pulse is switched off. The time taken for the protons to fully relax is measured in two ways. The first is the time taken for the magnetic vector to return to its resting state, and the second is the time needed for the axial spin to return to its resting state. The first is called T1 relaxation, and the second is called T2 relaxation. Hence, the time it takes for the hydrogen atoms to realign with the magnetic field, as well as the amount of energy released, changes depending on the environment and the chemical nature of the molecules. Physicians are able to give information about the difference between various types of tissues based on these magnetic properties.

Summarizing, MRI works and analyzes biological tissues in the following way:

1. a body organ or district is placed in a strong electromagnetic field;
2. a tomographic image is acquired (i.e., a slice);

3. each slice consists of voxels (i.e., volumetric picture element); a volume element represents a signal or color intensity value in a 3D space.

An MR examination is thus made up of a series of pulse sequences. Different tissues (such as fat and water) have different relaxation times and can be identified separately. By using a "fat suppression" pulse sequence, for example, the signal from fat will be removed, leaving only the signal from any abnormalities lying within it. Moreover, additional magnetic fields are used to produce 3-dimensional images that may be viewed from different angles. There are many forms of MRI, but diffusion MRI and functional MRI (fMRI) are two of the most common.

*Diffusion MRI.* This form of MRI measures how water molecules diffuse through body tissues. Certain disease processes (e.g., a stroke or tumor) can restrict this diffusion, so this method is often used to diagnose them. Diffusion MRI has only been around for about 15–20 years.

*Functional MRI.* In addition to structural imaging, MRI can also be used to visualize functional activity in the brain. Functional MRI measures changes in blood flow to different parts of the brain. It is used to observe brain structures and to determine which parts of the brain are handling critical functions. Functional MRI may also be used to evaluate damage from a head injury or Alzheimer's disease.

The traditional MRI unit is a large cylinder-shaped tube surrounded by a circular magnet. The patient will lie on a moveable examination table that slides into the center of the magnet inside a tunnel-shaped machine. Some MRI units, called short-bore systems, are designed so that the magnet does not completely surround the patient. Some newer MRI machines have a larger diameter bore which can be more comfortable for larger patients or patients with claustrophobia. Other MRI machines are open on the sides (open MRI); these open units are especially helpful for examining larger patients or those with claustrophobia. Newer open MRI units provide very high quality images for many types of exams. The different systems in medical imaging use magnetic field with different strength, ranging from open MRI units with magnetic field strength of 0.3 T to MRI systems with field strengths up to 1.0 T and whole-body scanners with field strengths up to 3.0 T (in clinical use). The MRI scan takes approximately 30–60 min, and it is important for the patient to stay as still as possible during the exam.

### 7.3.2.2 *When Using MRI*

In general, MRI scanners are particularly well-suited to image the nonbony parts or soft tissues of the body. Moreover, most diseases manifest themselves by an increase in water content, so MRI is a sensitive test for the detection of disease. The exact nature of the pathology can be more difficult to ascertain: for example, infection and tumor can in some cases look similar. A careful analysis of the images by a radiologist will often yield the correct answer. MR images of the soft tissue structures of the body (e.g., heart, liver, and many other organs) is more likely in some instances to identify and accurately characterize diseases than other imaging methods. This detail makes MRI an invaluable tool in the early diagnosis and evaluation of many focal lesions and tumors. Moreover, MRI has proven valuable in diagnosing a broad range of conditions, including cancer, heart and vascular disease, and muscular and bone abnormalities.

In the brain, MRI can differentiate between white matter and gray matter and can also be used to diagnose aneurysms and tumors. In particular, fMRI is used to observe brain structures and determine which areas of the brain "activate" (i.e., consume more oxygen) during various cognitive tasks. It is used to advance the understanding of brain organization and offers a potential new standard for assessing neurological status and neurosurgical risk.

In details, MR imaging of the body is performed to evaluate:

- organs of the chest and abdomen, including the heart;
- the liver, biliary tract, kidneys, spleen, bowel, pancreas, and adrenal glands;
- pelvic organs including the bladder and the reproductive organs such as the uterus and ovaries in females and the prostate gland in males;
- blood vessels (including MR angiography);
- lymph nodes.

Moreover, physicians can use an MRI investigation to help diagnose or monitor treatment for conditions such as:

- tumors of the chest, abdomen, or pelvis (Fig. 7.65);
- diseases of the liver, such as cirrhosis, and abnormalities of the bile ducts and pancreas;

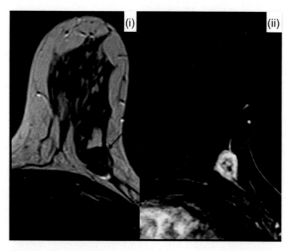

**FIG. 7.65** MRI of the breast. The round-shaped mass *(red arrow)* can be classified as highly suggestive of malignancy. *Modified from Marino, M.A., Riedl, C.C., Bernathova, M., Bernhart, C., Baltzer, P.A.T., Helbich, T.H., Pinker, K., 2018. Imaging phenotypes in women at high risk for breast cancer on mammography, ultrasound, and magnetic resonance imaging using the fifth edition of the breast imaging reporting and data system. Eur. J. Radiol. 106, 150–159.*

- inflammatory bowel disease such as Crohn's disease and ulcerative colitis;
- heart problems, such as congenital heart disease;
- malformations of the blood vessels and inflammation of the vessels (vasculitis);
- a fetus in the womb of a pregnant woman.

### 7.3.2.3 *Contrast Agent*

During an MRI, the patient may receive an injectable contrast, or dye, into the bloodstream just before the scan. The injection of contrast is most often done through an IV (intravenously, through a vein) that is placed in the back of patient's hand or the inside of her/his elbow. A saline solution will drip through the intravenous line to prevent clotting until the contrast material is injected at some point during the exam. Unlike contrast agents used in X-ray studies, MRI contrast agents do not contain iodine and, therefore, rarely cause allergic reactions or other problems. The contrast alters the local magnetic field, improving the quality of the images and providing more details in some instances. The contrast material most commonly used for an MRI exam contains a metal called gadolinium. Normal and abnormal tissue will respond differently to this contrast. Nephrogenic systemic fibrosis is currently a recognized, but rare, complication of MRI believed to be caused by the injection of high doses of gadolinium-based contrast material in patients with very poor kidney function. Careful assessment of kidney function before considering a contrast injection minimizes the risk of this very rare complication. There is a very slight risk of an allergic reaction if contrast material is injected. Such reactions are usually mild and easily controlled by medication. In addition, manufacturers of intravenous contrast indicate mothers should not breastfeed their babies for 24–48h after contrast medium is given. However, both the American College of Radiology (ACR) and the European Society of Urogenital Radiology note that the available data suggest that it is safe to continue breastfeeding after receiving intravenous contrast.

*Benefits.* MRI has the following advantages and disadvantages:

- it does not use ionizing radiation;
- it allows to discriminate between types of tissues (e.g., spleen and liver that X-rays do not differentiate), or between healthy and damaged tissues;
- it allows doctors to obtain functional and morphological information;
- it has a low spatial resolution (about 1 mm);
- scan times are much longer than other radiological techniques (e.g., a complete exam lasts 30–60 min).

*Possible risks.* There are no known biological hazards of MRI because, unlike X-rays and computed tomography, MRI uses radiation in the radiofrequency range which is found all around us and does not damage tissue as it passes through. Even if the magnetic field is not harmful, it may cause some medical device to malfunction. Pacemakers, metal clips, and metal valves can be dangerous in MRI scanners because of potential movement within a magnetic field. Metal joint prostheses are less of a problem, although there may be some distortion of the image close to the metal. MRI departments always check for implanted metal and can advise on their safety.

MRI has been used for scanning patients since the 1980s with no reports of any ill effects on pregnant women or their unborn babies. However, because the unborn baby will be in a strong magnetic field, pregnant women should not have this exam in the first 3–4 months of pregnancy unless the potential benefit from the MRI exam is assumed to outweigh the potential risks. Pregnant women should not receive injections of gadolinium contrast material except when absolutely necessary for medical treatment.

Jewelry and other accessories should be removed prior to the MRI scan. Because they can interfere with the magnetic field of the MRI unit, metal and electronic items are not allowed in the exam room. Dyes used in tattoos may contain iron and could heat up during an MRI scan, but this is rare. Tooth fillings and braces usually are not affected by the magnetic field, but they may distort images of the facial area or brain.

### 7.3.2.4 *CT Versus MRI*

An MRI is somewhat like an X-ray or a CT scan because it can provide images of internal body structures. It is more like a CT scan than an X-ray, and many people get the two scans confused because the equipment used for each is very similar. Plus, both an MRI and CT scan produce images of patient's bones, organs, and other internal tissues. The main differences between CT and MRI are:

- a CT scan uses X-rays, but an MRI uses magnets and radio waves. This means the patient is not exposed to radiation; in fact, no studies have linked MRIs to any harmful health effects. A CT scan uses radiation to make an image, and repeated exposure can be harmful;
- an MRI scan takes longer to perform (30–60 min, on average), whereas a CT scan is quick (around 5–10 min);
- an MRI (costing, on average, $1200 to $4000) is more expensive than a CT scan (which costs, on average, $1200 to $3200).
- unlike an MRI, a CT scanner does not show tendons and ligaments;
- MRI is better for examining the spinal cord and other soft tissues;
- a CT scanner is better suited to cancer, chest and lung problems (e.g., pneumonia), bleeding in the brain, especially after an injury;
- a brain tumor is more clearly visible on MRI;
- a CT scanner shows organ tear and organ injury more quickly, so it may be more suitable for trauma cases;
- broken bones and vertebrae are more clearly visible on a CT scan;
- CT scanner provides a better image of the lungs and organs in the chest cavity between the lungs;

- the contrast material used in MRI exams is less likely to produce an allergic reaction than the iodine-based contrast materials used for conventional X-rays and CT scanning;
- MRI provides a noninvasive alternative to X-rays, angiography, and CT for diagnosing problems of the heart and blood vessels.

## 7.3.3 Ultrasound Imaging

Ultrasound imaging, also called sonography, uses high-frequency sound waves to view inside the body. Because ultrasound images are captured in real time, they can also show movement of the body's internal organs as well as blood flowing through the blood vessels. Unlike X-ray imaging, there is no ionizing radiation exposure associated with ultrasound imaging.

In an ultrasound exam, a transducer (i.e., probe) is placed directly on the skin or inside a body opening. A thin layer of gel is applied to the skin so that the ultrasound waves are transmitted from the transducer through the gel into the body. The ultrasound image is produced based on the reflection of the waves off of the body structures. The strength of the sound signal and the time it takes for the wave to travel through the body provide the information necessary to produce an image. The most recent evolution is represented by the 3D technique, that, unlike the classical 2D image, it is based on the acquisition, by means of a special probe, of a "volume" of examined tissue. The volume under investigation is acquired and digitized in a few seconds or fractions of a second, after which it can then be examined:

- in 2D, with the analysis of infinite "slices" of the sample (on the three axes $x$, $y$, and $z$);
- in volumetric representation, with the investigation of the tissue or organ to be studied, which appears on the monitor as a solid that can be rotated on the three axes. In this way, its real aspect in the three dimensions is highlighted with particular clarity.

Ultrasound imaging is a medical tool that can help a physician evaluate, diagnose (Fig. 7.66), and treat medical conditions. It is usually used in the radiological, internist, surgical, and obstetric fields because it is a harmless, inexpensive technique and has a high diagnostic sensitivity on soft tissues. Common ultrasound imaging procedures include:

- abdominal ultrasound (to visualize abdominal tissues and organs);
- bone sonometry (to assess bone fragility);
- breast ultrasound (to visualize breast tissue);
- Doppler fetal heart rate monitors (to listen to the fetal heart beat);
- Doppler ultrasound (to visualize blood flow through a blood vessel, organs, or other structures);
- echocardiogram (to view the heart);

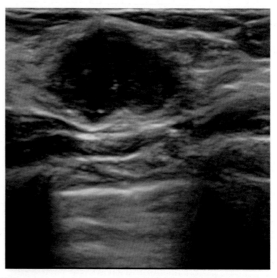

**FIG. 7.66** Ultrasound image of the breast; a 20-mm irregularly shaped mass and indistinct margins has been detected. *Modified from Marino, M.A., Riedl, C.C., Bernathova, M., Bernhart, C., Baltzer, P.A.T., Helbich, T.H., Pinker, K., 2018. Imaging phenotypes in women at high risk for breast cancer on mammography, ultrasound, and magnetic resonance imaging using the fifth edition of the breast imaging reporting and data system. Eur. J. Radiol. 106, 150–159.*

- fetal ultrasound (to view the fetus in pregnancy);
- ultrasound-guided biopsies (to collect a sample of tissue);
- ophthalmic ultrasound (to visualize ocular structures);
- ultrasound-guided needle placement (in blood vessels or other tissues of interest).

*Possible risks.* Ultrasound imaging has been used for over 20 years and has an excellent safety record. It is based on nonionizing radiation, so it does not have the same risks as X-rays or other types of imaging systems that use ionizing radiation. Although ultrasound imaging is generally considered safe when used prudently by appropriately trained health care providers, ultrasound energy has the potential to produce biological effects on the body. Ultrasound waves can heat the tissues slightly. In some cases, it can also produce small pockets of gas in body fluids or tissues. The long-term consequences of these effects are still unknown. Because of the particular concern for effects on the fetus, organizations such as the American Institute of Ultrasound in Medicine have advocated prudent use of ultrasound imaging in pregnancy. Furthermore, the use of ultrasound solely for nonmedical purposes such as obtaining fetal "keepsake" videos has been discouraged. Keepsake images or videos are reasonable if they are produced during a medically indicated exam, and if no additional exposure is required.

# 7.4 BIOCOMPATIBILITY AND CYTOCOMPATIBILITY ANALYSES

*Biocompatibility* is, by definition, the ability of a medical device or component materials to perform with an appropriate host response in a specific application. In a nutshell, the purpose of performing biocompatibility testing is to determine the fitness of a device for human use, and to see whether the use of the device can have

any potentially harmful side effects. Biological responses that are viewed as adverse and caused by a material in one specific application might not be regarded as such in a different situation. Enclosed in the general biocompatibility term are many aspects of cell behavior and function that deal with the appropriate cell response to the presence of a biomaterial. For instance, for bone prostheses, this implies that bone-forming cells (i.e., osteoblasts, see Section 6.1.2.1) do deposit a physiological bone ECM with all the natural constituents in balance. Each application needs to be tested in the proper settings that reflect the natural environment and the demands posed on the specific application. Although this will result in separate sets of test methods for each application, there are common parameters that should be considered (Kooten et al., 1997). Besides, even though a medical device is comprised of materials that are still known to be biocompatible, the device itself requires biocompatibility testing as well. For instance, a band-aid is made of at least three component materials: the adhesive, plastic, and gauze. Even though each of these materials is biocompatible if taken individually, testing of the band-aid itself is required in order to show the overall interaction effects. Biological testing is based upon in vitro and ex vivo test methods and in vivo animal models, so that the anticipated behavior when a medical device is used in human beings can be unfortunately judged only with caution, as it cannot be unequivocally concluded that the same biological response will also occur in our species. In addition, differences in the manner and extent of response to the same material between individuals indicate that some patients can have adverse reactions, even to well-established materials.

In a regulatory sense, biocompatibility is tested to determine the potential toxicity resulting from bodily contact with a material or medical device. Biocompatibility is, therefore, vital for medical devices. Inevitably, evaluating the biocompatibility of a device is a risk assessment exercise. There is no risk-free device or device material: the goal of device designers is to minimize risk while maximizing benefit to patients.

Biocompatibility testing is an important part of obtaining approval from the Food and Drug Administration (FDA or USFDA) and European Medicines Agency (EMA) to market a medical device in United States (US) and European Union (EU), respectively.

- The first step of the approval process in the United States is to confirm that a product is a medical device as defined by Section 201 (h) of the Federal Food, Drug, and Cosmetic Act (FFDCA). Because the FDA groups devices into three classes, the second step is therefore to classify the device. The classification is obtained from the FDA website. Of note, Class I devices have the lowest risk, and Class III the highest. For instance, introduction/drainage catheters and accessories are Class I devices, an endosseous dental implant abutment is classified as Class II, while an aortic stent is a Class III device. Novel medical devices or devices made with new, unfamiliar materials are automatically made Class III. The third step is to collect appropriate data. Most Class I devices are 510(k) exempt. A 510(k) is a premarket submission to be made to the FDA to demonstrate that the device to be marketed is at least as safe and effective, that is, substantially equivalent (SE) to a legally marketed device that is not subject to Premarket Approval (PMA). The submitter may not proceed to market the device until receiving an order declaring a device SE. Most Class II devices require a 510(k) application. Most Class III devices require a PMA, and for most PMA applications clinical trials are required. For premarket notification route, chemical characterization and biocompatibility test data are collected. Such data have to support

the claim that the device is safe and substantially equivalent to a legally marketed device. For PMA route, test data are collected to obtain an investigational device exemption (IDE). An IDE allows the device to be used in a clinical study in order to obtain safety and effectiveness data to support a PMA application. Of note, the majority of medical devices entering today's market were cleared by the FDA without clinical trials/data. In fact, >95% of medical devices used in the US were never tested in clinical trials. They were brought to market either under the 510(k) program, or they were on the market prior to 1976, when the safe medical device act was enacted. Nevertheless, the FDA is interested in the biocompatibility of raw materials and finalized medical devices.

- Medical devices cannot be placed on the EU market without conforming to the strict safety requirements of the EU legislation, collectively known as the Medical Devices Regulation (MDR). The current MDR (2017/745/EU) has brought Community legislation into line with technical advances, changes in medical science, and progress in law making. The MDR will replace the existing Medical Devices Directive (93/42/EEC) (MDD) and the Active Implantable Medical Devices Directive (90/385/EEC) (AIMDD). Published on May 5th, 2017, this new regulation will be enacted on May 25th, 2020. As the name suggests, it is no longer a directive, and all medical device companies have to be compliant with its requirements. Companies not following the new rules will no longer be allowed to sell their medical products in the EU market; currently approved medical devices will have a transitional period of three years to meet the new MDR requirements. The MDR maintains the division of medical devices into four different classes (Class I, IIa, IIb, and III, with Class III ranked the highest) depending on the kind and duration of contact with the human body. However, the Annex VIII of the MDR introduces classification changes in relation to certain devices. For example, surgical meshes and spinal disc replacement implants, or implantable devices coming into contact with the spinal column (except for screws, wedges, plates, and instruments), all active implantable devices and their accessories are now assigned to Class III. Of note, the higher the classification, the greater the level of assessment required by the Notified Bodies (NBs).

Because any medical device must meet or exceed the highest safety standards possible to ensure perfect product functioning, the best starting point in order to understand biocompatibility requirements is to know about the content of the ISO Standard 10993 entitled *Biological Evaluation of Medical Devices*, and prepared by Technical Committee ISO/TC 194 *Biological and Clinical Evaluation of Medical Devices*. A list of all parts in the ISO 10993 series can be found on the ISO website (https://www.iso.org/home.html). Part 1 provides an overview of biocompatibility and the suggested approach for risk mitigation from the perspective of materials and processing. Part 2 covers animal welfare requirements, while the remaining chapters dive deep into topics touching on risk mitigation, from sample preparation to animal studies and how to perform a toxicological risk assessment. Founded in 1947, the International Organization for Standardization (ISO) is an independent, nongovernmental organization in charge of promoting and setting worldwide guidelines for propriety, commercial, and industrial standards. A test method standard describes the test specimen to be used, the conditions under which it has to be tested, how many specimens and which kinds of controls have to be tested, and how the data must be analyzed. When a method has been standardized, it can be used in any other laboratory; in other words, the details are sufficient to ensure that different facilities will obtain the same results for the same samples. Testing strategies that comply with the ISO 10993 family are acceptable in the EU and most Asian countries. For products marketed in the US, the FDA has substantially adopted the ISO 10993 guideline as well.

The overall process of determining the biocompatibility of any medical device involves several stages. The first step in this process is the collection of data about the materials comprising the device, that is the chemical characterization of device components. The following stage is to perform in vitro screening, often on the device components only, and finally, to conduct confirmatory in vivo testing on the final device. Biological testing is probably the most critical issue in the biocompatibility evaluation. It is crucial to make sure that the finished device is challenged to ensure that any harmful effect can be expected when in use. When dealing with a material meant to be employed in a final product, its appropriateness should be assessed and documented. In fact, the required tests also depend on the use of the device, and the manner and duration of its interaction with the body. In test planning, it must be noted whether the device is a *surface device*, an *external communicating device*, or an *implant device*, and which tissue(s) the device will contact. For instance, implant devices interacting with the blood require more thorough testing than a surface device with an expected contact time with the body of a few days only. For each device category, certain effects must be considered and addressed in the regulatory submission for that device. It is worthy of note that ISO 10993-1: 2018 does not prescribe a specific battery of tests for any particular medical device. Rather, it provides a framework that can be used to design a biocompatibility testing program. The ISO materials biocompatibility matrix relies on the type and duration of body contact to categorizes devices, and also shows a list of potential biological effects. After completion of tests and collection of all data, it is recommended that an expert assessor interpret such data and results. This provides insight into whether additional tests need to be performed, or whether the existing data contain enough information for an overall biological safety assessment of the device.

The changes and adaptations of the ISO 10993-1: 2018 standard will impact the biological evaluation strategies of manufacturers of medical devices by putting more emphasis on chemical characterizations and sound toxicological evaluations. Thus, the era of the *tick-the-box of the flowchart* mentality for animal studies in the medical device business is to be replaced by systematic evaluation approaches that put more emphasis on the chemical composition of the device(s). On this note, the primary focus of in vitro cytotoxicity tests is to measure on cells in culture the consequences of leachable or secreted substances from a given biomaterial. However, lack of cytotoxicity cannot be equated with biocompatibility, because cell culture assays are not biocompatibility assays.

Once in vitro testing has been completed, in vivo biological trials can be done based upon the device's intended use. The goal of in vivo biocompatibility assessment of a medical device, or part thereof, is to determine its compatibility in the actual biological environment. Such testing can range from skin irritation to hemocompatibility and implantation testing. Turnaround time for tests can range from three weeks to greater than several months, depending on the specific test data needed. Subchronic or chronic implantation testing may last even longer (Keong and Halim, 2009). If an animal study is to be conducted, devices whose intended use is ex vivo (external communication) should be tested in such a way, and devices whose intended use is in vivo (implants) should be tested in an animal model mimicking as close as possible the clinical use conditions.

## 7.4.1 Cytotoxicity and Cytocompatibility Testing

A critical test in medical device biocompatibility testing is the assessment of in vitro cytotoxicity (i.e., the quality of being poisonous) according to the ISO 10993-5: 2009, entitled *Biological evaluation of medical devices Part 5: Tests for in vitro cytotoxicity*. Due to the general applicability of in vitro cytotoxicity tests and their widespread use in evaluating a wide range of devices and materials, it is the purpose of ISO 10993-5 to define a scheme for testing which requires decisions to be made in a series of steps, rather than to specify a single test. This should lead to the selection of the most appropriate test.

The various methods used and endpoints measured in cytotoxicity determination can be grouped into the following categories of evaluation:

- assessments of cell damage by morphological means;
- measurements of cell damage;
- measurements of cell growth;
- measurements of specific aspects of cellular metabolism.

Three categories of test are listed: *direct contact test*, *extract test*, and *indirect contact test*. The choice of one or more of such categories depends upon the nature of the sample to be evaluated, the potential site and the nature of the use. This choice then determines the details of the preparation of the samples to be tested, the preparation of the cultured cells, and the way in which the cells are exposed to the samples or to their extracts. Of note, a minimum of three replicates must be used for test samples and controls. Cytotoxic effects can be determined by either qualitative or quantitative means. Quantitative evaluation of cytotoxicity is preferable because qualitative means are more appropriate for screening purposes. Quantitative evaluation relies on the measurement of cell death, inhibition of cell growth, cell proliferation, or colony formation. The number of cells, amount of proteins, release of enzymes or vital dyes, reduction of vital dyes (e.g., (3-(4,5-dimethylthiazol-2-yl)-2, 5-diphenyltetrazoliumbromid, MTT), or any other measurable parameter may be quantified by objective means. MTT is widely regarded as the go-to assay for in vitro cytotoxicity testing, and is used regularly in academic and industrial R&D laboratories. The MTT assay has proven to be a reliable predictor of the in vivo biocompatibility performance, and found useful for the screening of new biomaterials or investigational compounds. The objective measure and response shall be recorded in the test report. Reduction of cell viability by >30% (i.e., cytotoxicity >70%) is considered a cytotoxic effect.

1. *Tests by direct contact* allows both the qualitative and the quantitative assessment of cytotoxicity. The direct contact procedure is recommended for low density materials, such as contact lens polymers. In this method, a piece of test material is placed directly onto cells growing in culture medium. The cells are next incubated so that leachable chemicals can diffuse from the test material into the culture medium and contact the cell layer. Possible reactivity of the test sample is indicated by malformation, degeneration, and lysis of cells around the test material. The step-by-step procedure for performing a direct contact test is reported in Fig. 7.67 and detailed in the figure legend.
2. *Tests on extracts* allow both the qualitative and the quantitative assessment of cytotoxicity (Fig. 7.68). Biocompatibility issues related to a medical device are most often caused by toxins that leach out of the materials into the surrounding tissues or body fluids. Accordingly, extracts of device materials (leachables) are often used in the laboratory to assessing the cytocompatibility. In such a kind of test, the extraction conditions (i.e.,

temperature and time) are at least as extreme as any condition the device or material will encounter during sterilization or clinical use. For devices that are susceptible to heat, an extraction condition of 37±1°C for 24±2h is considered reliable. *Extractants*, also called *extracting media*, are selected on the basis of the biological environment in which the test material is to be used. A physiological saline solution, and cell culture medium enriched with serum, best approximate the hydrophilic aqueous body fluids. Instead, vegetable oils are nonpolar, hydrophobic solvents used to mimic body lipids. Other suitable vehicles include purified water and dimethyl sulfoxide (DMSO). The standard surface area of the device is typically used to determine the volume of extract needed for each test performed. If the surface area cannot be determined due to the configuration of the device, a mass/volume of extracting fluid can be used. In either case, in order to enhance exposure to the extracting media, the device is cut into small pieces before extraction and extracts can be titrated soon after to yield a quantitative measurement of cytotoxicity. After preparation, the extracts are transferred onto a layer of cells and incubated. Following incubation, the cells are examined microscopically for malformation, degeneration, and cell lysis. A thorough explanation of the different steps to carry out extract tests is given in Fig. 7.68.

3. *Tests by indirect contact* by means of agar diffusion tests. It typically allow a qualitative assessment of cytotoxicity (Fig. 7.69). Such assays are appropriate for high density materials, such as elastomeric closures, and are not applicable in the case of leachables that cannot diffuse through the agar layer, or that may react with it. The use of indirect contact assays for the assessment of cytotoxicity shall be justified. In this method, a thin layer of nutrient-supplemented agar is placed over the cells in culture. The test

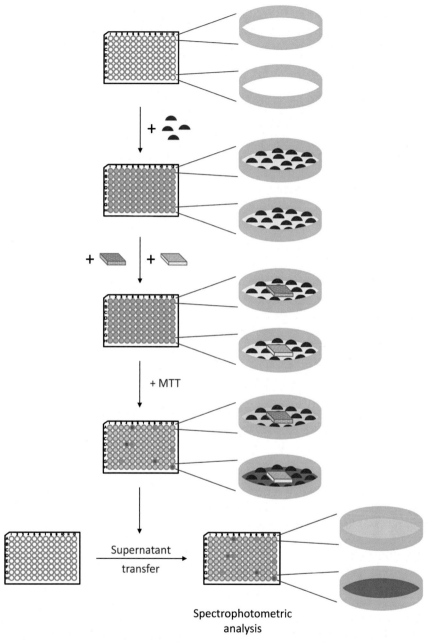

**FIG. 7.67** Direct contact assay carried out in a 96-well cell culture plate. Cells are plated in a multiwell culture plate and cultured in standard culture conditions. Different test materials (brownish parallelepipeds), previously sterilized, are carefully placed on the top of the cell layer in the center in every well, being careful that the specimen covers approximately one-tenth of the cell layer surface. Cells are next cultured under standard conditions in an incubator (e.g., for 24h at 37°C). In this way, the cells are in direct contact with the materials. At the end of the experiment, the cell viability is evaluated after the addition of a probe (e.g., MTT) that is reduced to a detectable product (pink to purple in figure) by living cells. Supernatants are moved to an empty multiwell plate and read by means of spectrophotometer. As the number of viable cells correlates with the color intensity determined by spectrophotometric measurements (quantitative assay), the more cytocompatible (or less cytotoxic) the material is, the more cells alive, and the stronger purple color we get.

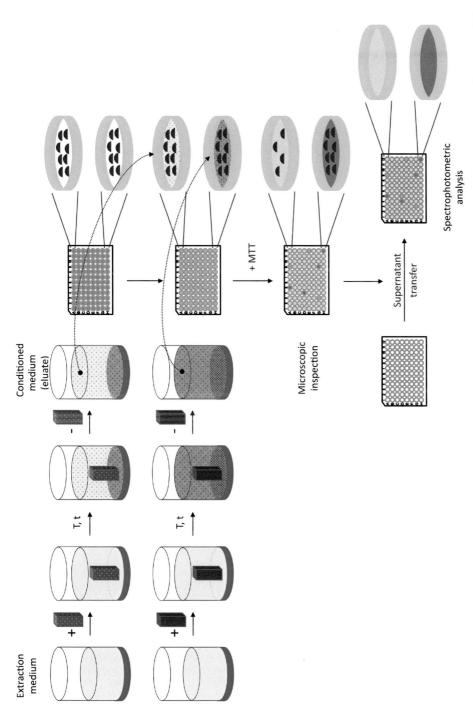

**FIG. 7.68** Extract test carried out in a 96-well cell culture plate. Extracts are obtained by incubating the different material specimens (brown parallel-epipeds), previously sterilized, in extracting media (e.g., cell culture medium enriched with serum) under predefined conditions (e.g., for 24 h at 37°C). Each conditioned medium is next applied to a cultured-cell monolayer. In this way, the cells are supplied with a fresh nutrient medium containing extractables (also called leachables) derived from the materials. The culture is then returned to the 37°C incubator and periodically removed for microscopic examination at designated times for as long as 3 days. Cells are observed for visible signs of toxicity, such as gross change in size or appearance, in response to the test materials (quantitative analysis). At the end of the experiment, the cell viability is evaluated after the addition of a probe (e.g., MTT) that is reduced to a detectable product (pink to purple in figure) by living cells. As the number of viable cells correlates with the color intensity determined by photometric measurements (quantitative assay), the more cytocompatible (or less cytotoxic) the material is, the more cells are alive, and the stronger purple color we get.

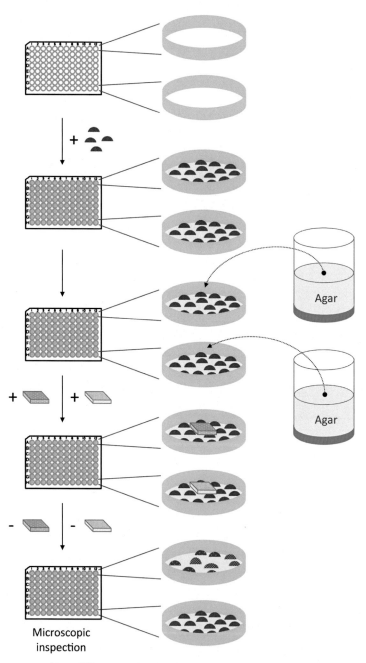

**FIG. 7.69** Indirect contact test (Agar diffusion method) carried out in a 96-well cell culture plate. Cells are plated in a multi-well culture plate and cultured in standard culture conditions until subconfluency. Culture medium is discarded from the vessel and replaced with fresh culture medium containing serum and melted agar to obtain a final mass concentration of agar of 0.5%–2%. The agar/culture medium mixture should be in a liquid state and at a temperature that is compatible with mammalian cells. Different test materials (brownish parallelepipeds), previously sterilized, are carefully placed over the solidified agar layer in the center of every well, being careful that the specimen covers approximately one tenth of the cell layer surface. The agar layer protects the cells from mechanical damage while allowing the diffusion of leachables from the test materials onto the cell layer. Cells are next cultured under standard conditions in an incubator (e.g., for 24–72h at 37°C). In this way, the cells are not in direct contact with the materials. At the end of the experiment, the cells are inspected by means of a microscope. A zone of malformed, degenerative, or lysed cells under and around the test material indicates cytotoxicity.

material, or an extract of the test material dried on filter paper, is placed on top of the agar layer, and the cells are next incubated. A zone of malformed, degenerative, or lysed cells under and around the test material indicates cytotoxicity. A more detailed explanation of the test procedure is presented in Fig. 7.69.

## 7.4.2 Hemocompatibility Testing

By definition, hemocompatible devices or device materials are those able to come into contact with blood without any appreciable, clinically-significant, adverse reactions such as thrombosis, hemolysis, and platelet, leukocyte, or complement activation, and/or other blood-associated adverse events. Materials used in blood contacting devices (e.g., intravascular catheters, hemodialysis sets, mechanical or tissue heart valves, vascular stents) must be assessed for blood compatibility to establish their safety. In practice, all materials are, to some extent, incompatible with the blood because they can either disrupt the blood cells (i.e., they give hemolysis) or activate the coagulation pathways (i.e., they are thrombogenic) and/or the complement system. The ISO 10993-4: 2017 standard specifies general requirements for evaluating the interactions of medical devices with blood. Some tests are listed below:

- the *hemolysis assay* is recommended for all devices or materials except those which contact intact skin or mucous membranes only. This kind of test allows to measure the damage to red blood cells (RBCs) when they are exposed to materials or to their extracts, and compares it to positive and negative controls;
- the *coagulation assay* measures the effect of the test item on human blood coagulation time. This test is recommended for all devices which are expected to be in contact with blood;

- the *Prothrombin Time assay* (PT) is a general screening test for the detection of coagulation abnormalities in the extrinsic pathway (see Section 6.2.1);
- the *Partial Thromboplastin Time assay* (PTT) detects coagulation abnormalities in the intrinsic pathway (see Section 6.2.1);
- the most common test for *thrombogenicity* is the in vivo method. For devices unsuited to this test method, ISO 10993-4 requires tests in each of four categories: coagulation, platelets, hematology, and complement system;
- the *complement activation testing* is recommended for implant devices that contact circulatory blood. This in vitro assay measures the complement activation (see Section 6.2.4) in human plasma as a result of exposure of the plasma to the test materials/device or extracts. The measure of complement actuation indicates whether a test item is capable of inducing a complement-induced inflammatory immune response in humans.

Other blood compatibility tests and specific in vivo studies may be required to complete the assessment of material-blood interactions, especially to comply with the ISO requirements.

## 7.4.3 Irritation Testing (Including Intracutaneous Reactivity)

According to ISO 10993-10, *Biological evaluation of medical devices—Part 10: Tests for irritation and skin sensitization*, such tests give an estimate of the local irritation potential of devices, materials, and/or their extracts, in a suitable animal model and using an appropriate site for application, such as skin, eye, or mucous membrane.

The route of exposure and duration of contact should be analogous to the anticipated clinical use of the device, but it is often better to exaggerate exposure conditions somewhat in order to establish a margin of safety for patients. To this end, suitable test to be performed are:

- the *primary skin irritation test*, which should be considered for topical devices that have external contact with intact or breached skin. The test material or the extract is applied directly to both intact and abraded sites on the animal skin. After a 24-h exposure, the material is removed, and the sites are scored for erythema and edema (i.e., redness and swelling);
- the *mucous membrane irritation test*, which is recommended for devices that will have externally communicating contact with intact, natural channels or tissues. These studies often use extracts rather than the material itself. Some common procedures include vaginal, cheek pouch, and eye irritation studies;
- the *intracutaneous reactivity test*, which shall be used to assess the localized tissue reaction to material extracts. This test is applicable when the determination of irritation by dermal or mucosal tests is inappropriate (e.g., when medical devices are implanted or are in contact with blood), and is recommended for devices that will have externally communicating or internal contact with the body or body fluids. In this test, extracts of the test material and blanks are injected intradermally. The injection sites are scored for erythema and edema. It reliably detects the potential for local irritation due to chemicals that may be extracted from a biomaterial.

*Sensitization testing* helps determining whether a material contains chemicals that may cause adverse local or systemic effects after repeated or prolonged exposure. These allergic or hypersensitivity reactions involve immunologic mechanisms. The evaluation of the sensitization potential may be performed using either specific chemicals from the test material, the test material itself, or (most often) test material extracts. Of note, the materials biocompatibility matrix reported in the ISO 10993-1: 2018 norm recommends sensitization testing for any class of medical devices.

## 7.4.4 Acute Systemic Toxicity Testing

The *systemic toxicity* is a potential adverse effect of the use of medical devices. Generalized effects, as well as organ and organ system effects, can result from the absorption, distribution, and metabolism of leachates from the device or its materials to parts of the body with which they are not in direct contact. According to the ISO 10993-11: 2017, the *acute systemic toxicity tests* shall be used where contact allows potential absorption of toxic leachables and degradation products, to estimate in an animal model the potential harmful effects of single or multiple exposures to medical devices, materials, and/or their extracts during a period of < 24 h. Such a test is typically carried out by using extracts of the device or device material. The extracts of the test material and negative control blanks are injected into animals (intravenously or intraperitoneally, depending on the extracting media), which are observed for toxic signs just after injection and at other points in time. The materials biocompatibility matrix recommends this test for all blood contact devices, but it may also be appropriate for any other device that gets into contact with internal tissues.

## 7.4.5 Subacute and Subchronic Toxicity Testing

Subacute and subchronic toxicity tests are given in the ISO 10993-11. These tests are carried out to determine the effects of either single or multiple exposures to, or contact with, medical devices, materials, and/or their extracts for a period not <24h and not >10% of the total lifespan of the test animal. Appropriate animal models are determined on a case-by-case basis. Such tests shall be waived if the available data about the chronic toxicity of the material(s) are sufficient to allow the subacute and subchronic toxicity to be evaluated. The reason for test waiving should be included in the overall biological evaluation report. These tests should be appropriate for the route and duration of contact.

If feasible, subacute, and subchronic systemic toxicity test protocols may be expanded to include implantation test protocols, in order to evaluate subacute and subchronic systemic and local effects.

## 7.4.6 Implantation Testing

The ISO 10993-6: 2016 specifies test methods for assessing local effects after implantation of biomaterials intended for use in medical devices to be implanted for short-term or long-term periods. Implantation studies are used to determine the biocompatibility of medical devices or materials that directly contact living tissues other than skin (e.g., implantable devices, sutures), and may be used to evaluate both absorbable and nonabsorbable materials. Again, to provide a reasonable assessment of safety, the implant study should closely approximate the intended clinical use. Local effects are evaluated by comparing the tissue response caused by a test sample to that caused by control materials used in medical devices whose clinical acceptability and biocompatibility are established. The device or material is implanted at an appropriate site in animal species suitable for the evaluation of the local response of living tissue to implanted material. The dynamics of biochemical exchange and cellular and immunologic responses may be assessed in implantation studies, especially through the use of histopathology. Histopathological analysis of implant sites increases greatly the amount of information obtained from these studies. Scoring relies on both microscopic and macroscopic parameters.

Of note, these implantation tests are not intended to evaluate or determine the performance of the test sample in terms of mechanical or functional loading. Moreover, the ISO 10993-6:2016 does not deal with systemic toxicity, carcinogenicity, teratogenicity or mutagenicity. However, the long-term implantation studies intended for the evaluation of local biological effects might provide insight into some of these properties.

## 7.4.7 Genotoxicity, Carcinogenicity, and Reproductive Toxicity Testing

The ISO 10993-3: 2014 specifies strategies and test methods for risk estimation, selection of hazard identification tests, and risk management, with respect to the possibility of the following potentially irreversible biological effects arising as a result of exposure to medical devices or materials thereof.

*Genotoxicity tests* allow to determine as to whether gene mutations, changes in chromosome structure, or other DNA or gene changes are caused by the test samples. Genotoxicity evaluations rely on a set of in vitro and in vivo tests to detect mutagens, which are substances that can directly or indirectly induce genetic damage. This damage can occur in somatic or germline cells, increasing the risk of cancer or inheritable defects, respectively. The most common test procedures for mutagenicity are the *Mouse Lymphoma Assay* and the *Ames test* that rely on the use of mammalian cells and bacteria, respectively, to detect DNA point mutations. A positive response in any of the in vitro tests requires additional in vivo assays or a presumption that the material is mutagenic. It is worthy of note that a strong correlation exists between mutagenicity and carcinogenicity.

*Carcinogenicity tests* are used to determine the tumorigenic potential of medical devices, materials, and/or extracts using multiple exposures for a major portion of the life span of a test animal (e.g., 18 months for mice, two years for rats, or years for dogs). Of note, carcinogenicity testing is expensive, highly problematic, and controversial.

*Reproductive toxicity tests* evaluate the potential effects of test materials and/or extracts on fertility, reproductive function, and prenatal and early postnatal development. They are often required for devices with permanent contact with internal tissues.

# References

Huaa, L., Chanb, Y.C., Wuc, Y.P., Wub, B.Y., 2009. The determination of hexavalent chromium (Cr6+) in electronic and electrical components and products to comply with RoHS regulations. J. Hazard. Mater. 163, 1360–1368.

Keong, L.C., Halim, A.S., 2009. In vitro models in biocompatibility assessment for biomedical-grade chitosan derivatives in wound management. Int. J. Mol. Sci. 10 (3), 1300–1313.

Kooten, T.G.V., Klein, C.L., Kohler, H., Kirkpatrick, C.J., Williams, D.F., Eloy, R., 1997. From cytotoxicity to biocompatibility testing in vitro: cell adhesion molecule expression defines a new set of parameters. J. Mater. Sci. Mater. Med. 8 (12), 835–841.

Munarin, F., Petrini, P., Gentilini, R., Pillai, R.S., Dirè, S., Tanzi, M.C., Sglavo, V.M., 2015. Micro- and nano-hydroxyapatite as active reinforcement for soft biocomposites. Int. J. Biol. Macromol. 72, 199–209.

# Further Reading

Achenbach, S., Marwan, M., Schepis, T., Pflederer, T., Bruder, H., Allmendinger, T., Petersilka, M., Anders, K., Lell, M., Kuettner, A., Ropers, D., Daniel, W.G., Flohr, T., 2009. High-pitch spiral acquisition: a new scan mode for coronary CT angiography. J. Cardiovasc. Comput. Tomogr. 3 (2), 117–121.

Agrawal, G., et al., 2017. Wettability and contact angle of polymeric biomaterials. In: Tanzi, M.C., Farè, S. (Eds.), Characterization of Polymeric Biomaterials. first ed., Woodhead Publ., Elsevier Ltd, Duxford, UK (Chapter 3).

Balci, M., 2005. Basic $^1$H- and $^{13}$C-NMR Spectroscopy. Elsevier Science, London, UK. ISBN: 9780444518118.

Barnes, J., 1992. High-Performance Liquid Chromatography. John Wiley and Sons, Chichester, West Sussex, UK.

Brunella, M.F., Serafini, A., Tanzi, M.C., 2017. Characterization of 2D polymeric biomaterial structures or surfaces. In: Tanzi, M.C., Farè, S. (Eds.), Characterization of Polymeric Biomaterials, first ed., Woodhead Publ., Elsevier Ltd, Duxford, UK (Chapter 1).

Bushberg, J.T., 2002. The Essential Physics of Medical Imaging. Lippincott Williams & Wilkins, Philadelphia, PA.

Chou, E.T., Carrino, J.A., 2007. Chapter 10—Magnetic resonance imaging. In: Pain Management, vol. 1., Elsevier, Amsterdam, NL, pp. 106–117.

De Nardo, L., Farè, S., 2017. Dynamico-mechanical characterization of polymer biomaterials. In: Tanzi, M.C., Farè, S. (Eds.), Characterization of Polymeric Biomaterials, first ed., Woodhead Publ., Elsevier Ltd (Chapter 9).

Detre, J.A., 2007. Magnetic Resonance Imaging. In: Neurobiology of Disease. Academic Press, Cambridge, MA, pp. 793–800.

Deyl, Z., 1984. Separation methods. In: New Comprehensive Biochemistry, vol. 8., Elsevier Sci. Publ. BV, Amsterdam, The Netherlands. ISBN: 0-444-805273.

Elmaoglu, M., Celik, A., 2012. MRI Handbook: MR Physics, Patient Positioning, and Protocols. Springer, New York, NY.

Giupponi, E., Candiani, G., 2017. Interaction of polymeric biomaterials with bacteria (static). In: Tanzi, M.C., Farè, S. (Eds.), Characterization of Polymeric Biomaterials. first ed., Woodhead Publ., Elsevier Ltd, Duxford, UK (Chapter 13).

https://www.iso.org/home.html.

Joint Commission (TJC), USA, https://www.jointcommission.org/sea_issue_47/.

Kalender, W.A., 2006. X ray computed tomography. Phys. Med. Biol. 51, 29–43.

Klug, H.P., Alexander, L.E., 1974. X-Ray Diffraction Procedures for Polycrystalline and Amorphous Materials, second ed. Wiley, New York.

Kumar, S., 2013. Physical and chemical characterization of biomaterials. In: Bandyopadhyay, A., Bose, S. (Eds.), Characterization of Biomaterials, first ed., Elsevier Pub., Kidlington, Oxford, UK. ISBN 978-0-12-415800-9D.

Lauterbur, P.C., 1973. Image formation by induced local interactions: examples employing nuclear magnetic resonance. Nature 242 (5394), 190–191.

Mäntele, W., Deniz, E., 2017. UV–VIS absorption spectroscopy: Lambert-beer reloaded. Spectrochim. Acta A Mol. Biomol. Spectrosc. 173, 965–968.

Marino, M.A., Riedl, C.C., Bernathova, M., Bernhart, C., Baltzer, P.A.T., Helbich, T.H., Pinker, K., 2018. Imaging phenotypes in women at high risk for breast cancer on mammography, ultrasound, and magnetic resonance imaging using the fifth edition of the breast imaging reporting and data system. Eur. J. Radiol. 106, 150–159.

Menard, K., Cassel, B., Thermomechanical Analysis. PerkinElmer, Inc. Technical Brochure "Basics of Thermomechanical Analysis with TMA 4000", www.perkinelmer.com/lab-solutions/resources/docs/TCH_TMA_4000.pdf.

Murphy, B., 2001. Fundamentals of Light Microscopy and Electronic Imaging. Wiley-Liss, John Wiley and Sons, New York, NY.

Musumeci, C., 2017. Advanced scanning probe microscopy of graphene and other 2D materials. Crystals 7 (7), 216. https://doi.org/10.3390/cryst7070216.

Oliveira, C.A., Candelária, I.S., Oliveira, P.B., Figueiredo, A., Caseiro-Alves, F., 2015. Metallosis: a diagnosis not only in patients with metal-on-metal prostheses. Eur. J. Radiol. 2, 3–6.

Pradhan, S., Rajamani, S., Agrawal, G., Dash, M., Samal, S.K., 2017. NMR, FT-IR and Raman Characterization of Biomaterials. In: Tanzi, M.C., Farè, S. (Eds.), Characterization of Polymeric Biomaterials, first ed. Woodhead Publ., Elsevier Ltd, Duxford, UK (Chapter 7).

Redondo, B., Gimeno, J.R., Pinar, E., Valdés, M., 2009. Unusual presentation of acute coronary syndrome. Bilateral coronary dissection after car accident. Am. J. Emerg. Med. 27 (8) 1024. e3-5.

Saunders, J., Ohlerth, S., 2011. CT physics and instrumentation—mechanical design. In: Veterinary Computed Tomography. first ed., John Wiley & Sons Ltd, Hoboken, NJ.

Scott, R.P.W., Cazes, J., 2001. Encyclopedia of chromatography. Marcel Dekker, Inc., New York

Socrates, G., 2001. Infrared and Raman Characteristic Group Frequencies: Tables and Charts. John Wiley & Sons, Ltd, Chichester, West Sussex, UK.

Sousa, A., Barrias, C.C., et al., 2017. In vitro interaction of polymeric biomaterials with cells. In: Tanzi, M.C., Farè, S. (Eds.), Characterization of Polymeric Biomaterials, first ed., Woodhead Publ., Elsevier Ltd, Duxford, UK (Chapter 12).

Stuart, B.H., 2005. *Infrared* Spectroscopy: *Fundamentals and Applications.* Analytical Techniques in the SciencesJohn Wiley & Sons, Ltd., Chichester, West Sussex, England.

Tanzi, M.C., 2017. Characterization of thermal properties and crystallinity of polymer biomaterials. In: Tanzi, M.C., Farè, S. (Eds.), Characterization of Polymeric Biomaterials, first ed., Woodhead Publ., Elsevier Ltd, Duxford, UK (Chapter 6).

Thomas, A.M.K., Banerjee, A.K., Busch, U., 2005. Classic Papers in Modern Diagnostic Radiology. Springer-Verlag, Heidelberg, Germany, pp. 70–72.

Wagner, M., 2018. Chapter 1, Introduction to Thermal Analysis. In: Thermal Analysis in Practice. Carl Hanser Verlag GmbH & Co. KG., Printed by Hubert & Co GmbH und Co KG BuchPartner, Gottingen, Germany.

Wunderlich, B., 1990. Thermal Analysis. Academic Press, Ltd, London, UK.

Yadav, L.D.S., 2005. Infrared (IR) Spectroscopy. In: Organic Spectroscopy. Springer, Dordrecht, pp. 52–106.

Yahia, L.'.H., Mireles, L.K., 2017. X-ray photoelectron spectroscopy (xps) and time-of-flight secondary ion mass spectrometry (ToF SIMS). In: Tanzi, M.C., Farè, S. (Eds.), Characterization of Polymeric Biomaterials, first ed. Woodhead Publ., Elsevier Ltd, Duxford, UK (Chapter 4).

Yoshioka, H., Schlechtweg, P.M., Kose, K., 2009. Chapter 3—Magnetic resonance imaging. In: Imaging of Arthritis and Metabolic Bone Disease. Elsevier, Amsterdam, NL, pp. 34–48.

# 8

# Advanced Applications

## 8.1 TISSUE ENGINEERING

### 8.1.1 Introduction

*Tissue engineering* (TE) can be defined as the use of a combination of cells, engineering materials, and suitable biochemical cues to repair damaged tissues and organs. This, however, is a simplistic definition, because the involved issues are much more complex than a simple assembling of the previously mentioned factors.

The first definition of TE is that attributed to R. Langer and J.P. Vacanti who described it as "an interdisciplinary field that applies the principles of engineering and life sciences toward the development of biological substitutes that restore, maintain, or improve tissue function or a whole organ" (Langer and Vacanti, 1993). However, there are contradictory opinions (Viola et al., 2003) and, in particular, the concept of functionality was introduced at a later time, when laboratory-made tissues began to be examined to see if they could function as natural tissues do (Guilak et al., 2003).

Functional TE can be described as "the production of functional tissues for research and applications through the elucidation of basic mechanisms of tissue development combined with fundamental engineering production processes."

Nevertheless, after different opinions were expressed concerning how precise and explicit this field definition needed to be, no formal definition was adopted.

TE is part of the broad field of regenerative medicine (RM), often referred to as Tissue Engineering and Regenerative Medicine (TERM) (Fisher and Mauck, 2013). It has been pointed out that TE emphasizes the starting materials and scaffolds used to create de novo tissue implants, whereas RM emphasizes endogenous tissue formation that may occur secondary to induction from the starting materials (Katari et al., 2015).

There are several strategies aiming at repairing damaged or pathological tissue and organs. A first distinction can be made between approaches that use cells (either injected with or without encapsulation and grown on scaffolds) and acellular approaches, where appropriate substrates of either of biological or synthetic origin are imbibed with molecules or factors able to attract the cells to the appropriate site to be restored. TE focuses on developing biomaterials able to deliver bioactive factors and/or cells to aid the healing response, and/or provide scaffolds able to support cell growth and promote appropriate tissue formation.

In this chapter, we mainly deal with the strategy of using a scaffold, taking into consideration the main problems related to the design and fabrication of the scaffolds.

## 8.1.2 Necessary Steps for Tissue Regeneration by Use of Scaffolds

The path necessary for the development of an engineered tissue before reaching clinical trial requires different steps, all of them of crucial importance. Fig. 8.1 illustrates, in a simplified form, all steps, which can be summarized as follows:

**(1)** Design criteria
**(2)** (a) Design and optimization of the scaffold; scaffold fabrication and characterization.
(b) Selection, supplying or isolation, and cultivation of the cells and/or bioactive factors
**(3)** Preparation of the construct scaffold ± bioactive factors
**(4)** Cell seeding and functional evaluation in vitro, subsequently in vivo, in the animal model
**(5)** Evaluation of the newly grown tissue

The different phases must be tailored to the specific approach, depending on the type of tissue to be reconstructed, the extent of the defect, and the pathology to be treated.

In any case, the design and preparation of an appropriate scaffold and the choice of the type of cells and their growth modalities play a crucial role.

## 8.1.3 The Scaffold and Materials

The scaffold must be designed in such a way as to present the shape and structure suitable for the growth of the regenerated tissue. To avoid rejection problems, cells of the same patient (autologous) are preferably used. To obtain satisfactory results, it is necessary to maintain the tissue-specific functions of the cells in culture, using appropriate environments (culture media) with active molecules, such as growth factors, while trying to simulate the physiological conditions in which the specific cells must operate, including the peculiar physical-mechanical stresses.

First, we need to consider a series of requirements necessary to optimize the performance of the scaffold, according to the specific tissue we want to restore.

Ideal features for a scaffold, in general, are:

**(a)** *high porosity*, with interconnected pores to allow cell growth, transport of nutrients, and elimination of waste products
**(b)** *biocompatibility* and *bioresorbability*, with degradation and resorption rates to be combined with tissue reconstruction
**(c)** *chemically suitable surface* for cell attachment, proliferation, and differentiation
**(d)** *mechanical properties* matching those of the tissues present at the implantation site

Admitted to be able to fabricate scaffolds meeting these requirements, some issues need to be taken into account:

– *Distribution of cells in the scaffold*. Typically, the scaffolds are manufactured then seeded with the cells after being placed in their liquid culture medium, which adsorbs on the scaffold, then diffuses inside. Although cells can easily adhere to and proliferate on the outermost part of the scaffold, their distribution across the entire scaffold is unevenly uniform due to restricted cell mobility and limited diffusive possibility of nutrients.

A solution could be represented by the direct inclusion of the cells inside the scaffold during its manufacturing, so as to control their distribution. Unfortunately, this is not feasible with most manufacturing processes, which involve the use of high temperatures or solvents and toxic chemicals that would damage or kill the cells.

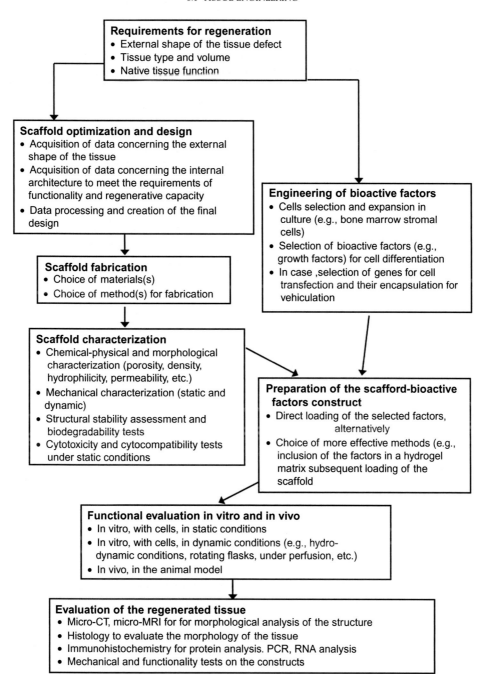

**Requirements for regeneration**
- External shape of the tissue defect
- Tissue type and volume
- Native tissue function

**Scaffold optimization and design**
- Acquisition of data concerning the external shape of the tissue
- Acquisition of data concerning the internal architecture to meet the requirements of functionality and regenerative capacity
- Data processing and creation of the final design

**Engineering of bioactive factors**
- Cells selection and expansion in culture (e.g., bone marrow stromal cells)
- Selection of bioactive factors (e.g., growth factors) for cell differentiation
- In case ,selection of genes for cell transfection and their encapsulation for vehiculation

**Scaffold fabrication**
- Choice of materials(s)
- Choice of method(s) for fabrication

**Scaffold characterization**
- Chemical-physical and morphological characterization (porosity, density, hydrophilicity, permeability, etc.)
- Mechanical characterization (static and dynamic)
- Structural stability assessment and biodegradability tests
- Cytotoxicity and cytocompatibility tests under static conditions

**Preparation of the scafford-bioactive factors construct**
- Direct loading of the selected factors, alternatively
- Choice of more effective methods (e.g., inclusion of the factors in a hydrogel matrix subsequent loading of the scaffold

**Functional evaluation in vitro and in vivo**
- In vitro, with cells, in static conditions
- In vitro, with cells, in dynamic conditions (e.g., hydro-dynamic conditions, rotating flasks, under perfusion, etc.)
- In vivo, in the animal model

**Evaluation of the regenerated tissue**
- Micro-CT, micro-MRI for for morphological analysis of the structure
- Histology to evaluate the morphology of the tissue
- Immunohistochemistry for protein analysis. PCR, RNA analysis
- Mechanical and functionality tests on the constructs

**FIG. 8.1**   Schematic view of the variables and activities involved in tissue engineering based on the use of scaffolds.

- *Vascularization and metabolic exchanges.* Even when the cells are well distributed into a scaffold of suitable design, blood supply is necessary for drainage of waste products and nutrient provision to the cells inside the scaffold. This can happen through the formation of new blood vessels from the surrounding vascularized tissue, but the process of neoangiogenesis takes time; meanwhile, the cells that lie deep may not survive.
- *Scaffold optimization.* To optimize the performance of the scaffold, it can be possible to design heterogeneous scaffolds by balancing the properties of different zones, so that a part of the scaffold presents a structure suitable to provide optimal mechanical properties, whereas another part offers optimized properties (microstructure, composition) to promote cell growth.
- *Distribution of growth factors and bioactive molecules.* The possibility of creating concentration gradients of biological cues is another key point for chemotactic response of cells, that is, for preferential cell migration to high concentrations of a mediator.
- *Use of different types of cells.* This strategy could be used to reconstruct complex and heterogeneous systems such as organs or parts of the human body in which different types of tissue coexist. Unfortunately, building such heterogeneous systems requires the ability to control and address both the distribution of cell types and their regenerative capacities, and the manufacture of complex scaffolds at the micro- and nanoscale. For now, these objectives are still out of reach.

### 8.1.3.1 *The Materials for the Preparation of the Scaffolds*

Both natural and synthetic materials have been tested. Except in the case of bone tissue, which requires totally or partially inorganic scaffolds, the material with which the scaffold is generally constructed is entirely polymeric. If it is true that preliminary results have confirmed that polymers of natural origin (collagen, chitin, hyaluronic acid and its derivatives, fibrin, alginates) are promising (see Chapter 4, Section 4.3), many of them may present problems of availability in large volumes. As a result, many researchers have turned to synthetic materials, in particular to biodegradable polymers.

Degradable materials must satisfy more strict requirements, in terms of their biocompatibility, than nondegradable materials. In addition to the potential problem of toxic contaminants leaching from the implant that must be eliminated before use, the potential toxicity of the degradation products and subsequent metabolites has to be taken into account. By consequence, only a limited number of starting reagents (monomers, cross-linkers, etc.) have been successfully applied to the preparation of biocompatible degradable biomaterials.

A list of requirements for a biodegradable material targeted for the preparation of scaffolds and delivery systems could include:

- will not evoke an excessive inflammatory response or toxic effects following implantation in the human body
- have an acceptable shelf life
- degradation products must be nontoxic and capable of being metabolized and excreted from the human body
- degradation time should "harmonize" with the process of healing/regeneration of tissues
- have appropriate mechanical properties for the chosen application, and the variation of mechanical properties during degradation should be compatible with the process of healing
- be easily processable
- be sterilizable (with an acceptable variation of the molecular weight)

To date, detailed toxicological studies in vivo and investigations of degradation rates and mechanisms have been published for only a small fraction of potentially useful hydrolytically (or enzymatically) unstable polymers.

## DEGRADATION MECHANISMS

To start, it is appropriate to make a distinction between degradation, erosion, and dissolution. *Degradation* is the process of chain breakage that changes the chemical structure; in the case of polymers of natural origin, such as proteins, it follows a conformational change that causes denaturation. *Erosion* involves a loss of mass, whereas *dissolution* consists of a loss of mass via solubilization.

Among polymers that undergo solubilization, we can cite dextran, polyvinyl alcohol, and polyethylene oxide, whereas polyacrylic acid and polyvinyl acetate undergo ionization before solubilization. Simple *hydrolysis* and *enzymatically induced hydrolysis* are the two main mechanisms for bioresorption of resorbable biomaterials. The first one occurs typically for synthetic polymers; the other one is characteristic of polymers of natural origin, such as polysaccharides and proteins (polyamides). Both the mechanisms involve cleavage of bonds susceptible to hydrolysis in the polymer backbone, causing fragmentation of the entire polymeric structure and producing low molecular weight oligomers that can be absorbed or excreted by the body (Gajjar and King, 2014).

Therefore, a typical resorption profile of a material consists of four steps:

- water absorption
- reduction of mechanical properties
- reduction of molar mass
- complete loss of weight

Initially, hydrolysis takes place in the amorphous regions of the polymer, converting long polymer chains into shorter fragments. This causes a reduction in molecular weight without a loss in physical properties as the matrix is still held together by the crystalline regions. Reduction in molecular weight is soon followed by a reduction in physical and mechanical properties

as more and more chains are converted to smaller fragments. At the later stages, the fragments are metabolized and converted to oligomers, which are either absorbed or excreted by the body.

The resorption process of a biomaterial is also classified according to its *erosion* mechanism. *Bulk erosion* is the mechanism when water diffuses rapidly into a polymer structure, leading to hydrolysis. The subsequent mass loss then occurs throughout the bulk of the material, as shown in Fig. 8.2. A characteristic behavior of bulk eroding polymers is the sudden and rapid loss of strength and structural integrity as the resorption continues over time.

Instead, in surface erosion, the mass loss occurs at the water/implant interface, causing resorption from the outer surface toward the interior while maintaining bulk integrity (see Fig. 8.2).

Sometimes the mechanism is complex, comprising multiple types of degradative mechanisms for the same material (a function not only of the chemical structure, but also the configuration, the conformation and the physical characteristics, such as hydrophobicity or hydrophilicity).

## SYNTHETIC BIODEGRADABLE POLYMERS

A small number of synthetic, degradable polymers have been used in medical implants and devices, which gained approval by the Food and Drug Administration (FDA)[1] for use in humans, among them are the polyesters listed in the following section.

Currently, the categories of synthetic biodegradable polymers taken into consideration are:

**(a)** *Polyesters.* Poly-$\alpha$-hydroxy acids:
polyglycolic acid (polyglycolide, PGA),
polylactic acid (polylactides: poly-L-lactic,

---

[1] The FDA does not approve polymers or materials *per se*, but only specific medical devices and drug delivery formulations.

**Bulk erosion**

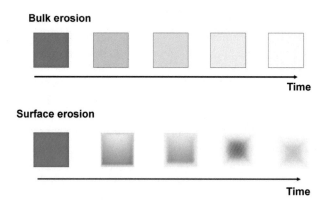

**Time**

**Surface erosion**

**Time**

FIG. 8.2    Bulk and surface erosion mechanisms. In bulk erosion, mass loss occurs throughout the bulk of the material, whereas in surface erosion, resorption occurs from the outer surface toward the interior while maintaining bulk integrity.

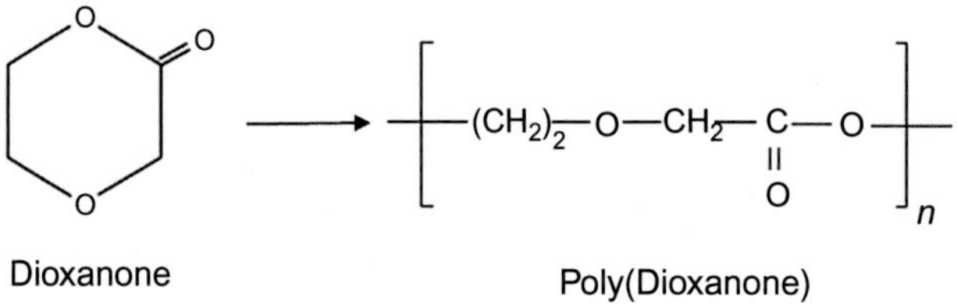

Dioxanone                         Poly(Dioxanone)

FIG. 8.3    Chemical structure of the cyclic monomer and polydioxanone (PDS). PDS is a poly-ether-ester, semicrystalline, with $T_g = -10°C$ to $-0°C$.

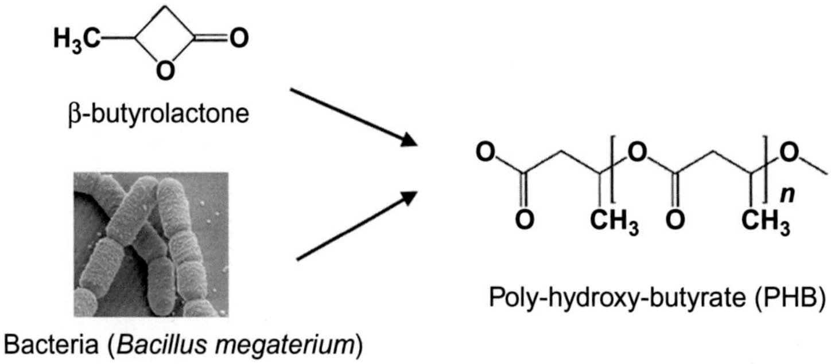

β-butyrolactone

Poly-hydroxy-butyrate (PHB)

Bacteria (*Bacillus megaterium*)

FIG. 8.4    Chemical structure of the cyclic monomer and poly-hydroxybutyrate, which is a bacterial polyester produced by many bacteria as an energy source, in particular by *Bacillus megaterium* (SEM image in the figure). It can also be produced by chemical synthesis from β-butyrolactone. PHB is semicrystalline, with $T_m = 160–180°C$, and $T_g = -5°C$ to 20°C.

PLLA, and poly D-L-lactic, PDLA), and a wide range of copolymers of lactic and glycolic acid. Also considered are polycaprolactone (PCL)[2] and a few others,

such as polyesters containing dioxanone (polydioxanone, PDS or PPDX or PDO, Fig. 8.3) and poly-hydroxybutyrate (PHB, Fig. 8.4).

In general, poly-α-hydroxy acids undergo bulk erosion as a result of hydrolysis.

[2] See also Chapter 4, Section 4.2.3.2

**TABLE 8.1** Properties of Bioresorbable and Bioerodible Polymers

| Polymer | Relative Mechanical Properties | Degradation Mechanism | Loss of Mechanical Properties (Months) | Mass Loss (Months) |
|---------|-------------------------------|----------------------|----------------------------------------|--------------------|
| PLLA | +++ | Bulk erosion | 9–15 | 36–48 |
| PDL/LLA 70/30 | ++ | Bulk erosion | 5–6 | 12–18 |
| PGA | +++ | Bulk erosion | 0.5–1 | 3–4 |
| PDLLA | + | Bulk erosion | 1–2 | 5–6 |
| PDLLA/PGA 50/50 | ++ | Bulk erosion | 1–2 | 3–4 |
| PCL | + | Bulk and surface erosion | 9–12 | 24–36 |

+++, good; ++, average; +, poor.
*Data from Hutmacher, D.W., 2000. Scaffolds in tissue engineering bone and cartilage. Biomaterials 21, 2529–2543.*

In water, the molecular weight of the polymer begins to decrease due to hydrolytic degradation after one day (PGA, PDLA) or after a few weeks (PLLA), but there is no mass loss until the molecular chains are reduced to a size that allows them to diffuse freely out of the polymer matrix. The mass loss is accompanied by the release of acidic products of degradation (glycolic, lactic acid and their oligomers) that, in vivo, may cause an intense inflammatory reaction. In particular, this occurs when the walls of the scaffold are thick, and there is insufficient capability of the surrounding tissue to dispose of the products of degradation, as a result of limited vascularization or poor metabolic activity.

PCL degrades more slowly than PLA and is useful in systems for controlled release of drugs (degradation time of 2–3 years). PDS shows a metabolic pathway similar to PGA, with degradation kinetics similar to PDL-LA.

For representative purposes, some data comparing mechanical properties and degradation time of poly-α-hydroxy acids and PCL are listed in Table 8.1.

Interestingly, PHB undergoes hydrolytic degradation with a *surface erosion mechanism*, releasing D(-)3-hydroxybutyric acid, normally present in the blood. The degradation rate is lower than that of synthetic polyesters. Copolymers can be prepared, for example, with valerolactone to obtain poly-hydroxybutyrate-co-hydroxyvalerate (PHBV), used in specialty packaging, orthopedic devices, and in controlled release of drugs. PHB and PHBV are part of the polyhydroxyalkanoates (PHAs) family (www.revolvy.com/page/Polyhydroxyalkanoates).

(b) *Polyorthoesters (POE)*. Over the last 30 years, polyorthoesters have evolved through four families, of which only the fourth (POE IV) has been shown to have all the necessary attributes to allow commercialization (Heller et al., 2002). To make a polyorthoester more hydrophilic, and therefore to increase the rate of hydrolysis of the polymer, short segments of glycolic or lactic acid are inserted into the backbone, as shown in Fig. 8.5. By varying the type of diols, it is possible to change the appearance

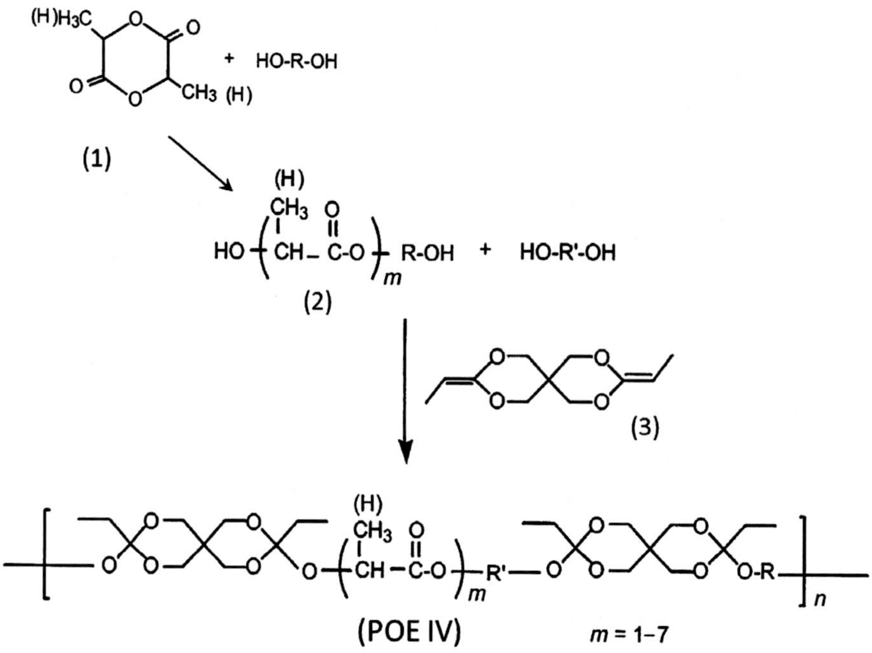

**FIG. 8.5** Chemical structure and synthetic path for polyorthoester IV (POE IV). (1) Glycolide and/or lactide are reacted with a diol to give short-chain polyglycolide and/or polylactide diols, (2) which react with diols and a diketene acetal (3) to produce POE IV. *From Heller, J., Barr, J., Ng, S.Y., Abdellauoi, K.S., Gurny, R., 2002. Poly(ortho esters): synthesis, characterization, properties, and uses. In: Domb, A., et al. (Eds.), Polyanhydrides and Poly(ortho esters). Adv. Drug Deliv. Rev. 54(7), 1015–1039.*

(from gel to solid), the degradation rate (from a few days to many months), sensitivity to pH, and the $T_g$ of the polymer.

The hydrolytic degradation of POE IV, that involves both surface and bulk erosion, produces α-hydroxyacids, propionic acid, and a mixture of diols and polyols. This POE and its block copolymers with polyethyleneglycol (PEG) can be successfully used in drug release applications.

(c) *Polyanhydrides.* These polymers have been investigated as bioresorbable and biocompatible biomaterials for short-term release of drugs since 1988 (Chasin et al., 1988). They are obtained from dicarboxylic acids (Fig. 8.6) by various synthetic methods and present, in general, a hydrophobic backbone with hydrolytically labile anhydride linkages; hydrolytic degradation can be controlled by control of the polymer composition (Kumar et al., 2002). Polyanhydrides are surface eroding due to their fast degradation, however they switch to bulk erosion once the device dimensions drop below a critical limit (Göpferich and Tessmar, 2002).

Depending on the starting monomers, aliphatic and aromatic polyanhydrides or copolymers, for example, with fatty acids can be obtained and cross-linked structures.

Aliphatic polyanhydrides were among the first materials to be investigated for the purpose of drug delivery. Some aliphatic polyanhydrides were studied for special applications, such as p(FA-SA), that is, copolymers derived from sebacic and fumaric acid, which were proposed for the

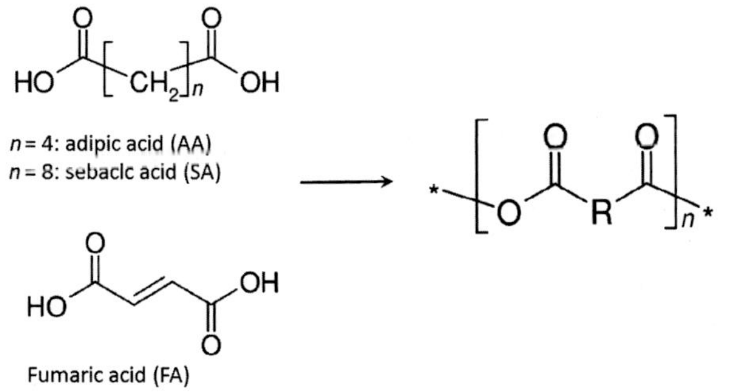

FIG. 8.6 Some dicarboxylic acids used in the synthesis of polyanhydrides (general structure on the right).

development of bioadhesive materials that interact with mucosal tissues. Aromatic polyanhydrides show high hydrophobicity and low degradation rates, whereas polyanhydrides derived from fatty acids were introduced with the aim to improve the mechanical properties (Gőpferich and Tessmar, 2002).

Aliphatic polyanhydrides of saturated diacid monomers are crystalline, with $T_m$ below 100°C, which degrade and are eliminated from the body within weeks; after polymerization, unsaturated polyanhydrides, such as the ones containing fumaric acid, still present double bonds that are available for secondary reactions to form a cross-linked matrix.

Polyester-anhydrides (prepared as di- and triblock copolymers with polycaprolactone, polylactic acid, and polyhydroxylbutyrate) and polyether anhydrides (prepared as di-, tri-, and brush copolymers of polyethylene glycol) have been investigated for tissue (bone) regeneration (Kumar et al., 2002).

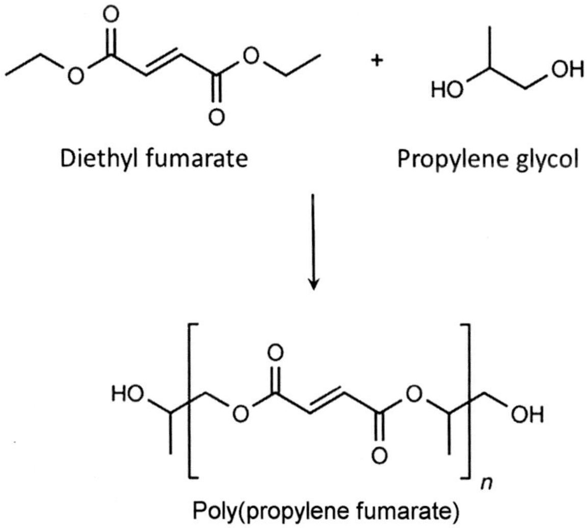

FIG. 8.7 Starting monomers and chemical structure of PPF.

**(d)** *Polypropylene fumarate (PPF)*. PPF is a linear and unsaturated polyester based on fumaric acid (Fig. 8.7) (Kasper et al., 2009). The degradation occurs via bulk erosion and leads to *fumaric acid*, a substance found in nature and eliminated through the tricarboxylic acid or Krebs cycle, and *1,2-propanediol* (or propylene glycol), used as diluent in pharmaceutical formulations.

In vitro, the rate of degradation varies from 84 to 200 days. The copolymer has an unsaturated double bond that can be used for further modification reactions (e.g., cross-links). It can be used in injectable formulations (which cross-link with the UV light, i.e., are photo-polymerizable)

**(e)** *Polycarbonates.*

*Aliphatic polycarbonates*: Poly(trimethylene carbonate) (PTMC; Fig. 8.8), shows elastomeric properties, with a $T_g$ of 40–60°C and low mechanical strength. It undergoes surface erosion with kinetics of degradation much more rapid in vivo than in vitro, as within in vivo environments hydrolysis is replaced by enzymatic degradation (Zhang et al., 2006). High MW PTMC and its copolymers with ε-caprolactone and D,L-lactide are useful for soft tissue engineering applications.

*Poly tyrosine-derived carbonates, poly(DTY)*: Other biodegradable polycarbonates can be synthesized, and, in particular, copolymers containing units of the amino acid tyrosine (i.e., tyrosine-derived polycarbonates) can be considered pseudo-polyamino acids. They are biocompatibile and retain good

mechanical resilience during degradation, like polycarbonates (Bourke and Kohn, 2003; Kohn et al., 2007).

Various studies have shown that the carbonate group hydrolyses more quickly, whereas the amide bond is stable in vitro. The hydrolysis of the carbonate group leads to alcohols and $CO_2$, so there is no release of acid groups, as is the case with poly-α-hydroxy acids. By varying the structure of the "Y" groups (Fig. 8.9), the mechanical properties, degradation rate, and cellular response can be varied.

In studies reported in the literature, polycarbonates with Y group = ethyl showed osteoconductive capacity and biocompatibility.

**(f)** *Polyphosphazenes*. This is a class of polymers displaying high biocompatibility. The degradation of the main chain, based on phosphorus and nitrogen (i.e., $-N=P-$), leads to biologically compatible products of hydrolysis (i.e., phosphates and ammonia). Some chemical structures are shown in Fig. 8.10.

Although most of the polyphosphazenes are stable in the presence of moisture, considerable efforts have focused on the synthesis of polymers that erode at biological pH to give innocuous products.

The discovery of the high immunostimulating activity of ionic polyphosphazenes has inspired extensive research in the area and moved interest in their commercial development. Polyphosphazenes have been investigated

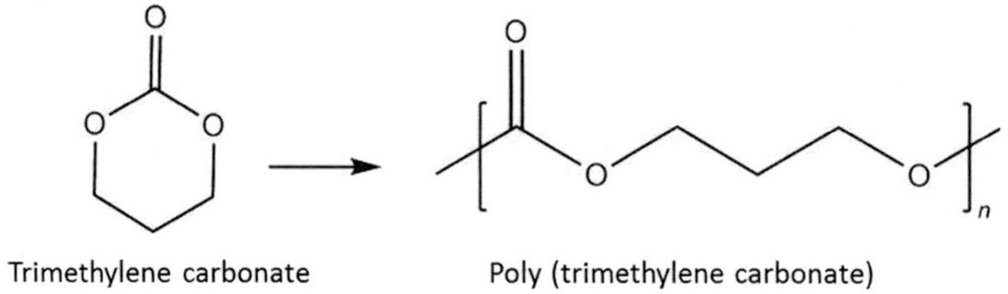

Trimethylene carbonate             Poly (trimethylene carbonate)

**FIG. 8.8**    Chemical structure of the monomer trimethylene carbonate and its polymer, PTMC.

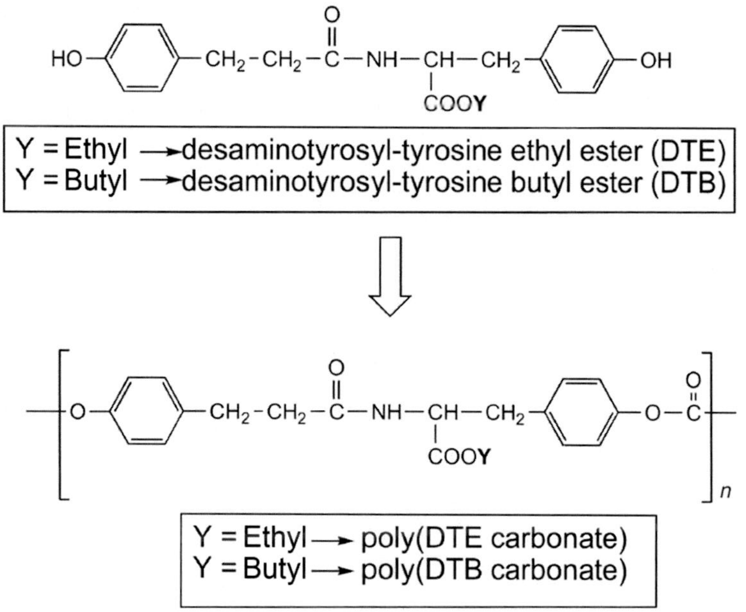

**FIG. 8.9** Chemical structures of poly(DTE carbonate) and poly(DTB carbonate). *From Fig. 3 in Kohn, J., Welsh, W.J., Knight, D., 2007. A new approach to the rationale discovery of polymeric biomaterials. Biomaterials 28(29), 4171–4177.*

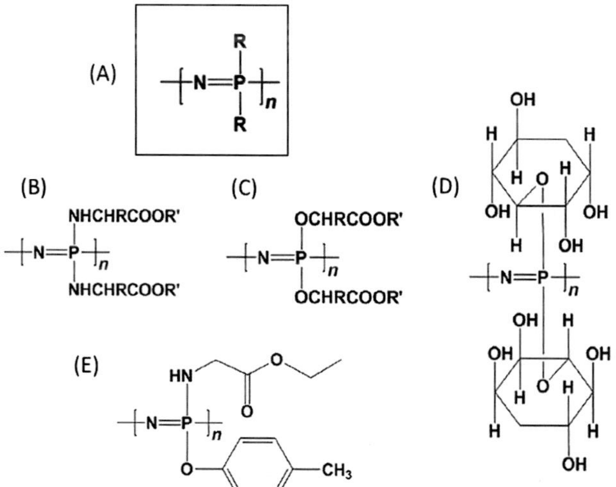

**FIG. 8.10** Some chemical structures of degradable and biocompatible polyphosphazenes. General structure of polyphosphazenes (A); with hydrolytically labile side groups such as amino acid esters (B), glycolate/lactate esters (C) or glucosyl groups (D); another possible biodegradable polyphosphazene (E).

for vaccine delivery vehicles, drug release, and scaffold for TE applications (Andrianov, 2009).

**(g)** *Biodegradable elastomers*

Biodegradable elastomers have been developed to overcome problems associated with biodegradable thermoplastic materials (Bruggeman et al., 2008), such as:

- rigid mechanical properties
- bulk degradation
- acidic degradation products (in the case of polyesters)

*Poly (polyol)sebacates.* This family of synthetic biodegradable polymers composed of structural units endogenous to the human metabolism, designated poly(polyol sebacate) (PPS) polymers, can be prepared from polyols and sebacic acid. Some examples are illustrated in Fig. 8.11.

Because of the multiple functionalities of the starting molecules, the obtained polymeric structures are cross-linked. Chemical, physical, and mechanical properties along with degradation rates can be tuned by varying the polyol type and stoichiometry of the reacting sebacic acid.

These thermoset networks exhibit tensile Young's moduli of 0.4–380 MPa with elongations at break from 11% to 200%, and $T_g$ ranging from 7°C to 46°C.

Elastomers such as poly(glycerol sebacate) (PGS) have shown to primarily degrade by surface erosion, retaining their structural integrity and shape stability during degradation in vivo.

Compared to PLGA, PPSs demonstrated similar in vitro and in vivo biocompatibility.

*Biodegradable Polyurethane (PU) Elastomers.* "Biostable" segmented polyurethanes have been described in Chapter 4 (Section 4.2.6), however, in

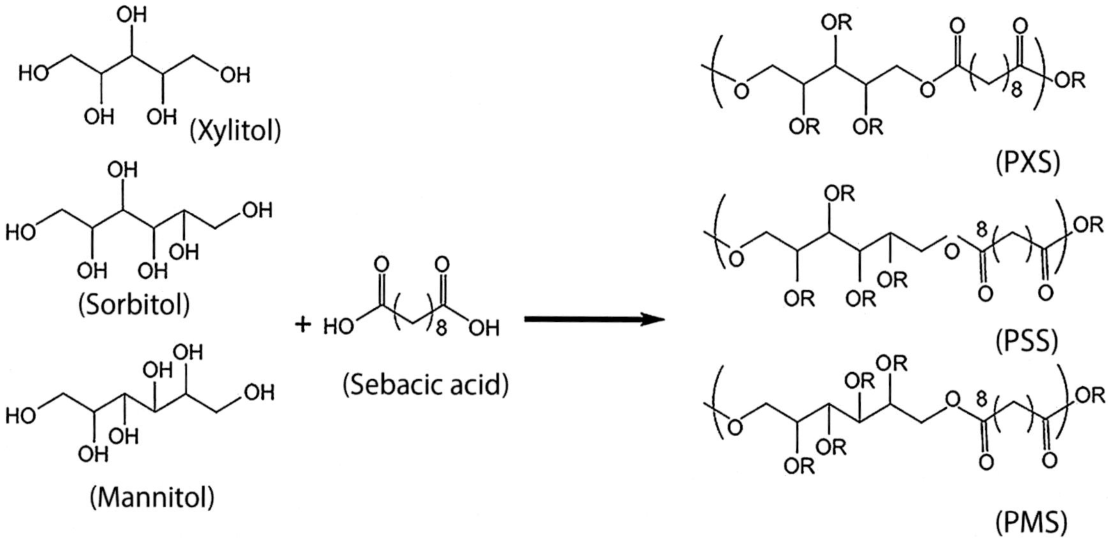

FIG. 8.11   Scheme for the synthesis of some polyol-based polysebacates. Xylitol, sorbitol, and mannitol are polymerized with sebacic acid in different stoichiometries, yielding poly xylitol sebacate (PXS), poly sorbitol sebacate (PSS), and poly mannitol sebacate (PMS). Simplified chemical structure of the polymers are shown. *Modified from Bruggeman, J.P., de Bruin, B.-J., Bettinger, C.J., Langer, R., 2008. Biodegradable poly(polyol sebacate) polymers. Biomaterials 29, 4726–4735.*

recent years, there has been increased interest in developing biodegradable PU elastomers for use as tissue scaffolds and drug delivery systems (Guelcher, 2008). To obtain a structure that is unstable in the physiological environment and degrades releasing molecules tolerable by the metabolism of the human body, therefore free from toxic effects, macrodiols of the polyester type (e.g., polylactide-, polyglycolide-, polycaprolactone-diols) and hydrophilic polyether diols, such as polyethylene glycol (PEG), can be selected (Rechichi et al., 2008). In particular, by use of PEG, hydrophilic, and together with polyester macrodiols, hydrophobic, it is possible to obtain a higher degradation rate and better biocompatibility.

Biodegradable PURs are designed to undergo hydrolytic or enzymatic degradation to noncytotoxic decomposition products in vivo. Although the degradation mechanism is not fully understood, it is hypothesized that ester linkages hydrolyze yielding α-hydroxyacids degradation products, as well as urethane and urea fragments with terminal acid groups, which have been theorized to catalyze further degradation (Guelcher, 2008).

**(h)** *Hydrogels.* This class of polymeric materials is described in Chapter 1 (Section 1.2.5), and nondegradable acrylic hydrogels are presented in Chapter 4 (Section 4.2.2.5). Degradable hydrogels have mainly been designed and synthesized for applications in drug delivery and, more recently, as TE scaffolds; they may degrade and dissolve by hydrolysis or enzymolysis of main chain, side chain, or cross-linker bonds (Peppas and Hoffman, 2012).

Most investigated hydrolytically degradable hydrogels contains PEG blocks and can be either:

– Triblock copolymers of A-B-A structure where A (or B) may be PLA, PLGA, or other hydrophobic polyesters that form hydrophobic blocks, and B (or A) may be PEG, a hydrophilic block. These structures form hydrogels held together by hydrophobic forces and degrade into endogenous metabolites such as lactic or glycolic acids, whereas the PEG blocks are then excreted through the kidneys.
– Cross-linked and degradable PEG-based gels prepared from acrylate- or methacrylate-terminated block copolymers that include PEG as a hydrophilic block. These gels may have the A-B-A triblock structure of (methacrylate)–PEG–methacrylate, which is photo-polymerized and later degrades by hydrolysis of the ester bonds linking PEG to the methacrylate cross-links.
– Enzymatically degradable PEG-based hydrogels, which may include peptide blocks (e.g., fibrin peptide cross-linking blocks) that act as a substrate for a naturally occurring enzymes, thus degrading and dissolving by proteolysis.

### 8.1.4 Methods for Scaffold Fabrication

Many techniques have been developed to fabricate three-dimensional (3D) porous architectures, and some of these techniques have gained much attention due to their versatility. This section points out techniques actually available to fabricate scaffolds for TERM applications to improve the function and architecture of the scaffold for a specific application. In fact, in the body, cells and tissue are organized in a 3D architecture; hence, scaffolds have to be fabricated by a different methodology to facilitate cell distribution and guide their growth into the 3D space.

The techniques for producing scaffolds can be divided into two categories: nondesigned

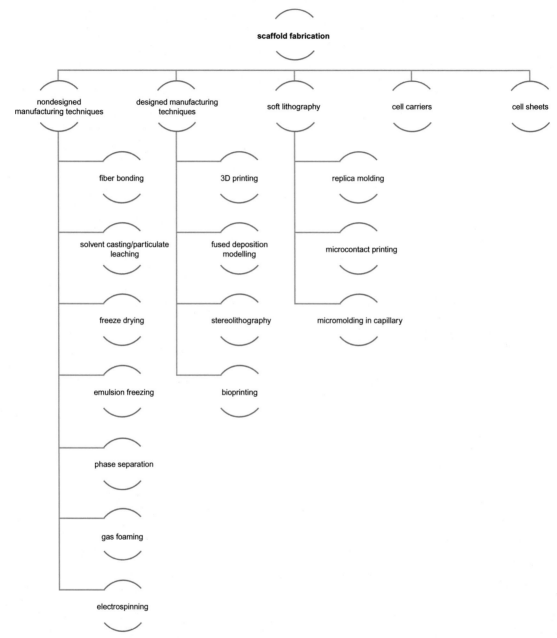

**FIG. 8.12** Possible approaches and techniques for the production of scaffolds for regenerative medicine and tissue engineering applications.

manufacturing techniques and designed manufacturing techniques (Fig. 8.12). Nondesigned manufacturing techniques include solvent casting/particulate leaching (SC/PL), freeze drying, phase separation, gas foaming, electrospinning, or a combination of these techniques. Designed manufacturing techniques include additive manufacturing (AM) processes, such as

**TABLE 8.2** Pros and Cons of Main Techniques for Scaffold Fabrication

| Technique | Advantages | Disadvantages |
| --- | --- | --- |
| Fiber bonding | High surface to volume ratio<br>High porosity | Poor mechanical property |
| Solvent casting/particulate leaching | Control over porosity and pore size | Limited mechanical properties<br>Possible solvent residual |
| Freeze drying | High temperature and separate leaching step not required | Small pore size<br>Long processing time |
| Phase separation | Control over porosity and pore size | Difficult to control precisely scaffold morphology |
| Gas foaming | Free of harsh organic solvents<br>Control over porosity and pore size | Limited mechanical property<br>Inadequate pore interconnectivity |
| Electrospinning | Control over porosity, pore size and fiber diameter | Limited mechanical property<br>Pore size decrease with fiber thickness |
| Additive manufacturing | Excellent control over geometry, porosity | Limited polymer type |

fused deposition modeling, 3D printing, stereolithography, and the more recently developed technique for hydrogel printing and combined printing of hydrogel and cells.

The fabrication technique for TE scaffolds is directly related to the bulk and surface properties of the polymer and the proposed function of the scaffold. Although each technique has its advantages and disadvantages (Table 8.2), the appropriate selection of the methodologies must satisfy the requirements of the specific tissue to be repaired or engineered. The different techniques here described allow for appropriate scaffold design by controlling the pore size, pore interconnectivity, and porosity so as to be suitable for nutrient diffusion and cell attachment.

### 8.1.4.1 Nondesigned Manufacturing Techniques

#### FIBER BONDING

The earliest scaffolds were nonwoven fiber meshes, made of nonbonded fibers with low mechanical properties. To overcome this problem, a fiber bonding technique (Fig. 8.13) for scaffold fabrication was developed by Mikos et al., aimed at binding the fibers together at the intersection points. The scaffold obtained by fiber bonding provides mechanical stability and allows tissue ingrowth. One of the main advantages of using the fiber is its large surface area, which is suitable for scaffold applications. Therefore, it provides more surface area for cell attachment and sufficient space for regeneration of extracellular matrix. The disadvantages of this technique are represented by low control of the porosity and pore size, lack of availability of suitable solvents, and required appropriate melting temperature of polymers. Two different approaches are reported in literature, reported here as examples for producing scaffolds with this technique. Similar approaches can be developed for other biocompatible polymeric fibers.

Synthetic polymer (PLLA) is dissolved in chloroform, then a nonwoven mesh of PGA fiber is added. Subsequently, solvent is removed by evaporation and, as a result, a composite material consisting of nonbonded PGA fiber embedded in PLLA matrix is formed. Fiber bonding occurs during post-treatment at a temperature above the melting temperature of PGA. As a result, PLLA matrix of the composite is removed by dissolution in methylene chloride agent

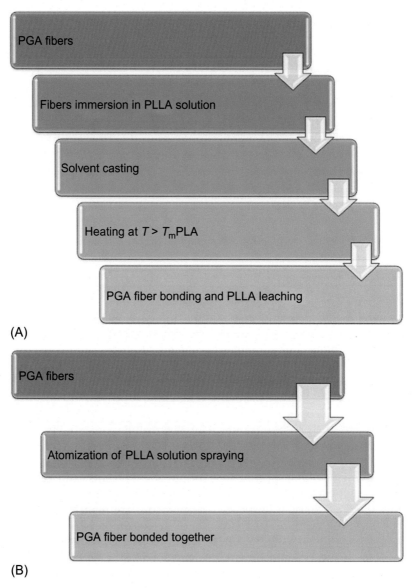

**FIG. 8.13**　Scheme of two different proposed approaches in fiber bonding technique: (A) PLLA solution is added to PGA fibers; (B) PLLA is sprayed onto PGA fibers.

because PGA is insoluble in this solvent (Figs. 8.13A and 8.14A top). Hence, this process yields the scaffolds of PGA fibers bonded together by heat treatment. In the second approach (Figs. 8.13B and 8.14A bottom), a non-woven fiber mesh is sprayed with an atomized PLLA or PLGA solution. The polymer solution builds up on the PGA fibers and bonds them at the interconnection points. The produced scaffold allows cell adhesion onto PLLA while the mechanical properties are related to PGA fibers (Fig. 8.14B).

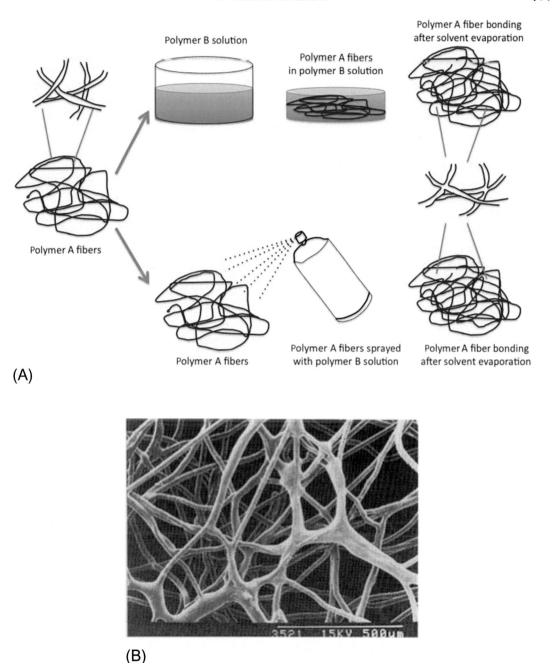

**FIG. 8.14** (A) Sketch of two different proposed approaches in fiber bonding technique: (top) Polymer B solution (e.g., PLLA) is added to polymer A fibers (e.g., PGA); (bottom) polymer B solution (e.g., PLLA) is sprayed onto polymer A fibers (e.g., PGA); (B) Example of scaffold obtained by fiber bonding. *From Mikos, A.G., Bao, Y., Cima, L.G., Ingber, D.E., Vacanti, J.P., Langer, R., 1993. Preparation of poly(glycolic acid) bonded fiber structures for cell attachment and transplantation. J. Biomed. Mater. Res. 27(2), 183–189.*

## SOLVENT CASTING/PARTICULATE LEACHING

To overcome the drawbacks related to the fiber bonding technique, a SC/PL technique has been developed. SC/PL is one of the most common methods used for the preparation of porous scaffolds. The SC/PL method is totally based on the dispersion of a porogen (e.g., salt, sugar, gelatin, and wax) in a polymeric solution or powdered materials, followed by the porogen leaching. Porogen can be ground into small particles that have a size adequate for pores in the scaffold. In fact, the porogen acts as placeholder for pores and interconnection of the pores in the scaffolds. Highly porous scaffolds with porosity up to 95% and pore diameter up to 500 µm can be prepared by using this technique (Fig. 8.15A). Porous constructs of synthetic biodegradable polymers could be prepared; for example, SC/PL is studied for PLLA and PLGA scaffolds, but it could be applied to any other polymer that is soluble in a solvent, such as chloroform or methylene chloride.

The main objective of this technique is the realization of bigger pore size and increased pore interconnectivity. The main advantage of this technique is its simplicity, versatility, and ease of controlling the pore size and geometry (Fig. 8.15B). Pore geometry is controlled by the selection of the shape of the specific porogen agent, and pore size is controlled by sieving the porogen particle to the specific dimensional range. One of the main drawbacks of this technique is that it can only produce thin scaffolds or membranes up to 3 mm thick; it is very difficult to design the scaffolds with accurate and controlled pore interconnectivity due to the difficulty in leaching all the porogen from the polymer.

However, leaching of salt crystals, typically sodium chloride, often results in poor interconnection and irregular pore shape. To overcome these drawbacks, different modifications of this technique have been proposed, such as leaching of microspheres and the partial fusion of particles before casting, respectively, to overcome poor pore shape and to improve interconnection.

A variation of this traditional approach consists in pouring the NaCl porogen particles with the appropriate size into a mold, and then a polymer solution is cast into the porogen-filled mold. After the evaporation of the solvent, the salt crystals are leached away using water to form the pores of the scaffold.

Another possible approach based on SC/PL consists of the use of porogen agents different from NaCl crystals. In fact, the shape of the pore can influence cells' adhesion and proliferation; for that reason, the possible use of spherical particles to be used as porogen agents has been investigated. Among possible alternative materials to NaCl, ice, sucrose, gelatin, and paraffin in the shape of microspheres have been proposed. As examples, the approaches based on the use of paraffin and gelatin porogen agents are described in the scheme reported in Fig. 8.16 and in the following section.

The first step is the preparation of the microspheres to be used as a porogen agent:

- *Paraffin microspheres.* Paraffin microspheres can be obtained as reported in Fig. 8.17A. Paraffin is melted at 65°C and poured into a glass beaker containing water and polyvinylpyrrolidone kept at 70°C on a magnetic stirrer. The emulsion is then vigorously stirred, and paraffin microspheres are solidified by rapid cooling, obtained by adding ice-cold water. The microspheres are then washed with deionized water, dried at room temperature, and kept in a desiccator until use.
- *Gelatin microspheres.* Gelatin microspheres are prepared by dissolving gelatin in water (Fig. 8.17B) at a very high concentration, and the solution is added to a beaker containing soybean oil heated to 60°C under stirring. After a few minutes, the emulsion is cooled down to 15°C in an ice bath, and gelatin droplets are dehydrated by adding cold acetone. Microspheres can be then removed from the

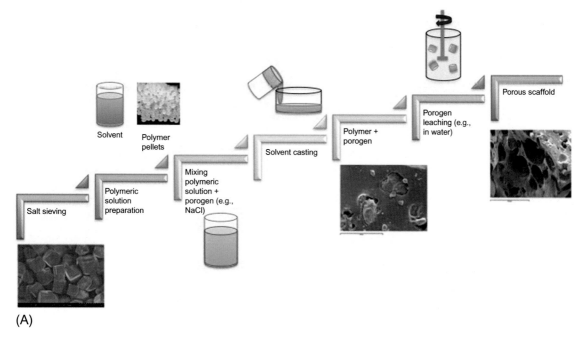

(A)

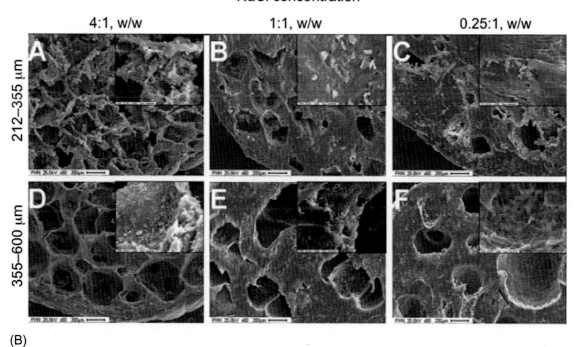

(B)

**FIG. 8.15** (A) Scheme of the different steps in a solvent casting/particulate leaching technique. (B) The effect of different NaCl crystal sizes on scaffold pore dimension. PCL-HA scaffolds examined by SEM prepared by varying NaCl size and concentration. (A, B, C) Scaffold prepared using NaCl porogen particles with 212–355 μm size. (D, E, F) Scaffold prepared with 355–600 μm particles. The insert in each image represents high magnification (1000 ×) of the images (magnification 60 ×). (A and D) Scaffolds prepared with 4:1 NaCl concentration; (B and E) scaffolds prepared with 1:1 NaCl concentration; (C and F) scaffolds prepared with 0.25:1 NaCl concentration. *Modified from Yu, H., Matthew, H.W., Wooley, P.H., Yang, S.Y., 2008. Effect of porosity and pore size on microstructures and mechanical properties of poly-epsilon-caprolactone-hydroxyapatite composites. J. Biomed. Mater. Res. Part B Appl. Biomater. 86(2), 541–547.*

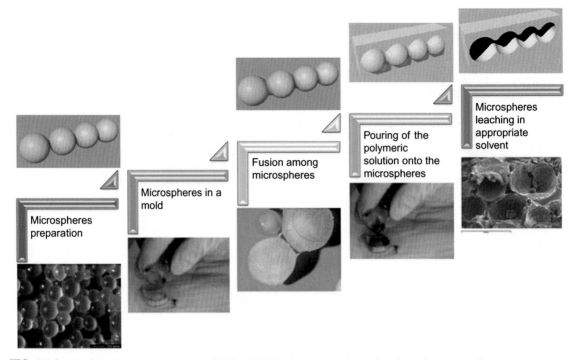

**FIG. 8.16**    Steps for the preparation of scaffold by SC/PL when microspheres fused together are used as a porogen agent.

soybean oil, washed several times with acetone, and dried.

After microsphere preparation, paraffin and gelatin microspheres are sieved to obtain different ranges of particle dimensions and leveled in a glass mold to the thickness desired for the scaffold. Then, to achieve paraffin adhesion, the mold containing the microspheres is placed in an oven at 40–50°C until satisfactory adhesion of microspheres is observed. A similar procedure can be used for gelatin microspheres, but the molds have to be kept in an oven for up to 96 h with appropriate relative humidity at about 50°C. The polymer with which the scaffold can be obtained (e.g., PLA, PLA-TMC) is dissolved in an appropriate solvent (e.g., acetone, chloroform) at the desired concentration. Polymer solution is then cast into the mold containing the paraffin or gelatin microspheres. After vacuum drying, specimens are punched out from the sheets

(Fig. 8.18A) before dissolving the porogen by washing the scaffolds several times with hexane (for paraffin) or deionized water (for gelatin). Scaffolds prepared with this approach can exhibit higher pore interconnection (Fig. 8.18B) than using the previously described SC/PL methods, due to the partial fusion among the microspheres and the amount of microspheres placed in the mold. Moreover, the pore dimension can be selected by sieving the microspheres in the appropriate dimension range.

### FREEZE DRYING

A freeze drying method is very useful for fabricating porous scaffold. Sublimation is the main aspect to be considered in this technique. The main steps of the process are reported in Fig. 8.19A; in particular, a solution with the desired polymer concentration is prepared, then lyophilization is involved to remove the

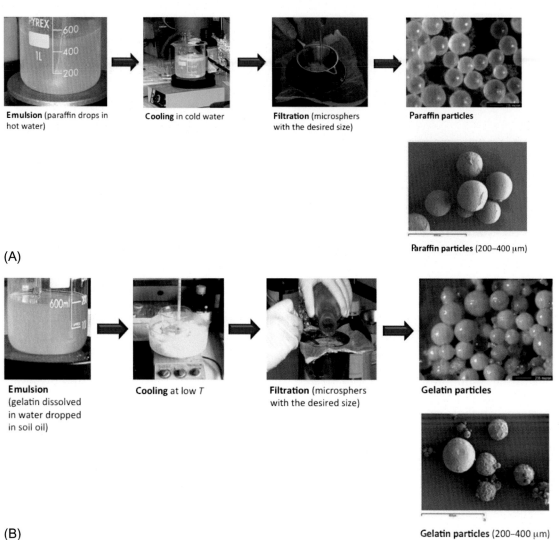

**Emulsion** (paraffin drops in hot water)

**Cooling** in cold water

**Filtration** (microsphers with the desired size)

**Paraffin particles**

**Paraffin particles** (200–400 μm)

(A)

**Emulsion** (gelatin dissolved in water dropped in soil oil)

**Cooling** at low *T*

**Filtration** (microsphers with the desired size)

**Gelatin particles**

**Gelatin particles** (200–400 μm)

(B)

**FIG. 8.17** Scheme of the steps for the preparation of (A) paraffin microspheres and (B) gelatin microspheres.

solvent once the solution is frozen in a mold. By controlling the rate of freezing and the pH of the solution, the pore size is controlled as well; in particular, small pores are a result of a fast freezing rate. The advantage of this technique is that no need of high temperature or pressure is needed. This technique is applied to a number of different polymers including natural polymers (e.g., silk fibroin, collagen, gelatin) and synthetic polymers (e.g., PGA, PLA, PLGA).

A modification of this technique is the combination of emulsion and freeze drying (Fig. 8.19B); this technique is based on immiscibility, such as with water and oil. The polymer is dissolved into a solvent and then a nonsolvent is added. The solution is then mixed to form a homogeneous emulsion mixture. This mixture is poured into

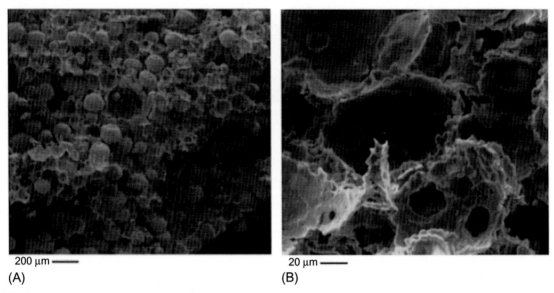

200 µm ———        20 µm ———
(A)                                            (B)

**FIG. 8.18**  Examples of polymeric scaffold obtained by SC/PL when microspheres are fused together and polymeric solution is poured onto them. (A) Microspheres embedded in the polymer (after solvent casting step); (B) polymeric porous scaffold after microsphere leaching.

a mold and quenched under low temperature or by using liquid nitrogen. Freeze drying allows the sublimation of both solvent and nonsolvent, so that porous scaffolds are produced. The pores are highly interconnected, which is advantageous for nutrient supply, metabolic waste clearance, cells ingrowth, and vascularization.

## PHASE SEPARATION

Thermally induced phase separation (TIPS) has become popular to fabricate scaffolds for TE. Scaffold fabrication using a phase separation technique requires a temperature change that separates the polymeric solution in two phases; one polymeric solution has a low polymer concentration (polymer lean phase), and the other one has a high polymer concentration (polymer rich phase). The separation of liquid-liquid phase occurs by lowering the temperature of the solution, and two solid phases are formed by quenching (Fig. 8.20). The

solvent is removed by extraction, evaporation, and sublimation processes, forming a porous scaffold.

By altering the types of polymer and solvent, polymer concentration, and phase separation temperature, different types of porous scaffolds with micro- and macrostructured morphology can be produced (Fig. 8.21). Depending on the thermodynamics and kinetic behavior of the polymer solution under certain conditions, TIPS can be a complicated process. As seen in Fig. 8.20, there are two curves, the bimodal curve that represents the thermodynamic equilibrium of liquid-liquid demixing and the spinodal curve. When the temperature of a solution is above the binodal curve, the polymer solution is homogeneous. The maximum point, at which both the binodal and the spinodal curves merge, is the critical point of the system. The area under the spinodal curve is the unstable region, and the area located in the zone between the binodal and spinodal curves is the metastable region.

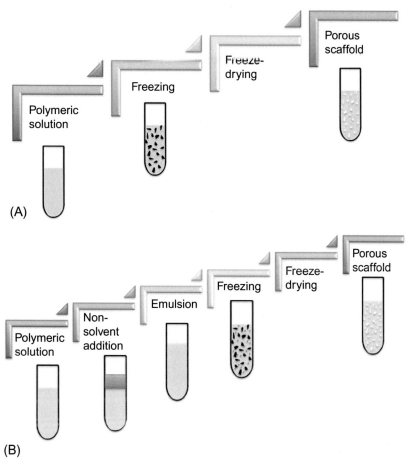

**FIG. 8.19** Scheme of the different steps in (A) freeze drying and (B) emulsion/freeze drying technique.

## GAS FOAMING

To eliminate the use of organic solvents and/or high temperature in the pore-making process, a technique involving gas as a porogen can be used. In fact, the residues of organic solvent that remain after the previously described processes can damage cells and nearby tissues. Different approaches are reported in literature, all based on the use of gas or porogen agents that can create gas as a blowing agent. However, also in this case, it is not possible to incorporate cells during the scaffold production.

*Approach A.* In this approach, a gas foaming technique uses high-pressured carbon dioxide gas for the fabrication of highly porous scaffolds. The porosity and porous structure of the scaffolds depend upon the amount of gas dissolved in the polymer. The process started with the formation of solid discs of synthetic materials (e.g., PGA, PLLA, or PLGA) by compression molding. Then, the polymer discs are placed in a chamber and exposed to carbon dioxide at high pressure ($P = 800\,psi$) to saturate the polymer with gas.

**FIG. 8.20** Phase diagram of a typical ternary solution. The thermodynamic equilibrium boundaries are represented by the binodal and the spinodal curve. At high temperatures (above the binodal curve), the polymer solution is in the one-phase region, and the solution is homogeneous. Upon decreasing the temperature, the solution reaches the binodal curve, and demixing into a polymer-rich and polymer-lean phase occurs. The area between the binodal and spinodal curves is the metastable region, where a (quite slow) mechanism of nucleation and growth takes place, leading to (relatively) large pores. Conversely, the separation mechanism, which takes place in the spinodal region and is called spinodal decomposition, is very fast and usually leads to the formation of a network of small interconnected pores.

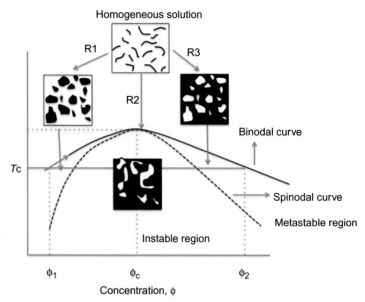

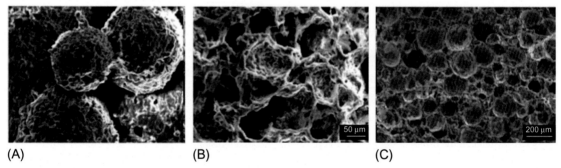

**FIG. 8.21** SEM images of PU scaffolds obtained (A) at the metastable zone in the left side of the phase diagram, (B) at the instable zone, and (C) at the metastable zone in the right side of the phase diagram.

After that, pressure is rapidly decreased to atmospheric pressure. Under this condition, dissolved $CO_2$ becomes unstable and minimizes the free energy; as a result, pore nucleation occurs. These pores cause significant expansion of polymeric volume and a decrease in polymeric density. The disadvantage is mainly the low pores' interconnection, especially on the surface of the scaffold, where a compact skin is formed. Although the fabrication method requires no leaching step (as, for example, in SC/PL) and uses no harsh chemical solvents, the high pressure involved in porous scaffold formation does not allow the incorporation of cells or bioactive molecules. Moreover, cell migration throughout the scaffold could be difficult due to the unconnected pore structure. *Approach B.* To produce an open pore morphology using this technique, both a gas foaming and particulate leaching technique can be used together using two salts,

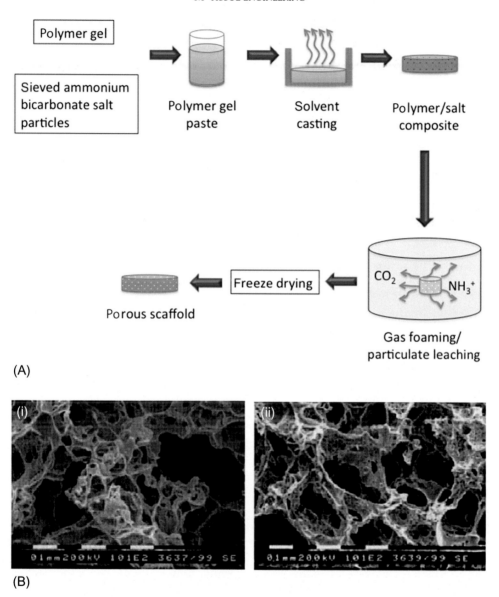

FIG. 8.22 (A) Scheme of the preparation of porous scaffolds by gas foaming/particulate leaching technique; (B) example of scaffolds obtained using this technique: (i) surface and (ii) cross-section. *Modified from Yoon, J.J., Park, T.G., 2001. Degradation behaviours of biodegradable macroporous scaffolds prepared by gas foaming of effervescent salts. J. Biomed. Mater. Res. 55(3), 401–408.*

ammonium bicarbonate and citric acid (Fig. 8.22A). Ammonium bicarbonate salt particles (in the range 100–200 or 300–500 μm) and a polymer (e.g., poly(D,L-lactic-*co*-glycolic acid), PLGA) are mixed together, then compact disks are obtained by compression or solvent casting, immersed in an acidic aqueous solution (i.e., aqueous solution of citric acid), so that the salt particles are then removed by leaching. At this step, the

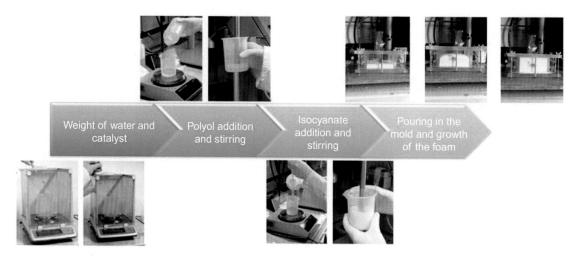

FIG. 8.23    Scheme of the preparation of polyurethane porous scaffolds by gas foaming.

porogen agent reacts with the solvent, and $CO_2$ and $NH_3^+$ are released, promoting pore formation (Fig. 8.22A). This combination produces a porous scaffold with an open, interconnected morphology (Fig. 8.22B) without the use of any organic solvents. *Approach C.* If polyurethane is used as a material to fabricate a porous scaffold by gas foaming, a different approach has to be considered, as already described in Chapter 3, Section 3.2.2.1. Polyurethane foams can be synthesized with a one-step bulk polymerization process, by adding the expanding agent (i.e., water) to a premixed quantity of polyols with a catalyst (e.g., Fe-acetyl-acetonate). The last step is the addition of the diisocyanate (e.g., prepolymer methylene diphenyl diisocyanate, MDI) that causes the starting of the foaming reaction. The reaction between water and prepolymer MDI produces carbon dioxide ($CO_2$), which is the porogen agent that creates the porous structure. The procedure is schematically reported in Fig. 8.23:

- the appropriate amount of water is added to the polyol-catalyst mixture
- the weighed amount of polyols is added to the mixture and mixed to homogenize the reaction mixture
- the appropriate quantity of polymeric MDI is then added, the reaction mixture is appropriately stirred, and poured into a closed mold to allow the expansion of the foam into a fixed volume

The mold is firmly closed, and the expanding reaction is allowed to take place at room temperature. After 72h, the foam is extracted from the mold and the compact skin on the surface (Fig. 8.24A) is cut, so that the porous polyurethane scaffolds are obtained (Fig. 8.24B).

This technique is applicable to polyurethane and is an easy and low-cost technique that allows obtaining scaffolds with interconnected pores; the dimension of the pores can be set by varying the quantity of water used as an expanding agent and the reaction mixture quantity poured into the mold.

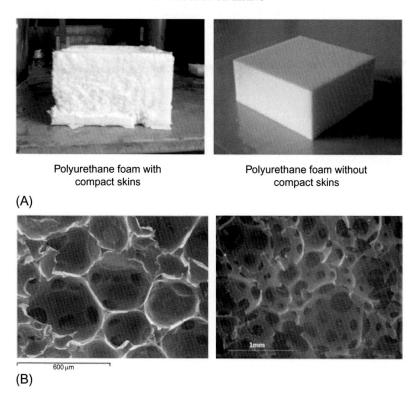

Polyurethane foam with compact skins

Polyurethane foam without compact skins

(A)

(B)

FIG. 8.24    (A) Polyurethane foam after the extraction from the mold with compact skin on the surface (on the *left*) and polyurethane foam after skin removal (on the *right*); (B) SEM images of polyurethane foams obtained by gas foaming.

## ELECTROSPINNING

Electrospinning (see Chapter 3, Section 3.2.4) is a method to create porous scaffolds composed of fibers. To process solutions or melt polymers into continuous fibers with diameters ranging from nanometers to submicrometers, electrospinning is a highly versatile technique. Many biocompatible polymers are used for electrospinning, either synthetic (e.g., PGA, PLGA, PCL) or natural ones (e.g., silk fibroin, collagen, chitosan, gelatin). Moreover, the process does not require the use of coagulation chemistry or high temperature for fiber generation and bonding. The ultrafine fibers in the obtained electrospun mats have high surface-to-volume ratios and porosity higher than 90%, and mimic the natural extracellular matrix of different body tissues.

This process (Fig. 8.25) is controlled by high intensity electric field (voltage around 10–15 kV) between two electrodes having electric charges of opposite polarity. The polymer solution is loaded into a syringe with a metal capillary connected to one electrode, and the other is connected to the metal collector. Subjected to the generated electric field, the polymer solution, loaded in a syringe, is expelled at a constant rate through a metal capillary, so that the polymeric jet travels in the field, becoming thin fibers and deposited onto the conductive fiber collector. For obtaining different thicknesses of fiber mat, fiber size, and orientation, process parameters have to be considered (e.g., polymer concentration, solvent, ejection rate, applied voltage, distance between collecting plate and capillary).

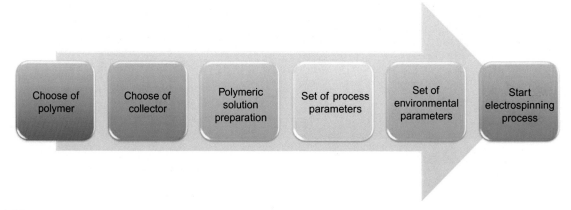

FIG. 8.25  Scheme of the preparation of electrospinning nanofibrous scaffolds.

One of the main advantages of electrospinning is that it can produce the scaffold with structural features suitable for cells' colonization, proliferation, and subsequent tissue organization. In fact, it is possible to produce electrospun fibers with designed orientation, high aspect ratio, high surface area, and having control over pore geometry (Fig. 8.26). These characteristics are favorable for better cell growth for in vitro and in vivo studies and applications, because they directly influence the cell adhesion, cell expression, and transportation of oxygen and nutrients to the cells. Electrospun scaffolds are widely used in RM, TE, wound dressing, and drug release membranes, because of the possibility of selecting and adjusting the combination of proper component ratios, properties of electrospun scaffolds for customizing the set-up, and to obtain scaffolds with the desired function.

### 8.1.4.2 Designed Manufacturing Techniques

Designed manufacturing techniques are also called additive manufacturing (AM) techniques (see Chapter 3, Section 3.7) and represent more advanced techniques for scaffold fabrication. AM is an efficient approach for the production of scaffolds with a desired property; another advantage of the AM technique is to produce the parts with highly reproducible architecture and compositional variations. AM has advantages compared to nondesigned manufacturing techniques (Section 8.1.4.1); for example, it can control matrix architecture (e.g., size, shape, pores interconnectivity, orientation), yielding a biomimetic structure. Moreover, it has the ability to control the mechanical properties, biological effects, and degradation kinetics of the obtained scaffolds.

In the next section, some techniques already described in Chapter 3 will be discussed. Hence, here some examples of materials and scaffolds that can be obtained with the most used techniques.

*Fused Deposition Modeling* (*FDM, Chapter 3, Section 3.7.3.3*). FDM printing for the fabrication of scaffold is based on the use of thermoplastic polymers, such as polylactic acid, polycaprolactone, polyvinyl alcohol, and biodegradable polyurethane, with adequate rheological properties. An advantage of FDM-based techniques is its high resolution, which allows production of complex scaffolds. Moreover, FDM offers good mechanical strength and the option to obtain gradient porous scaffolds to mimic

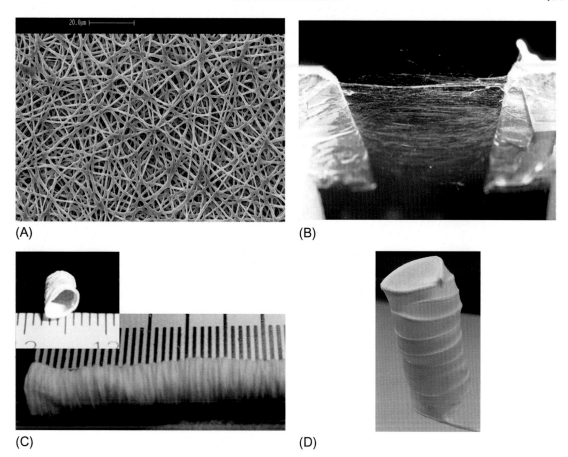

**FIG. 8.26** (A) SEM image of electrospun polyurethane with random orientation; (B) macroimage of oriented polycaprolactone (PCL) electrospun nanofibers obtained by using a parallel plates collector; (C) macroimage of tubular electrospun structure in silk fibroin as a scaffold for vascular vessel regeneration; (D) macroimage of 3D helical-designed electrospun scaffolds.

complex tissues. The main drawback is the limited thermoplastic material options with good melt viscosity properties for extrusion.

*Precision Extruding Deposition (PED)* represents a variation of FDM in which polymeric pellets are used instead of filaments. In fact, in FDM, filament fabrication represents a time-consuming step that can, in the case of biodegradable polyesters (e.g., PLA, PCL), possibly decrease in molecular weight, hence a variation in degradation rate. PED consists of a mini-extruder mounted on a high-precision positioning system (Fig. 8.27A). PED can be used with

bulk materials in pellet form, avoiding most of the material preparation steps in a filament-based system. This configuration opens up the opportunity for the use of a wider range of materials, making the PED a viable manufacturing process for composite scaffold materials (Fig. 8.27B).

*Low-temperature deposition manufacturing (LDM)* was proposed as an alternative technique to FDM because this process can better preserve bioactivities of scaffold materials because of its nonheating liquefying processing of materials. Natural biopolymers, such as collagen type, gelatin, sodium alginate, and chitosan, can be successfully printed

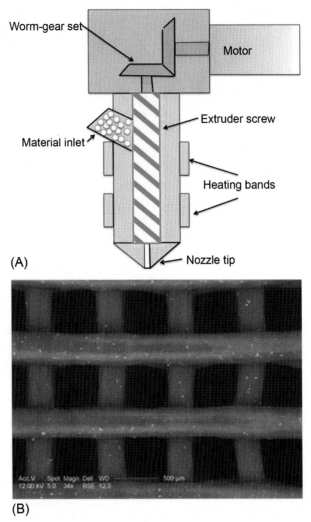

(A)

(B)

**FIG. 8.27** (A) Schematic of mini-extruder in precision extrusion deposition (PED) system; (B) SEM image of melt-blended PCL-HA scaffold fabricated by PED. *Modified from Shor, L., Güçeri, S., Wen, X., Gandhi, M., Sun, W., 2007. Fabrication of three-dimensional polycaprolactone/hydroxyapatite tissue scaffolds and osteoblast-scaffold interactions in vitro. Biomaterials 28(35), 5291–5297.*

without compromising their bioactivities. Synthetic polymers, such as poly(lactic-co-glycolic acid), polyurethane, poly(D,L-lactide), and poly(L-lactic acid), can be dissolved in an appropriate solvent and processed into a scaffold. In addition, inorganic particles, such as nano-hydroxyapatite, can be incorporated into the polymeric solution enhancing the biological and mechanical properties of the scaffolds. The LDM system builds scaffolds layer by layer on a platform in a low-temperature environment ($T < 0°C$) so that the printed layers are frozen on the platform (Fig. 8.28A). After the forming process, the frozen scaffolds are freeze dried to remove the solvent, obtaining the porosity in the scaffold (Fig. 8.28B).

*Stereolithography (SLA, Chapter 3, Section 3.7.3.1).* SLA is based on the principle of photopolymerization, in which free radicals

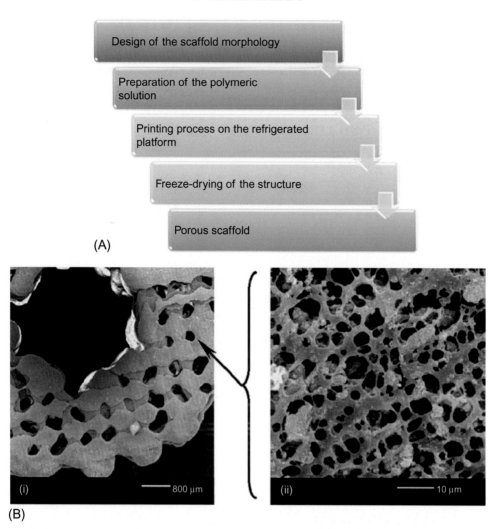

**FIG. 8.28** (A) Steps of low-temperature deposition manufacturing (LDM) system; (B) SEM image of melt-blended PCL-HA scaffold fabricated by PED. *Modified from Xiong, Z., Yan, Y., Wang, S., Zhang, R., Zhang, C., 2002. Fabrication of porous scaffolds for bone tissue engineering via low-temperature deposition. Scripta Mater. 46(10), 771–776.*

are released after the interaction between the photoinitiator and UV light (Fig. 8.29A). An important parameter in SLA is the thickness of the cured layer, which can change depending on the energy of UV light to which the polymer is exposed. It is important that the biomaterial is FDA-approved for human use. SLA stands out as a versatile technique, as the drug

and photopolymer can be mixed together prior to printing, so that the drug is entrapped in the polymeric matrix. Moreover, SLA has a higher resolution (Fig. 8.29B) compared to other AM techniques. Also, heating is minimized during printing, so thermolabile drugs can be used. The main disadvantage is that there are not many photopolymers available

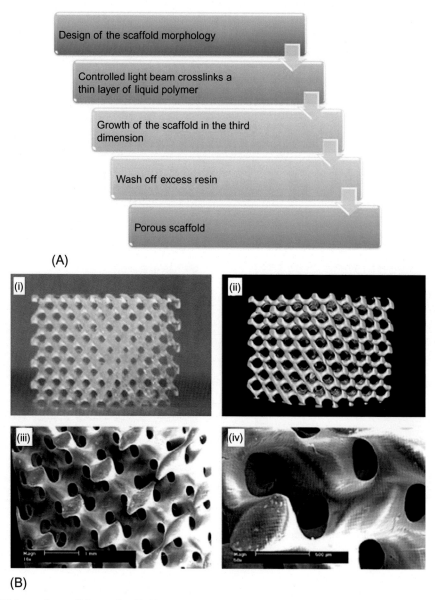

**FIG. 8.29** (A) Steps of stereolithography (SLA) system; (B) macroimage (i), micro-CT visualization (ii), and SEM images (iii and iv) of a PCL scaffold built by SLA. *Modified from Elomaa, L., Teixeira, S., Hakala, R. Korhonen, H., Grijpma, D.W., Seppälä, J.V., 2011. Preparation of poly(ε-caprolactone)-based tissue engineering scaffolds by stereolithography. Acta Biomater. 7, 3850–3856.*

for scaffolds because many of them are not considered safe for human applications. Polymers such as poly(ethylene glycol), PEG, poly(ε-caprolactone), PCL, poly(trimethylene carbonate), PTMC, poly(propylene fumarate), PPF, poly(D,L-lactide), and PDLLA have been used and functionalized with photoreactive groups like methacrylate (MA) for use in TE.

Materials such as PEGDA hydrogels have been used with cells with some good results, but they are not degradable, whereas the low elasticity and low glass transition temperatures of TMC, PCL, and PPF are not ideal for use in scaffolds.

## 3D BIOPRINTING

Recently, in the context of TERM applications, the definition of biofabrication as a research field was updated as "the automated generation of biologically functional products with structural organization from living cells, bioactive molecules, biomaterials, cell aggregates such as micro-tissues, or hybrid cell-material constructs, through bioprinting or bioassembly and subsequent tissue maturation processes".

Whereas AM methods (see Chapter 3) have demonstrated different degrees of success in fabricating 3D scaffolds to accommodate cells that can develop new tissues, most of them are incapable of simultaneously depositing biomaterials and cells. This disadvantage limits the possibility to mimic cell distributions in native tissues, particularly when multiple tissue interfaces are to be developed. Three main technologies have demonstrated the ability to incorporate cells during the process of AM into a biomaterial carrier:

– laser-induced forward transfer;
– inkjet printing (both thermal and piezoelectric);
– robotic dispensing.

The combination of biological molecules, comprised of bioactive factors, particles, matrixes, or living cells, and biomaterials takes the name of *bioink*. Due to the incorporation of biological elements within the polymer, foresight has to be given during the printing process and for the chosen polymer ink. In addition to this, the polymer viscosity has to be evaluated to not cause high shear stresses that could cause damage to cells embedded in the ink. Moreover, as the polymer solution contains biological material, it is important to also consider the

biological behavior response in terms of viability of the cells encapsulated in the gel.

Three-dimensional bioprinting represents a process in which the *bioink* (Table 8.3) is deposited simultaneously following a defined 3D pattern and shape through a bottom-up assembly. This technology opened new doors and made significant progresses toward recapitulating the complexity of human tissues. The combination of computer-aided design/computer-aided manufacturing systems with 3D bioprinting technology is expected to eventually enable the conversion of medical images into tissue constructs for patient-personalized tissue repair. Hence, the hydrogel matrix properties are key parameters that need to be optimized to enable a high-resolution printing process and cell survival, as well as to provide defined cell microenvironments for the targeted tissue-engineering approach.

Extrusion-based approaches are among the most popular techniques in 3D bioprinting. This technique is based on a "through contact" printing process, and it is usually used to produce scaffolds in which the constituent material shows thermoresponsive behavior. Moreover, it is based on the dispersion of a fluid through a mechanic (by screw or piston) or pneumatic control system,

**TABLE 8.3** Examples of Bioinks That Can be Used in 3D Bioprinting Approaches

| Material | Printing Mode |
| --- | --- |
| Alginate | Extrusion-based |
| | Inkjet-based |
| Gelatin | Laser-based |
| | Extrusion-based |
| Silk | Extrusion-based |
| Hyaluronic acid | Extrusion-based |
| Fibrin | Inkjet-based |
| Collagen | Extrusion-based |

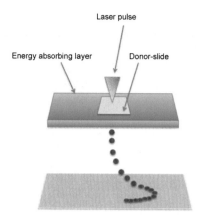

Laser pulse

Energy absorbing layer          Donor-slide

**FIG. 8.30**    Laser-induced forward transfer technique.

coupled with an automatic robotic system, that allows for the precise filament deposition during the printing process. The most widely used bioprinting techniques can be grouped into three categories: laser-assisted, extrusion-based, and inkjet-based systems.

## LASER-INDUCED FORWARD TRANSFER

Among laser-based systems, stereolithography and two photon polymerization are two techniques that rely on a laser source to induce the polymerization of the material for the fabrication of precise patterns in a volume that contains the polymer. Laser-induced forward transfer (LIFT) technology is part of the same category, with the difference that the laser source used in this technique is needed to induce local evaporation of an energy-absorbing layer (i.e., donor side covered with a laser energy-absorbing layer and a layer of cell-containing bioink) that, in turn, generates high gas pressure propelling the bioink droplets from the donor slide toward a collector side (Fig. 8.30). Because it's a nozzle-free approach, it is not affected by clogging issues; it has a good resolution, requires small material volumes, and allows for the precise deposition of materials and high densities of cells without negatively affecting viability or cellular function.

## INKJET SYSTEM

In inkjet printing, drop on demand is the most used inkjet printing mode, where the material is ejected from the cartridge in a shape of separated single droplets. In the biofabrication field, it is defined as dispensing through a small orifice and precise positioning of very small volumes (e.g., 1–100 picolitres) of bioink on a substrate by an actuator that can be thermal, piezoelectric, or electrostatic; the first two are most commonly used. For thermal inkjet (Fig. 8.31A), small volumes of the printing fluid are vaporized by a microheater to form the pulse that expels droplets from the print head. The generated heat and resulting evaporation result in stress for the deposited cells. In piezoelectric inkjet printing (Fig. 8.31B), no heating is used, but a direct mechanical pulse is applied to the fluid in the nozzle by a piezoelectric actuator, which causes a shockwave that forces the bioink through the nozzle. Inkjet printing has successfully been applied for accurate deposition of cells. One of the main restrictions of inkjet technology is the low upper limit of the viscosity for the ink (about $0.1\,Pa\,s^{-1}$). This is due to the fact that small droplets of this ink are deposited onto a substrate at high velocity, and the low viscosity promotes spreading of the droplet on the surface.

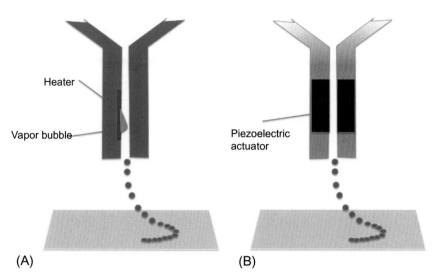

**FIG. 8.31** Scheme of inkjet printer: (A) thermal inkjet printer; (B) piezoelectric inkjet printer.

### ROBOTIC DISPENSING

An alternative approach for the design and fabrication of organized 3D hydrogel constructs is based on dispensing systems. Hydrogels loaded with cells are inserted in disposable plastic syringes and dispensed onto a platform. Different systems can be used for dispensing the bioink (Fig. 8.32): pneumatic, piston-driven, or screw-driven. Rather than single droplets, robotic dispensing yields larger hydrogel strands. To maintain the shape of the constructs after printing, hydrogels with higher viscosities are often used. Fabrication speed using robotic dispensing is significantly higher compared to the other systems, and anatomically shaped constructs have successfully been fabricated. *Piston-driven* deposition provides more direct control over the flow of the hydrogel from the nozzle due to the delay of the compressed gas volume in the pneumatic systems. *Screw-based* systems are beneficial for the dispensing of hydrogels with higher viscosities.

Cells have been deposited with high viability and no notable effects on differentiation capacity using both pneumatic and piston driven systems. Screw extrusion can generate larger pressure drops at the nozzle, which can potentially be harmful for embedded cells.

### SOFT LITHOGRAPHY

Topographic and chemical surface patterning of a substrate is recognized as a powerful tool for regulating cell function; controlling the scale and pattern would significantly help the development of specific cell-regulating cues in different biomedical applications such as TE, implants, cell-based biosensors, microarrays, and cell biology. In particular, for the regeneration of some tissues (e.g., skeletal muscle), cell orientation is very important to guide cells during the formation of the desired tissue. Cells can be induced to align in a specific direction following physical or chemical stimuli. Soft lithography, named for its use of soft, elastomeric elements in pattern formation, represents a nonphotolithographic strategy for carrying out micro- and nanofabrication. It provides a convenient, effective, and low-cost method for the formation and manufacturing of micro- and nanostructures. The name *soft lithography*

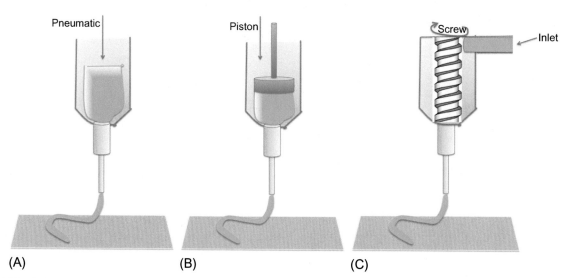

**FIG. 8.32** Scheme of inkjet printer: (A) pneumatic driven dispenser; (B) piston-driven dispenser; (C) screw-driven dispenser.

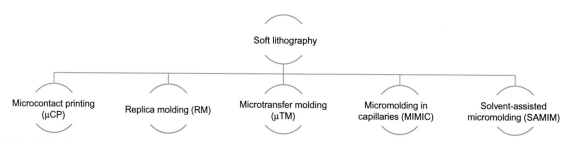

**FIG. 8.33** Classification of soft lithographic techniques.

does not cover one specific method but a group of techniques with the common feature that, at some stage of the process, an elastomeric ("soft") material is used to create the chemical structures. Five techniques can be primarily identified in soft lithography, as shown in Fig. 8.33. In each of them, the starting point is a topographically structured master (usually in silicon); this master is traditionally produced by photolithography. Then, the stamps are formed by casting the elastomeric polymer (usually polydymethylsiloxane, PDMS) over the topographic master, producing the corresponding replica (Fig. 8.34).

*Microcontact printing ($\mu CP$)* is an efficient method for pattern transfer, in which a conformal contact between the stamp and the surface of the substrate is the key to its success. In most of the applications, the PDMS stamps are used after an oxygen/air plasma treatment to facilitate the adsorption of the ink to the stamp surface. Specific functionalization of the stamp surface can offer additional possibilities. Once the PDMS stamp is obtained, the following step is the "inking" of the stamp. The ink is the molecular layer that is transferred during the printing process from the stamp surface to the substrate (Fig. 8.35A). Prior to transfer,

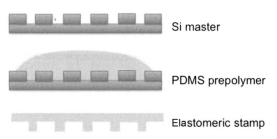

**FIG. 8.34** Fabrication of the PDMS stamp from the mold obtained by photolythography.

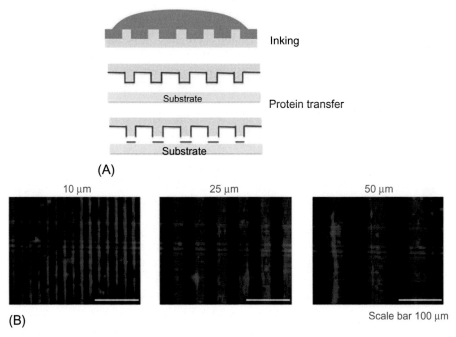

**(A)**

**(B)**

**FIG. 8.35** Microcontact printing technique: (A) step of the process: inking step followed by transfer of protein on the substrate; (B) images at fluorescence microscope of fibronectin stripes obtained by micro-CP onto a biodegradable polymeric substrate (PLA-TMC).

the stamp with the inking solution is incubated and then the excess solution is dried; after that, the stamp and the substrate are put in conformal contact, applying a light pressure to ensure the transfer of the ink from the stamp to the substrate (Fig. 8.35B).

*Replica molding (RM)* is an efficient method for the duplication of the information (i.e., shape, morphology, and structure) present in the surface of a mold. The polymer can be dissolved in an appropriate solvent, or UV- or thermally curable prepolymers can be used, as they do not contain solvent (Fig. 8.36). A possible shrinkage of less than 3% on curing can be obtained; the cured polymers, therefore, possess almost the same dimensions and topographies in the PDMS mold. The fidelity of this process is largely determined by van der Waals interactions, wetting, and kinetic factors such as filling of the mold.

## 8.1.5 The Cell Types

The production of an engineered tissue in vitro requires the use of cells to populate matrices and produce matrix resembling that of the native tissue (Howard et al., 2008). The main successes in this field have come from the use of primary cells taken from the patient and used in conjunction with scaffolds to produce tissue for reimplantation. However, this strategy has limitations, because of the invasive nature of cell collection and the potential for cells to be in a diseased state. Therefore, attention has become focused upon the use of stem cells, including embryonic stem (ES) cells, bone marrow mesenchymal stem cells (BM-MSCs), and umbilical cord-derived mesenchymal stem cells (UC-MSCs). One of the critical steps of using stem cells for RM is the ability to control the differentiation of the cells to the desired tissue lineages.

Before describing the pros and cons of stem cells, we consider cell lines and primary cells.

### 8.1.5.1 Cell Lines

Cell lines are able to proliferate "in vitro" for long periods and can be cryopreserved in a cell bank. Most of the fundamental phenotypic changes occur initially after the initial isolation from the tissue of origin, and the cell line is thus more homogeneous, stable, and reproducible than a heterogeneous population of primary cells.

These types of cells are named "immortalized" and are very useful for cytotoxicity and cytocompatibility in vitro assays (described in Chapter 7, Section 7.3.1). Cell lines are cost-effective, easy to use, provide an unlimited supply of material, and bypass ethical concerns associated with the use of animal and human tissue; they also provide a pure population of cells, thus consistent samples and reproducible results. However, despite being a powerful tool, care must be paid in interpreting the results because cell lines are genetically manipulated and may not adequately represent primary cells.

The downside of most continuous cell lines is that they do not represent the real situation "in vivo." They have spread uncontrollably all over the world, creating immense potential for misidentification and the introduction of contaminants. It is therefore advisable that continuous cell lines be acquired from internationally recognized cell and tissue banks, such as the American Type Culture Collection (ATCC; https://www.lgcstandards-atcc.org/?geo_country=it).

### 8.1.5.2 Primary Cells and Tissue Cultures

Primary cultures isolated from animals or humans represent heterogeneous populations compared, for example, to the type of cell and state of differentiation. Each isolated sample is unique, and it is impossible to reproduce it exactly. The de-differentiation process starts from the moment the cells are separated from the parent tissues, and therefore a primary cell culture is a system dynamic, in constant state of modification. Primary cell cultures require complex nutrient media, supplemented with animal serum and other undefined components and are difficult to standardize. It is therefore necessary to rely on supply from institutions or research centers of proven experience.

Primary cells are cells isolated directly from living tissue (e.g., biopsy) and established for growth in vitro. These cells have undergone very few population doublings and are therefore more representative of the main functional component of the tissue from which they are derived in comparison to cell lines.

Before in vivo studies, mouse or rat cells can be used to refine doses and reduce the number of animals required for preclinical toxicology. Human cells can be used to determine accuracy of extrapolating human data from an animal model (www.thermofisher.com).

### 8.1.5.3 Stem Cells

The tissues of the human body are composed of various cell types, each with specific characteristics. When the embryo reaches the blastula

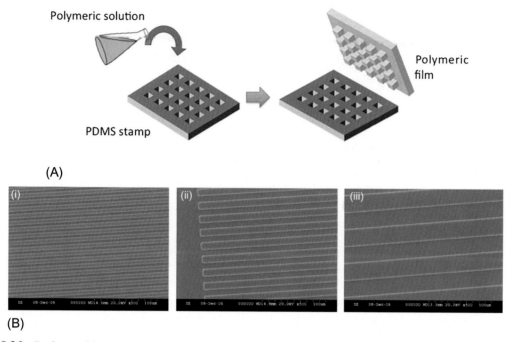

**FIG. 8.36** Replica molding technique: (A) step of the process: pouring of the polymeric solution onto the PDMS stamp followed by solvent casting and extraction of the microstructured polymer from the PDMS stamp; (B) images at SEM of microstructured substrates in PLA-TMC: (i) depth $= 1\,\mu m$, width $= 5\,\mu m$; (ii) depth $= 1\,\mu m$, width $= 10\,\mu m$; (iii) depth $= 1\,\mu m$, width $= 25\,\mu m$.

stage during development, each of its cells is called *totipotent* because it is able to give rise to any of the 254 different cell types, from which all the adult tissues will derive. Continuing the development, these first become *pluripotent* (or *multipotent*) and finally *unipotent*, that is, specialized cells.

The transition from pluripotent to unipotent cells does not occur for all cells during embryonic development. Totipotent cells, also called *embryonic stem cells*, are able to give rise to all the tissues of the human organism.

The tissues that supply stem cells are: umbilical cord blood, embryos in the early stages of development, tissues of the fetus, and more recently, placenta and amniotic fluid.

The adult tissues suppliers of stem cells are: bone marrow, blood, adipose tissue, nervous system, muscle, and liver tissues.

If the ethical problems related to the use of embryonic cells make it difficult to use for TE, it is desirable to use adult stem cells (mesenchymal), which offer significant advantages compared to primary autologous cells, often already endowed with poor proliferative abilities in a healthy individual.

*Embryonic Stem Cells (ES).* ES cells could allow the production of type-matched tissues for individual patients, either through stem cell banking or by the use of therapeutic cloning. ES cells have the ability to be maintained for long (theoretically indefinite) culture periods, therefore potentially providing large amounts of cells for tissues that could not be derived directly from a tissue source. Proof of the true pluripotent nature of ES cells is teratoma formation. The use of stem cells will therefore require a method to ensure differentiation, by in vivo

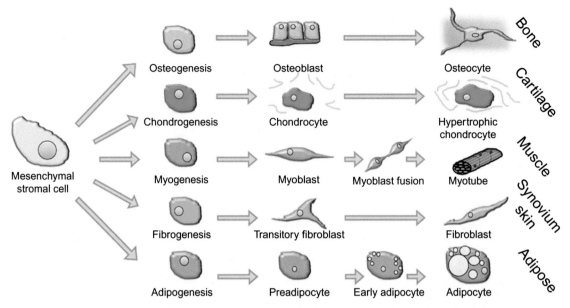

**FIG. 8.37** The cells and tissues that arise from bone marrow-derived mesenchymal stromal cells. *From Fig. 1 in Hardy, R., Cooper, M.S., 2011. Glucocorticoid-induced osteoporosis—a disorder of mesenchymal stromal cells? Front. Endocrinol. 2, 24. doi:https://doi.org/10.3389/fendo.2011.00024.*

demonstration of an absence of teratoma formation (Howard et al., 2008).

*Bone Marrow-derived Mesenchymal Stem Cells (BMSCs).* The MSC cell population can be isolated as a fraction of the adherent bone marrow colony-forming units; they are fibroblastic and can be differentiated to the osteogenic and other lineages, as shown in Fig. 8.37.

BMSCs are generally isolated from bone marrow aspirates harvested from the superior iliac crest of the pelvis in humans, although other sources such as the tibial and femoral marrow compartments and the thoracic and lumbar spine are also available (Tae et al., 2006). Although they represent a minor fraction of the total nucleated cell population in marrow, MSCs can be plated and enriched using standard cell culture techniques.

*Umbilical Cord-derived Mesenchymal Stem Cells (UC-MSCs).* Since the discovery that umbilical cord blood contains MSCs that can undergo multilineage differentiation, much research has been focused on determining their applications. The analysis of their gene expression profile reveals similarities to BM-MSCs, with an ability to differentiate into adipocytes, osteoblasts, hepatocytes, and neuronal-like cells (Howard et al., 2008). This source of stem cells may provide a large pool of material, which can be purified using noninvasive techniques. In addition, as the differentiating potential of BMSCs may decrease with age, an alternative such as UC-MSCs can be of huge value.

*Adipose Tissue-derived MSC (ADSCs).* Adipose tissue stromal cells contain adipocyte progenitors at various stages of maturity, including the stromal-vascular (SV) cells, a class of fat-resident-committed mesenchymal progenitors, exhibiting at least a bipotent differentiation potential, because they can differentiate into adipocytes or chondrocytes (Tae et al., 2006).

Stem cells from adipose tissue, variously referred to as processed lipoaspirate (PLA) cells, have been shown to have similar differentiation potentials. There is a little difference between cells from marrow and fat in terms of yield, growth kinetics, cell senescence, multilineage differentiation capacity, and gene transduction efficiency. The utility of these cells in therapeutic applications may then depend on the availability of tissue specimens and the ease of in vitro expansion. However, works on the osteogenic and chondrogenic potentials of adipose-derived MSCs showed inferior capacity for osteogenesis or chondrogenesis compared with bone marrow-derived MSCs.

### 8.1.5.4 Induced Pluripotent Stem Cells

The possibility to create stem cells from nonembryonic cells through the process of reprogramming lifted the ethical limitation of embryonic stem cells and is driving stem cell research in a new direction.

The first induced pluripotent stem cells (iPSCs) were generated in 2006 (Takahashi and Yamanaka, 2006) from mouse embryonic fibroblasts in the lab of Dr. Shinya Yamanaka, work that earned him the Nobel Prize in Physiology or Medicine in 2012 (along with John B. Gurdon) "for the discovery that mature cells can be reprogrammed to become pluripotent".

Although many stem cell researchers dream of iPSCs being used to directly treat disease, that path has not been clear. With only one clinical trial to date, Yamanaka is instead focusing efforts on establishing an iPSC cell bank. However, the clinical utility of iPSCs, right now, is not as exciting as their role in research and their usefulness in modeling disease.

The method of reprogramming has changed significantly in the last decade. Although the original iPSC "cocktail" was quite simple, recent work has focused on further refining and simplifying this process (Fisher and Mauck, 2013). Instead of ectopic expression of several transcription factors, an alternative approach is to express specific micro-RNAs (miRNAs).

Another major issue limiting translation of iPSC technology has been the need for viral infection to introduce the necessary transcription factors for reprogramming, which raises regulatory concerns due to genomic integration or mutagenesis and is relatively inefficient. Alternatively, synthetic mRNAs can be used directly to reprogram differentiated cells into iPSCs. This approach was much more efficient (two orders of magnitude higher than viral transfection).

Others have tested the hypothesis that differentiated cells can be reprogrammed in situ, abrogating the need for in vitro manipulation. However, some caution must be taken in this process, as there remain several issues that could limit their eventual clinical potential. For instance, although some studies have focused on the in vivo applications of these cells in situations of organ repair, issues have arisen related to the immunogenicity of autologous cell-derived iPSCs that are reprogrammed in vitro and reimplanted.

The main question remains: how similar are iPSCs to ESCs, and are the differences between iPSCs and ESCs substantial?

### 8.1.5.5 Endothelial Cells and Neovascularization

It is known that the need for capillary vasculature and the difficulty in obtaining are the greatest impediments for most cellular approaches for tissue regeneration. Two of the main approaches researched for the induction of vascularization in 3D matrices are:

- the design of a tissue implant grown in vitro up to complete maturation of the vascular network
- the design of an implant bioactivated by suitable angiogenic modulators able to induce vascular regeneration in situ

Whichever approach is pursued, it is necessary to consider the "milieu" (culture medium) in which the cells responsible for vascular regeneration, that is, the endothelial cells (ECs), are naturally exposed.

The adhesion, proliferation, and reorganization of ECs in capillary structures is allowed in a permissive 3D context, which is able to offer all the soluble and insoluble signals necessary for revascularization.

The cells of the vascular endothelium reside in a complex microenvironment, characterized by a slow but continuous turnover due to degradation and resynthesis phenomena. It is in this environment that cells *read* the immobilized signals in the matrix and *feel* gradients of soluble signals. The induction of angiogenesis is mediated by a large number of growth factors and cytokines produced by the cells and by the surrounding host tissue. When the neovascularization process is induced in vivo, endothelial cells proliferate, migrate, and are able to invade the surrounding tissues. All the stimuli that induce ECs to be activated to pass from a quiescent state to the state of dynamic remodeling that leads to a new capillary bed are not yet completely known.

## 8.1.6 Dynamic Cell Culture and Bioreactors

The term "bioreactor" was originally used to indicate the fermentation chambers used for bacteria culture. In the last decade, it is used to define devices very different from one another, but all able to perform bioprocesses (i.e., biochemical and/or biological processes) "under closely monitored and tightly controlled environmental and operating conditions (e.g., pH, temperature, pressure, nutrient supply and waste removal)."

Bioreactors are devices that provide the nutrients with a transport system to the cells and enable the efficient removal of toxic or inhibitory products of cellular metabolism (Riboldi et al., 2017).

In the case of TE, the requirements for a bioreactor are:

- generation of flow patterns of culture medium for efficient and uniform cell seeding on 3D scaffolds

- increase in the transport of nutrients and gases from the culture medium to the engineered constructs
- physical stimulation of scaffold/cells constructs during their development
- online control of the chemical-physical parameters of the culture (pH, gas concentration, metabolites)

Many improvements in the use of bioreactors for the in vitro development of biological tissues are associated with the innovations in the biomaterials field. In fact, manufacturing or surface modification techniques, such as stereolithography, peptide functionalization, 3D printing and bioprinting, and microfluidic approaches allow the development of innovative scaffolds, which produce effective 3D support for cell culturing traditionally performed in a bidimensional (2D) environment.

In particular, the creation of flow patterns allows simulation of physiological conditions, maintaining constant concentrations of chemical components in the culture medium, and facilitating mass transfer in engineered constructs. Moreover, the generation of specific mechanical stimuli (compression, traction, flexion, torsion, and shear) reproduces the stresses that the specific tissue undergoes in vivo as much as possible, and promotes cellular orientation and the production of extracellular matrix, thereby stimulating the growth and maturation of the tissue in vitro (Riboldi et al., 2017).

Cyclic mechanical tensile stresses applied by the use of a bioreactor are particularly useful for the in vitro formation of those tissues formed under the same stimulus, for example, skeletal muscle tissue, cardiac tissue, and tendons and ligaments. An example of this type of bioreactor is shown in Fig. 8.38. Similarly, tissues that are physiologically subjected to compression stress, such as cartilage and bone, proved to be very sensitive to compressive stimulation in bioreactors, possibly also combining a hydrodynamic perfusion or other complex stimuli.

# 8.2 FUNDAMENTALS OF BIOTECHNOLOGY

Since the commencement in 1970s when Professor Berg's (Nobel Prize in Chemistry in 1980) experiments in genetic engineering had early success, modern biotechnology has become an integral feature of the public's consciousness, hopes, and fears. Issues such as novel diseases, genetically engineered foods, cloning, bioterrorism, and biosafety impact and interact with the public's feelings and instincts, which in turn necessitate the comprehension and conveyance of the fundamentals of biotechnology in an easy and simplistic manner (Doelle et al. (2009)).

For centuries, humans have used microorganisms to produce drinks and foods without understanding the microbial processes underlying their production. Brewing and baking bread (Fig. 8.39) are typical examples of processes that fall within the concept of biotechnology that, in this practical case, is the use of yeast to produce the desired item. Such traditional processes usually utilize the living organisms in their wild-type form (or slightly modified through breeding), whereas the more modern form of biotechnology will generally involve a more advanced modification of biological systems or organisms. This is because there are few drawbacks associated with the use of naturally occurring animals, plants, and microorganisms for the production of desired products, that are in the purity of the living stocks, the production of undesired and sometimes toxic by-products, the inability to withstand harsh biochemical processes/treatments, and generally high production costs.

The pressing problems that human beings have faced in the beginning of the 21st century are that water and nutrient deficits (especially proteins), environmental pollution, paucity of raw materials and energy resources, the necessity to generate new environmentally friendly

**FIG. 8.38**  Bioreactor used for the cyclic tensile mechanical stimulation of engineered muscle tissue constructs. A cyclic tensile stress, designed to mimic the features of the physiological development of muscle tissue in vitro, produced a tissue structure organization closer to the native one as compared to the constructs cultured in static conditions. *From Fig. 1 in Candiani, G., Riboldi, S.A., Sadr, N., Lorenzoni, S., Neuenschwander, P., Montevecchi, F.M., Mantero, S., 2010. Cyclic mechanical stimulation favors myosin-heavy chain accumulation in engineered skeletal muscle constructs. J. Appl. Biomater. Biomech. 8(2), 68–75.*

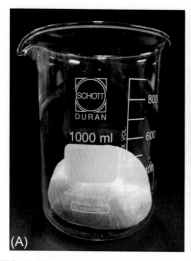  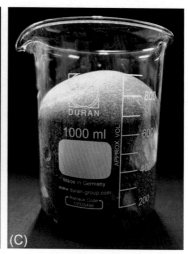

**FIG. 8.39** (A) Unleavened dough from wheat flour, (B) after rising, (C) and backing. By fermentation, the yeast species *Saccharomyces cerevisiae*, commonly named baker's yeast, converts fermentable carbohydrates to carbon dioxide ($CO_2$) and ethanol ($CH_3CH_2OH$). $CO_2$ is the major gas responsible for dough leavening over time: as the dough rises and proofs, $CO_2$ is released and bubbles become entrapped *(arrows)*.

materials, and development of new diagnostics and therapeutics cannot be solved in traditional ways. Therefore, it is necessary to engage in principally new, emergent technologies to support human life and increase its quality (Matyushenko et al., 2016). Over the past three decades, biologists have increasingly applied the principles of physics, chemistry, and mathematics to gain precise knowledge of how cells make these substances at the molecular level. By combining this newly gained knowledge with the methods of engineering, what has emerged is the concept of biotechnology, sometimes also referred to as biotech, which embraces all of the previously mentioned disciplines. Biotechnology can be defined in many ways. In the broadest terms, *biotechnology* means any technological application that uses biological systems, living organisms, or derivatives thereof to make or modify products or processes for a specific use, as reported in Art. 2 of the United Nations' Convention on Biological Diversity (CBD; www.cbd.int/). Yet, in 2002, the Organization for Economic Cooperation and Development (OECD; http://www.oecd.

org/) developed both a single definition of biotechnology and a list-based definition of different types of biotechnology techniques. The single definition defined biotechnology as "the application of science and technology to living organisms, as well as parts, products, and models thereof, to alter living or nonliving materials for the production of knowledge, goods, and service," and covered all modern biotechnology but also many traditional or borderline activities. For this reason, the OECD recommended the single definition always to be accompanied by the list-based definition, which functions as an interpretative guideline to the former. The list was indicative rather than exhaustive and was expected to change over time, as data collection and biotechnology activities evolve. In 2008, OECD member countries decided to begin work on revising and updating the list-based definition of biotechnology techniques, which now reads as follows:

– *DNA/RNA*: genomics, pharmacogenomics, gene probes, genetic engineering, DNA/RNA sequencing/synthesis/amplification, gene expression profiling, and use of antisense technology.

- *Proteins and other molecules*: sequencing/ synthesis/engineering of proteins and peptides (including large molecule hormones), improved delivery methods for large molecule drugs, proteomics, protein isolation and purification, signaling, identification of cell receptor.
- *Cell and tissue culture and engineering*: cell/ tissue culture, TE (including tissue scaffolds and biomedical engineering), cellular fusion, vaccine/immune stimulants, embryo manipulation.
- *Process biotechnology techniques*: fermentation using bioreactors, bioprocessing, bioleaching, biopulping, biobleaching, biodesulfurization, bioremediation, biofiltration, and phytoremediation
- *Gene and RNA vectors*: gene therapy, viral and nonviral vectors.
- *Bioinformatics*: construction of databases on genomes, protein sequences, modeling complex biological processes, including systems biology.
- *Nanobiotechnology*: application of the tools and processes of nano/microfabrication to build devices for studying biosystems and applications in drug delivery, diagnostics, etc.

Looking at things in a different way, biotechnology can be broadly divided into two major branches:

- Nongene biotechnology, which deals with whole cells, tissues, or even individual organisms.
- Gene biotechnology, which involves gene manipulation, cloning, etc.

Biotechnology is a very controversial subject, and each type of biotech has to deal with its own ethical issues.

The most important classification of biotechnologies is their division by area of application. In this regard, one of the most common approaches is a "color" method that was first proposed by R.R. Colwell in 2003 (Matyushenko et al., 2016). The most developed segments are referred to as (Fig. 8.40):

- *Red* or *medical biotechnology*, which involves the application of biological techniques to product research and development in health care and medicine, such as the production of drugs and diagnostics with the help of cellular technology and genetic engineering. On a global perspective, medical biotechnology is recognized for its potential to stimulate the economy as well as reduce health inequities. The main applications are at a *diagnostic* level, such as in *prediagnosis* (i.e., to screen for and detect the predisposition for diseases in individuals), *diagnosis*, and *prognostication* (i.e., better prediction of outcomes for particular diseases, or the effects of therapy), and a *therapeutic* level, such as in *prophylaxis* (i.e., the treatment given to prevent a disease) and *therapy* (i.e., the treatment intended to relieve or heal a disease).
- *Green biotechnology* or *agrifood biotechnology* approaches and applications include creating new plant varieties of agricultural interest (e.g., pest-resistant grains), and producing biofertilizers and biopesticides. This area of biotech is based exclusively on transgenics (i.e., Genetically Modified Organisms (GMOs)), so-called because they have a foreign extra-gene or genes inserted into their genome. The consequences of such deliberate genetic modification(s) are yet uncertain and debated. For centuries, species have been crossed to produce new varieties with better properties. But is it the same as the genetic crossing of species? There is still an additional problem of contamination: GMO crops are spread easily and, if they prove to be harmful to mankind, as someone has unfoundedly claimed so far, it will be a hard endeavor to distinguish GMOs from the natural plants.
- *Blue* or *marine biotechnology*, rarely mentioned, encompasses processes in aquatic environments. Blue biotech is

concerned with the exploration and exploitation of diverse aquatic organisms, such as those from Earth's oceans, lakes, rivers, and streams, in order to develop new products. Because aquatic life has adapted to thrive in extreme environmental conditions, the exploration of such great biodiversity could enable the development of new pharmaceuticals or other products that can withstand extreme conditions, and which consequently have high economic value. For instance, one example is the use of wound dressings coated with chitosan (see Chapter 4, Section 4.3.2.3);

- *Grey* or *environmental biotechnology* pertains to applications/technologies related to the environment. Grey biotechnology processes can be split up into two main branches, which are biodiversity maintenance and contaminants removal. Biodiversity maintenance is achieved through the studying the molecular biology and performing genetic analysis of populations/ species that constitute the ecosystem(s), the comparison and classification of the species, the use of cloning techniques to preserve species, and application of genome storage technologies. Pollutants removal or bioremediation, for instance, uses microorganisms and plants to isolate and dispose of different substances, such as heavy metals and hydrocarbons.

- *White biotechnology*, which relies on industrial processes such as the production of new chemicals or the development of new automotive fuels. Growing concerns about the dependence on mineral oil and the awareness that the world's oil supplies are not limitless are additional factors prompting the biotechnology industries to explore nature's richness in a search of methods to replace petroleum-based synthetics (Frazzetto, 2003). White biotech thus concentrates on the production of energy from renewable resources and biomasses. Starch from corn, potatoes, sugar cane, and wheat is already

used to produce bioethanol as a substitute for gasoline. And because plastics have rapidly replaced wood and metals in many consumer items, buildings, and furniture, one of the first goals on white biotech's agenda is the production of biodegradable plastics. Rising consumer demand for green chemicals, particularly in North America and Europe who are willing to pay a higher price or a green premium for numerous renewable products, is expected to push manufacturers to employ white biotechnology processes as opposed to traditional chemical manufacturing. White biotech aims to use living cells, such as yeasts, molds, bacteria, and enzymes to synthesize products that are easily degradable, and require less energy and create less waste during their production process. In this regard, a product that could benefit greatly from white biotech is paper. In fact, much of the cost and considerable pollution involved in the paper-making process is caused by "krafting," a method for removing lignin from the wood substrate.

In recent years, there has been a growing interest in biotechnology and subdisciplines. Readers who still doubt the incredible hopes and enthusiasm in biotech should refer to the wide variety of scholarly journals specifically dealing with the subject, such as *Nature Biotechnology, Trends in Biotechnology, Current Opinion in Biotechnology, Biotechnology Advances, Plant Biotechnology Journal, Biotechnology for Biofuels, Frontiers in Bioengineering and Biotechnology, Biotechnology and Bioengineering, Microbial Biotechnology, Critical Reviews in Biotechnology, Journal of Animal Science and Biotechnology, Applied Microbiology and Biotechnology, Journal of Industrial Microbiology and Biotechnology, BMC Biotechnology*, and *Journal of Biotechnology*, to cite only a few. Of note, this is not a comprehensive and exhaustive list but rather a personal selection of some amongst the most renowned journals devoted to cutting-edge biotechnology research.

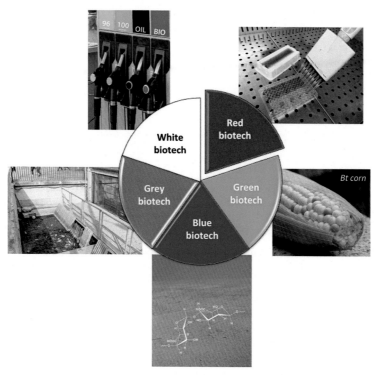

**FIG. 8.40** Biotechnology classification by colors. Some case-specific applications are reported in the figure. Clockwise: one example of *red* biotech is the enzyme-linked immunosorbent assay (ELISA); Bt corn, a widely known crop genetically modified to express one or more proteins from the bacterium *Bacillus thuringiensis* that are poisonous to certain insect pests, is an example of *green* biotech application; alginate, a polysaccharide extracted from brown algae and used for cosmetic products, is useful in *blue* biotech; *grey* biotechnology primarily deals with wastewater treatment systems by means of special reactors in which natural principles have been harnessed to treat waste; biofuels such as bioethanol and biodiesel are examples of *white* biotech.

To understand the molecular basis of biotechnology, first we will provide you with the basics of cell functioning. On this ground, the following sections cover the fundamental understanding of molecular biology and genetic engineering to provide readers with a clear view of the scientific principles and how these have been used to advance biotechnology.

## 8.2.1 Nucleic Acids: From Structure to Function

Genetic material is stored in the form of DNA, which is the hereditary material in all living cells. In humans, the nucleus (see Chapter 6, Section 6.1.1) of each cell contains $3 \times 10^9$ bp, collectively known as the *genome*. Arrayed along this molecule are an estimated 35,000 genes. A *gene* is the molecular unit of hereditary, also defined as a sequence on the DNA polynucleotide strand. Of note, the precise order of the nucleotides arranged on a DNA stretch determines the genetic instructions it encodes.

The way genes are activated is called *gene expression*. Because DNA is housed within the nucleus, while protein synthesis takes place in the cytoplasm, there must be some sort of intermediate messenger, called messenger RNA (mRNA), that leaves the nucleus and manages protein synthesis elsewhere: mRNA is a single-stranded (ss) nucleic acid that carries a copy of the genetic code for a single gene out of the

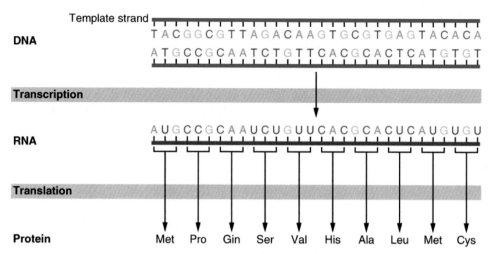

**FIG. 8.41**   The genetic code. DNA holds all of the genetic information necessary to build proteins. The nucleotide sequence of a gene is transcribed to give RNA, which, in turn, is translated into a protein. *From https://opentextbc.ca/anatomyandphysiology/.*

nucleus, and it is used to produce proteins in the cytoplasm (Fig. 8.41). Therefore, protein synthesis begins from genes. Because each gene provides the code necessary to synthesize a particular protein, not all genes are copied all the time. In fact, gene expression is controlled individually for each gene, so that only the genes whose products are needed at a specific time are expressed.

Recall that proteins are linear polymers composed of many amino acids (see Chapter 1, Section 1.6.1.1). The sequence of nucleotides in a gene, which is its A, T, C, and G order (see Chapter 1, Section 1.6.5.1), translates to a polyamino acid sequence (i.e., a protein). Similar to the way in which the three-letter code r-a-t signals the image of a rat, three DNA nucleotides in a row (i.e., a triplet) codes for a specific amino acid. For instance, the DNA triplet GCA specifies the amino acid Ala (Fig. 8.42). In this way, a gene, which is composed of multiple triplets in a unique sequence, provides the code to build a particular protein, with multiple amino acids in the proper sequence to give rise to its primary structure (see Chapter 1, Section 1.6.2).

Have you ever had to transcribe something? Maybe someone left a message on your voice mail, and you had to write it down on paper. As in this example, transcription is a process in which information is rewritten. Gene expression begins with *transcription* (Fig. 8.41), that is the synthesis of a mRNA strand complementary to the gene of interest. Such process is so-called because the DNA sequence transcribed a similar nucleotide alphabet specific to the RNA. The mechanism of transcription has parallels to that of DNA replication, in that a double-stranded (ds) DNA region unwinds and the two strands separate, but during transcription only that small portion of dsDNA is split apart. Unlike DNA replication in which both strands are copied, only one strand is transcribed (Fig. 8.41); the transcription process relies on Watson-Crick base pairing, and the resultant ssRNA is the reverse-complement of the original ssDNA sequence.

Like translating a book from one language into another, the codons on the mRNA must be translated into the amino acid alphabet of proteins. *Translation* is the process of decoding an mRNA sequence into an amino acid chain. The environment where translation takes place is the ribosome (see Chapter 6, Section 6.1.1).

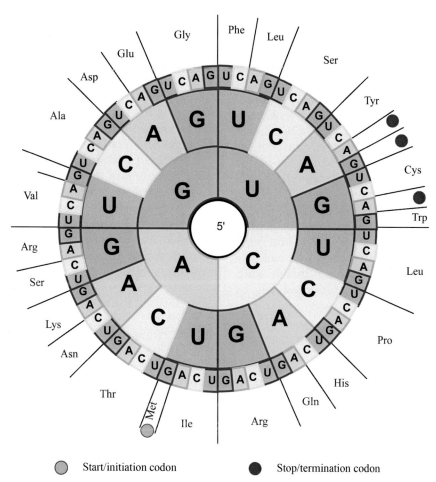

**FIG. 8.42** The RNA codon chart details the various nucleotide combinations that create the 20 known amino acids. To decode the codon, move from the center circle towards the periphery. The starting codon for mRNA is invariably AUG. Stop codons, also called termination codons, are nucleotide triplets within mRNA (UAA, UAG, and UGA) that signal a termination of translation into proteins and do not code for any amino acid. Of note, the DNA codon table is obtained by substituting T in place of U in the RNA codon table, which is exactly identical to it.

This flow of genetic information has become known as the Central Dogma of Molecular Biology, and was first described by Professor Crick (Nobel Prize in Physiology or Medicine in 1956) as one-way traffic: "DNA makes RNA, and RNA makes protein," meaning that the flow is unidirectional and irreversible. The Dogma was next revised in 1970 (Crick, 1970) and is now represented by major milestones:

- DNA can be copied to DNA through a process called *DNA replication*.
- DNA information can be copied into RNA through a process called *transcription*.
- RNA information can be translated into protein through a process named *translation*.
- RNA can be copied into RNA through a process called *RNA replication*.
- RNA can be copied back into DNA through a process called *reverse transcription*.

The unknown and impossible transfers describe a protein being copied from a protein, the synthesis of RNA and DNA using the primary structure of a protein as a template.

In a cell, the DNA does not usually exist by itself. Instead it associates with specialized proteins that help organize it to give its final three-dimensional (3D) structure. The complex of DNA with histones and other structural proteins is called *chromatin*. Because histones are highly alkaline proteins, dsDNA wrapped around them to give nucleosomes (see Chapter 1, Section 1.6.5.2) in a way resembling a thread wrapped around a spool. Of note, chromatin is decondensed for most of the cell life, meaning that it exists in long, thin strings that look like squiggles under the microscope. In this state, the DNA can be accessed relatively easily by the cellular machinery that reads and copies (i.e., transcribes) the DNA, through a process fundamental in allowing the cell to grow and function.

But chromatin can also condense. In fact, when the cell is about to divide, chromatin folds into characteristic formations, that are called *chromosomes*. Each chromosome contains a single piece of dsDNA along with the aforementioned packing proteins. Worth noting that, if comparing the length of chromosomes to that of naked DNA, the packing ratio of DNA in chromosomes is approximately 10,000:1. The DNA constituting the human genome is distributed into individual chromosomes, which are 23 pairs and a total full complement of 46 chromosomes. Every somatic cell, that is, any cell of a living organism other than the reproductive cells called *gametes*, in physiological conditions has two copies of each. Such human cells are called *diploid* because they carry two complete sets of chromosomes: one set of 23 chromosomes comes from the father and the other one from the mother. The total number of chromosomes is called the *chromosome number*, whereas *ploidy* refers to the number of complete sets of chromosomes in a cell. In this regard, the number of chromosomes in a single set is called the *haploid number*, given by a small letter $n$.

As opposed to somatic cells, gametes, that is sperm and ova, are *haploid* cells. They combine through the fertilization event to give rise to the *zygote*, the first diploid cell (i.e., it displays $2n$ chromosomes). The chromosomes in each pair, one of which comes from the sperm (i.e., it is paternal) and one from the egg (i.e., it is maternal), are said to be *homologous*. Individual chromosomes are always depicted with the short p arms (p stands for *petite*, the French word for small) at the top and the long q arms (q stands for *queue*) at the bottom (Fig. 8.43). A chromosome consists of two identical DNA sequences, called *sister chromatids*, which are joined at the *centromere*. Centromere placement is also be used to identify the gross morphology, or shape, of chromosomes.

Each chromosome carries many genes. The term *genetic locus*, or simply *locus* (plural loci), describes the position of any gene on a given chromosome (Fig. 8.43). Under a light microscope and using a suitable staining, one can observe DNA bands on a chromosome, which are visible in different shades of gray. These bands indicate higher or lower concentrated bp sequences of a certain kind. To describe the position of these bands and the corresponding genes, the locus starts with the chromosome number first, followed by either p or q to refer to the chromatid arm. For instance, the chromosomal locus of a typical gene might be written 3p21, which reads "three p two one" and means that we are pointing to chromosome three, p arm, region two, subregion one.

An *allele* is a variant form of a given gene. In other words, alleles are genes with the same function, however with slight variations in them that gives rise to different observable traits. Diploid organisms, such as humans, have two alleles at each genetic locus, with one allele inherited from each parent. Each pair

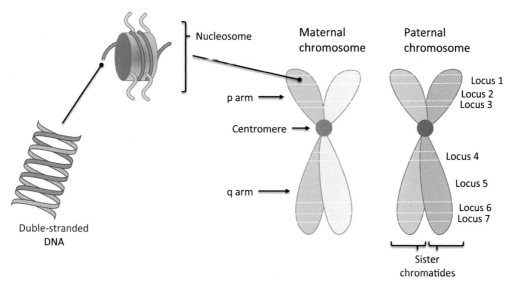

**FIG. 8.43** Idiogram of a generic autosome pair. During fertilization, each parent contributes one chromosome to each pair of homologous chromosomes, so the offspring receives half of its chromosomes from the mother and half from the father, through the egg and the sperm cell, respectively. Chromosomes are an organized package of double-stranded (ds) DNA that wraps around highly alkaline proteins, called histones, to give nucleosomes. Chromosomes and chromatids occur as rod- or thread-shaped structures in the cell nucleus.

of alleles represents the *genotype* of a specific gene: genotypes are described as homozygous if there are two identical alleles at a specific locus, and as heterozygous if the two alleles differ. Alleles contribute to the organism's *phenotype*, which is the outward appearance of the organism.

Images of the individual chromosomes are conventionally arranged into a standardized format known as *karyogram*. The *karyotype* describes the number, appearance, and physical characteristics of chromosomes in the cell nucleus, in other words, what they look like under a light microscope after Giemsa staining. The karyogram is a microphotograph of all chromosomes represented in a standard format. On this photograph, all chromosomes are sorted and rearranged in pairs by size and position of their centromeres. Basically, the human karyogram is the photo of the whole set of all 22 identical autosome pairs plus the

two gender-specific chromosomes (XX/XY) (Fig. 8.44).

To divide, a cell must complete several important tasks: it has to grow, copy its genetic material, and physically split into two identical daughter cells. Cells perform these functions in an organized, predictable series of steps that make up the *cell cycle*. From the very definition of cell cycle, it is a cycle rather than a linear path (Fig. 8.45). In fact, at the end of each go-round, the two daughter cells can start the exact same process over again from the beginning.

When eukaryotic cells divide, genomic DNA must be equally partitioned into both daughter cells. To accomplish this, the DNA becomes highly compacted into the classic chromosomes that can be seen by means of a light microscope (Fig. 8.44). Once a cell has divided, its chromosomes uncoil again. In eukaryotes, the cell cycle consists of a long preparatory period, called interphase (I), during which the cell grows and

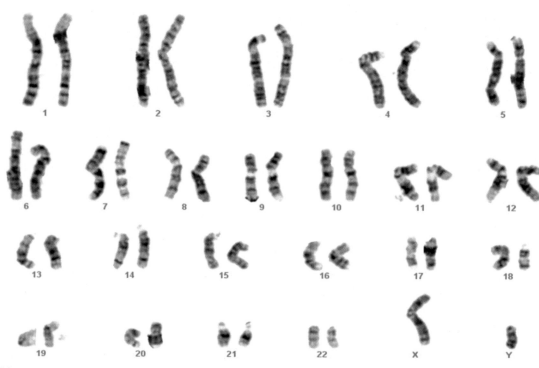

**FIG. 8.44**   Human male karyogram (karyotype 46 XY). The picture shows the whole set of all 22 autosome pairs plus the two gender-specific chromosomes (XY in males). *From Kasasbeh, F.A., Shawabkeh, M.M., Hawamdeh, A.A., 2011. Deletion of 18p syndrome. Lab. Med. 42(7), 436–438.*

makes a copy of its DNA. The I phase is divided into first gap (G$_1$), synthesis (S), and G$_2$ phases (Fig. 8.45). The prefix inter- means between, reflecting that interphase takes place between one mitotic (M) phase and the next one. During G$_1$, the cell grows physically larger, copies organelles, and makes the molecular building blocks it will need in later steps. In S phase, the cell synthesizes a complete copy of the nuclear DNA. Finally, in G$_2$, the cell grows more, makes proteins and organelles, and begins to reorganize its contents in preparation for the M phase. Of note, G$_2$ ends when M begins. The first portion of the M phase is called *karyokinesis*, or nuclear division: the cell segregates its duplicated chromosomes, and it is divided into five substages named prophase, prometaphase, metaphase, anaphase, and telophase

(not shown in Fig. 8.45). The end portion of the M phase, called *cytokinesis*, or cell motion, is the physical process of cell division that leads to the separation of the cytoplasmic components into the two indentical daughter cells.

## 8.2.2 Genetic Engineering

*Genetic engineering*, sometimes referred to as genetic modification or *recombinant DNA technology*, is a broad term relating to the direct manipulation of the genetic composition of cells, or an organism, by artificial means, often involving the transfer of specific traits, or genes, from one organism into an animal or plant of the same or an entirely different species, to obtain enhanced and desired characteristics or products.

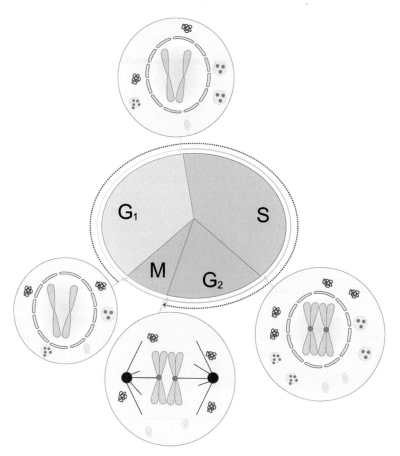

**FIG. 8.45** The cell cycle. A cell on the path to cell division proceeds through a series of precisely timed and carefully regulated stages of growth, DNA replication, and division that produce two identical daughter cells (i.e., a cell clone). The cell cycle consists of the interphase (I) *(blue dotted arrow)* and the following mitotic (M) phase *(yellow solid arrow)*. The most schematic possible view of a cell at the transition between two consecutive stages of the cell cycle is depicted in the figure. During the I phase, the cell undergoes normal growth processes while preparing for cell division. The first stage of I is called the first gap ($G_1$) because little change is visible from a microscopic observation. However, during this stage, the cell is extremely active at the biochemical level. Of note, an unreplicated chromosome contains one double-stranded (ds) DNA molecule. Throughout I, the nuclear DNA remains in a semicondensed chromatin configuration, as opposed to what is shown in the figure for clarity. Along the S phase, the DNA replication leads to the formation of identical dsDNA molecules (i.e., a sister chromatid pair) that are firmly attached to each other at the centromere. In the $G_2$ phase, the cell synthesizes proteins necessary for the chromosome segregation. Some cell organelles are duplicated, and the cytoskeleton is dismantled to provide resources for the mitotic phase.

Genetic engineering has manyfold applications and the potential to deal with major aspects of life, for instance, improving health, enhancing food resources, and resistance to divergent adverse environmental effects. In the past century, recombinant DNA technology was just imaginary: that desirable features could be improved in living bodies by controlling the expressions of target genes. However, in the recent era, genetic engineering has demonstrated unique impacts in bringing advancements to human life. By virtue of this new technology, crucial proteins required for health problems and dietary purposes can be produced safely, affordably, and sufficiently (Khan et al. 2016b). Manipulation of an organism's genome is carried out either through the introduction of one or several new genes and regulatory elements, or by decreasing or blocking the expression of endogenous genes through their recombination.

Considering the fact that each human cell contains approximately two meters of DNA, a small tissue sample contains many kilometers of nucleic acids. It is therefore relatively easy to isolate a sample of DNA from a collection of cells, but picking a specific gene within such a sample is like looking for a needle in a haystack. However, recombinant DNA technology has made it possible to isolate one gene or any other DNA stretch, enabling scientists to determine its nucleotide sequence, study transcripts, mutate it in highly specific ways, and reinsert the modified sequence into the same or another living entity.

In biology, a clone is a group of individual cells or organisms descended from one progenitor. The members of a clone family are genetically identical, because (somatic) cell replication produces identical daughter cells at each cell division. The use of the word *clone* has been extended to recombinant DNA technology, which has provided scientists with the ability to produce many copies, or DNA clones, of a single DNA fragment, such as a gene. In practice, the procedure is carried out by inserting a DNA stretch into another DNA molecule, called vector or carrier, and then allowing this molecule to replicate in a host bacterium. The most commonly used vector is a plasmid, which is a circular DNA molecule originated from bacteria, and engineered ad hoc. In nature, plasmids are not part of the bacterial genome, but they carry genes that provide the host bacterium with useful features, such as drug resistance, mating ability, and toxin production. Of note, they are small enough to be conveniently manipulated to carry extra DNA spliced into them. The process of introducing a foreign gene(s) into a vector is called *gene*

*cloning* (Fig. 8.46). DNA-cutting enzymes specific for the target DNA sequence, which are called restriction nucleases, are used to obtain a specific gene from an organism's genome. The DNA fragment of interest is next joined to the desired plasmid vector by means of an enzyme called ligase. The DNA sequence that has been pasted into the vector is called *insert*, whereas the plasmid now containing the insert is called a *chimera*. Through the process of *transformation*, this is introduced into a recipient cell, such as an easy-to-grow bacterium. The microorganism is next plated and grown in culture (i.e., in vitro): this single cell is expanded exponentially to generate a large number of bacteria, each of which contains copies of the original chimera. Because the plasmid vector also carries a gene that codes for antibiotic resistance that allows the microorganisms to survive in the presence of a specific antibiotic such as ampicillin, the bacteria that take up the chimera are selected on ampicillin-containing nutrient plates: bacteria without the chimera will die, whereas each bacterium carrying the chimera live and reproduce and gives rise to a small, dot-like group, or *colony*, of identical bacteria (i.e., a bacterial clone). This is because, as they duplicate, they replicate the chimera and pass it on to the offspring, making copies of the DNA insert they contain. In this way, clones containing the relevant DNA fragment are selected to be harvested.

What is the point of making many copies of a DNA sequence in a vector? In some cases, scientists need lots of DNA copies to carry out experiments or to build new chimeras. In other cases, the piece of DNA encodes a useful protein, and the bacteria are used to synthesize it (Fig. 8.46). Once the target protein has been synthesized, it is either secreted or retained by cells. In the former case, the culture medium is harvested, and the protein of interest is readily made available for the specific application. Instead, in the latter case, the bacteria are harvested and split open to release it. Unfortunately, the bacteria do contain

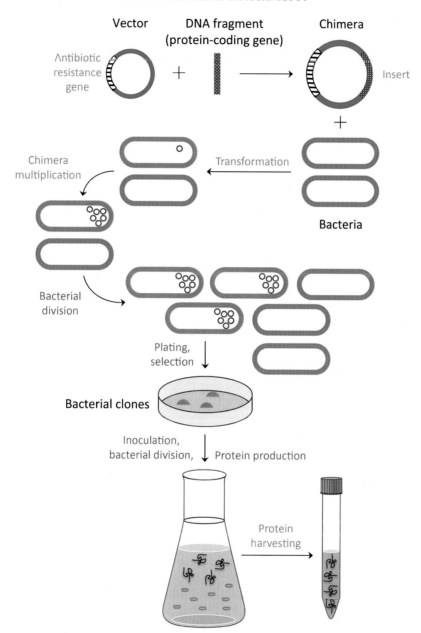

**FIG. 8.46** DNA cloning or molecular cloning is a set of experimental methods that are used to assemble recombinant DNA molecules and to direct their replication within a host microorganism. In a conventional molecular cloning experiment, the gene to be cloned is obtained from an organism of interest by treating in the test tube genomic DNA with enzymes to generate smaller DNA fragments. Such fragments are next combined with plasmid vectors to generate chimeras, which are introduced into host microorganisms, such as bacteria, by a process called transformation. The chimeric plasmids are replicated along with the host DNA and the bacteria themselves. Bacteria carrying the chimera(s) are grown up and selected. A bacterial clone can be grown in a liquid culture medium to produce a large amount of target protein that is finally harvested.

many proteins other than the target protein. Because of this, the protein of interest must be purified, that is, it needs to be separated from the other cell contents by means of biochemical techniques. A nice example of use of gene cloning for therapeutic purposes is given by patients suffering from type I diabetes. They must regularly inject themselves with insulin from another source in order to prevent dangerously high levels of blood glucose. For many years, all of the insulin used in diabetics was from the pancreases of pigs and cows that had been slaughtered for meat. In the 1980s, scientists developed new techniques to produce human insulin in *Escherichia coli* (*E. coli*) bacteria. This recombinant insulin is now the primary treatment used. Other examples of protein therapeutics include recombinant human growth hormone (rhGH or hGH), which is administered to patients who are unable to synthesize it, and tissue plasminogen activator (tPA) (see Chapter 6, Section 6.2.1), which is used to treat strokes and prevent blood clots. Why use bacteria to produce a target protein? Because bacteria are cheap, easy-to-grow, clonal, multiply quickly, are relatively easy to transform, and can be stored at −80°C almost indefinitely.

The following points highlight the top three applications of genetic engineering.

### 8.2.2.1 Creating Genetically Modified Organisms

People have been altering the genomes of plants and animals for many years using traditional breeding techniques. Artificial selection for desired traits has resulted in a variety of different organisms, ranging from sweet corn to hairless cats. But this artificial selection, in which organisms that exhibit specific traits are chosen to breed subsequent generations, has been limited to naturally occurring genetic modifications. In recent decades, however, advances in the field of genetic engineering have allowed for precise control over the genetic changes introduced into animals, plants, or single-celled life forms to give rise to *genetically modified organisms* (GMO), or *transgenic organisms*. By definition, this is any creature whose genetic material has been altered using genetic engineering techniques, or "any living organism that possesses a novel combination of genetic material obtained through the use of modern biotechnology," as defined in the Cartagena Protocol on Biosafety to the Convention on Biological Diversity. GMOs are intended to produce many medications, food and other goods, and are widely used in scientific research.

Due to the relative ease of manipulating extra-genomic DNA in bacteria, they were the first organisms to be genetically modified in the laboratory, as shown in Section 8.2.2. Genes from a wide variety of organisms have thus been pasted in a plasmid vector, that was inserted into bacteria for storage and expression.

Within the field known as *pharming*, a portmanteau of pharmaceutical and farming, intensive research has been conducted to develop transgenic animals for the production of biotherapeutics. Pharming refers to the generation of a GMO through the insertion of genes that code for protein therapeutics into host animals or plants that would otherwise not biosynthesize such products. Pharming is a useful alternative to traditional pharmaceutical development because GM livestock and plants are relatively inexpensive to produce and maintain. One of the first mammals successfully engineered for the purpose of pharming was a sheep named Tracy, born in 1990 and created by Professor Wilmut's group at The Roslin Institute in Scotland. Through DNA injection, a zygote (i.e., a single-celled fertilized embryo) was genetically engineered to produce milk containing large quantities of the human enzyme alpha-1 antitrypsin, a therapeutic protein used to treat cystic fibrosis and emphysema. In 1997, Wilmut and colleagues generated another pharmed sheep named Polly, a Poll Dorset clone made from nuclear transfer and using a fetal fibroblast nucleus genetically engineered to express the human factor IX (see Chapter 6, Section 6.2.1). This clotting factor naturally occurs in healthy humans, but it is absent in people suffering from hemophilia. Because of this, their blood does not clot normally. Hemophilics thus require replacement therapy with a therapeutic protein. Polly, along with two other sheep born in 1997 and engineered to produce human factor IX, represent a major advance in this regard. The first pharmed agent produced by animals to gain approval by the FDA and the European Medicines Agency (EMA) for therapeutic use was the recombinant human antithrombin marketed as ATryn, a plasma protein with anticoagulant properties used to treat hereditary antithrombin-deficient patients who have to undergo surgical or childbirth procedures. This biotherapeutic is isolated and purified from the milk of GM goats. Of note, one genetically modified goat can produce the same amount of antithrombin in a year as 90,000 blood donations.

### 8.2.2.2 *Organism Cloning*

*Organism cloning*, sometimes also called *reproductive cloning*, refers to the procedure of creating a new multicellular organism genetically identical to another one (i.e., a clone). However, because of the random DNA mutations that may occur during development or induced by the environment where they grow up in, clones may not look exactly the same or behave in the same way.

In the late 1990s, Dolly the sheep was the first mammal to be cloned from an adult cell. Dolly was part of a series of experiments carried out at The Roslin Institute by Professor Wilmut's team that were trying to develop a better method for producing GM livestock. Scientists at Roslin also wanted to learn more about how cells change during development and whether a specialized cell, such as an

endothelial cell or a neuron, could be used to make a whole new animal. Due of the nature of the research, the team was made up of many different professionals, including basic scientists, embryologists, surgeons, veterinarians, and farm staff. Dolly was cloned from a cell taken from the mammary gland of a 6-year-old Finn-Dorset sheep and an egg cell from a Scottish Blackface sheep (Fig. 8.47), and was born in July 1996. Dolly's white face was one of the first signs that she was a clone, because if she was genetically related to her Scottish Blackface surrogate mother, she would have had a black face. Because Dolly's genome came from a mammary gland cell, she was named after the country singer Dolly Parton. In February 2003, Dolly was euthanized because of the progressive lung disease and severe arthritis. A Finn-Dorset such as Dolly has a life expectancy of about 12 years, but Dolly lived only six and a half years. Dolly's premature death supported the view of some scientists who claim that all cloned animals are born with health problems (Williams, 2003). In response, The Roslin scientists did not believe there was a connection with Dolly being a clone and revealed that other sheep in the same flock died of the same disease. After her death, The Roslin Institute donated Dolly's body to the National Museum of Scotland, where she has become one of the most popular exhibits.

Dolly the sheep was generated using a cloning method called Somatic Cell Nuclear Transfer (SCNT), in which the nucleus of an egg cell is removed and replaced with the nucleus of an adult donor cell.

Gland epithelial (diploid) cells were harvested from the udder of a Finn-Dorset ewe (i.e., the nuclear donor) and placed in culture dishes with very low nutrient concentration, so that they stopped dividing. In other words, they switched off their active genes and arrested in $G_1$ phase (Wilmut et al., 1997).

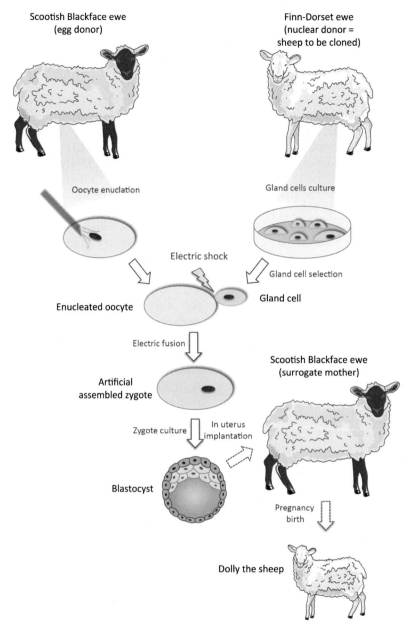

**FIG. 8.47**    The strategy used to create Dolly the sheep. The detailed explanation of each step of the creation process is given in the following text.

Meanwhile, an unfertilized egg (haploid) cell was taken from a Scottish Blackface ewe (i.e., the egg donor). Its nucleus was sucked out by means of a micromanipulator, leaving the egg cell enucleated but containing all the cellular machinery necessary to grow an embryo.

One gland cell was selected and placed next to the recipient enucleated oocyte. The application of an appropriately timed electric pulse caused them to fuse together.

After about six days, the artificially-assembled zygote developed into a blastocyst that was surgically implanted in the uterus of another Scottish Blackface ewe (i.e., the surrogate mother).

After gestation, the pregnant Scottish Blackface ewe gave birth to offspring that was a Finn-Dorset lamb genetically identical to the original gland cell donor. Because 99.9% of the cell DNA is contained in the nucleus in the form of chromosomal DNA, with the remaining 0.1% of DNA found in mitochondria, the resulting sheep shared almost exactly the same genome as the original gland cell.

The breakthrough represented by Dolly is one of the major scientific achievements of the last century because it overturned the concept that the differentiated status of a somatic cell is irreversible (Loi et al., 2013). In fact, the generation of Dolly the sheep showed that the nuclear genes of such a mature differentiated somatic cell are still capable of reverting to an embryonic totipotent state, creating a cell that can then go on to develop into any part of an animal. In the case of Dolly, she was the only successful case out of 434 attempted fusions of oocytes and donor cells that were taken from cultures of mammary glands. Even if cloning of adult nuclei has been become much more efficient, there will still be hazards to humans. For instance, the donor cells could suffer mutations in situ from radiation, chemicals, and/or aging during the lifetime of the donor. It is even not unusual that mutations also arise in the donor cells during cell culture (McKinnell and Di Berardino, 1999).

### 8.2.2.3 Gene Therapy

According to the FDA definition, *gene therapy* is the administration of genetic material to modify or manipulate the expression of a gene product or to alter the biological properties of living cells for therapeutic use (https://www.fda.gov/). Gene therapy is an experimental technique that uses nucleic acids to treat, or sometimes prevent, diseases. In the future, this technique may allow physicians to treat any disorder by inserting a gene into a patient's cells instead of using drugs or surgery. Researchers are testing several gene therapy approaches including:

- replacing a mutated gene that causes disease with an healthy copy of the same gene;
- inactivating, or "knocking out," a mutated gene that is functioning improperly;
- introducing a new or modified gene into the body to help fight a disease.

Although gene therapy is potentially a promising treatment option for a number of diseases, such as inherited disorders, some types of cancer, and certain viral infections, this therapeutic approach remains risky and is still under study to make sure that it will be safe and effective for patients. In light of that, gene therapy is currently being tested only for patients who lack other treatment options. These include conditions that, in the absence of treatment, cause severe disability or early death. Yet, despite that, a very recent brief by the Massachusetts Institute of Technology (MIT)'s New Drug Development Paradigms Initiative (NEWDIGS) projected that around 40 gene therapy technologies would be approved by the end of 2022, and 45% of them would be cancer treatments, 34% for the treatment of orphan diseases, 17% for common diseases, and 4% for extremely rare diseases (i.e., they affect fewer than 1 in 100).

Currently, the only way to receive gene therapy is to participate in a clinical trial. Gene therapy clinical trials are research studies that help physicians in determining whether an approach is safe and effective. As of August 2018, the number of gene therapy clinical trials approved worldwide is 2805 (http://www.abedia.com/wiley/).

Virtually all cells in the human body contain genes, making them potential targets for gene

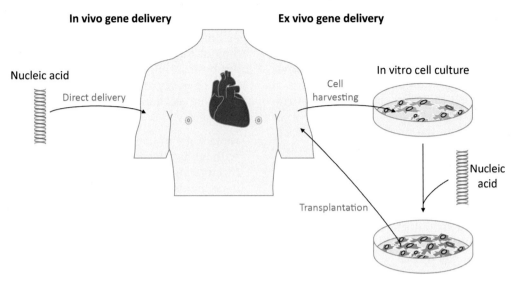

**FIG. 8.48**   In vivo and ex vivo gene transfer strategies. In in vivo gene transfer (left), the genetic material is delivered directly to the patient, and the genetic modification takes place in situ. In ex vivo gene transfer (right), the cells of interest are removed from the patient and established in cell culture. They are genetically modified and then readministered to the patient.

therapy. Depending on the ploidy, such cells can be divided into two major categories: somatic cells and germline cells (eggs or sperm). Although it is possible to treat both of them, to date, human gene therapy has been directed at somatic cells only, whereas gene delivery to germline remains controversial; this is because the effects will be passed onto offspring. Somatic gene therapy can be realized through two strategies (Fig. 8.48):

– *ex vivo*, which means that the cells are modified outside the patient's body and then transplanted back in again. In an ex vivo delivery approach, the cells are typically harvested from the patient's own body (i.e., autologous cells, that give autotransplantation or autograft) or, less frequently, from another healthy individual of the same species called *donor* (i.e., homologous cells, that give allograft, allogeneic transplant, or homograft). The cells are next treated in vitro with the gene therapy agent, expanded, and transplanted into the patient. Such modified cells further replicate and spread in the body. The ex vivo strategy

ideally allows the transfer of a gene(s) to a specific cell subpopulation without affecting off-target cells and organs. A major drawback of this approach is that not just any kind of cell can be withdrawn from individuals (e.g., nondividing cells such as neurons, cardiomyocytes);

– *in vivo*, which involves the direct injection of the gene therapy agent into the patient's body. Depending on various factors, in vivo gene therapy can be administered intravenously, injected into the muscles, or injected or infused into an organ or bodily structure.

One of the major hurdles in the development of gene therapy is the delivery of the effector to and into the target cell. Gene delivery is the process by which foreign nucleic acids are transferred to host cells. *Transfection* is the process of deliberately introducing nucleic acids (either DNA or RNA) into eukaryotic cells, eliciting the desired effect(s). This process typically involves opening transient pores or holes in the cell membrane to allow the uptake of nucleic acids. The introduced nucleic acids may exist in the cells transiently, such that they are only expressed for a limited period of time and do not undergo replication. Conversely, transfection may be stable and nucleic acids integrate into the genome of the recipient cell and duplicate when the host genome replicates.

Gene delivery can be achieved using *physical methods* and *gene delivery vectors*.

1. *Physical methods* employ physical forces to permit the nucleic acids to cross the cell membrane without the support of any carrier agent. For this reason, they are also known as carrier-free gene delivery strategies or methods. The main physical procedures for naked nucleic acids delivery are:
   - *naked nucleic acids injection* or *hydrodynamic delivery*, and *microinjection*. The nucleic acids delivery through a rapid injection of a relatively large volume of solution, that is, the application of controlled hydrodynamic pressure to blood vessels, is a way to enhance endothelial and parenchymal cell permeability (Fig. 8.49). It is roughly defined as the direct-pressure injection of a solution into a cell through a glass capillary. The method has been in existence almost as long as there have been microscopes to observe the process (King, 2004). This method is the most affective delivery method because simple, efficient, and versatile (Suda and Liu, 2007). Hydrodynamic tail vein injection had its inception in the late 1990s with investigations into intravascular injection of plasmid DNA solutions in whole animals. On the other hand, microinjection is an effective and reproducible method for introducing exogenous molecules into a defined cell populations. The concentration of nucleic acids and the timing of the experiment can be finely tuned in order to minimize some problems associated with overexpression (Rose, 2007);
   - *sonoporation*, or *ultrasound-mediated gene transfer*, employs ultrasound to deliver nucleic acids (Lentacker et al., 2014). This technique produces some disruption of the plasma membrane by using ultrasound waves of 1–3 MHz with an intensity of 0.5–2.5 W cm$^{-2}$ by acoustic cavitation. Ultrasound waves aided with air-filled microbubbles enhanced gene delivery efficiency (Fig. 8.50). Microbubbles cavitate, oscillate, and break up upon absorption of ultrasound waves and produce localized shocks in the form of a high-velocity microjet that interrupts the nearby membrane, resulting in defects that allow nucleic acid transfer. The frequency and the intensity of ultrasonic waves and the type of target tissue influence the absorption rate (Tabassum Khan, 2017);
   - *electroporation*, or *gene electrotransfer*, involves the delivery of a voltage (current) across the cell surface, which results in the spontaneous creation of pores in the plasma membrane. When the voltage is applied and the pores open up, current travels through the cell. Recalling that nucleic acids are anionic at pH 7, it is clear that they migrate in the same direction as the electron flow in the applied current. Instead, when the current is turned off, the pores spontaneously close, and allow trapping some nucleic acids within the cells. Effective electroporation depends on electric field parameters, electrode design, and the tissues or cells being targeted. Although almost any tissue can be electroporated, including muscle, skin, heart, liver, lung, and vasculature, no single combination of such variables leads to greatest efficacy in every situation (Young and Dean, 2015) (Fig. 8.51);
   - *magnetofection*, or *magnet-assisted transfection*, mediates gene transfer in vitro and in vivo using supramagnetic nanoparticles coated with nucleic acids in the presence of an external magnetic field (EMF). Mechanistically, magnetofection enhances gene delivery by guiding and

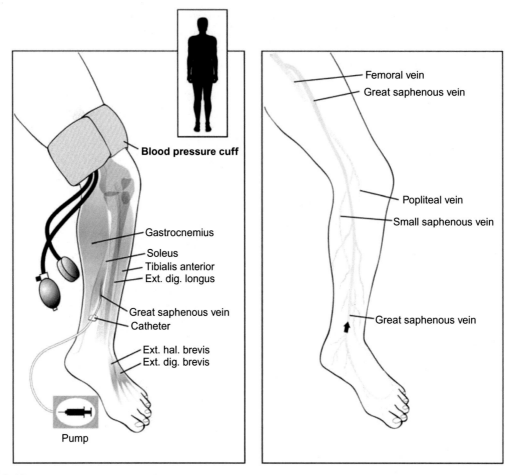

**FIG. 8.49**    Hydrodynamic limb vein (HLV) delivery procedure. A plasmid DNA solution is infused into a limb vein (anterograde). Gene transfer is facilitated by the placement of a proximal tourniquet, which allows a transient increase in vascular pressure upon injection. The plasmid is only extravasated in the areas displaying increased pressure, thus limiting gene transfer to the isolated limb. The complete procedure can be accomplished in 5–10 min. *From Herweijer, H., Wolff, J.A., 2006. Gene therapy progress and prospects: hydrodynamic gene delivery. Gene Ther. 14(2), 99–107.*

maintaining nucleic acid-loaded particles in close contact with the target, thus increasing endocytic uptake of such particles. After intravenous injection, further enhancement comes from magnetic field-facilitated extravasation of the particles into surrounding tissues (Krötz et al., 2003) (Fig. 8.52). Unfortunately, magnetofection has some drawbacks:

particles <50 nm are not suitable for magnetic targeting, while particles too large in size ($\varnothing > 5\,\mu m$) are less prone to pass through the blood capillaries. The blood flow rate also affects the transfection efficacy of this method; for instance, a flow rate of about $20\,cm\,s^{-1}$ in the human aorta makes the transfection tricky. The external magnetic flux density and gradient

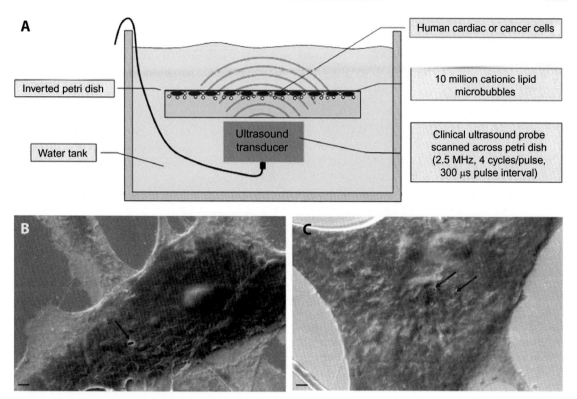

FIG. 8.50 (A) Schematic of an experimental setup used for cell sonoporation. (B and C) Representative scanning electron microscopy (SEM) images of cancer cells after ultrasound treatment (4.8 MPa peak). *Red arrows* highlight cell membrane holes induced by sonoporation. Bars = 2 μm. *Modified from Bhutto, D.F., Murphy, E.M., Priddy, M.C., Centner, C.C., Moore Iv, J.B., Bolli, R., Kopechek, J.A., 2018. Effect of molecular weight on sonoporation-mediated uptake in human cells. Ultrasound Med. Biol. 44, 2662–2672.*

decreases at a distance from the magnetic pole, which also affects the transfection efficacy (Jinturkar et al., 2011);

— *gene gun*, also called *biolistic gene transfer* (short for biological ballistics), or *particle bombardment system*, is the method of directly shooting nucleic acids into cells using a device called a gene gun. The biolistic gene transfer relies on the use of nontoxic, subcellular-sized (Ø = 0.5–5 μm) particles made of a heavy metal, such as tungsten or gold, which are first coated with nucleic acids. The nucleic acids/metal particles are next loaded onto one side of a plastic bullet, or microprojectile (Fig. 8.53).

A pressurized inert gas, usually helium, provides the force for the gun. The macroprojectile is abruptly stopped at the end of the shaft, but the particles coated with nucleic acids emerge from the gun with great speed and force. Efficient gene transfer necessitates fine optimization of the procedure to attain good penetration, yet minimizing tissue/cell damage. The parameters that affect the gene transfer efficiency are the size and the quantity of microspheres, the microspheres-to-nucleic acids ratio, and the bombardment force that relies on the gene gun instrumentation. Although minimal surgery is required for

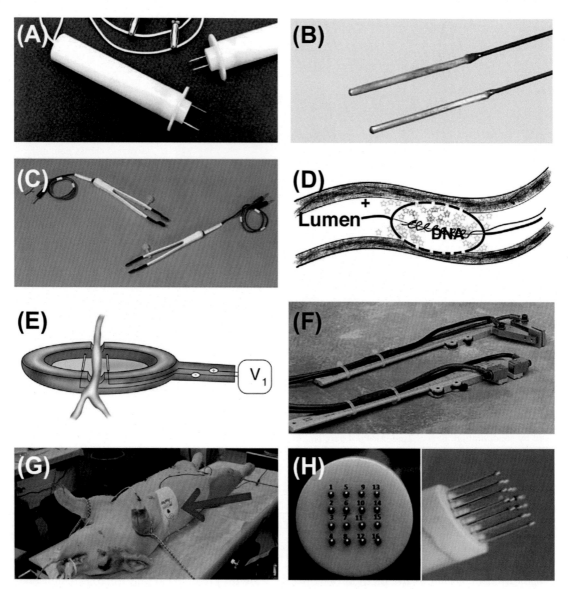

FIG. 8.51 Examples of electrodes for electroporation. (A) Penetrating, two-needle arrays. (B) Nonpenetrating parallel needles (Genetrode electrodes, Genetronics, San Diego, CA, USA). (C) Plate electrodes (Tweezertrodes, BTX, Hollister, MA, USA). (D) Cartoon of a balloon catheter-based electrode for delivery of DNA and electroporation. (E) Spoon electrode for vascular electroporation. (F) Caliper-mounted plate electrodes. (G) Conformable defibrillator pads for electroporation *(arrow)*. (H) Multielectrode array. *From Young, J.L., Dean, D.A., 2015. Nonviral Vectors for Gene Therapy—Physical Methods and Medical Translation, pp. 49–88.*

efficient gene transfer to the internal organs, and the gene gun deliver nucleic acids to a narrow regions, this technique suffers from very low penetration depth when delivered from the body surface. For this reason, most applications of the gene gun are limited to exposed tissues, including skin and muscles. To date, DNA

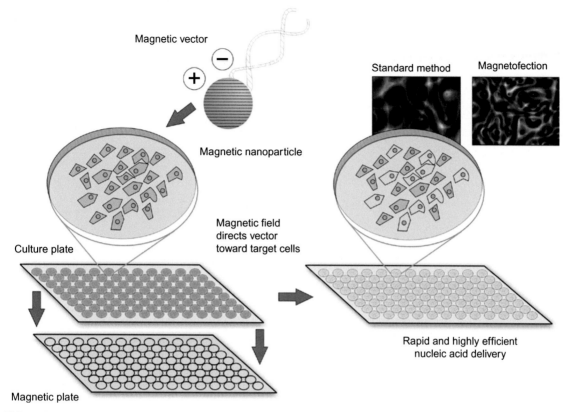

**FIG. 8.52** Magnetofection procedure applied to cells in culture. *From Jinturkar, K.A., et al., 2011. Challenges in Delivery of Therapeutic Genomics and Proteomics, pp. 83–126.*

vaccination is the most common application of biolistic gene transfer because the technique allows delivering small amounts of DNA sufficient to induce immune response against the gene product. In fact, this technique is superior to other physical methods for DNA vaccination (Alsaggar and Liu, 2015).

2. *Gene delivery vectors* permit the nucleic acids to reach the cells and cross the cell membranes by means of a carrier. The main classes of gene delivery vectors are:

  — the *viral vectors*. Viruses have evolved to become highly efficient at nucleic acid delivery to specific cell types while avoiding immunosurveillance by an infected host. These properties make viruses attractive vehicles for gene therapy. The term *transduction* is used to describe virus-mediated gene transfer into eukaryotic cells. However, the term *transfection* is sometimes used as well t refer to viral-mediated gene transfer. Several types of viruses, including retroviruses, adenoviruses (AdVs), adeno-associated viruses (AAVs), and herpes simplex viruses (HSVs), have been modified in the laboratory for use in gene therapy applications (Robbins and Ghivizzani, 1998). Significant improvements in vector engineering, delivery, and safety have

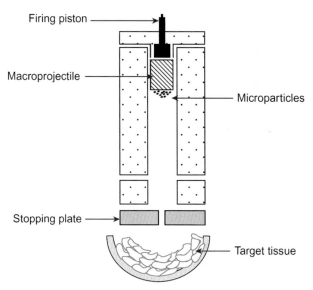

**FIG. 8.53**    Gene gun technique for nucleic acids delivery to cells in culture. *From Jinturkar, K.A., et al., 2011. Challenges in Delivery of Therapeutic Genomics and Proteomics, pp. 83–126.*

placed viral vector-based therapy at the forefront of modern gene medicine (Lundstrom, 2018). Different viral vectors have their own unique pros and cons. Retroviral vectors can permanently integrate into the genome of the infected cell but require mitotic cell division for transduction. Adenoviral vectors can efficiently deliver genes to a wide variety of dividing and nondividing cell types, but immune elimination of infected cells often limits gene expression in vivo. HSVs can deliver large amounts of exogenous DNA, but cytotoxicity and maintenance of transgene expression are a hindrance. AAVs are used to infect many nondividing and dividing cell types but have a limited nucleic acid-carrying capacity.

A number of preclinical studies have demonstrated therapeutic and prophylactic efficacy in animal models and clinical trials.

- The *chemical vectors*, or *nonviral vectors* for gene delivery, are chemicals that are able to interact with nucleic acids to make

nanoparticles and microparticles suitable to mimic some viral functions required for gene transfer (Pichon et al., 2010). Transfection by means of nonviral vectors comprises three main steps: complexation of nucleic acids, interaction of the resulting particles with the cell membrane and their entry into the cell, followed by release of nucleic acids into the cytosol, and intracellular transport into the nucleus, if needed for expression. However, unlike viral analogues that have evolved means to overcome cellular barriers and immune defense mechanisms, chemical gene carriers consistently exhibit significantly reduced transfection efficiency as they are hindered by numerous extra- and intracellular obstacles (Mintzer and Simanek, 2009). The development of nonviral vectors for gene delivery is a fascinating subject that is certain to advance rapidly over the next few years (Candiani, 2016). Chemical carriers can be characterized into three types:

- inorganic particles. One of the cheapest and oldest material used for gene delivery is calcium phosphate (CaP) (Khan et al., 2016a). The principle of CaP coprecipitation involves mixing nucleic acids with calcium chloride ($CaCl_2$) in a buffered saline/phosphate solution to generate a calcium-phosphate-DNA coprecipitate (Fig. 8.54). The suspension is next added in vitro to the cells to be transfected that take up some of the precipitate by endocytosis. This technique is easy to master, is an effective way to transfect many cell types. However, CaP coprecipitation is prone to variability due to the inherent sensitivity of this technique to slight changes in pH, temperature, and buffer salt concentrations. Besides, it can be cytotoxic to many cell types, especially to primary cells. In addition, it is unsuitable for in vivo transfer of nucleic acids to whole animals, and shows relatively poor transfection efficiency as compared to the other chemical vectors;

- *lipid-based gene delivery vectors.* A cationic lipid consists of a positively charged head group and one or two hydrophobic tails. The cationic head group governs the interaction between the lipid and the phosphate backbone of nucleic acids, and facilitates their condensation. They self-assemble with nucleic acids to produce nanometric or microscale complexes called *lipoplexes.* Even if the complexation step may seem very simple in concept, it determines the behavior and the transfection potential of lipoplexes. Consequently, the concentration, temperature, environment, and kinetics of mixing all are factors that should be considered carefully in any protocol of lipoplex formation (Tros de Ilarduya et al., 2010). Anionic, neutral, or cationic lipoplexes can be obtained depending on the charge ratio of the cationic lipids used to complex a given amount of anionic nucleic acids. It is worthy of note that a slight excess of positive charge confers lipoplexes with higher transfection efficiency (Pezzoli et al., 2013);

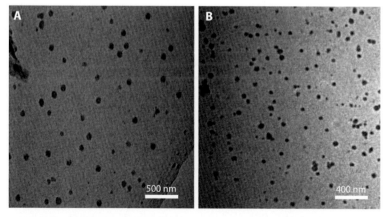

**FIG. 8.54** Transmission electron micrographs (TEM) of (A) plasmid DNA-loaded and (B) void calcium phosphate (CaP) nanoparticles. *From Bisht, S., Bhakta, G., Mitra, S., Maitra, A., 2005. pDNA loaded calcium phosphate nanoparticles: highly efficient non-viral vector for gene delivery. Int. J. Pharm. 288(1), 157–168.*

- *polymer-based gene delivery vectors.* Cationic polymers used for gene delivery include natural DNA-binding proteins such as histones, synthetic polypeptides, poly(ethylene imine) (PEI), poly-L-lysine (PLL), chitosan, and many others (Pezzoli and Candiani, 2013). Of note, as opposed to cationic lipids, cationic polymers are devoid of a hydrophobic domain. They can be combined with and condense nucleic acids to form small-sized complexes, called polyplexes. This is crucial for gene transfer, as small particles may be favorable for improving transfection efficacy, especially in vivo. Three approaches to design and optimize polymers are being pursued. The first relies on the rational design of the chemical structure to take into account identified biological barriers to gene transfer. The second approach is based on the systematic modification of the chemical structure of the polymer to establish structure-activity relationships. Considering the different gene delivery applications of polymers, the design of the proper polymer becomes a sophisticated task (Tros de Ilarduya et al., 2010). The molecular weight, the degree of branching, or the surface charge have to be adjusted to produce stable complexes, yet simultaneously generate systems with the desired release properties (Malloggi et al., 2015).

Over the past decade, significant progress has been made in the understanding of the cellular pathways and underlying mechanisms involved in lipoplex- and polyplex-mediated transfection (Fig. 8.55). The general mechanisms of delivery involve the following steps: before complexes are taken up by cells, serum in the surrounding milieu can break apart complexes, causing nucleic acids to be released and degraded. Lipoplexes and polyplexes bind to the cell surface by nonspecific, electrostatic interactions between the cationic complexes and the anionic cell-surface proteoglycans (Ruponen et al., 2001), and enter the cells by endocytosis, or fuse with the plasma membrane. Once within the cells, nucleic acids need to be released in order to become active. Finally, nuclear DNA uptake is fundamental for transcription (Tros de Ilarduya et al., 2010).

## 8.2.3 Polymerase Chain Reaction

A multitude of biotechnological techniques used in basic research as well as in clinical diagnostics on an everyday basis depend on DNA polymerases and their inherent ability to replicate DNA strands with astoundingly high fidelity (Aschenbrenner and Marx, 2017). Most methods for keeping track of DNA sequences in a genome use an enzymatic reaction known as Polymerase Chain Reaction (PCR). The PCR is an enzymatic process first introduced by Mullis in 1983. The impact of his pioneering work on the field of molecular biology and biotechnology earned him a Nobel Prize in Chemistry after a decade. As the inventor himself stated, "The PCR can generate one hundred billions of similar molecules in one afternoon. The reaction is easy to execute. It requires no more than a test tube, a few simple reagents, and a source of heat."

The PCR enables the production of large amounts of specific DNA sequences (e.g., genes), called *templates*, from a small quantity of the original molecule by means of a simple enzymatic reaction. From then onward, the easiness, speed and cost-effectiveness of such technique has led to its widespread use in many downstream applications such as DNA cloning, diagnosing diseases, and paternity tests among others.

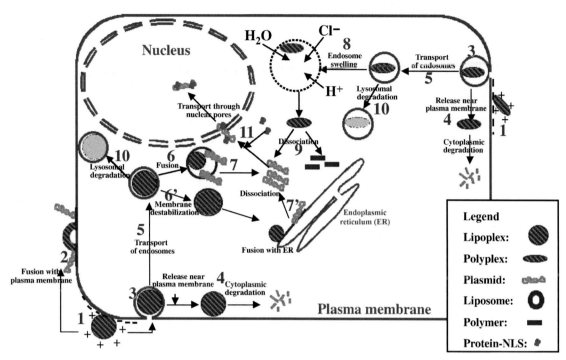

**FIG. 8.55** Summary of the steps involved in lipoplex- and polyplex-mediated gene delivery. *From Elouahabi, A., Ruysschaert J.-M., 2005. Formation and intracellular trafficking of lipoplexes and polyplexes. Mol. Therapy 11, 336–347.*

The five core components needed to set up a PCR fill in a test tube with a total volume of 20–100 µL. They are (Fig. 8.56):

– *a DNA template,* which is the target dsDNA sequence to be copied and amplified cycle after cycle;
– *a pair of primers,* which are synthetic, single-stranded oligonucleotides that are specifically designed to be complementary to (i.e., able to recognize and match with) either end of the ssDNA template. One primer is referred to as *reverse* because it binds to the sense strand of the template (i.e., the one that goes from 5′→3′ in the sequence read from left to right), whereas the other one that binds to the antisense DNA strand (from 3′→5′) is

called *forward.* A pair of *reverse-forward* primers with specificity for target regions on the template is required for the reaction to start;
– *a DNA polymerase,* which is the enzyme in charge of the synthesis of new copies of the template. This enzyme is able to add single nucleotides to the 3′-OH of each primer in a way that they are complementary to the ssDNA template, thus creating a replica of the original DNA sequence. Because the enzymatic reaction occurs during high temperature conditions, a thermostable DNA polymerase is needed to avoid the inactivation of the enzyme every thermal cycle. For this purpose, an enzyme known as *Taq*

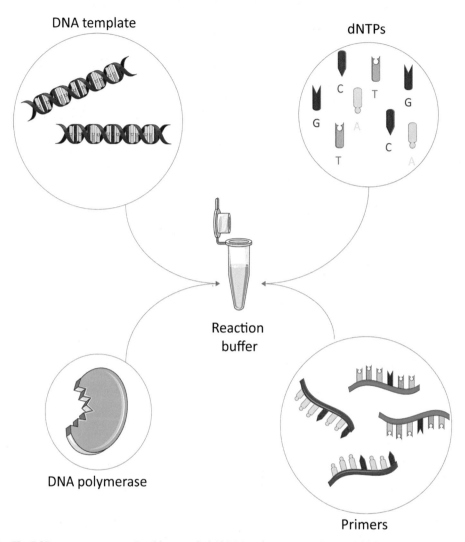

**FIG. 8.56**   The PCR core components. Double-stranded (ds) DNA fragments to be amplified, called templates, are inserted in the reaction buffer together with free deoxynucleotide triphosphates (dNTPs), which are the building blocks of the newly synthesized strands. Pairs of forward-reverse primers, which are single-stranded (ss) oligonucleotides complementary to specific regions of the template, are essential for the enzymatic reaction to start. The DNA polymerase is the enzyme responsible for the replication of the template.

polymerase, so called because isolated from thermophilic bacterium *Thermus aquaticus* who thrive at temperatures of 50–80°C, is nowadays widely used. However, due to the poor replication fidelity of *Taq* polymerase,

another enzyme called *Pfu* polymerase, obtained from the hyperthermophilic archaeon *Pyrococcus furiosus*, is sometimes preferred. In fact, the specific feature of *Pfu* polymerase is the 3′→5′ exonuclease proofreading activity, which is the ability to detect, remove, and replace mispaired nucleotides that have been introduced by mistake during the template copying;

- *four different kinds of deoxynucleotide triphosphates (dNTPs)*. Single nucleotides must be available in the reaction buffer as building blocks to generate copies of the template. During the replication of the DNA template, the polymerase catalyses the reaction between the 3′-OH of the last nucleotide of the strand and the 5′-phosphate group of the nucleotide that has to be added to the novel strand. The two nucleotides are joined together by a condensation reaction, that is, the formation of a phosphodiester bond and the release of pyrophosphate (*PPi*);
- *the reaction buffer*, which provides the polymerase with a suitable environment, in terms of pH and ionic composition, to copy the template. For instance, the buffer contains $Mg^{2+}$ ions that serve as cofactors of the polymerase enzyme.

During a PCR run, the template is exponentially amplified through subsequent thermal cycles, by heating and cooling (i.e., thermal cycling) all the ingredients. At the beginning of each cycle, the newly synthesized DNA molecules will act as the templates for the next reaction, thus giving rise to the so-called chain reaction. The amplification is carried out in a laboratory apparatus called a *thermocycler*, also known as *thermal cycler*, which raises and lowers the temperature of the sample(s) in a holding block in discrete, preprogrammed steps, allowing for thermal denaturation and reannealing of samples, and the polymerization to take place. Such thermal changes have to be fast and exacting for high-quality results. A PCR experiment is typically run for 25–40 thermal cycles, also called *rounds*, each one

consisting of three consecutive phases (Fig. 8.57):

- the *denaturation* or *melting* phase. The temperature is first raised to 95°C for 1 min. This step allows the denaturation of dsDNA template because the high temperature causes the hydrogen bonds between the bases in the dsDNA to break and the two strands to separate;
- the *annealing* phase, during which the temperature is lowered to 50–60°C for 40–50 s to favor the annealing of the two primers to their complementary sequences on each respective ssDNA template. When each primer stably matches with the template, polymerase starts polymerizing a copy strand;
- the *elongation* or *extension* phase. The temperature is raised for 2 min to 72°C, which is the optimal working temperature for the DNA polymerase. Once activated, the polymerase starts to synthesize new DNA strands in the 5′→3′ direction by adding available dNTPs to the 3′-OH of the previous nucleotide. Under optimal conditions, polymerase works at a 1,000 nt $min^{-1}$ addition rate.

The three-step cycle is repeated many times to obtain a fair amount of amplified DNA fragments called *amplicons*. Because each template gives rise to two identical amplicons after every cycle, the number of amplicon copies produced ($X_n$) at the end of each thermal cycle is calculated as follows:

$$X_n = X_0 \times 2^n$$

where $X_0$ is the initial number of DNA molecules, and $n$ is the number of thermal cycles undertaken. For instance, after 30 thermal cycles we get $\approx 10$ billion amplicons from one molecule of dsDNA.

The amplification efficiency strongly depends on the choice of primers and the operating parameters.

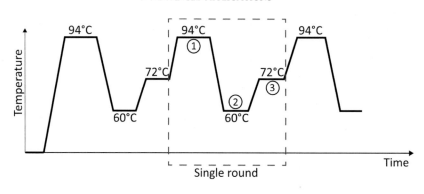

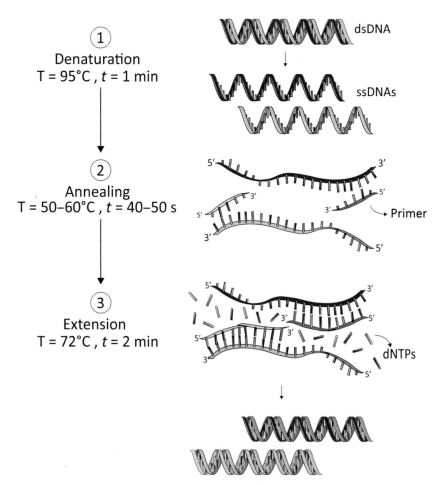

**FIG. 8.57** Time-temperature graph showing the three consecutive thermal phase of every PCR round (upper panel). First, the temperature is raised to 94°C (1, melting phase), next lowered to 50–60°C (2, annealing phase), and raised again to 72°C (3, extension phase). Schematic representation of what happens during a PCR round (lower panel). During melting, the double-stranded (ds) DNA template breaks apart, giving rise to two single-stranded (ss) DNA molecules (1). When the temperature is lowered to the optimal annealing temperature for primers, reverse and forward primers hybridize with the ssDNA (2). Newly synthesized DNA templates (i.e., amplicons) are produced during the extension phase (3), as the DNA polymerase adds deoxynucleotide triphosphates (dNTPs) to the 3′-OH of both primers.

# References

Alsaggar, M., Liu, D., 2015. Physical methods for gene transfer. In: Nonviral Vectors for Gene Therapy—Physical Methods and Medical Translation. Academic Press, pp. 1–24.

Andrianov, A.K. (Ed.), 2009. Polyphosphazenes for Biomedical Applications. John Wiley & Sons, Inc, Hoboken, NJ.

Aschenbrenner, J., Marx, A., 2017. DNA polymerases and biotechnological applications. Curr. Opin. Biotechnol. 48, 187–195.

Bourke, S.L., Kohn, J., 2003. Polymers derived from the amino acid L-tyrosine: polycarbonates, polyarylates and copolymers with poly(ethylene glycol). Adv. Drug Deliv. Rev. 55, 447–466.

Bruggeman, J.P., de Bruin, B.-J., Bettinger, C.J., Langer, R., 2008. Biodegradable poly(polyol sebacate) polymers. Biomaterials 29, 4726–4735.

Candiani, G., 2016. Non-Viral Gene Delivery Vectors. Humana Press/Springer, New York City, NY.

Chasin, M., Lewis, D., Langer, R., 1988. Polyanhydrides for controlled drug delivery. Biopharm. Manuf. 1, 33–35.

Crick, F., 1970. Central dogma of molecular biology. Nature 227 (5258), 561–563.

Doelle, H.W., Rokem, S., Berovic, M., 2009. Biotechnology—Volume I: Fundamentals in Biotechnology. UNESCO—Encyclopedia Life Support Systems (UNESCO-EOLSS).

Fisher, M.F., Mauck, R.L., 2013. Tissue engineering and regenerative medicine: recent innovations and the transition to translation. Tissue Eng. Part B Rev. 19 (1), 1–13.

Frazzetto, G., 2003. White biotechnology. EMBO Rep. 4 (9), 835–837.

Gajjar, C.R., King, M.W., 2014. Resorbable Fiber-Forming Polymers for Biotextile Application. SpringerBriefs in Materials, Springer Nature Switzerland AG, Cham. http://dx.doi.org/10.1007/978-3-319-08305-6_2.

Gőpferich, A., Tessmar, J., 2002. Polyanhydride degradation and erosion. Adv. Drug Deliv. Rev. 54, 911–931.

Guelcher, S.A., 2008. Biodegradable Polyurethanes: Synthesis and Applications in Regenerative Medicine. Tissue Eng. Part B Rev. 14 (1), 3–17.

Guilak, F., Butler, D.L., Goldstein, S.A., Mooney, D.J. (Eds.), 2003. Functional Tissue Engineering. Springer-Verlag, New York. ISBN 0-387-95553-4.

Heller, J., Barr, J., Ng, S.Y., Abdellauoi, K.S., Gurny, R., 2002. Poly(ortho esters): synthesis, characterization, properties and uses. Adv. Drug Deliv. Rev. 54 (7), 1015–1039.

Howard, D., Buttery, L.D., Shakesheff, K.M., Roberts, S.J., 2008. Review. Tissue engineering: strategies, stem cells and scaffolds, J. Anat. 213, 66–72.

Jinturkar, K.A., Rathi, M.N., Misra, A., 2011. Gene delivery using physical methods. In: Challenges in Delivery of Therapeutic Genomics and Proteomics, Elsevier, Burlington, MA, pp. 83–126.

Kasper, F.K., Tanahashi, K., Fisher, J.P., Mikos, A.G., 2009. Synthesis of poly(propylene fumarate). Nat. Protoc. 4, 518–525.

Katari, R., Peloso, A., Orlando, G., 2015. Tissue engineering and regenerative medicine: semantic considerations for an evolving paradigm. Front. Bioeng. Biotechnol. 2, 1–6. 57.

Khan, M.A., Wu, V.M., Ghosh, S., Uskoković, V., 2016a. Gene delivery using calcium phosphate nanoparticles: optimization of the transfection process and the effects of citrate and poly(l-lysine) as additives. J. Colloid Interface Sci. 471, 48–58.

Khan, S., Ullah, M.W., Siddique, R., Nabi, G., Manan, S., Yousaf, M., Hou, H.W., 2016b. Role of recombinant DNA technology to improve life. Int. J. Genomics.

King, R., 2004. Gene delivery to mammalian cells by microinjection. Methods Mol. Biol. 245, 167–174.

Kohn, J., Welsh, W.J., Knight, D., 2007. A new approach to the rationale discovery of polymeric Biomaterials. Biomaterials 28 (29), 4171–4177.

Krötz, F., Wit, C.d., Sohn, H.-Y., Zahler, S., Gloe, T., Pohl, U., Plank, C., 2003. Magnetofection—a highly efficient tool for antisense oligonucleotide delivery in vitro and in vivo. Mol. Ther. 7 (5), 700–710.

Kumar, N., Langer, R.S., Domb, A.J., 2002. Polyanhydrides: an overview. Adv. Drug Deliv. Rev. 54, 889–910.

Langer, R., Vacanti, J.P., 1993. Science 260 (5110), 920–926. http://dx.doi.org/10.1126/science.8493529.

Lentacker, I., De Cock, I., Deckers, R., De Smedt, S.C., Moonen, C.T.W., 2014. Understanding ultrasound induced sonoporation: definitions and underlying mechanisms. Adv. Drug Deliv. Rev. 72, 49–64.

Loi, P., Czernik, M., Zacchini, F., Iuso, D., Scapolo, P.A., Ptak, G., 2013. Sheep: the first large animal model in nuclear transfer research. Cell. Reprogram. 15 (5), 367–373.

Lundstrom, K., 2018. Viral vectors in gene therapy. Diseases 6(2).

Malloggi, C., Pezzoli, D., Magagnin, L., De Nardo, L., Mantovani, D., Tallarita, E., Candiani, G., 2015. Comparative evaluation and optimization of off-the-shelf cationic polymers for gene delivery purposes. Polym. Chem. 6 (35), 6325–6339.

Matyushenko, I., Sviatukha, I., Grigorova-Berenda, L., 2016. Modern approaches to classification of biotechnology as a part of NBIC-technologies for bioeconomy. Br. J. Econ. Manag. Trade 14 (4), 1–14.

McKinnell, R.G., Di Berardino, M.A., 1999. The biology of cloning: history and rationale. Bioscience 49 (11), 875–885.

Mintzer, M.A., Simanek, E.E., 2009. Nonviral vectors for gene delivery. Chem. Rev. 109 (2), 259–302.

Peppas, N.A., Hoffman, A.S., 2012. Hydrogels. In: Ratner, B.D., Hoffman, A.S., Schoen, F.J., Lemons, J.E. (Eds.), Biomaterials Science: An Introduction to Materials in Medicine, third ed. Academic Press, Cambridge, MA, pp. 171–172 (Chapter I.2.5).

Pezzoli, D., Candiani, G., 2013. Non-viral gene delivery strategies for gene therapy: a "ménage à trois" among nucleic acids, materials, and the biological environment. J. Nanopart. Res. 15(3).

Pezzoli, D., Kajaste-Rudnitski, A., Chiesa, R., Candiani, G., 2013. Lipid-based nanoparticles as nonviral gene delivery vectors. In: Nanomaterial Interfaces in Biology. pp. 269–279.

Pichon, C., Billiet, L., Midoux, P., 2010. Chemical vectors for gene delivery: uptake and intracellular trafficking. Curr. Opin. Biotechnol. 21 (5), 640–645.

Rechichi, A., Ciardelli, G., D'Acunto, M., Vozzi, G., Giusti, P., 2008. Degradable block polyurethanes from nontoxic building blocks as scaffold materials to support cell growth and proliferation. J. Biomed. Mater. Res. 84A, 847–855.

Riboldi, S.A., Bertoldi, S., Mantero, S., 2017. In vitro dynamic culture of cell-biomaterial constructs. In: Tanzi, M.C., Farè, S. (Eds.), Characterization of Polymeric Biomaterials, first ed. Woodhead Publ., Elsevier Ltd., Duxford (Chapter 4).

Robbins, P.D., Ghivizzani, S.C., 1998. Viral Vectors for Gene Therapy. Pharmacol. Ther. 80 (1), 35–47.

Rose, D.W., 2007. Genetic manipulation of mammalian cells by microinjection. Cold Spring Harb. Protoc. 2007 (10). pdb.prot4754-pdb.prot4754.

Ruponen, M., Rönkkö, S., Honkakoski, P., Pelkonen, J., Tammi, M., Urtti, A., 2001. Extracellular glycosaminoglycans modify cellular trafficking of lipoplexes and polyplexes. J. Biol. Chem. 276 (36), 33875–33880.

Suda, T., Liu, D., 2007. Hydrodynamic gene delivery: its principles and applications. Mol. Ther. 15 (12), 2063–2069.

Tabassum Khan, N., 2017. Non-viral mediated physical approach for gene delivery. Drug Design. Open Access 06(02).

Tae, S.-K., Lee, S.-H., Park, J.-S., Im, G.-I., 2006. Mesenchymal stem cells for tissue engineering and regenerative medicine. Biomed. Mater. 1, 63–71.

Takahashi, K., Yamanaka, S., 2006. Induction of pluripotent stem cells from mouse embryonic and adult fibroblast cultures by defined factors. Cell 126, 663–676.

Tros de Ilarduya, C., Sun, Y., Düzgüneş, N., 2010. Gene delivery by lipoplexes and polyplexes. Eur. J. Pharm. Sci. 40 (3), 159–170.

Viola, J., Lal, B., Grad, O., 2003. The Emergence of Tissue Engineering as a Research Field. October 14. The National Science Foundation, Arlington, VA. https://www.nsf.gov/pubs/2004/nsf0450/start.htm.

Williams, N., 2003. Death of Dolly marks cloning milestone. Curr. Biol. 13 (6), R209–R210.

Wilmut, I., Schnieke, A.E., McWhir, J., Kind, A.J., Campbell, K.H.S., 1997. Viable offspring derived from fetal and adult mammalian cells. Nature 385 (6619), 810–813.

Young, J.L., Dean, D.A., 2015. Electroporation-mediated gene delivery. In: Nonviral Vectors for Gene Therapy—Physical Methods and Medical Translation, pp. 49–88.

Zhang, Z., Kuijerb, R., Bulstra, S.K., Grijpma, D.W., Feijena, J., 2006. The in vivo and in vitro degradation behavior of poly(trimethylene carbonate). Biomaterials 27, 1741–1748.

# Further Reading

Altomare, L., Farè, S., 2008. Cells response to topographic and chemical micropatterns. J. Appl. Biomater. Biomech. 6 (3), 132–143.

Altomare, L., Riehle, M., Gadegaard, N., Tanzi, M.C., Farè, S., 2010a. Microcontact printing of fibronectin on a biodegradable polymeric surface for skeletal muscle cell orientation. Int. J. Artif. Organs 33 (8), 535–543.

Altomare, L., Gadegaard, N., Visai, L., Tanzi, M.C., Farè, S., 2010b. Biodegradable microgrooved polymeric surfaces obtained by photolithography for skeletal muscle cell orientation and myotube development. Acta Biomater. 6 (6), 1948–1957.

Bhutto, D.F., Murphy, E.M., Priddy, M.C., Centner, C.C., Moore Iv, J.B., Bolli, R., Kopechek, J.A., 2018. Effect of molecular weight on sonoporation-mediated uptake in human cells. Ultrasound Med. Biol. 44, 2662–2672.

Bisht, S., Bhakta, G., Mitra, S., Maitra, A., 2005. pDNA loaded calcium phosphate nanoparticles: highly efficient non-viral vector for gene delivery. Int. J. Pharm. 288 (1), 157–168.

Candiani, G., Riboldi, S.A., Sadr, N., Lorenzoni, S., Neuenschwander, P., Montevecchi, F.M., Mantero, S., 2010. Cyclic mechanical stimulation favors myosin heavy chain accumulation in engineered skeletal muscle constructs. J. Appl. Biomater. Biomech. 8 (2), 68–75.

Cha, C., Soman, P., Zhu, W., Nikkhah, M., Camci-Unal, G., Chen, S., Khademhosseini, A., 2014. Structural reinforcement of cell-laden hydrogels with microfabricated three dimensional scaffolds. Biomater. Sci. 2, 703–709.

Chen, G., Ushida, T., Tateishi, T., 2002. Development of biodegradable porous scaffolds for tissue engineering. J. Mater. Sci. Eng. C 17, 63–69.

Elomaa, L., Teixeira, S., Hakala, R., Korhonen, H., Grijpma, D.W., Seppälä, J.V., 2011. Preparation of poly (ε-caprolactone)-based tissue engineering scaffolds by stereolithography. Acta Biomater. 7, 3850–3856.

Elouahabi, A., Ruysschaert, J.-M., 2005. Formation and intracellular trafficking of lipoplexes and polyplexes. Mol. Ther. 11 (3), 336–347.

Gunatillake, P.A., Adhikari, R., 2003. Biodegradable synthetic polymers for tissue engineering. Eur. Cell Mater. 5, 1–16.

Hardy, R., Cooper, M.S., 2011. Glucocorticoid-induced osteoporosis—a disorder of mesenchymal stromal cells? Front. Endocrinol. 2, 24. http://dx.doi.org/10.3389/fendo.2011.00024.

Hejazi, F., Mirzadeh, H., Contessi, N., Tanzi, M.C., Faré, S., 2017. Novel class of collector in electrospinning device for the fabrication of 3D nanofibrous structure for large defect load-bearing tissue engineering application. J. Biomed. Mater. Res. A 105 (5), 1535–1548.

Herweijer, H., Wolff, J.A., 2006. Gene therapy progress and prospects: Hydrodynamic gene delivery. Gene Ther. 14 (2), 99–107.

Hutmacher, D.W., 2000. Scaffolds in tissue engineering bone and cartilage. Biomaterials 21, 2529–2543.

Kasasbeh, F.A., Shawabkeh, M.M., Hawamdeh, A.A., 2011. Deletion of 18p Syndrome. Lab. Med. 42 (7), 436–438.

Kaur, G., Dufour, J.M., 2012. Cell lines Valuable tools or useless artifacts. Spermatogenesis 2 (1), 1–5. http://dx.doi.org/10.4161/spmg.2.1.19885.

Klotz, B.J., Gawlitta, D., Rosenberg, A.J.W.P., Malda, J., Melchels, F.P.W., 2016. Gelatin-methacryloyl hydrogels: towards biofabrication-based tissue repair. Trends Biotechnol. 34 (5), 394–407.

Konta, A.A., Garcia-Pina, M., Serrano, D.R., 2017. Personalised 3D printed medicines: which techniques and polymers are more successful? Bioengineering 4, 79.

Malda, J., Visser, J., Melchels, F.P., Jüngst, T., Hennink, W.E., Dhert, W.J.A., Jü Groll, J., Hutmacher, D.W., 2013. 25th Anniversary article: engineering hydrogels for biofabrication. Adv. Mater. 25, 5011–5028.

Melchels, F., Feijen, J., Grijpma, D., 2009. A Poly(D,L-Lactide) resin for the preparation of tissue engineering scaffolds by stereolithography. Biomaterials 30, 3801–3809.

Mikos, A.G., Bao, Y., Cima, L.G., Ingber, D.E., Vacanti, J.P., Langer, R., 1993. Preparation of poly(glycolic acid) bonded fiber structures for cell attachment and transplantation. J. Biomed. Mater. Res. 27 (?), 183–189.

Priya, V.S.V., Roy, H.K., jyothi, N., Prasanthi, N.L., 2016. Polymers in Drug Delivery Technology, Types of Polymers and Applications. Sch. Acad. J. Pharm. 5 (7), 305–308.

Shor, L., Güçeri, S., Wen, X., Gandhi, M., Sun, W., 2007. Fabrication of three-dimensional polycaprolactone/hydroxyapatite tissue scaffolds and osteoblast-scaffold interactions in vitro. Biomaterials 28 (35), 5291–5297.

Sughanthy Siva, A.P., Ansari, M.N.M., 2015. A review on bone scaffold fabrication methods. Int. Res. J. Eng. Technol. 2 (6), 1232–1238.

Sultana, N., 2015. Scaffold fabrication protocols. In: Sultana, N., Hassan, M.I., Lim, M.M. (Eds.), Composite Synthetic Scaffolds for Tissue Engineering and Regenerative Medicine. SpringerBrief in Materials, Springer, New York, NY. www.regenerativemedicine.net/Tissue.html.

Xiong, Z., Yan, Y., Wang, S., Zhang, R., Zhang, C., 2002. Fabrication of porous scaffolds for bone tissue engineering via low-temperature deposition. Scripta Mater. 46 (10), 771–776.

Yoon, J.J., Park, T.G., 2001. Degradation behaviors of biodegradable macroporous scaffolds prepared by gas foaming of effervescent salts. J. Biomed. Mater. Res. 55 (3), 401–408.

Yu, H., Matthew, H.W., Wooley, P.H., Yang, S.Y., 2008. Effect of porosity and pore size on microstructures and mechanical properties of poly-epsilon-caprolactone- hydroxyapatite composites. J. Biomed. Mater. Res. B Appl. Biomater. 86 (2), 541–547.

# Index

Note: Page numbers followed by *f* indicate figures, *t* indicate tables, and *np* indicate footnotes.

Printed in the United States
By Bookmasters